第四届结构工程新进展国际论坛文集

The 4th International Forum on Advances in Structural Engineering

# 混凝土结构与材料新进展

# Progess in Concrete Structures and Materials

吴智深　吴刚　缪昌文　主编

Editors in Chief: Zhishen Wu, Gang Wu & Changwen Miao

中国建筑工业出版社

China Architecture & Building Press

**图书在版编目（CIP）数据**

混凝土结构与材料新进展/吴智深，吴刚，缪昌文主编．—北京：中国建筑工业出版社，2010.11

ISBN 978-7-112-12538-8

Ⅰ.①混… Ⅱ.①吴…②吴…③缪… Ⅲ.①混凝土结构-国际学术会议-文集②混凝土-建筑材料-国际学术会议-文集 Ⅳ.①TU37-53 ②TU528-53

中国版本图书馆CIP数据核字（2010）第196202号

责任编辑：赵梦梅　刘婷婷
责任设计：李志立
责任校对：王　颖　王雪竹

**混凝土结构与材料新进展**

Progess in Concrete Structures and Materials

吴智深　吴刚　缪昌文　主编

Editors in Chief：Zhishen Wu，Gang Wu & Changwen Miao

*

中国建筑工业出版社出版、发行（北京西郊百万庄）
各地新华书店、建筑书店经销
北京红光制版公司制版
北京盛通印刷股份有限公司印刷

*

开本：787×1092毫米　1/16　印张：28¾　字数：718千字
2010年11月第一版　2010年11月第一次印刷
定价：**78.00**元

ISBN 978-7-112-12538-8
(19821)

# 前　言 Preface

当前，中国正在从事世界上最大规模的土木工程建设。中国土木工程（包括结构工程）持续的、大规模的建设，为结构工程技术的发展、学科的建设、人才的培养、队伍的壮健，特别是自主创新能力的提升，提供了历史性的机遇。正是基于这种时代背景，由中国建筑工业出版社、同济大学《建筑钢结构进展》编辑部、香港理工大学《结构工程进展》（Advances in Structural Engineering）编委会联合主办，东南大学土木工程学院、混凝土及预应力混凝土结构教育部重点实验室、江苏省建筑科学研究院有限公司和高性能土木材料国家重点实验室联合承办的第四届"结构工程新进展国际论坛（The 4th International Forum on Advances in Structural Engineering）"在江苏南京进行。

本届论坛主题是"混凝土结构与材料新进展"。目前，我国每年的混凝土用量约为13～15亿立方米，钢筋用量约为2000万吨，用量接近全世界总用量的50%，钢筋混凝土仍将是我国今后相当长时期内主要工程结构材料，同时，在世界范围内，如何应对既存的大量混凝土结构的耐久性问题也是土木工程师们共同关注的问题。如何通过新材料的快速推广、预应力等技术的创新应用、结构体系的优化设计、智能化方法的合理引入等，推动混凝土及混凝土结构的高性能化，提高混凝土结构的安全性、耐久性是当前的主要研究热点。在此背景下，本论坛选择"混凝土结构与材料新进展"为主题具有重要意义。论坛荣幸地邀请到了25位特邀报告专家，报告主题涵盖了混凝土结构与材料的各个领域，阐述了该领域内的最新发展信息，向与会者提供了一个与专家学者互动并获取宝贵经验的机会。

感谢论坛特邀报告人，他们不仅在大会上做了精彩的报告，而且部分特邀报告人还奉献了精心准备的论文，使得本书得以顺利出版。

感谢论坛自由投稿作者以及参加本次论坛的所有代表，正是大家的积极参与和配合，才使得本次论坛能够顺利举行。

感谢中国建筑工业出版社、同济大学《建筑钢结构进展》编辑部、香港理工大学《结构工程进展》编委会对本次论坛的指导、支持和帮助。

感谢东南大学土木工程学院、混凝土及预应力混凝土结构教育部重点实验室、江苏省建筑科学研究院有限公司和高性能土木材料国家重点实验室对本次论坛成功主办的努力和付出。

# 目　录 Contents

# 第 1 章 Chapter 1

# 基于病害研究的大跨径 PC 箱梁桥设计探讨*

吕志涛，潘钻峰，王景全

（东南大学　混凝土及预应力混凝土结构教育部重点实验室，江苏南京 2100961）

**提　要**：近二十年来，很多大跨径预应力混凝土箱梁桥在运营一段时间后，出现了跨中下挠和梁体开裂的两大病害，而且变形和裂缝常常是"并发症"，严重影响了这些桥梁的正常使用性能和耐久性，甚至引起结构安全问题。当前，导致这些病害的原因还没有完全明确。本文基于大跨径预应力混凝土箱梁病害的调查研究，从箱梁桥的箱梁构造设计、箍筋设置、纵向预应力设计等方面对箱梁桥进行了研究分析，对导致箱梁结构长期下挠的主要成因在机理上进行了解释，研究了斜裂缝出现后箱梁剪切变形对总挠度的影响。基于收缩徐变及预应力效应的不确定分析，对体外备用预应力束在大跨径预应力混凝土箱梁桥中的应用和设计进行了研究。结果表明，高跨比设置过小、腹板下弯束的取消或腹板下弯束设置不足是产生箱梁开裂下挠病害的主要原因；出现斜裂缝之后薄壁箱梁的剪切变形对主梁下挠的影响不可忽视。

**关键词**：箱梁桥；下挠；斜裂缝；高跨比；腹板下弯束；剪切变形；备用束

# ISSUES IN DESIGN OF LONG-SPAN PC BOX GIRDER BRIDGES BASED ON INVESTIGATION ON BRIDGE DISEASES

Z. T. Lü, Z. F. Pan, J. Q. Wang

(Key Laboratory of R. C. & P. C. Structures, Ministry of Education, Southeast University, Nanjing, 210096, China.)

**Abstract**: In the past twenty years, excessive deflections and cracks in the box girders have commonly appeared in some long-span prestressed concrete box girder bridges after a period of operation, reducing the serviceability and durability of bridges. However, the causes of excessive deflections and cracks are still not clear. Based on the investigation on the preceding two problems, the long-span PC box girder bridges are studied from several aspects, such as the high-span ratio, longitudinal prestress design, composition design and configuration of stirrups. Moreover, the cause mechanism of excessive deflection problem is explained, and the impact of shear deformation on structural deflection is studied after the formation of inclined cracks. In addition, the efficiency of different layouts of spare

* 基金项目："十一五"交通部科技攻关重大专项（200631800001），江苏交通科学研究项目（09Y012）

tendons is compared for improving the deflection problems, and the determination of spare tendons is performed through the uncertainty analysis of creep, shrinkage and prestress effects. Research indicates that the smaller high-span ratio and the cancellation or insufficiency of web's bent-down tendons are the main causes for the preceding two problems, and the effects of shear deformation and development of inclined cracks on the deflection could not be ignored.

**Keywords**: box girder bridge, deflection, inclined crack, high-span ratio, web's bent-down tendons, shear deformation, spare tendon

## 1.1 引言

近二十年来，一些大跨径预应力混凝土箱梁桥在运营一段时间后，普遍出现"腹板开裂"和"跨中下挠"等质量通病。而且裂缝和下挠均随时间延续而不断发展，直接影响着结构的安全性与正常使用性能，对结构的耐久性造成很大危害，这些问题一直困扰着桥梁设计工程师和工程管理人员。对于桥梁病害的原因，多次在全国桥梁学术会议上有所探讨，归结起来，主要包括如下因素：桥梁超载，预应力损失过大，施工不当，混凝土的收缩、徐变，梁空间作用，温度效应等。近年来，很多学者认为纵向预应力数量不足是产生下挠的主要原因。一些学者基于荷载平衡思想，对预应力抵消恒载的作用效应进行了研究，提出了一些设计方法，如"恒载零弯矩设计方法"[1,2]，"合理成桥状态设计方法"[3]等，特别是零弯矩设计方法在国内多座桥梁中得到了应用，其基本思想是配置足够多的纵向预应力使梁处于接近轴压状态，从而达到减小长期变形发展的目的。对于 200m 跨以下的预应力混凝土桥梁，按零弯矩设计方法增大预应力配筋量可以做到接近零弯矩的状态，但对于大跨径桥梁，由于恒载比例增加，且受到截面尺寸的限制，体内难以布置足够的预应力筋来抵消恒载，所以大跨径箱梁桥的病害问题还无法真正解决。

本文根据对已出现病害的大跨径箱梁桥的结构特性的调查和分析，从高跨比、腹板及配筋设计、纵向预应力束设计、下挠原因推演等方面对箱梁桥进行了研究分析，并研究了混凝土梁斜向开裂后，剪切变形对总挠度的影响。针对上述质量通病，提出可通过设置后期备用束（体内束或体外束，或者两者并用）得以改善，通过运营期内定期检测，在必要时张拉备用钢束，改善桥梁内力及变形状态，而备用束量的设计可根据长期变形置信界限值与确定性计算值的差异的来确定。

## 1.2 箱梁构造及箍筋的合理设计

### 1.2.1 高跨比的设计

众所周知，结构变形不仅与所受到的作用力有关，结构自身的刚度对变形影响也很大，高跨比是影响结构刚度的一个重要因素。目前，箱梁桥的跨度和宽度均趋于增加，而箱梁构造尺寸却因要求兼顾轻型美观的原因设计得越来越小，从而增加了箱梁桥的正常使用性能（裂缝和变形）对各种荷载效应的敏感性。为限制挠度的过大发展，以往研究往往偏重于预应力筋用量的增加，尽量多地抵消恒载作用，而对箱梁结构构造的研究偏少。高跨比是影响主梁受力的重要参数之一，其值增大对桥梁结构的影响主要体现在两个方面：适当增加梁高，可增加主梁的刚度，改善截面的受力状态；大跨径箱梁桥中的预应力效应很大程度上是利用预应力的偏心弯矩来抵消恒载的，适当增大高跨比，在预应力筋用量不变的情形下，纵向预应力筋的偏心距增大，增加了预应力利用效率，而恒载增加十分有限。所以适当增加高跨比，是一种有效抑制后期下挠病害的有效方法。本文对国内外近80座连续刚构的墩顶及跨中梁高的高跨比进行了统计，如图1所示。从中可以看出，目前对于采用普通混凝土或高性能混凝土的连续刚构，墩顶截面跨高比通常为1/15～1/20，跨中截面为1/40～1/65。而对于采用轻质高强混凝土或部分钢结构的连续刚构，其恒载效应减小，高跨比可适当减小，由图1看出，部分梁段采用轻质混凝土或钢结构后，墩顶高跨比一般可小于1/20，跨中高跨比可做到1/70以下。高跨比的下降是桥梁上部结构轻型化的表现，但高跨比太小，主梁刚度得不到保证，此外，腹板的受剪面积减小，截面主拉应力增大，从而增加了后期病害发生的几率。表1中出现病害的大跨径箱梁桥，其高跨比均相对较小。澳大利亚的Getway桥其采用的高跨比相对较大，目前还没有文献报道其下挠过大问题。

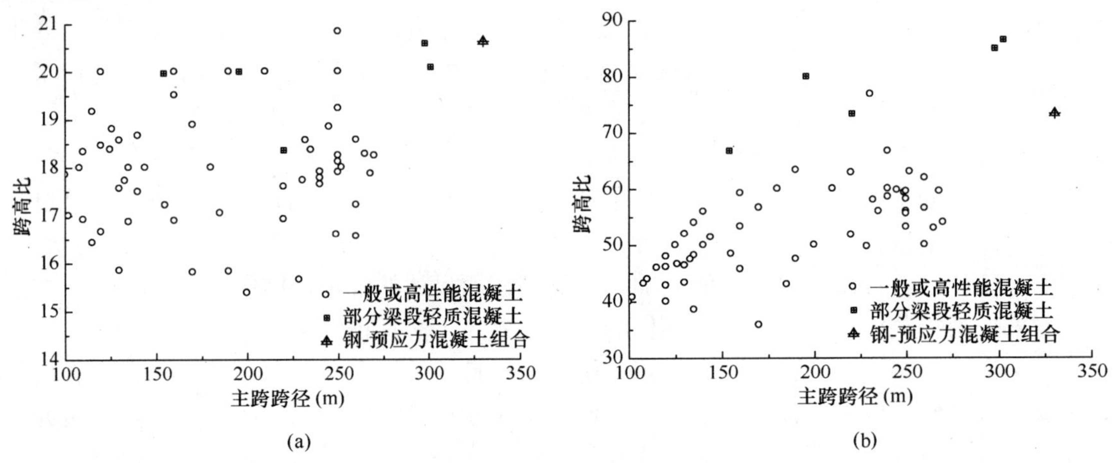


图1 预应力混凝土连续刚构跨高比统计

(a) 根部主梁跨高比；(b) 跨中主梁跨高比

**国内外几座大跨径箱梁桥的病害调查** **表1**

| 跨径组合（m） | 国家 | 桥型 | 悬臂束 | 高跨比 | | 腹板斜裂缝 | 下挠量（cm） | 测量时成桥年数 |
|---|---|---|---|---|---|---|---|---|
| | | | | 根部 | 跨中 | | | |
| 140+240+140 | 中国 | 连续刚构 | 直线束 | 1/17.8 | 1/60 | — | 31.7 | — |
| 162+3×245+162 | 中国 | 连续刚构 | 直线束 | 1/18.9 | 1/59.8 | 有 | 30.5 | 7 |
| 150+270+150 | 中国 | 连续刚构 | 直线束 | 1/18.2 | 1/54 | 有 | 26 | 7 |
| 65+125+180+110 | 中国 | 连续刚构 | 直线束 | 1/18 | 1/60 | 有 | — | — |
| 105+4×160+105 | 中国 | 连续刚构 | 直线束 | 1/20 | 1/53.3 | 有 | 22 | 10 |

续表

| 跨径组合（m） | 国家 | 桥型 | 悬臂束 | 高跨比 | | 腹板斜裂缝 | 下挠量（cm） | 测量时成桥年数 |
|---|---|---|---|---|---|---|---|---|
| | | | | 根部 | 跨中 | | | |
| 99＋195＋99（Parrotts Ferry） | 美国 | 连续刚构（轻骨料混凝土） | — | 1/20 | 1/80 | 有 | 63.5 | 12 |
| 72＋241＋72 | 帕劳 | 带铰 T 构 | 直线束 | 1/17 | 1/65.8 | 有 | 139 | 18 |
| 100＋220＋100（Stovset） | 挪威 | 连续刚构（中跨部分轻质混凝土） | — | 1/18.3 | 1/73.3 | — | 20 | 8 |
| 94＋301＋72（Stolma） | 挪威 | 连续刚构（中跨部分轻质混凝土） | — | 1/20.1 | 1/86 | — | 9.2 | 3 |
| 主跨 260（Getway） | 澳大利亚 | 连续刚构 | — | 1/16.6 | 1/50.0 | — | — | — |

以某主跨跨径 268m 的预应力混凝土连续刚构为例，分析高跨比对结构长期变形的影响。该桥箱梁顶板宽 16.4m，底板宽 7.5m。根部梁高为 15m（$h_1/L=1/17.9$），跨中梁高为 4.5m（$h_2/L=1/59.6$），梁底线形按曲线 1.6 次抛物线布置，箱梁截面如图 2 所示，悬臂束部分采用了下弯束，节段划分及预应力布置见图 3。中跨合龙束 15 对，在合龙时，先张拉了 12 对合龙束，预留了 3 对后期底板张拉束（Z6，Z10 和 Z14），预期成桥一年后进行张拉，以期改善桥梁的运营后的内力和变形状况。

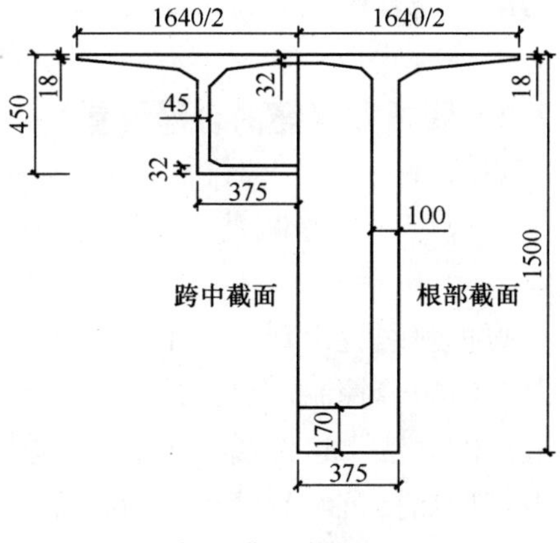

图 2　箱梁典型截面图（cm）

现以墩顶处梁高作为变化参数来研究该桥的长期性能，分别取 17m（$h_1/L=1/15.8$）和 14m（$h_1/L=1/19.1$），相应的，跨中梁高改变为 5m（$h_2/L=1/53.6$）和 4m（$h_2/L=1/67$），梁底曲线仍按 1.6 次抛物线布置。计算长期变形时，混凝土收缩徐变模式采用 CEB-FIP78 模型。成桥后跨中挠度的发展如图 4 所示。

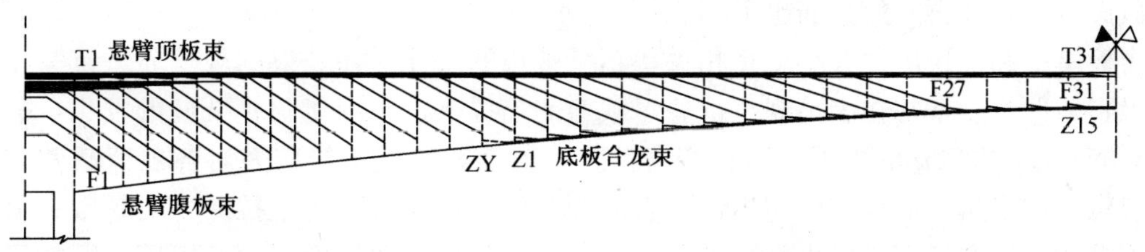


图 3　带有下弯束的纵向预应力配束方案

从图 4 可以看出，将墩顶及跨中的高跨比从原先的 1/17.9、1/59.6 分别增加至 1/15.8、1/53.6 后，30 年后的跨中挠度从 17cm 减少至 7.6cm；如将墩顶及跨中的高跨比分别减小至 1/19.1、1/67 后，30 年后的跨中挠度为 25.5cm 左右，所以适当控制高跨比是抑制长期变形过大的一个有效措施。也有学者提出，墩顶截面对主梁变形影响较大，为

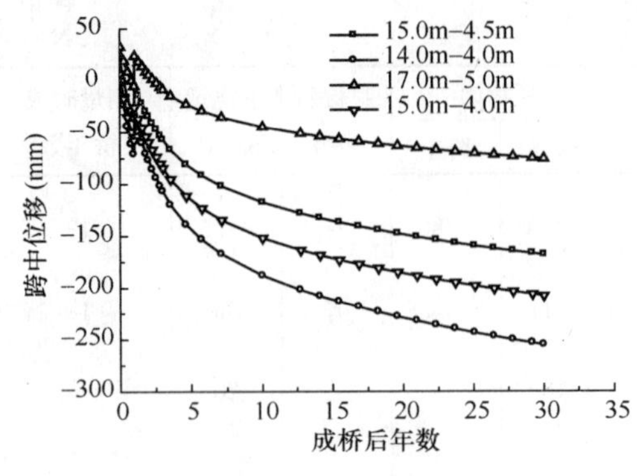


图 4 不同高跨比下的跨中挠度发展

减轻自重，可适当减小跨中截面高跨比，其实不然。本文分析了保持墩顶高跨比不变（$h_1/L=1/17.9$），将跨中截面减小至 4.0m（$h_2/L=1/67$），跨中长期变形的计算结果如图 4 所示。从中可以看出，将跨中截面减小 0.5m 后，30 年后的跨中挠度为 20.8cm，比原高跨比下的 17cm 有所增加。这是因为，减小跨中截面高度后，减小了中跨合龙束的偏心距，从而降低了其抵消恒载效应的效率。目前，新建的几座大跨径的预应力混凝土连续刚构，汉源大树大渡河大桥（133m＋255m＋133m）、柏溪金沙江大桥（140m＋249m＋140m）、庙子坪岷江大桥[4]（125m＋220m＋125m）墩顶高跨比分别为 1/15.9、1/16.6 和 1/16.3，均比表 1 中已出现病害桥梁的高跨比要大。

### 1.2.2 腹板和箍筋的合理配置

如前所述，由于对桥梁轻型、超柔和美观的追求，现代箱梁桥的箱壁越来越薄，特别是腹板尺寸日趋减薄，从而削弱了箱梁的抗裂和抗剪能力，使混凝土箱梁更易出现裂缝。

腹板斜裂缝出现后，不易完全闭合，不像弯曲裂缝在使用阶段多数情况下可以闭合，所以对斜裂缝控制需更加严格。不少设计人员，认为箱梁腹板斜裂缝问题严重，可采用多配箍筋的办法。这混淆了抗裂与抗剪承载力两者的概念。应当认识到，提高截面的抗剪承载力与增加抵抗斜裂缝出现荷载能力（提高受剪抗裂性）并不是同一回事。抗裂性能不好不等于抗剪承载力差，加密箍筋的做法的确对提高抗剪承载力有利，但箍筋并不能防止腹板斜裂缝的出现，提高不了斜截面的抗裂性。况且，箍筋增多，减小了混凝土截面，更不利于抗裂。当然，多配箍筋，对斜裂缝出现之后的开展具有抑制作用。所以，在箱梁满足抗剪承载力和构造的要求下，无需多加配置箍筋。一味地增大配箍量，可能导致腹板混凝土浇筑的不密实，反而影响了施工和混凝土浇筑质量。而且，在抗剪承载力中，若混凝土抗剪截面过小，加再多的箍筋也无用。

腹板厚度和主拉应力大小息息相关，箱梁腹板的尺寸往往是根据规范通过抗剪承载力验算得到。但是，常常在满足规范抗剪承载力的前提下，并不能防止裂缝的发生，只能控制斜裂缝宽度在 0.2mm 以下。腹板变薄后，一方面，腹板主拉应力对各项荷载效应变得更为敏感，如果某一荷载变化超出设计者的预期值，可能会引起主拉应力的超限；另一方面，钢筋的布置变得困难，也影响了混凝土浇筑质量。合理地增大腹板厚度，既能提高抗剪承载力，又能有效地改善箱梁的抗裂性能，便于施工中的钢筋布置，尤其预应力在腹板上的锚固。

还应提出，前些年，不少设计人员在设计纵向预应力束时，取消腹板下弯束，通过下式计算截面的抗剪能力：

$$\tau=\tau_{\alpha}+0.2\sigma_{x}+0.4\sigma_{y} \tag{1}$$

即认为桥梁截面上的抗剪能力由混凝土结构本身的抗剪能力 $\tau_\alpha$、纵向预应力产生的截面正应力 $\sigma_x$ 和竖向预应力产生的竖向压应力 $\sigma_y$ 三部分组成，因此认为，只要多配竖向预应力，就可解决斜截面的抗裂问题，也不必考虑 $\sigma_x$、$\sigma_y$ 和 $\tau_\alpha$ 之间的相互关系。这种想法至今仍有不少市场，以致在梁腹板中大量配置竖向预应力束，又多又密。实际上，竖向预应力束的作用是与截面混凝土尺寸有关联的，不能无节制地扩大 $\sigma_y$ 的作用。更有甚者，还将竖向预应力束配置在绝大部分梁段上，这种做法浪费很大。

同样地，在设计箍筋时，也是不加考虑箍筋的抗剪承载力部分与混凝土截面抗剪承载力部分的相互关系。在一些设计中，也是又粗又密地配置箍筋，如 Φ16@120，以致大大增加了施工难度，影响了混凝土的密实度，并且，也是全梁段多配箍筋，甚至在梁跨中区段也配得又粗又密（实际上有的只需按构造配置箍筋），浪费十分惊人。

## 1.3　腹板下弯束的设置

大跨径预应力混凝土梁桥一般采用悬臂浇筑的施工方法，预应力束的布置主要依据施工和成桥阶段的受力状态确定。传统纵向预应力配束方案是根据梁在荷载作用下的弯矩包络图设计的，配置了悬臂预应力束（顶板直线束、腹板下弯束）和连续预应力束等。自 20 世纪 80 年代末，曾有专家建议取消下弯束，悬臂预应力束只采用顶板直线束（如图 5 所示）。该方案的优点是腹板内预应力管道较少，给箱梁施工带来了极大方便。但是，相应地需要布置大量的竖向预应力束，来限制主拉应力。这种直线式的配束方案，不仅导致工程费用的增加，而且经历近二十年的工程实践，很多大跨径混凝土梁桥出现了病害，其中最突出的问题是箱梁腹板出现斜裂缝和跨中挠度的不断增大。目前出现腹板裂缝病害的大跨径箱梁桥，大都采用了直线式的配束方案或腹板下弯束相对顶板直线束的用量较小，如表 1 所示。文献[5]调查结果显示，采用直线式配束方案的箱梁桥出现腹板斜裂缝的桥梁比例高达 95%，比采用部分下弯束的箱梁桥腹板开裂的比例高出 21%。因此，笔者认为纵向预应力下弯束的取消或腹板下弯束设置过少，是导致病害出现的主要原因之一。因为采用下弯束限制主拉应力及斜裂缝的出现是最为有效的。当然，采用部分下弯束的箱梁桥出现斜裂缝的比例也不低，可能与腹板下弯束数量及布置方式有很大关系，所以腹板下弯束的设计还存在优化改进的地方。

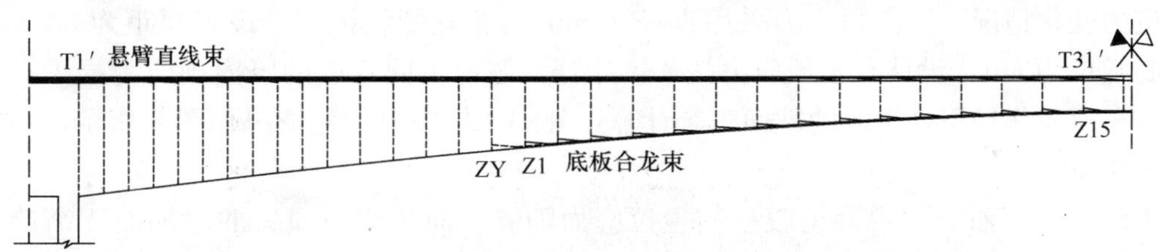


图 5　直线式纵向预应力配束方案

腹板斜裂缝的出现主要是由于主拉应力超限，理论上讲，可以通过纵向预压应力与竖向预应力提供的竖向预压应力的乘积来限制主拉应力的大小[6]。但实际效果不佳，这是由于受梁高的限制，竖向预应力束太短，预应力损失较大，实际预应力有效值与设计值相差较大，而且存在预应力空白区。文献[7]测得的竖向预应力损失高达 50%。文献[8]对衡阳东阳渡湘江大桥的竖向预应力损失进行了现场测试，部分实测的总损失达到了初始张拉力的47％左右；由于锚具变形和钢筋回缩变形值一定，竖向预应力筋越短，该项损失越大，实测的锚具变形和钢筋回缩引起的损失占预应力总损失的53.7％。此外，竖向预应力孔道较窄，压浆存在漏浆或灌浆不饱满问题，导致对抗剪截面的削弱，使腹板主应力增大。从钱江三桥[9]随机抽检35根的结果来看，无浆占71.42％，不饱满占11.42％，开孔流水的占40％。竖向预应力筋存在的这些问题使得其作用很难达到设计要求。

主拉应力与剪应力的大小也密切相关，可以通过减小竖向剪应力来限制主拉应力的大小。剪应力主要由箱梁截面的剪力大小决定，而腹板下弯束能提供预剪力，从而减小了箱梁截面的剪应力。仍以图2所示的主跨为268m的连续刚构为例，在自重作用下，该连续刚构主跨半跨截面的剪力在墩顶处最大，跨中位置处接近为零。如将其悬臂阶段的部分纵向预应力束（腹板下弯束，F1～F27）改为图5所示的直线式布置方案，图3和图5两种预应力配束方案提供的预剪力如图6所示。

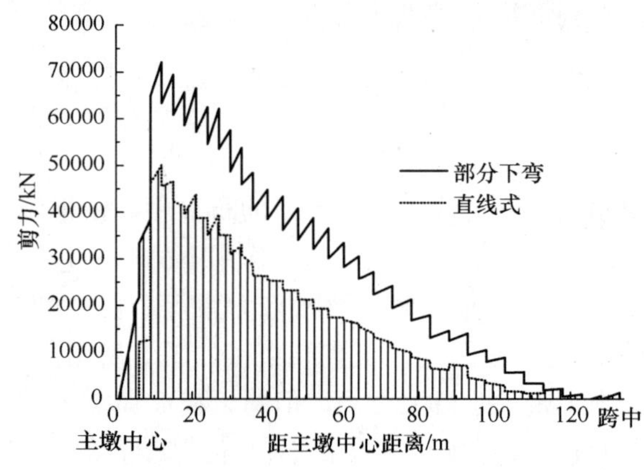


图6 纵向预应力作用下的中跨半跨剪力图

图6表明，部分纵向预应力采用下弯束后，提供了较高的预剪力，这样能够平衡较多的恒载剪力，从而减小了箱梁截面中的剪应力。因此，为抵抗腹板主拉应力，设置腹板下弯束是比较有效的方式。在跨中区段，恒载产生的剪力较小，所以该区段无需配置竖向预应力束，即使配置竖向预应力束，该区域梁高较小，竖向预应力损失较大，效果也不佳。

本文选取图2所示的连续刚构主跨区段距主墩距离为73m截面（近$L/4$截面）进行分析，比较两种配束方案下腹板的主拉应力状况，该截面尺寸如图7所示。选取的计算点A，B，C分别位于上翼缘与腹板交接处、箱梁几何中心、腹板与下翼缘交接处，考虑在不同的竖向预应力损失的情况下，计算上述两种纵向预应力的配束方案下的该截面腹板的主拉应力变化情况。竖向预应力采用直径32mm的精轧螺纹钢筋，设置间距为50cm，张拉控制吨位为60t，设计有效预应力按48t计算。假定不同程度的竖向预应力损失，在恒载和汽车荷载作用下，选取的截面三个计算点处的主拉应力变化情况如图8所示，图中负值代表拉应力。

由图8可以看出，混凝土腹板的主拉应力随着竖向预应力损失的增加而逐渐趋于受拉，且增长速度较快。比较直线式和下弯式配束方案，采用下弯束后，有效地改变了主拉应力的大小，即使竖向预应力不发挥作用，主拉应力值也很小。所以通过下弯束的合理设置，可控制主拉应力大小；竖向预应力筋在剪应力较小、高度较小的梁段完全可以取消。

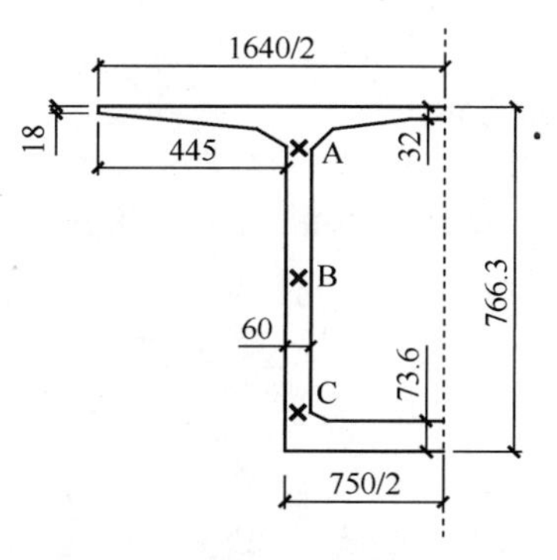


图 7　箱梁计算截面（cm）

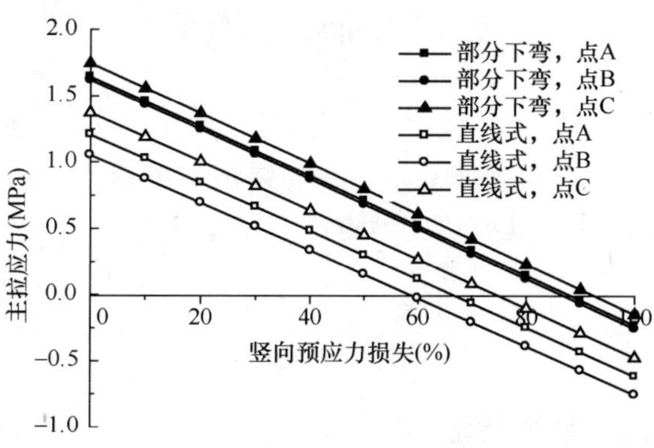


图 8　主拉应力计算结果

## 1.4　剪切变形对挠度的影响

### 1.4.1　斜向开裂后的剪切变形计算

如前所述，大跨径箱梁桥的下挠过大和箱梁开裂常常是“并发症”。不能总用“综合”因素来分析跨中挠度过大的问题，我们认为，对待此类质量通病，仅从反复计算（箱梁空间作用、混凝土收缩徐变、日照、温差作用等等，甚至预应力松弛超过预期值及预应力损失过大）以及施工工艺方面去分析是远远不够的。混凝土开裂后，结构的下挠变化的机理变得异常复杂，下挠现象也会逐步恶化。混凝土开裂不但本身会引起主梁刚度的下降，还会对截面特性、混凝土收缩徐变效应及预应力效应产生影响，引起内力重分布，继续加剧裂缝的发生和开展。

主梁刚度由弯曲刚度和剪切刚度组成，通常，梁的挠度主要由弯曲刚度（弯矩）控制，但是，对于薄腹箱梁，剪切变形及其引起的下挠是不可忽略的，特别是在斜裂缝出现及发展之后，剪切变形对薄腹箱梁的挠度影响会更大。混凝土开裂前，梁的工作处于弹性阶段，其剪切刚度可以根据弹性力学的方法计算。

$$K_v = GA_v = 0.417E_cA_v \tag{2}$$

式中，$G$ 为剪切模量；$A_v$为抗剪面积，对于矩形梁，$A_v$取 0.83A，对于 T 梁，近似取腹板面积。然而，当混凝土出现斜裂缝后，剪切变形的影响不能再用式（1.2）计算，因为开裂后的混凝土泊松比对第二应力方向已失去作用，横向作用到剪切裂缝上的力由抗剪钢筋来承担。Timoshenko[10]认为，对于 $l/h\geqslant 10$ 的梁，在弹性阶段可以忽略剪切变形对挠度的影响，但出现斜裂缝后，剪切刚度对总挠度的影响不可忽略，且随时间推移，影响将不断增加。本文第一作者在 20 世纪 60 年代中期一批抗剪性能试验研究中，就有过这种定性的认识。德国著名学者 Leonhardt[11]也曾指出，当梁的跨高比 $l/h\leqslant 12$ 时，混凝土开裂后，在跨高比不同，截面及配筋不同的梁中，由剪切变形可为弯曲变形的 0.2～3 倍。本文作者[12]借助于变角桁架模型，量化分析了斜裂缝出现后的瞬时剪切变形及长期剪切变形，并与三根薄腹梁的试验结果进行了比较。

变角桁架模型的剪切刚度的计算公式为：

$$K_{\mathrm{v}}=\frac{\rho_{\mathrm{v}} n_{\mathrm{v}} E_{\mathrm{c}} A_{\mathrm{v}} \cot^2\alpha}{1+4\rho_{\mathrm{v}} n_{\mathrm{v}}\left(1+0.39\cot^2\alpha\right)^2} \tag{3}$$

式中，$\rho_{\mathrm{v}}$为配箍率；$n_{\mathrm{v}}$为箍筋弹模与混凝土弹模的比值；$\alpha$ 为斜压杆倾角。故桁架模型在$V_{\mathrm{t}}$作用下，其剪切变形为：

$$\delta_{\mathrm{v}}=\int_l \frac{V_{\mathrm{t}}\overline{V}}{K_{\mathrm{v}}}\mathrm{d}x \tag{4}$$

## 1.4.2 试验验证

### 1.4.2.1 试验设计

本次试验设计了 3 根长度为 4m 的高强混凝土 T 形梁，梁的计算跨径为 3.6m，箍筋有两种形式：$\phi8@150$（$\rho_{\mathrm{v}}=0.67\%$）和 $\phi8@200$（$\rho_{\mathrm{v}}=0.5\%$），3 根薄腹梁的纵筋均为 6$\Phi$22+2$\phi$8。各梁的设计参数如表 2 所示，加载方式为两点集中力加载，各梁的截面配筋及加载方式如图 9 所示。为了研究斜裂缝出现后的剪切变形对跨中挠度的影响，重点量测了在箍筋应力屈服前，混凝土薄腹梁的挠度变化。在梁的支座、加载点及跨中均布置了百分表，量测梁的挠度变化。

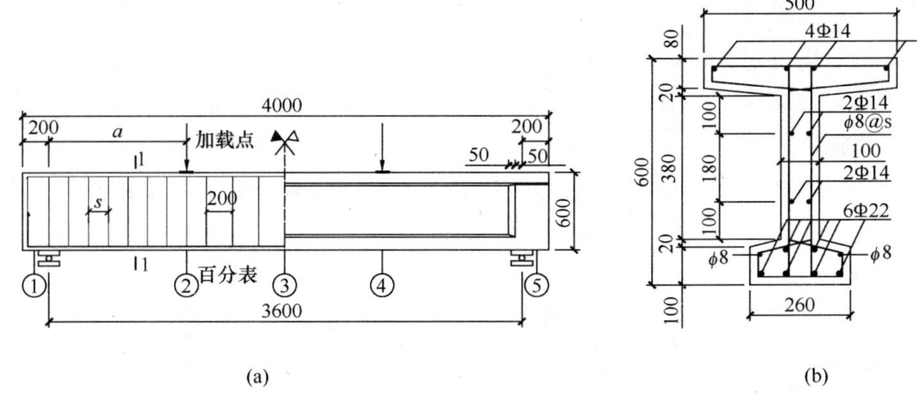


图 9 试验梁设计

（a）加载方式及测点布置图；（b）薄腹梁的配筋示意图

试验梁设计参数 表 2

| 编号 | $f_{\mathrm{cu}}$ (MPa) | $h_0$ (mm) | $a$ (mm) | $a/h_0$ | 纵筋 | | 箍筋 | | | $V_{\mathrm{u}}$ (kN) |
|---|---|---|---|---|---|---|---|---|---|---|
| | | | | | $f_{\mathrm{sy}}$ (MPa) | 配筋 | $f_{\mathrm{vy}}$ (MPa) | $s$ (mm) | $\rho_{\mathrm{v}}$ (%) | |
| L1 | 72 | 546 | 1050 | 1.923 | 373 | 62Φ2+2Φ8 | 360 | 150 | 0.67 | 497 |
| L2 | 72 | 546 | 1050 | 1.923 | 373 | 6Φ22+2Φ8 | 360 | 200 | 0.5 | 425 |
| L3 | 72 | 546 | 800 | 1.465 | 373 | 6Φ22+2Φ8 | 360 | 150 | 0.67 | — |

### 1.4.2.2 变形比较

本文重点测量了 3 根梁的挠度在箍筋及纵筋屈服前随荷载的变化，3 根试验梁的挠度随荷载的变化见图 10～图 14。观测完梁挠度随荷载发展的变化之后，将薄腹梁加载至破坏，L1 和 L2 的破坏荷载见表 2 中的 $V_{\mathrm{u}}$，两者的破坏形态均为剪压破坏。L3 梁加载至

500kN 后并未有破坏的征兆，由于试验设备的限制，L3 并没有加载至破坏。

在桁架模型中，如果不考虑混凝土对抗剪的贡献（受压区混凝土的抗剪、骨料的咬合作用和纵筋的销栓作用等），计算出的承载力偏保守。相应的，出现斜裂缝之后总剪切变形的计算应包括两个部分：一是考虑在斜裂缝刚刚出现时，即 $V_c$ 作用下的剪切变形，按式（1.2）计算；另一方面，斜裂缝出现之后，剪力传递机理改变，此式剪切变形的计算采用前述的变角桁架模型，变角桁架承受的剪力为 $V_t=V-V_c$，按式（1.3）计算。对应于不同的斜裂缝形式，斜裂缝出现的剪力可根据文献[13]计算。

3 根试验梁挠度随荷载变化的计算值与试验值的比较如图 10～图 14 所示。图 10～图 14 中弯曲变形 J，弯曲变形 A 分别表示根据中国混凝土及预应力混凝土桥涵设计规范 JTGD 62—04[14]和美国混凝土协会 ACI318R-08[13]方法计算出的弯曲变形。

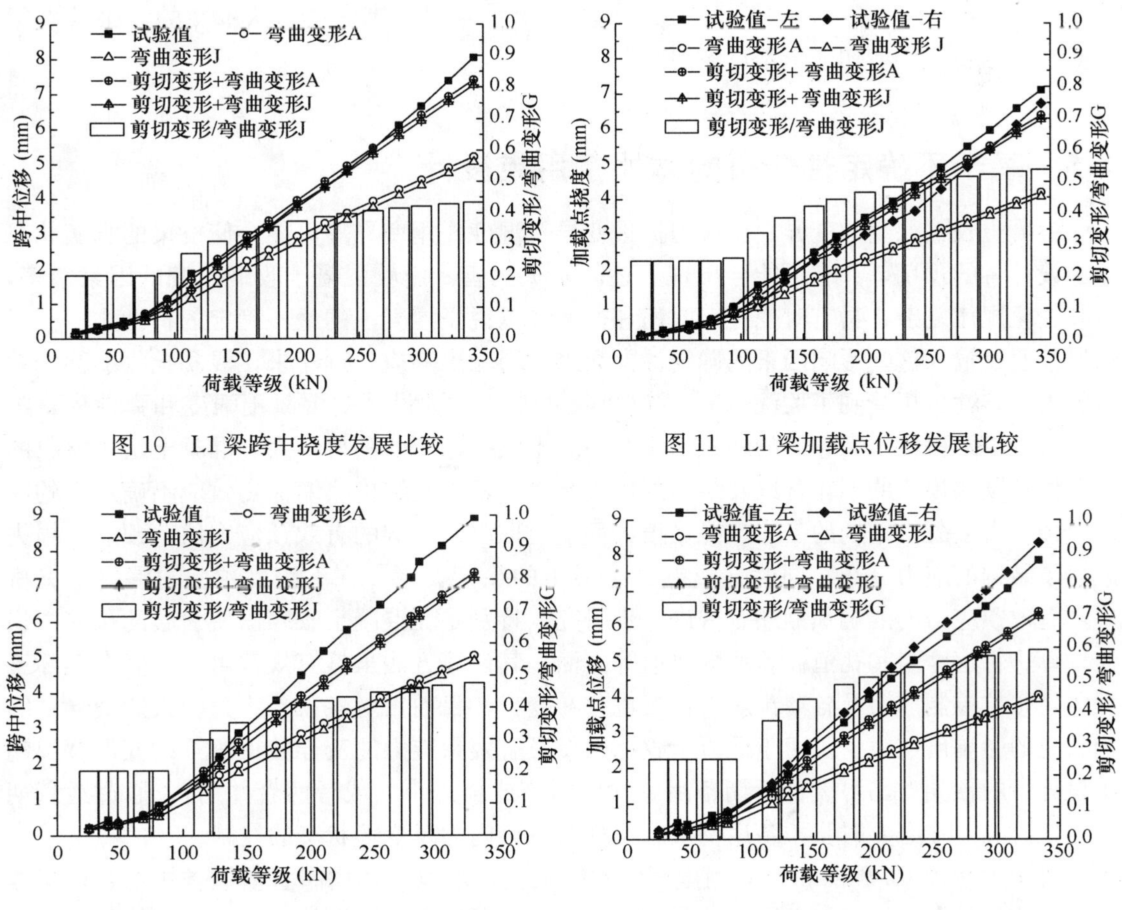


图 10　L1 梁跨中挠度发展比较

图 11　L1 梁加载点位移发展比较

图 12　L2 梁跨中挠度发展比较

图 13　L2 梁加载点位移发展比较

由图 10～图 14 可以看出，采用 JTGD 62—04 方法计算出的钢筋混凝土梁弯曲变形稍小于 ACI 规范的计算值，但均小于实测的变形值。这说明，如果忽略出现斜裂缝之后的剪切变形，会有低估变形的危险，这对保证混凝土结构的正常使用性能不利。比较图 10 和图 12 可以看出，随着配箍率的减少，剪切变形与弯曲变形的比值稍有增加。比较图 11

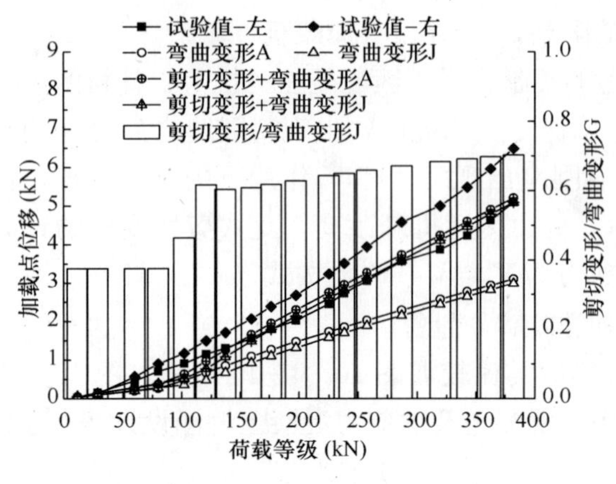


图 14 L3 梁加载点位移发展比较

和图 14 看出，随着剪跨比的减小，剪切变形所占总变形的比重有较大增加。从剪切变形与弯曲变形比值柱形图可以看出，随着荷载等级的增加，剪切变形所占的比重会越来越大。图 14 中，剪切变形与弯曲变形比值柱形图在 120kN 时有突变，那是因为 L3 梁的剪跨比较小，斜裂缝先于弯曲裂缝出现，故剪切变形比重会突然增大；随着荷载增加，弯曲裂缝出现之后，弯曲刚度较小，故剪切变形比重下降。我们认为，薄腹箱梁斜向开裂后的剪切变形是下挠过大的主要原因之一。从本文的 3 根薄腹梁试验也可以看出，剪切变形不可忽略。

## 1.5 基于不确定性分析的体外备用束设计

混凝土收缩徐变的变异、预应力损失变异及温度作用变异等可能会使桥梁的真实状况超出设计者的预期，亦有可能会出现上述的常见病害。影响混凝土收缩徐变的因素众多，且其变化规律复杂，具有时变性，收缩和徐变可以说是混凝土的最不确定的特性之一。以往研究通过输入这些影响因素的确定性参数来获得输出响应，近年来，混凝土收缩徐变效应的不确定性分析受到了关注[15~17]，这也很正常，因为设计的混凝土强度和弹性模量都已经考虑了随机性，而它们的随机性没有混凝土收缩徐变的随机性大。混凝土组成材料的变异性、预测模式的不完善以及结构的环境条件等，都是结构收缩徐变效应不确定性的原因。预应力混凝土结构的工作状况在很大程度上取决于结构的有效预应力值，预应力损失改变了结构的应力分布，同时又影响到混凝土的徐变变形。在大跨度预应力混凝土梁桥中，有效预应力的准确预测还有待进一步研究，特别是在混凝土收缩徐变引起的预应力损失计算、管道摩阻和孔道偏差系数取值等方面，其计算方法虽然有规范可循，但在实际工程中，张拉控制应力及各项预应力损失也存在变异，目前针对各项预应力损失的变异对结构状况的影响研究很少。由于混凝土收缩徐变及有效预应力的变异使得桥梁的变形状况超出设计者的预期时，可设置后期备用束（体内束或体外束，或者两者并用），通过运营期内定期检测，在必要时张拉备用钢束，实现对桥梁内力及变形状态的控制。

对于大跨径 PC 箱梁桥，受到截面尺寸的限制，难以在体内布置后期备用束，这里考虑设置体外备用束方法。后期体外备用束存在三种布置方案，即体外顶板束、体外底板束和体外折线束，如图 15（a）。为提高体外预应力束的利用效率，需认识到何种布置方案及位置对控制桥梁下挠过大最为有效。假定一单位力作用于桥梁跨中，其产生的弯矩如图 15（b）所示，而三种体外预应力束布置方案产生的弯矩图如图 15（c，d，e），由弯矩图乘可知，当顶板束、底板束的张拉端，折线束的转向块设置在单位力作用下主梁弯矩图的反弯点 A 附近时，这时体外束布置位置最优，其改善下挠的效率最高。以图 2 和图 3 所示的连续刚构为例，按上述三种方案进行布置一对 25$\Phi$15.24 的体外预应力束，如图 15（a），张拉控制应力为 1209MPa。在该三种体外束布置方案下，跨中的上拱值见表 3。

三种体外束布置方案下的跨中上拱值　表 3

| 方　案 | 顶板束 | 底板束 | 折线束 |
|---|---|---|---|
| 上拱（mm） | 11 | 6 | 11 |

从表 3 可以看出，布置体外顶板束和折线束对改善长期下挠病害最为有效。此外，顶板束及折线束产生的剪力（如图 16 所示）与恒载下的剪力方向相反，体外顶板束和折线束也能减小腹板的主拉应力。所以，设计体外备用束时，优先考虑采用顶板束和折线束。

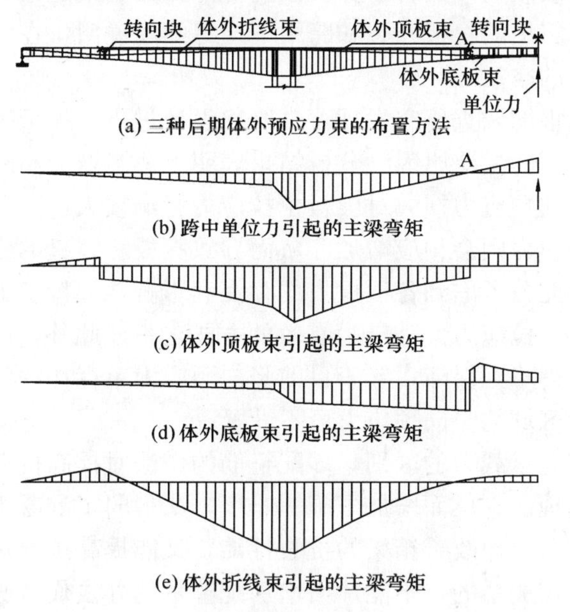


图 15　三种体外束布置方案下的主梁弯矩

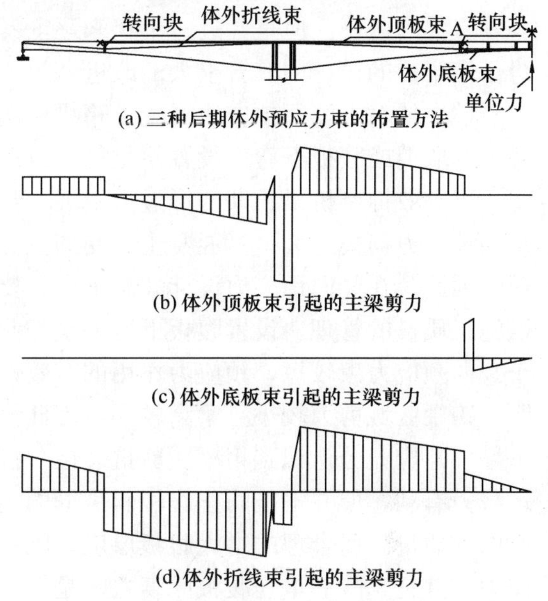


图 16　三种体外束布置方案下的主梁剪力

本文选取了收缩模型不确定性系数 $\Psi_1$、徐变模型不确定性系数 $\Psi_2$、环境相对湿度 $RH$、混凝土强度 $f_{cu}$、自重 $k_{G1}$、二期恒载 $k_{G2}$、预应力张拉控制应力 $\sigma_{con}$、管道每米局部偏差 $k$ 和预应力筋与塑料波纹管道的摩擦系数 $\mu$（$k$ 和 $\mu$ 的均值和变异系数是根据大量实桥和室内试验结果进行统计而得到），共 9 个随机变量进行不确定分析，基于现行桥规中的收缩徐变模型[14]，以成桥后的跨中挠度相对成桥时的增量为输出响应，进行统计分析，得到其均值及具有 95%置信水平的区间，如图 17 所示。模型中考虑了汽车荷载的准永久值效应对桥梁长期变形的影响。

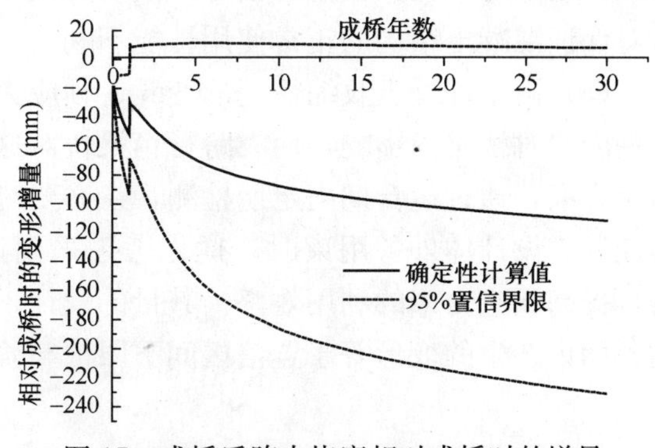


图 17　成桥后跨中挠度相对成桥时的增量

根据上述长期变形不确定性分析可知，10 年后，具有 95%保证率的跨中长期变形增量区间的下限值与确定性计算值的差异约为 10cm，30 年后该差异为 12cm 左右。可布置 11 对 25$\Phi$15.24 的体外顶板束，来控制 30 年后桥梁的长期变形。但是，如果所有的顶板束均布置在图 15（a）位置 A 时，预应力集中锚固将使得顶板局部应力过大，这时顶板束的锚固位置可基于图 15（a）的 A 点逐步向左右相邻节段分散，使体外顶板束产生的变形等于 12cm 左右。当然，根据早期的变形差值，所需的备用束的量会减小，目前该方面工作正在进一步研究中。

## 1.6 结论

（1）预应力混凝土箱梁桥（连续梁和连续刚构）具有不少突出的优点，在桥梁建设中有着广泛的应用。但是，近二十年来逐渐暴露出的腹板开裂和跨中下挠等问题说明，在大跨径箱梁桥的设计中有着很大的改进和提高的空间。

（2）从对大跨径预应力混凝土箱梁桥病害的调查研究发现，高跨比设计过小，腹板下弯束的取消或腹板下弯束设置不足是导致病害的主要原因。分析结果表明，适当增加高跨比，可有效抑制桥梁长期变形过大发展。竖向预应力损失对腹板主拉应力影响很大，一旦竖向预应力损失过大，其混凝土承受的主拉应力则会相应增加，从而导致腹板斜裂缝的出现。因此，在竖向预应力设计和施工中，要充分考虑到各项损失的存在和影响，应特别加强施工质量的管理。设置腹板下弯束是控制主拉应力大小最为有效的方法之一。此外，由于竖向预应力束较短，预应力作用的有效性得不到保障，鉴于目前竖向预应力束存在的问题、跨中区域剪力较小且梁高较小，因此该区域的竖向预应力可不设置。

（3）应充分认识到箱梁受剪抗裂性与抗剪承载力的区别，多配箍筋的做法对提高抗剪承载力有利，但并不能防止腹板斜裂缝的出现。它反而增加了施工难度，又削弱了混凝土截面和质量。而合理的增大腹板厚度，既能有效地改善箱梁的抗裂性能，又能提高抗剪承载力。并且，配筋量与腹板厚度之间是有一定关系的，不能片面增加配箍量的办法代替腹板厚度的值。

（4）目前国内大跨径预应力混凝土梁桥的质量通病是腹板斜向开裂和挠度过大，而且变形和裂缝常常是“并发症”，我们认为，薄腹箱梁斜向开裂后的剪切变形是下挠过大的主要原因之一。从本文的 3 根薄腹梁试验也可以看出，剪切变形不可忽略。混凝土梁出现斜裂缝之后，剪力传递机理发生改变，可用变角桁架模型来计算剪切变形，瞬时剪切变形的计算结果与本文试验吻合较好。按桥规[14]计算的变形存在低估斜向开裂梁变形的危险，这对保证混凝土结构的正常使用性能不利。

（5）由于混凝土收缩徐变的变异、预应力损失变异及温度作用变异等，很难准确计算这些因素对桥梁的影响，而大跨径箱梁桥对这些因素较为敏感，所以可设置后期备用体外预应力束，通过运营期内定期检测，在必要时张拉备用钢束，实现对桥梁内力及变形状态的控制。设计体外备用束时，优先考虑采用体外顶板束和折线束，并找出最优的布置方案，提高预应力束的利用效率。其量的设计，可根据时变效应不确定性分析的结果，使后期备用束产生的变形等于置信区间下限值与确定性计算值的差值。

## 参考文献

[1] 上官兴，郭圣栋等. 预应力混凝土连续梁桥的"恒载零挠度设计"新理念. 全国既有桥梁加固、改造与评价学术会议论文集，2008，65-72.

[2] 王法武，石雪飞. 大跨径预应力混凝土梁桥长期挠度控制研究. 公路，2006，8：72-76.

[3] 刘钊，戴玮等. 大跨度预应力混凝土梁桥的合理成桥状态设计方法. 桥梁建设，2010，1：40-44.

[4] 王勇，蒋劲松. 连续刚构设计构思的探讨. 西南公路，2003(1)：25-27.

[5] 王国亮，谢峻等. 在用大跨度预应力混凝土箱梁桥裂缝调查研究. 公路交通科技，2008，25(8)：52-56.

[6] 杨高中，杨征宇等. 连续刚构桥在我国的应用与发展. 公路，1998(6)：1-7.

[7] 黄豪，唐小兵等. 竖向预应力作用效果的数值模拟与预应力损失的试验研究. 武汉理工大学学报，2007，31(5)：922-924.

[8] 汪剑. 大跨径预应力混凝土箱梁桥非荷载效应及预应力损失研究. 长沙：湖南大学，2007.

[9] 施颖，郑建群. 从设计层面探讨预应力混凝土连续梁桥裂缝控制. 重庆交通学院学报，2005，24(4)：13-18.

[10] S. Timoshenko 著，萧敬勋译. 材料力学. 天津：天津科学技术出版社，1989.

[11] F. Leonhardt 著. 钢筋混凝土结构裂缝与变形的验算. 北京：水利电力出版社，1983.

[12] 吕志涛，潘钻峰. 斜向开裂混凝土梁的瞬时及长期剪切变形研究. 建筑科学与工程学报，2010，27(2)：1-9.

[13] American Concrete Institute. Building code requirements for structural concrete. ACI 318-08 and "Commentary." ACI 318R-08，Farmington Hills，Mich，2008.

[14] JTG D62-2004. 公路钢筋混凝土及预应力混凝土桥涵设计规范.

[15] MADSEN H O，BAZANT Z P. Uncertainty analysis of creep and shrinkage effects in concrete structures. ACI Journal，1983，82(2)：116-127.

[16] 潘钻峰，吕志涛等. 苏通大桥连续刚构收缩徐变效应的不确定性分析. 工程力学，2009，26(9)：67-73.

[17] 潘钻峰，吕志涛. 基于不确定分析的大跨径 PC 箱梁桥的后期备用束设计. 建筑科学与工程学报，2010，27(3).

# 第 2 章 Chapter 2

# 混凝土与预应力混凝土结构的受剪设计与评估

# SHEAR DESIGN AND EVALUATION OF REINFORCED AND PRESTRESSED CONCRETE STRUCTURES

Michael P. Collins and Evan C. Bentz
University of Toronto, 35 St. George Street, Toronto, Canada.

**Abstract**: Deficiencies in the shear design of concrete structures are inherently more dangerous than deficiencies in flexural design because shear failures can occur without prior warning and with no possibility for redistribution of internal forces. While assessing the shear capacity of a reinforced concrete structure accurately is critically important for public safety, the traditional techniques available for this task are open to dispute. This paper will summarize recent research aimed at understanding the basic mechanisms of shear transfer and will introduce simple, rational, general and accurate procedures for evaluating shear capacity. Examples will be given of the use of these procedures in evaluating the safety of existing structures and in diagnosing the cause of structural failures.

**Keywords**: Shear, size effect, finite elements analysis, footings, strength evaluation, safety

## 2.1 INTRODUCTION

The infrastructure of modern civilization depends to a large extent upon reinforced concrete, which is mouldable "synthetic rock" reinforced with steel bars. In structures made from this material the concrete primarily looks after the compressive stresses while the reinforcing bars are designed to carry the tensile stresses. Like wood, and unlike steel, this material does not have the same strength properties in different directions. Such anisotropic materials are particularly susceptible to what are called shear failures. Thus a reinforced concrete member subjected to shear may develop diagonal cracks and if the member does not contain an appropriate amount of shear reinforcement, these cracks can result in the sudden failure of the member. Shear reinforcement, traditionally called stirrups, links together the flexural tension and flexural compression sides of a member and ensures that the two sides act as a unit. Shear failures involve the breakdown of this linkage and typically involve the opening of a major diagonal crack. See Figure1. Avoiding such failures, which can be catastrophic, is the objective of shear design.

Figure 1　Shear failures 1955 and 2006

To introduce the concepts which will be discussed in this paper consider the simple example of a reinforced concrete beam spanning between three columns as shown in Figure2. As the uniformly distributed load applied to this beam is increased the concrete will first crack on the top face of the beam over the central supporting column where the bending moment has its highest value. At somewhat higher loads the concrete will crack on the bottom face at the location where the moments causing tension on this face have the highest value. To prevent these bending cracks from opening there must be appropriate quantities of longitudinal reinforcement near the top face and the bottom face of the beam. The initial flexural cracks at maximum moment locations will be vertical but as the load is increased new flexural cracks will form at other locations along the beam and as they spread in from the outer faces of the beam these new cracks will become inclined. These inclined "diagonal" cracks are called shear cracks because they form in regions where the shear force is high and the direction of the cracks is dictated by the sense of the shear force. See Figure 2. To prevent these diagonal cracks from opening and causing a shear failure there must be appropriate quantities of transverse shear reinforcement provided along the length of the beam. Thus, the flexural design of the beam can be thought of as arriving at an arrangement of longitudinal reinforcement so that the flexural strength at all sections appropriately exceeds the bending moments caused by the loads. On the other hand, the shear design consists of choosing an arrangement of stirrups so that the shear strength of the

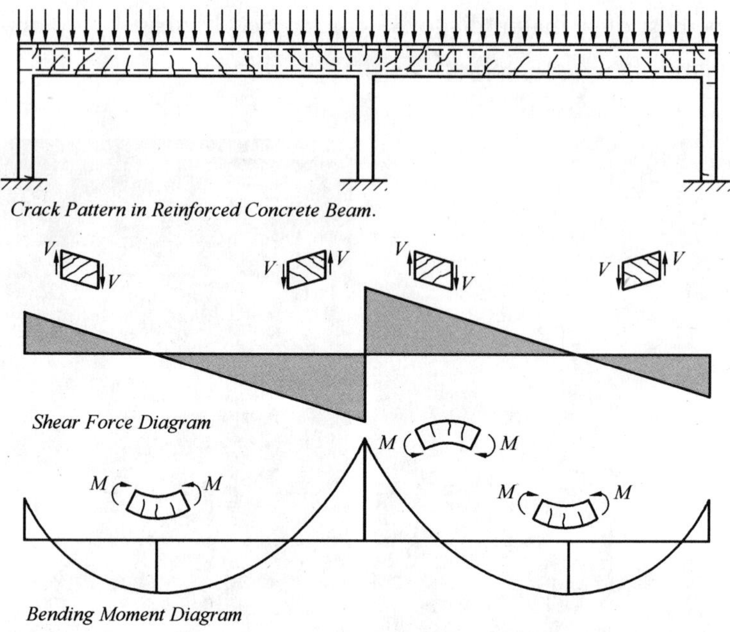


Figure 2 Influence of shear forces and bending moments on crack patterns

beam at all sections appropriately exceeds the shear forces caused by the loads.

In designing for flexure, engineers have available a simple, general, rational method called the "plane sections" theory, see Figure 3, capable of predicting not only the flexural strength, but also the complete load-deformation response of reinforced concrete sections. Because of this, there is little disagreement between different international design codes as to the flexural strength of a given reinforced concrete section or the quantities of reinforcement needed to ensure ductile flexural behaviour. There is, however, substantial disagreement as to the magnitude of the shear strength of structural elements and the reinforcement requirements needed to ensure ductile shear response. Further, rather than the simple, general, behavioural theory available for flexural design, the building code procedures for shear design typically consist of a collection of complex, restricted empirical equations for shear strength. In view of the disparity between the state-of-the-knowledge in flexure, and the state-of-the-knowledge in shear, it is perhaps not surprising that while failures of reinforced concrete structures due to deficiencies in flexural design are extremely rare, failures due to deficiencies in shear design occur much more frequently.

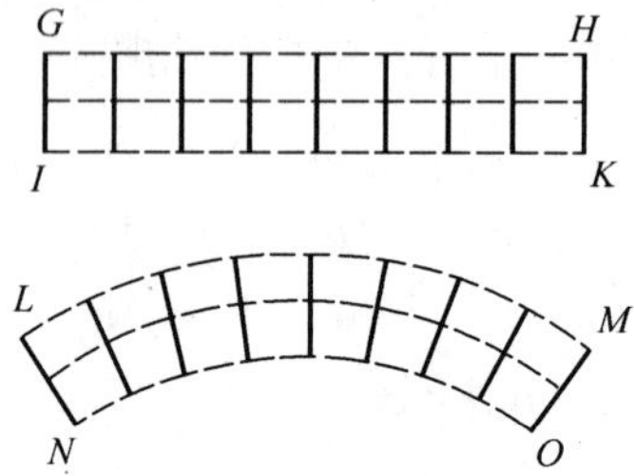


Figure 3 Robert Hooke's 1678 observation that plane sections remain plane

As an illustration of the inconsistency between some current shear design provisions, consider the results of the four slab-strip specimens shown in Figure 4. Two of these specimens were large (L) with thicknesses comparable to that of the 2006 failed bridge slab

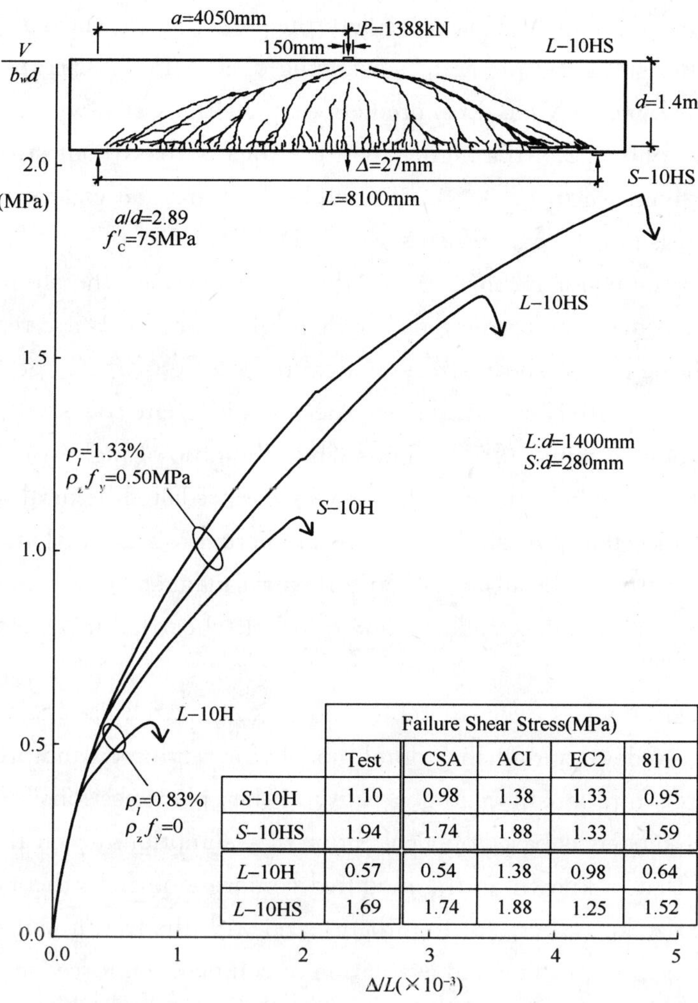


| | Failure Shear Stress(MPa) | | | | |
|---|---|---|---|---|---|
| | Test | CSA | ACI | EC2 | 8110 |
| S-10H | 1.10 | 0.98 | 1.38 | 1.33 | 0.95 |
| S-10HS | 1.94 | 1.74 | 1.88 | 1.33 | 1.59 |
| L-10H | 0.57 | 0.54 | 1.38 | 0.98 | 0.64 |
| L-10HS | 1.69 | 1.74 | 1.88 | 1.25 | 1.52 |

Figure 4　Four high strength concrete slab strips failing in shear.
Note *L*-10H is predicted to fail in flexure at a shear stress of 1. 26 MPa

shown in Figure 1 while the other two were small (S) with thickness similar in size to traditional laboratory specimens. For each pair of specimens, one contained no shear reinforcement (as was the case for the bridge slab) and 0. 83% of longitudinal reinforcement while the other contained 0. 50 MPa of shear reinforcement. To avoid flexural failures, the specimens with shear reinforcement contained somewhat more longitudinal reinforcement (1. 33%). All four specimens failed in shear. The table in Figure 4 compares the observed failures shear stresses with the failure shear stresses predicted by current Canadian (CSA), American (ACI), European (EC2), and British, (8110) codes. Note that for each slab strip, there are considerable differences between the four code predictions. Thus for the large specimen without stirrups, L-10H, the highest predicted failure shear stress is 2. 56 times the lowest predicted failure shear stress. Further, the codes give very different predictions as to how the failure shear stress will change as the depth is made larger or shear reinforcement is added. The ACI code predicts the same failure shear stress for the thick slabs as for the thin slabs and predicts that adding 0. 50 MPa of stirrups will increase the failure shear stress by 0. 50 MPa. EC2 predicts that the thick slab with stirrups will have a failure shear stress only 0. 27 MPa higher than the thick slab without stirrups. The British code predicts this increase to be 0. 88 MPa while the Canadian code predicts 1. 20 MPa. The observed increase in the experiments was 1. 12 MPa.

In contrast to the major inconsistencies that exist between the shear strength predictions of these four codes, the predictions for flexural failure loads are remarkably consistent. If the magnitude of the shear stress at flexural failure of the large specimen without shear reinforcement, L-10H, is calculated, the four different codes all predict a value of 1. 26 MPa within plus or minus 0. 5%. Thus while the ratio of highest to lowest shear failure load can reach 2. 56, the ratio of highest to lowest predicted flexural failure load is only about 1. 01. The uncertainty in shear strength prediction can mean that while an engineer might predict a structure will fail in a ductile flexural manner (e. g. 1. 26 MPa flexure $<$ 1. 38 MPa shear), it may in fact collapse due to brittle shear failure, see Figure 1 and Figure 4.

During the last 30 years a considerable amount of research has been conducted worldwide with the aim of developing behavioural models for reinforced and prestressed concrete in shear comparable in rationality and generality to the "plane sections" theory for flexure. One group of such models is based on a collection of assumptions about the behaviour of reinforced concrete that is known as the modified compression field theory (MCFT). This paper will summarize key aspects of this theory and will illustrate the use of models based on this theory in the shear design and evaluation of complex reinforced and prestressed concrete structures.

## 2. 2 THE MODIFIED COMPRESSION FIELD THEORY (MCFT)

As the name implies, the MCFT is a modified version of the compression field theory, CFT. The CFT, in turn, is built upon the classic truss analogy of Ritter and Mörsch. The first widely used theoretical model for shear response of reinforced concrete was presented by Emil Mörsch in his pioneering 1907 textbook Der Eisenbetonbau. Mörsch assumed that after the concrete cracks, the shear would be carried by diagonal compressive stresses in the concrete and by tensile stresses in the shear reinforcement, See Figure 5. While he recognized that the angle of inclination of the diagonal compres-

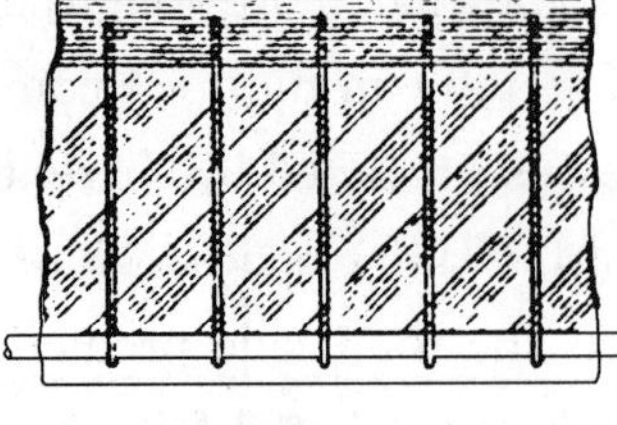


Figure 5 Emil Mörsch and the 45° truss model for shear

sive stresses was not necessarily 45°, he believed that it was mathematically impossible to determine the angle of inclination of these stresses and, hence, recommended 45° a s a conservative approximation. Mörsch calculated the shear stress, $v$, in the web of a beam as $v = V/(b_w jd)$ where $b_w$ is the web width, and $jd$ is the flexural lever arm. Expressing the amount of web reinforcement as $\rho_z = A_v/(b_w s)$, where $A_v$ is the area of the shear

reinforcement spaced at a distance s, and assuming that the shear reinforcement yields, Mörsch's approach predicts that the shear stress that can be resisted by a member is equal to $\rho_z f_y$. If the angle $\theta$ was known, the shear strength expression becomes $\rho_z f_y \cot\theta$. Various suggestions have been made as to appropriate values for $\theta$ with the recent EC2 shear provisions suggesting that $\cot\theta$ can be taken as large as 2.5, (i. e. $\theta = 21.8°$). Rather than using a value greater than unity for $\cot\theta$, the ACI code has traditionally adjusted for the conservative nature of Mörsch's equation by adding a concrete contribution, $v_c$, so that the ACI predicted shear strength becomes $v_c + \rho_z f_y$ Where $v_c$ is taken as equal to the shear stress to cause diagonal cracking, usually taken as $2\sqrt{f'_c}$ in psi units.

Herbert Wagner's tension field theory (1928) dealing with the post-buckling shear capacity of thin-skinned aluminum panels in aircraft structures developed very similar equations to those of Mörsch for the post-cracking shear capacity of reinforced concrete. However, rather than assuming the angle of principal tension was 45°, Wagner assumed that the angle of principal tensile stress in the buckled thin skin would equal the angle of principal tensile strain in the skin. By examining the strains of the longitudinal and transverse compression rib reinforcement, Wagner could determine this angle. When the same approach was applied to cracked reinforced concrete, first in torsion (Mitchell and Collins 1974) and later in shear, (Collins 1978), very promising results were obtained. The current equations of the Compression Field Theory are summarized in Figure 6. These equations enable the stress-strain response of an element subjected to shear and biaxial stresses to be predicted. Note that there are three equations of equilibrium which can be elegantly summarized with a Mohr's circle of stress, three equations of geometric compatibility shown by a Mohr's circle of strain, and three stress-strain relationships for the reinforcement and concrete behaviour. Note that the strains are average strains measured over base lengths long enough to include several cracks.

Equation 10 in Figure 6 was developed by testing reinforced concrete panels in a specially developed membrane-element tester which permitted loading in pure shear. The equation was later verified from experiments conducted in two larger testing machines.

It can be seen from Eq. 10 that as the principal tensile strain in the cracked concrete, $\varepsilon_1$, increases, the maximum compressive stress that can be resisted, $f_{2max}$, reduces substantially, see Figure 7 and Figure 8.

While the CFT provided reasonable predictions for members with significant amounts of shear reinforcement, it was not capable of predicting the response of members not con-

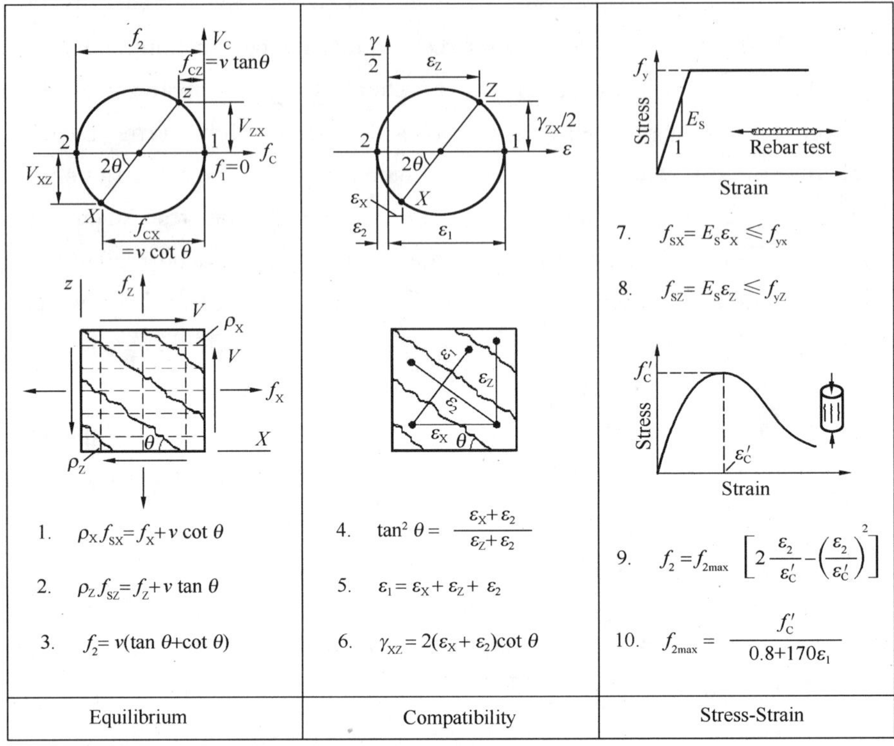


Figure 6 The Compression Field Theory

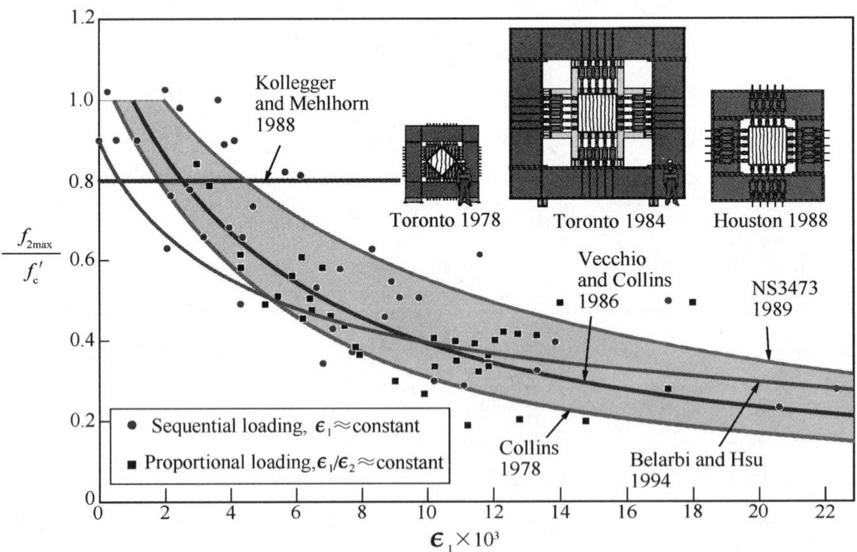


Figure 7 Reduction in crushing strength with principal tensile strain

taining shear reinforcement. The element shear tests which had been conducted to find the relationships between principal compressive stress and principal tensile and compressive strain, also showed that the Mohr's circles of stress did not go through he origin as assumed by the CFT, see top right plot in Figure 9. By accounting for the influence of these

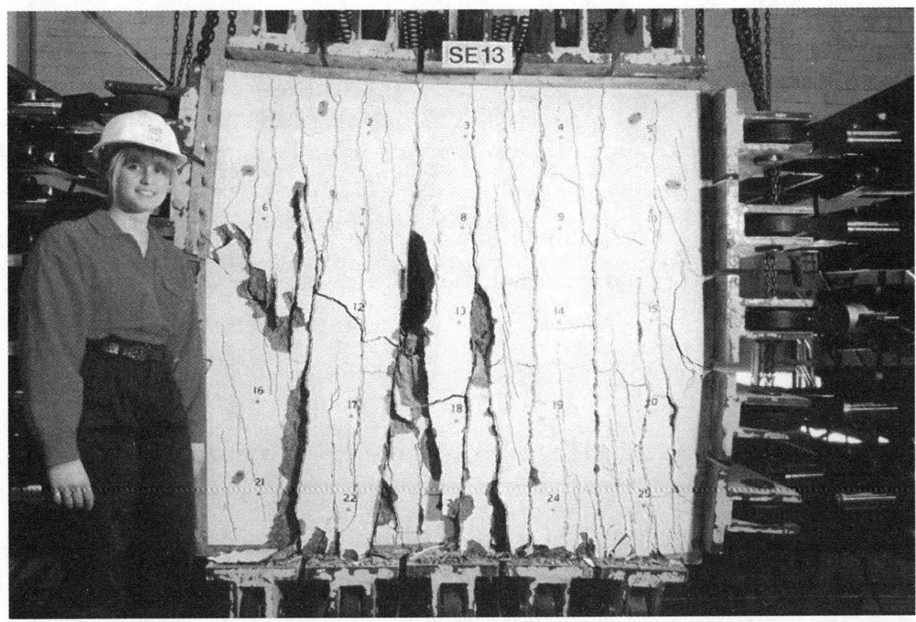


Figure 8 reinforced concrete element subjected to pure shear failing in compression with $f_{2\max}$ of 24 MPa and a cylinder strength of $f'_c = 80.5$MPa

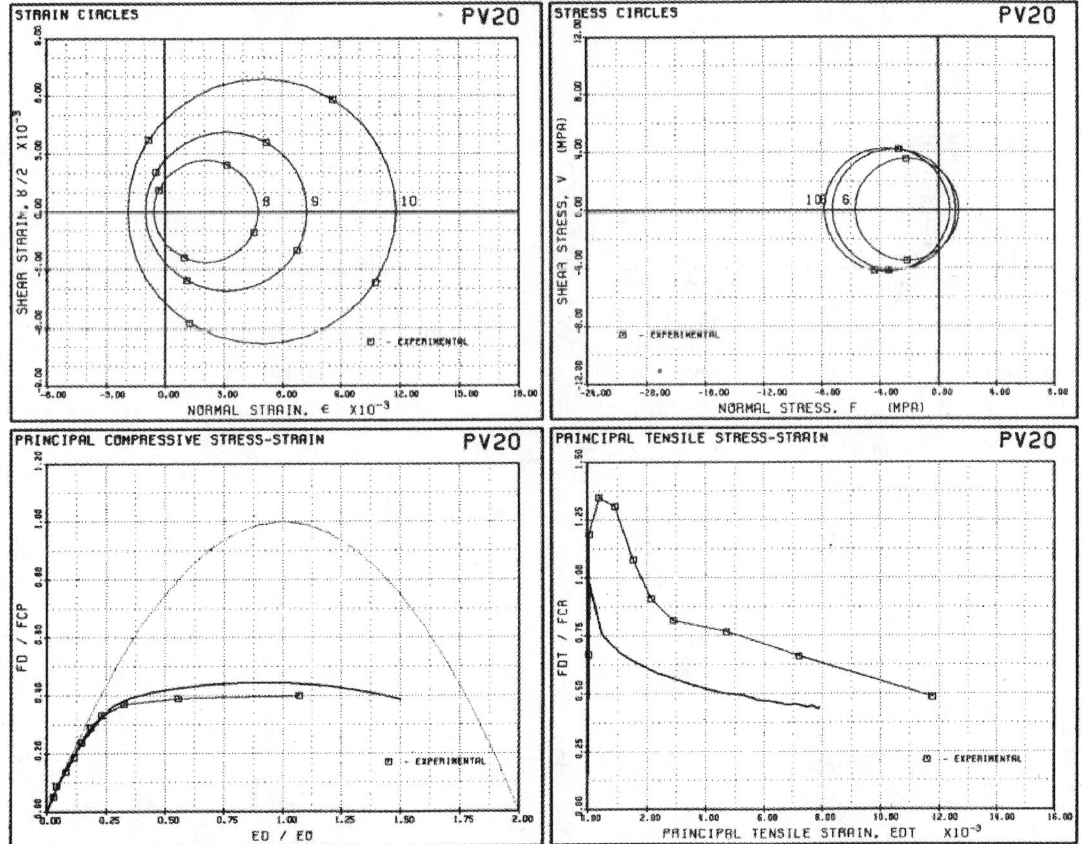


Figure 9 Results from specimen PV20 tested in the membrane element tester

average tensile stresses in the cracked concrete, the compression field theory was changed to the modified compression field theory, MCFT, enabling a much wider range of predictions to be made.

Figure 10 shows the 15 equations of the current MCFT. Across the width of the figure the equations are divided into three sets: five equilibrium equations, five geometric conditions and five stress-strain relationships. In the upper part of the figure the equations deal with average stresses, average strains and the relationships between average stresses and average strains. Average stresses (e. g. $f_{sx}$) and average strains (e. g. $\varepsilon_x$) correspond to stresses and strains averaged out over lengths long enough to damp out the local variations that occur at cracks and between cracks. The bottom five relationships in the figure are concerned with stresses at a crack (e. g. $f_{sxcr}$), the crack width and the maximum shear stress that can be transmitted across the crack. The term $v_{ci}$ stands for shear stress transmitted across the crack interface. The two small inset stress-strain figures represent the standard bilinear stress-strain relationship assumed for reinforcing bars and the uniaxial compressive stress-compressive strain relationship obtained from a standard cylinder test. The average stress-average strain relationships for the cracked concrete given by Equation 13 and Equation 14 in Figure 10 are the result of the panel testing and the early shell ele-

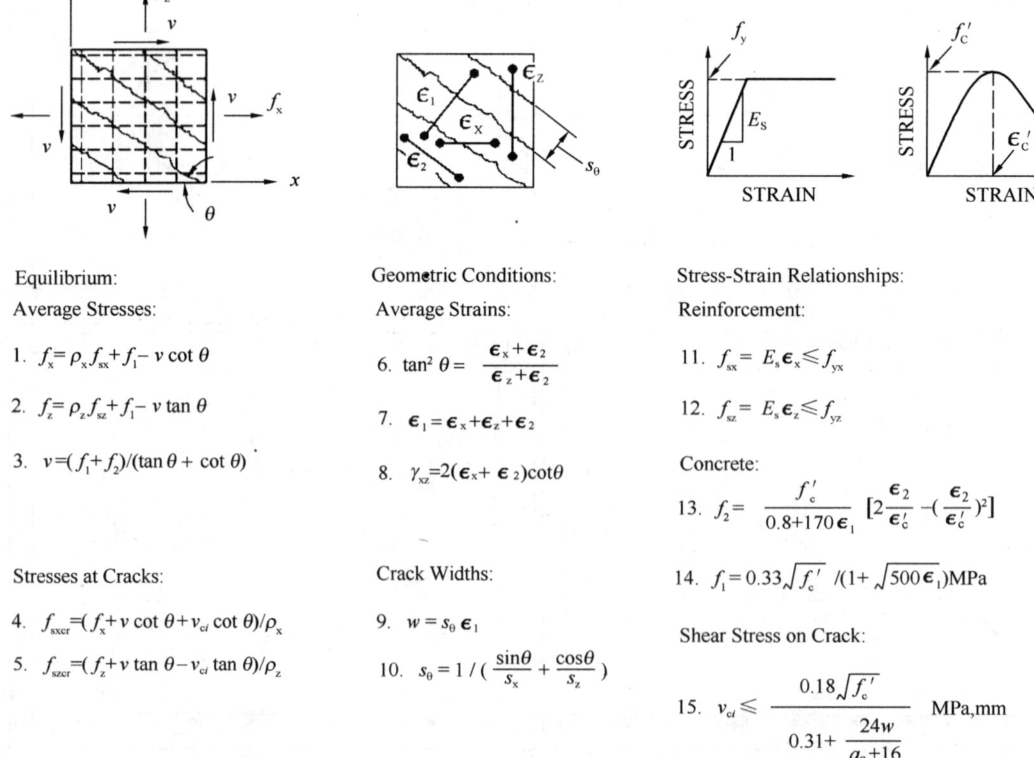


Figure 10 The equations of the Modified Compression Field Theory

ment testing program. The relationship used for determining the ability of the crack surfaces to transmit the interface shear stresses, Eq. 15 in Figure 10, was derived from the aggregate interlock experiments of Walraven (1981). Note that this shear stress limit is a function of the crack width, $w$, the maximum aggregate size, $a_g$, and the concrete cylinder strength, $f_c$

While solving these 15 equations by hand is very tedious, it is straightforward using an appropriate computer program such as Membrane-2000 available on the web (Bentz 2010). An example of the ability of the MCFT to predict the shear response of reinforced concrete elements is shown in Figure11. All six elements plotted in the figure contained about 3% of longitudinal reinforcement, had similar concrete strengths and were loaded in pure shear. The specimens had different amounts of transverse reinforcement. The SE specimens were tested in Toronto while the A and B specimens were tested in Houston (Pang and Hsu 1995). It can be seen that as the amount of transverse reinforcement was increased, the post-cracking shear stiffness and the shear strength of the elements increased, but the ductility of the elements decreased. It can be seen that there is excellent agreement between the lines representing the response predicted by the MCFT and the points showing the measured response.

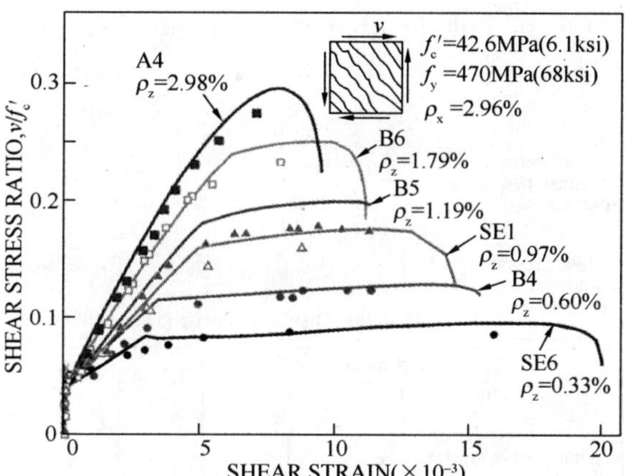


Figure 11　Predicted and observed response of six elements in pure shear

## 2.3　MCFT ANALYSES OF BEAMS AND COLUMNS

The MCFT is a procedure to predict the load-deformation response of a reinforced concrete element subjected to bi-axial stresses, $f_x$, $f_z$ and $v$. To use the theory to predict the load-deformation response of a reinforced concrete beam such as that shown in Figure 12, the beam could be represented as a two dimensional grid of elements with the response of each element being predicted by the MCFT. This is the basis of non-linear finite element programs such as VecTor2 (Vecchio 1989, 2010) which have been developed at the University of Toronto over the past 25 years. If such a program is used to analyze the beam it will be found that in zones extending for a distance of about d away from the point loads and the reactions there will be significant vertical compressive stresses in the concrete. These clamping stresses will enhance the shear strength of the elements in these zones making it probable that the shear failure will occur outside of these zones. For beams with short shear spans the zones with significant clamping stresses will overlap and the shear strength of the beam will be considerably increased. It is important to recognize that in these "disturbed regions" the shear stress distribution over the depth of the beam is influenced by the distribution of the clamping stresses and near the loads and reactions, plane sections do not remain plane.

Outside of the disturbed regions discussed above it is appropriate to assume that plane sections remain plane and that the clamping stresses are negligible. With these two assumptions a beam cross-section can be modeled as a stack of biaxially stressed elements with the response of each element being predicted by the MCFT. This is the basis of program Response-2000 (Bentz 2000, 2010) which can be used to predict the shear stress distribution over the depth of the beam and the complete load-deformation response of concrete sections subjected to shear, flexure and axial load, see Figure 12. If only the shear strength of a beam cross-section is required then the web of the beam can be approximated by just one biaxial element located at mid-depth and the shear stress on the element can be assumed to be $V/(b_w d_v)$ where $b_w$ is the web width and $d_v$ is the flexural lever arm which can be taken as 0.9$d$. The longitudinal strain, $\varepsilon_x$, at mid-depth of the beam can be found from the calculated strain in the longitudinal flexural reinforcement and the assumption that plane sections remain plane. For a given value of $\varepsilon_x$ the failure shear stress can then be calculated from the MCFT as the sum of two terms, $V_c$ and $V_s$, see Figure 12. The first term depends on the ability of the cracks to transmit shear stress, characterized by the coefficient $\beta$, while the second term is a function of the amount of shear reinforcement, $\rho_z f_y$, and the angle of inclination of the principal compressive stresses, $\theta$. Both $\beta$ and $\theta$ depend on the strain $\varepsilon_x$ and the equivalent crack spacing $s_{xe}$. This MCFT-based sectional design model for shear will be explained in more detail in the next section of the paper.

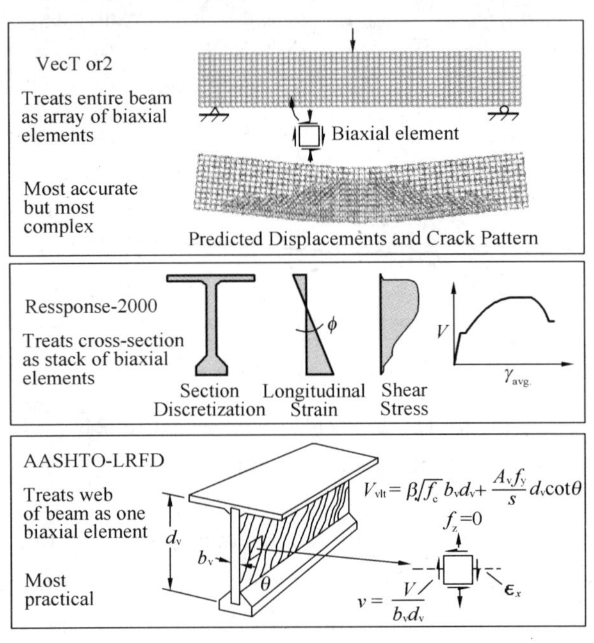


Figure 12 Levels of Approximation in MCFT Analyses

A good illustration of the use of these three different shear analysis tools is given in the ACI paper by Collins, Bentz and Kim (2002) titled "Shear Strength of Circular Reinforced Concrete Columns". In Table 1 of this paper, the experimental failure loads of 15 circular columns tested at Toronto are compared with the predicted strengths from AASH-

TO, Response and TRIX, an earlier version of VecTor2, along with predictions from ACI and predictions from an expression recommended by Kowalsky and Priestley (2000) at the University of California, San Diego (UCSD). The key part from this table is reproduced in Figure 13 below. It can be seen that the accuracy of the MCFT methods improves as the complexity of the method increases with the average experimental shear strength to predicted shear strength ratios and the coefficients of variation going from 1.08 and 12.8% for AASHTO to 1.01 and 6.0% for TRIX. Figure 14 shows the ability of Response-2000 to accurately predict the full shear-shear strain response of column SC3 from Figure 13.

| Investigator | Section | Loading Arrangement | L/D | Specimen |
|---|---|---|---|---|
| Aregawi | 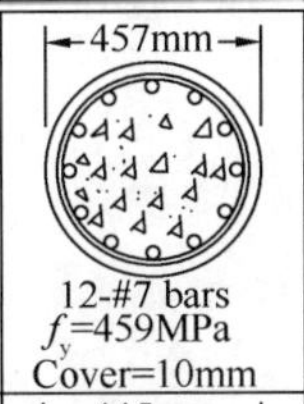 457mm<br>12-#7 bars<br>$f_y$=459MPa<br>Cover=10mm | 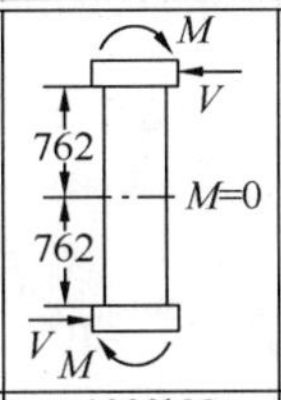 M, V<br>762<br>M=0<br>762<br>V, M | 3.33 | EB1<br>WB1<br>EB2<br>WB2 |
| Khalifa | 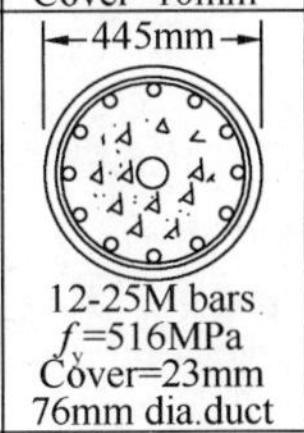 445mm<br>12-25M bars<br>$f_y$=516MPa<br>Cover=23mm<br>76mm dia.duct | 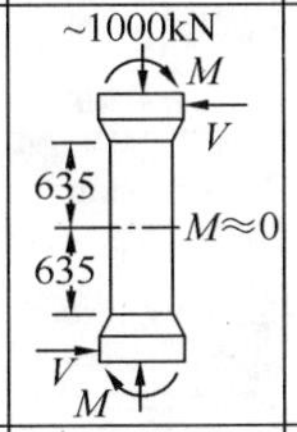 ~1000kN<br>M, V<br>635<br>M≈0<br>635<br>V, M | 2.85 | SC0<br>SC1<br>SC2<br>SC3<br>SC4+ |
| Kim | 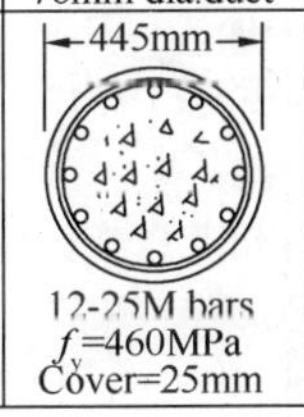 445mm<br>12-25M bars<br>$f_y$=460MPa<br>Cover=25mm | 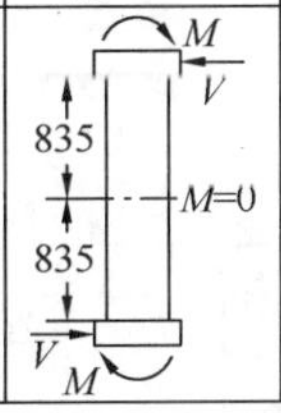 M, V<br>835<br>M=0<br>835<br>V, M | 3.75 | YJC Control<br>YJC200R<br>YJC150R<br>YJC100R<br>YJC200W<br>YJC100W |

| Observed Shear (kN) | Failure Ratios: Experimental/Computed | | | | |
|---|---|---|---|---|---|
| | ACI 11.3.1 | UCSD 2000 | AASHTO 2000 | R2k 1.0.1 | TRIX 97 |
| 363 | 1.30 | 0.84 | 1.04 | 1.06 | 1.01 |
| 467 | 1.47 | 0.98 | 1.19 | 1.15 | 0.99 |
| 348 | 1.38 | 0.91 | 1.08 | 1.20 | 0.97 |
| 437 | 1.51 | 1.03 | 1.20 | 1.24 | 1.07 |
| 326 | 1.72 | 0.84 | 1.37 | 1.14 | 1.03 |
| 324 | 1.36 | 0.77 | 1.26 | 1.23 | 1.03 |
| 478 | 1.11 | 0.80 | 1.02 | 1.04 | 0.98 |
| 578 | 1.04 | 0.82 | 0.99 | 0.99 | 1.06 |
| 455 | 1.12 | 0.77 | 1.03 | 0.96 | 0.87 |
| 212 | 1.45 | 0.95 | 1.14 | 1.21 | 0.97 |
| 323 | 1.15 | 0.87 | 0.95 | 0.97 | 0.98 |
| 411 | 1.33 | 1.03 | 1.11 | 1.10 | 1.05 |
| 479 | 1.25 | 1.01 | 1.07 | 1.05 | 1.07 |
| 315 | 1.07 | 0.99 | 0.88 | 0.93 | 0.95 |
| 434 | 0.98 | 1.05 | 0.86 | 0.94 | 1.11 |
| Average | 1.28 | 0.91 | 1.08 | 1.08 | 1.01 |
| C.O.V(%) | 16.1 | 113 | 12.8 | 10.1 | 6.0 |
| min | 0.98 | 0.77 | 0.86 | 0.93 | 0.87 |
| max | 1.72 | 1.05 | 1.37 | 1.24 | 1.11 |

Figure 13　Example of relative quality of shear strength predictions at different levels of approximation

Response-2000 can determine the flexural, shear and axial response of a reinforced or prestressed concrete section. For a prismatic member in which the section properties are constant along the length, the program can determine the deflected shape of the member by integrating the curvatures, shear strains and axial strains along the member length. Examples of such predictions for two large girders are given in Figure 15. These girders were tested as part of a program related to evaluating the shear safety of over 500 highway bridges in the state of Oregon. Currently there is a two billion dollar program underway to strengthen many of these bridges. As part of a project to develop an assessment methodology to identify the critical bridges, 44 full-sized bridge girders were loaded to failure at Or-

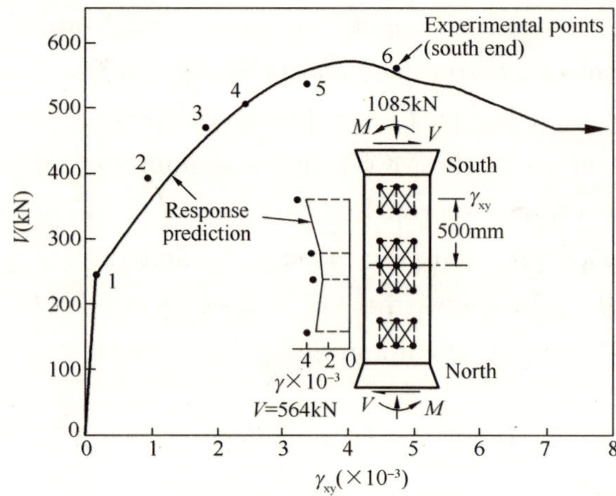


Figure 14 Example of load-deformation prediction of a column

egon State University under the direction of Professor Chris Higgins (Higgins et al. 2004). It was concluded that what was called the AASHTO-MCFT and Response-2000 provided the most accurate means of assessing shear capacity. For the 31 girders not subjected to fatigue loading, the AASHTO LRFD method gave an average test to predicted ratio of 1.10 with a coefficient of variation of 7.7% while Response-2000 gave an average of 0.98 with a coefficient of variation of 6.5%.

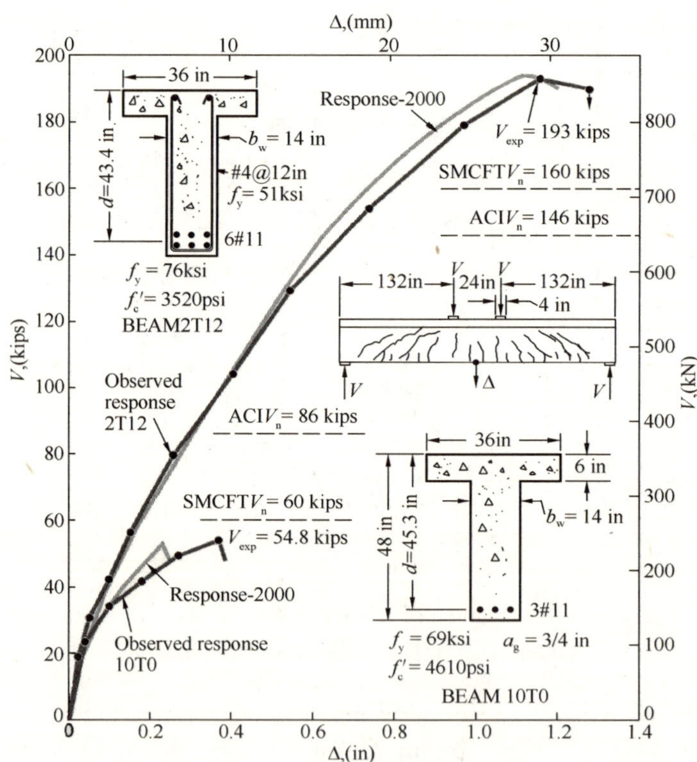


Figure 15 Example of load-displacement predictions for two bridge girders

## 2.4 CODE-BASED MCFT SHEAR STRENGTH EQUATIONS

The Canadian CSA code includes a shear design procedure based on the simplified MCFT (Bentz et al. 2006, Bentz and Collins 2006) which is similar to the AASHTO LR-

FD method discussed above. The unfactored shear strength, $V_n$, of a section as calculated by the Canadian code can be determined from the following equations where the unfactored failure shear stress $v_n$ is defined as $V_n/(b_w d_v)$ where $d_v$ can be taken as $0.9d$:

$$v_n = \beta\sqrt{f'_c} + \rho_z f_y \cot\theta \leqslant 0.25 f'_c \tag{1}$$

The aggregate interlock parameter $\beta$ primarily depends on the width of the cracks and size of the aggregate $a_g$. As crack width depends on both the average tensile strain in the cracked concrete and the crack spacing it is not surprising that the expression for $\beta$ derived from the MCFT has a strain term and a crack spacing term. The equation is

$$\beta = \frac{0.4}{1+1500\varepsilon_x} \cdot \frac{1300}{1000+s_{xe}} \tag{2}$$

The longitudinal strain at mid-depth, $\varepsilon_x$, can be conservatively taken as one half of the tensile strain in the flexural tensile reinforcement. Allowing for the tension in the longitudinal reinforcement caused by the shear and assuming that there is no axial load or prestressing gives

$$\varepsilon_x = v_n \frac{1+M/(V\cdot d_v)}{2E_s\rho_l} \tag{3}$$

where $M/V$ is the ratio of bending moment to shear at the section being considered and $\rho_l$ is the geometric ratio of the area of longitudinal flexural tension reinforcement to the shear area. That is: $\rho_l = A_s/(b_w d_v)$.

If the member contains more than the specified minimum amount of shear reinforcement (i. e. $\rho_z f_y \geqslant 0.06\sqrt{f'_c}$) then it can be assumed that the crack spacing will be well controlled and hence $s_{xe}$ can be taken as 300 mm which reduces the crack spacing term in Eq. (2) to unity. If the member contains only concentrated longitudinal reinforcement then the spacing of vertical cracks near mid-depth of the member, $s_x$, is assumed to be equal to $0.9d$. To allow for the influence of aggregate size the effective crack spacing, derived from Eq. (15) in Figure 10, is taken as:

$$s_{xe} = \frac{35 s_x}{15 + a_g} \geqslant 0.85 s_x \tag{4}$$

For high strength concrete the cracks will go through the aggregate rather than around the aggregate particles leading to smoother crack surfaces with less aggregate interlock capacity. To account for this, if $f'_c$ exceeds 70 MPa the term $a_g$ in Eq. (4) is taken as zero. As $f'_c$ goes from 60 MPa to 70 MPa, $a_g$ is linearly reduced to zero. As an additional allowance for the low aggregate interlock capacity of high strength concrete the term $\sqrt{f'_c}$ in Eq. (1) is not allowed to exceed 8MPa.

The MCFT predicts that the angle of the principal compressive stress, $\theta$, at shear failure for members with shear reinforcement depends primarily upon the longitudinal strain, $\varepsilon_x$ as:

$$\theta = 29^\circ + 7000\varepsilon_x \tag{5}$$

## 2.5 EXAMPLE CALCULATIONS USING MCFT SECTIONAL MODEL

As an example of the use of the MCFT procedure, the failure shear stress of the large specimen with no shear reinforcement described in Figure 4, will be calculated. Because of the beneficial effects of the clamping stresses in the vicinity of the load or the reaction, the critical section for shear is taken to be located at a distance of $d_v$ from the face of the load or the support. For the simply supported beam shown in Figure 4 the section $d_v$ from the load has the higher moment and hence is more critical. At this section the $M/V$ ratio is $4050-75-0.9\times1400 = 2715$ mm.

As the concrete cylinder strength is 75 MPa the term $a_g$ is taken as zero and $\sqrt{f'_c}$ in Eq. (1) is taken as 8 MPa. From Eq. (4) $s_{xe}= 2940$ mm and hence Eq. (1) becomes:

$$v_n = 1.056/(1+1500\varepsilon_x)\text{MPa} \tag{1a}$$

while Eq. (3) becomes

$$\varepsilon_x = v_n \frac{1+2715/(0.9\cdot 1400)}{2\cdot 200000\cdot (0.0083/0.9)} = \frac{v_n}{1169} \tag{3a}$$

Substituting Eq. (3a) into Eq. (1a) and solving the quadratic we find that the predicted value for $v_n$ is 0.599 MPa and the corresponding value of $\varepsilon_x$ is $0.511\times10^{-3}$. Note that the CSA failure shear stresses listed in Figure 4 are 90% of the value calculated above because the shear stresses in the table are calculated as $V/(b_w d)$.

As a further example, consider the large specimen in Figure 3 with stirrups. For this specimen, the crack spacing can be taken as 300 mm and hence Eq. (1) becomes

$$v_n = 3.20/(1+1500\varepsilon_x)+0.5\cot\theta \tag{1b}$$

Because there is 1.33/0.83 more longitudinal reinforcement in this specimen Eq. (3a) becomes

$$\varepsilon_x = v_n/1870 \tag{3b}$$

If $\varepsilon_x$ is assumed to be 0.001, Eq. (5) gives $\theta$ as 36° and then Eq. (1b) predicts the failure shear stress to be 1.968 MPa. For this value of $v_n$ Eq. (3b) predicts $\varepsilon_x$ to be $1.05\times10^{-3}$ indicating that the estimated value of $\varepsilon_x$ was reasonably accurate. Convergence is reached when the estimated value of $\varepsilon_x$ is $1.034\times10^{-3}$ giving a calculated failure stress of 1.937 MPa and an angle of 36.2°. The appearance of this section as it failed is shown in Figure 16. In this figure the shear slip on the wide failure crack can be seen from the displacements of the grid of horizontal and vertical lines spaced 100 mm apart. The crack widths shown in the photograph were measured when the load was at 87% of the failure value. This slab with stirrups displayed crack widths prior to failure that were ten times larger than those measured for the companion slab without stirrups. Note that ACI code assumes that the failure crack shown in Figure 14 will be at 45° to the longitudinal axis, while the EC2 calculation assumes it will be at 22°. It can be seen in Figure 14 that near mid-depth, the angle of the failure crack is about 32o which is not far from the value predicted by the CSA equation.

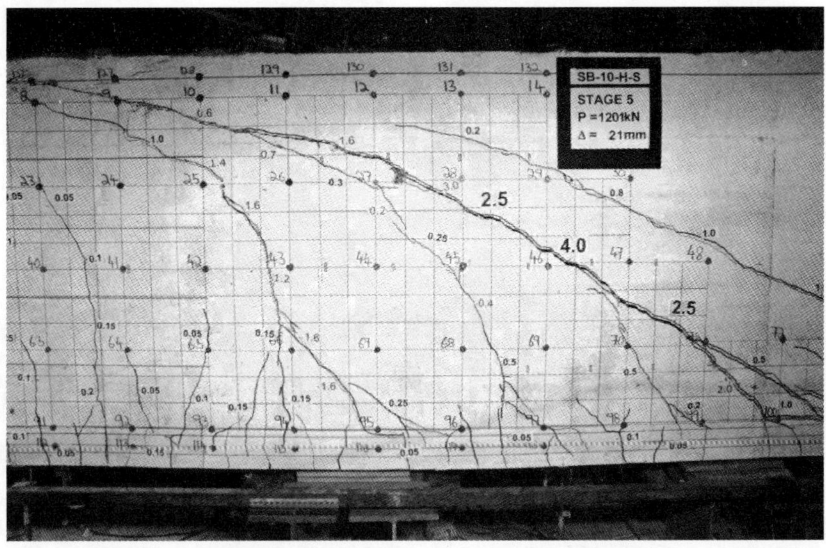


Figure 16　Failure of Slab Strip L—10HS

## 2.6 EFFECT OF AXIAL LOAD AND PRESTRESSING ON SHEAR STRENGTH

In the code-based MCFT method, axial load and prestressing is predicted to impact the shear strength of reinforced concrete by lowering the longitudinal strain in the member at mid-depth and thus narrowing the cracks. The full equation for $\varepsilon_x$ given in the CSA code is:

$$\varepsilon_x = \frac{M/d_v + V - V_p + 0.5N - A_p f_{p0}}{2(E_s A_s + E_p A_p)} \tag{6}$$

Where $N$ is the axial load (tension positive), $V_p$ is the vertical component of prestressing force, $A_p$ is the area of prestressed reinforcement on the flexural tension side, and $f_{p0}$ is the stress in this steel when the strain in the surrounding concrete is equal to zero.

To demonstrate the use of these provisions, suppose the beam without links in Figure 15 were designed in the 1950's by the traditional ACI shear provisions of that time. Since then it has become clear that those provisions produce designs that are insufficiently conservative in terms of shear strength. Thus it may be desired to externally post-tension this beam to improve the shear strength with 2 tendons at mid-depth of 7—15 mm strands with an effective prestressing force of $1277 \times 2 = 2550$ kN of compression. As this is external prestressing, it will be treated as an axial force in Eq. 6 as $N = -2550$ kN which will lower $\varepsilon_x$ and thus increase the shear capacity. Assuming $\varepsilon_x = 0.10 \times 10^{-3}$, $V_r$ is calculated as 462 kN. Eq. 6 evaluates to a strain of $0.16 \times 10^{-3}$. Convergence is obtained when $\varepsilon_x = 0.125 \times 10^{-3}$ with a shear strength of 447 kN. Thus the inclusion of the external post-tensioning is predicted to have almost doubled the shear strength of this member.

Figure 17 shows a comparison of a set of 4 recent experiments to the predictions of the CSA method, EC2, the ACI code, and BS8110, the existing British Standard. As can be seen, the CSA provisions provide the best estimate of the impact of prestressing. This work is explained in the recent doctoral thesis of Liping Xie (2009).

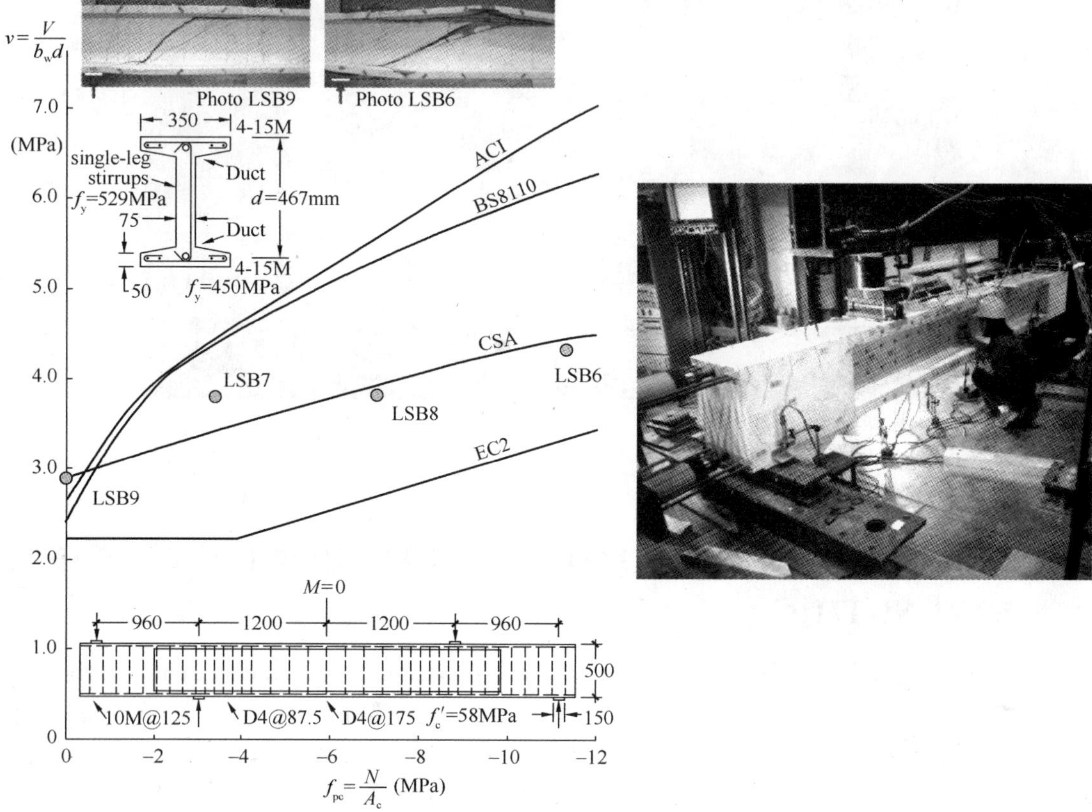


Figure 17 Effect of prestressing on shear strength

## 2.7 THE SIZE EFFECT IN SHEAR

The MCFT indicates that large members without transverse shear reinforcement would fail at lower shear stresses than geometrically similar smaller members. This is important because the so called "basic expression for shear strength of members without shear reinforcement" in the ACI code was developed from experiments on small beams. The MCFT predicts that this "size effect in shear" would quickly diminish if transverse shear reinforcement were added to the member. Figure 18 shows the predicted shear capacities of two series of web elements. In one series the spacing of the diagonal cracks is assumed to be 300 mm while for the second it is assumed to be 2000 mm. The smaller crack spacing assumption would be appropriate for traditional laboratory sized specimens without shear reinforcement while the 2000 mm would be appropriate for very deep slabs without shear reinforcement. If there is no transverse reinforcement the element with the wider crack spacing has only half the strength of that with the smaller crack spacing. However once a minimum amount of transverse reinforcement is provided, the difference in strength is much less significant. Further in an actual beam adding stirrups would significantly reduce the spacing of the diagonal cracks. Analyses such as these were used to arrive at a minimum shear reinforcement amount beyond which the size effect could be neglected. For the AASHTO LRFD specifications this minimum was initially set in terms of the quantity $\rho_v f_y$ which was not to be less than $0.083\sqrt{f'_c}$ in MPa units.

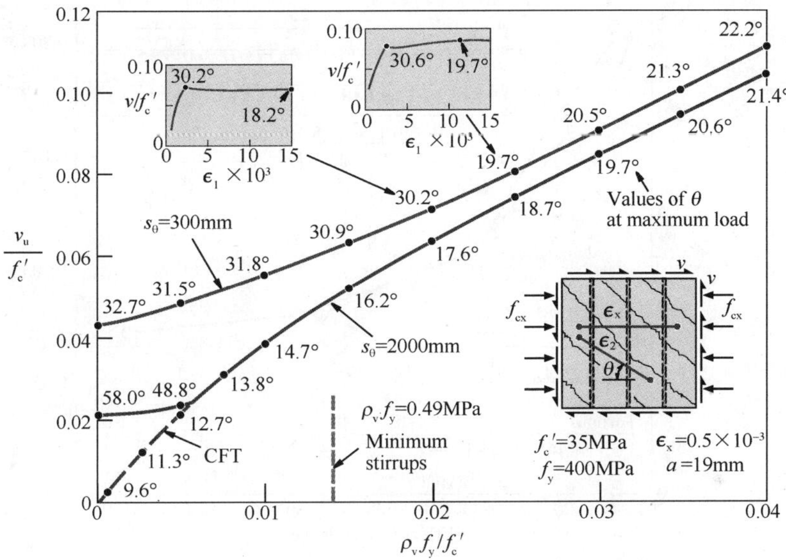


Figure 18　Effect of amount of transverse reinforcement on size effect in shear

In the last 30 years a number of major projects studied how the failure shear stress of members with less than 1% of longitudinal reinforcement varied as either the width or depth of the test specimens were changed while keeping all other variables constant. The results from two of the larger of these test series (Shioya 1989, Lubell et al. 2004) are summarized in Figure 19. It can be seen that there is a consistent and significant decrease in failure shear stress as the depth of the specimens is increased. In these size effect series it was noted that the spacing of flexural cracks at mid-depth of the specimens was typically about 0.5d, hence for the same reinforcement tensile stress, the crack widths at mid-depth increased as depth of the specimens increased. The ACI expressions for the shear strength of members without shear reinforcement were based on experiments of members with depths less than about 350 mm and, hence, did not account for this reduction in failure shear stress that happens as member depth increases. Because of this, a number of major concrete structures designed over the last 50 years may have considerably lower factors of safety against shear failure than assumed by the original designers. Evaluating the safety of these structures and identifying those in most urgent need of repair is a major challenge for the structural engineering profession. For example, shown in Figure 20 is an actual beam in a major industrial facility which is about 5 metres deep and contains considerably less than the currently specified minimum quantity of shear reinforcement. Does this beam need to be strengthened in shear? Figure 20 compares this actual beam with the largest shear specimen ever tested (Shioya 1989), along with some of the largest members tested at the University of Toronto which contained small amounts of transverse reinforcement. Also shown is the size of the traditional shear tests (6″×12″) upon which the design provisions used for the beam were based.

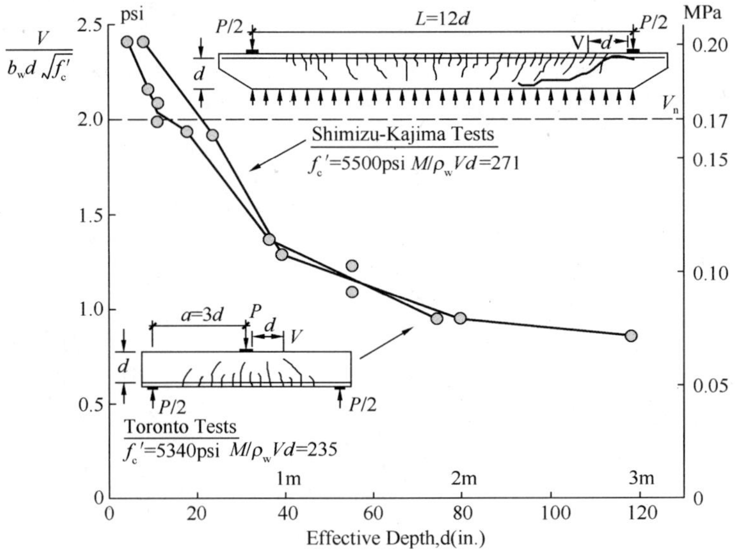


Figure 19 Size effect in shear

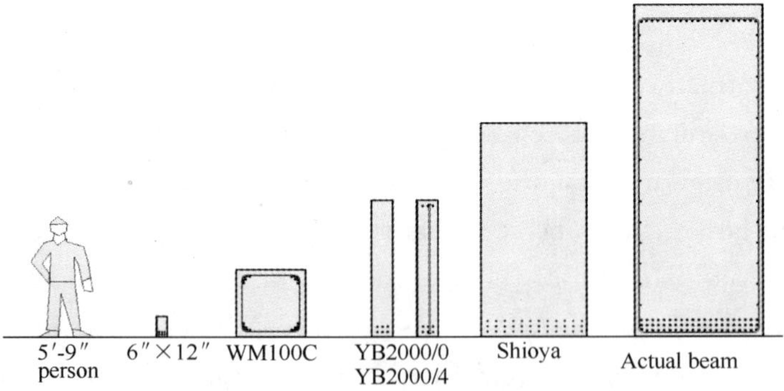


Figure 20 Relative size of beams

Figure 21 compares the experimental results of beams containing less than the CSA specified minimum shear reinforcement with the predictions from the CSA code and Response-2000. It can be seen that while the code assumes no benefits of any shear reinforcement less than the minimum, the experiments and the MCFT predictions show that there can be significant increases in shear capacity even for very small quantities of shear reinforcement though there is, of course, still a significant size effect for these members.

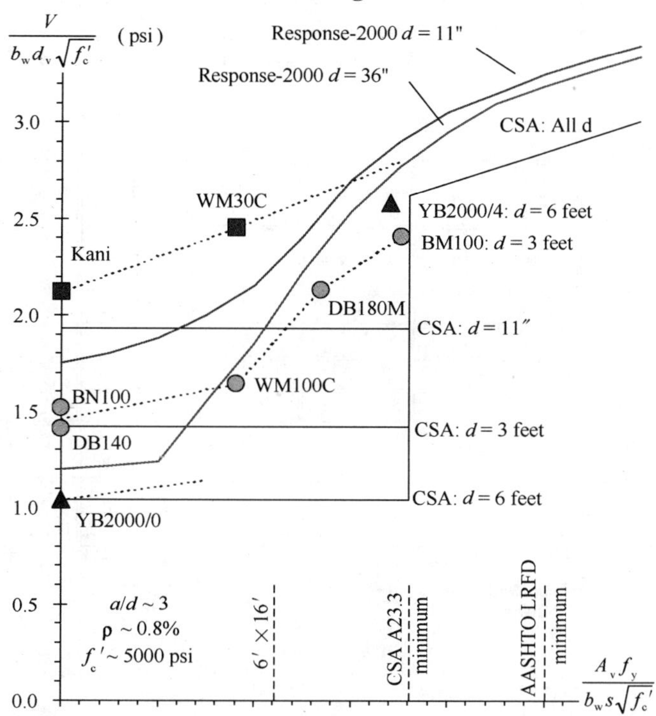


Figure 21　Effect of small quantities of transverse reinforcement on size effect in shear

## 2.8 SHEAR CAPACITY OF TRANSFER SLABS

Figure 22 shows three situations where the traditional ACI shear design procedures may result in members with inadequate safety. The thick one-way slab in the condominium building transfers closely-spaced wall loads from the upper stories to more widely-spaced walls below, the large strip footings transfer wall loads to the soil, while the thick roof slab of the subway tunnel carries high soil pressures across the subway opening. The thickness of each element has been chosen using ACI 318-05 so that $\phi V_c$ just exceeds $V_u$ and hence shear reinforcement is not required.

In order to understand at a more fundamental level the mechanisms involved in the shear failure of critical members such as the transfer slab discussed above, ten large-scale and ten geometrically-similar, small-scale, shear-critical reinforced concrete slab-strip specimens were tested. The key results of this work were summarized in a 2007 ACI paper by Sherwood, Bentz and Collins titled "Effect of Aggregate Size on Beam-Shear Strength of Thick Slabs." Additional information on the acoustic emission imaging techniques that were used to track crack propagation in a number of the specimens is given in a 2007 paper

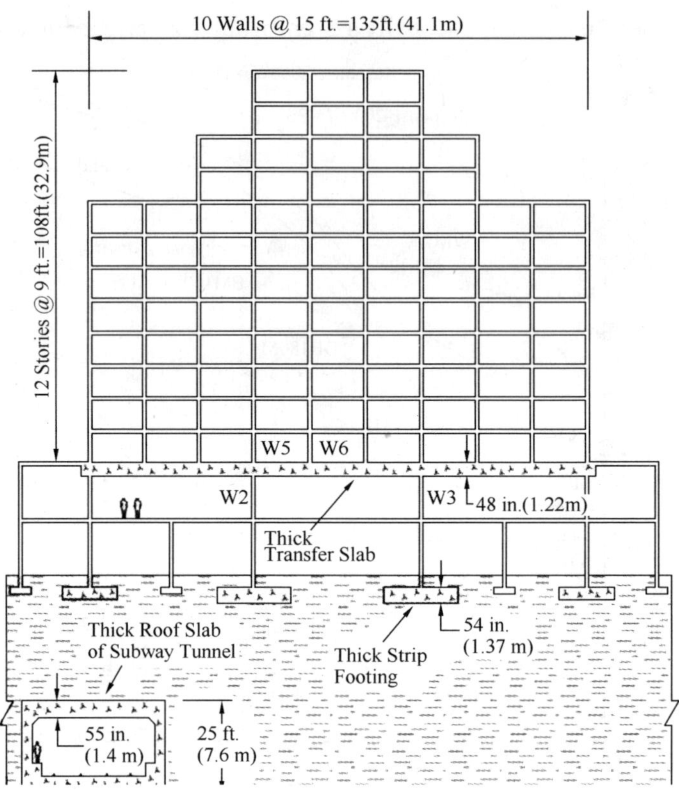


Figure 22 Structures using thick transfer slab

by Katsaga, Sherwood, Collins and Young. The large-scale specimens had an effective depth of 1400 mm, which was large enough to permit very detailed measurements to be made. See Figure 23. Apart from depth ($d$=1400mm or 280mm) the prime variable was the maximum aggregate size (10, 20, 40 and 50mm). Figure 23 shows that increasing the maximum aggregate size from 10mm to 50mm made the path of the shear crack rougher and increased the failure shear stress for the large specimens by 33%. The figure also shows that the CSA equations predict the failure shear stresses for all four specimens reasonably well. For the large specimen with small aggregate, both ACI and the 2004 Eurocode (EC2) give unconservative predictions.

The specimens shown in Figure 23 were designed so that the results could be compared with previous experiments conducted at Toronto. Figure 24 compares the failure shear stresses for specimens made from normal-strength concrete with the failure shear stresses for specimens made from high-strength concrete. It can be seen that, except for small specimens, no significant increase of shear strength results from the use of high-strength concrete for these members which do not have shear reinforcement. It can also be seen that the CSA standard predicts these results very well. As shown in the inset figures in Figure 24, the high-strength concrete members such as H5 have cracks which pass through the aggregate resulting in smoother crack surfaces less capable of transmitting shear stresses.

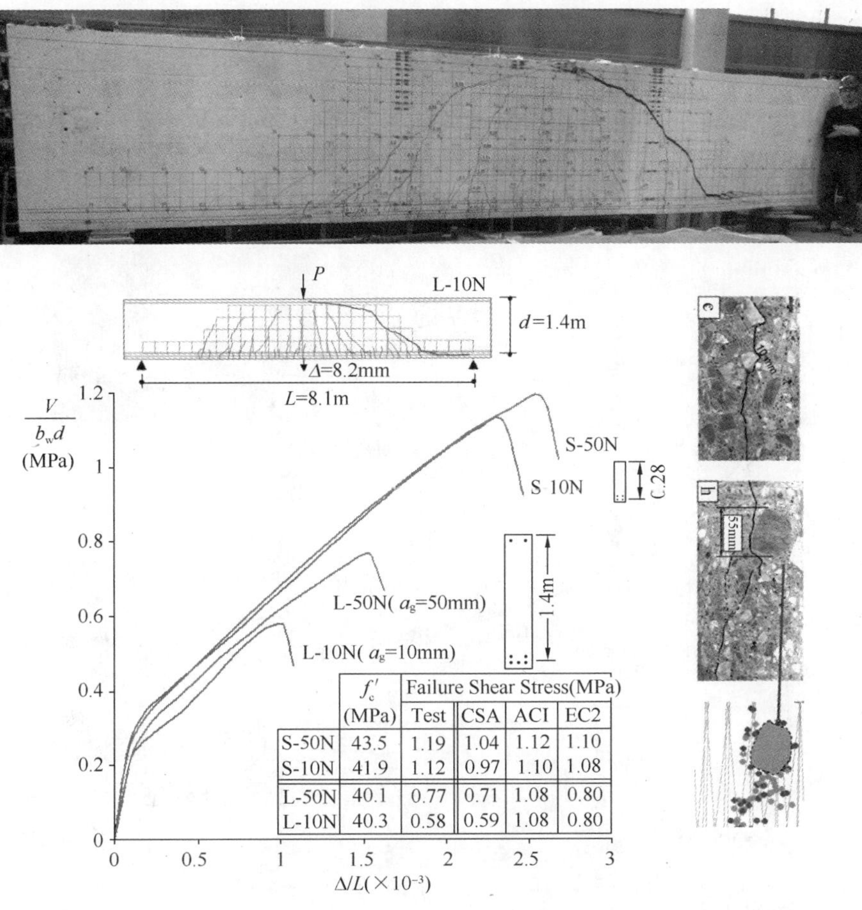


|  | $f_c'$ (MPa) | Failure Shear Stress(MPa) Test | CSA | ACI | EC2 |
|---|---|---|---|---|---|
| S-50N | 43.5 | 1.19 | 1.04 | 1.12 | 1.10 |
| S-10N | 41.9 | 1.12 | 0.97 | 1.10 | 1.08 |
| L-50N | 40.1 | 0.77 | 0.71 | 1.08 | 0.80 |
| L-10N | 40.3 | 0.58 | 0.59 | 1.08 | 0.80 |

Figure 23　Effect of Aggregate size and member depth on shear strength

Figure 24　Size effect for high strength and normal strength concrete

While the CSA shear provisions predict the influence of size and concrete strength very well, the ACI provisions, which were formulated in the 1960's, and the EC2 provisions, which were formulated in the mid 1990's, do not account well for either the effect of size or the effect of concrete strength. See Figure 25.

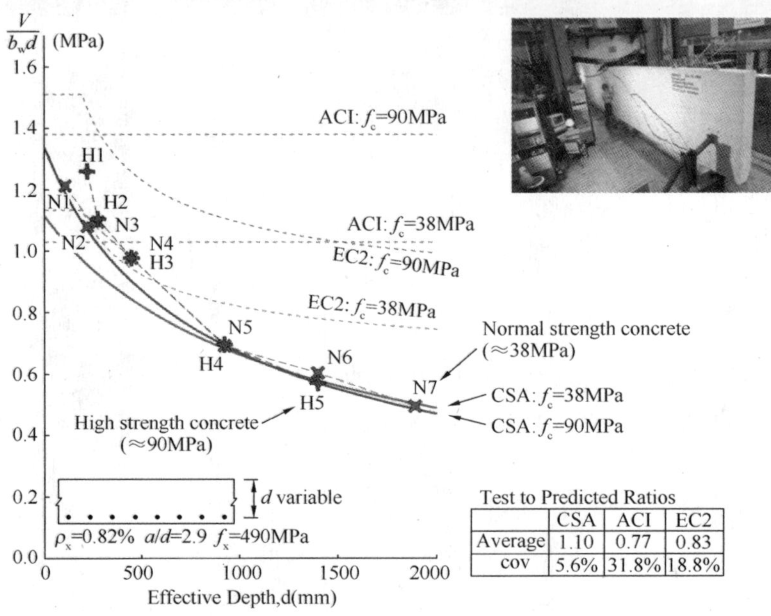


| | CSA | ACI | EC2 |
|---|---|---|---|
| Average | 1.10 | 0.77 | 0.83 |
| cov | 5.6% | 31.8% | 18.8% |

Figure 25 Size effect predictions of ACI and EC2 codes

## 2.9 MCFT SHEAR STRENGTH FOR DISTURBED REGIONS

In regions of members close to applied point loads or significant changes in geometry, plane sections will not remain plane, the shear stress distribution will be significantly different from that assumed by Mörsch, and the clamping stresses will have a significant influence on shear strength. In these regions, strut-and-tie models more accurately describe the flow of forces. In these models, the flow of compressive stresses in the concrete is represented by concrete struts while the tension reinforcement is represented by ties. Such strut-and-tie models are given by the CSA, EC2 and ACI codes. The principal difference between these three implementations concerns the expressions used for predicting the crushing strengths of the struts. The CSA code uses compatibility conditions to calculate the principal tensile strains in the concrete struts where they cross or intersect with tension ties and uses equation 10 of Figure 6 to define the crushing strength. Assuming that the compressive strain in the concrete at crushing of the strut is 0.002, and that the tensile strain in the reinforcing steel tension tie is $\varepsilon_s$, the principal tensile strain in the strut can be determined from Equations 4 and 5 from Figure 6 as:

$$\varepsilon_1 = \varepsilon_s + (\varepsilon_s + 0.002)\cot^2\theta_s \tag{7}$$

and the crushing strength of the strut is given by:

$$f_{cu} = \frac{f'_c}{0.8 + 170\varepsilon_1} \leqslant 0.85 f'_c \tag{8}$$

Thus if a tie has a tensile strain of 0.002 and the angle between the strut and the tie is 25° the crushing stress of the strut is predicted to be 0.23$f'_c$ By contrast, EC2 and ACI assume that the crushing stress of the strut depends only on the concrete strength (about 0.5$f'_c$ for EC2 and ACI). In addition, the ACI code requires that the angle between the strut and tie not be less than 25°.

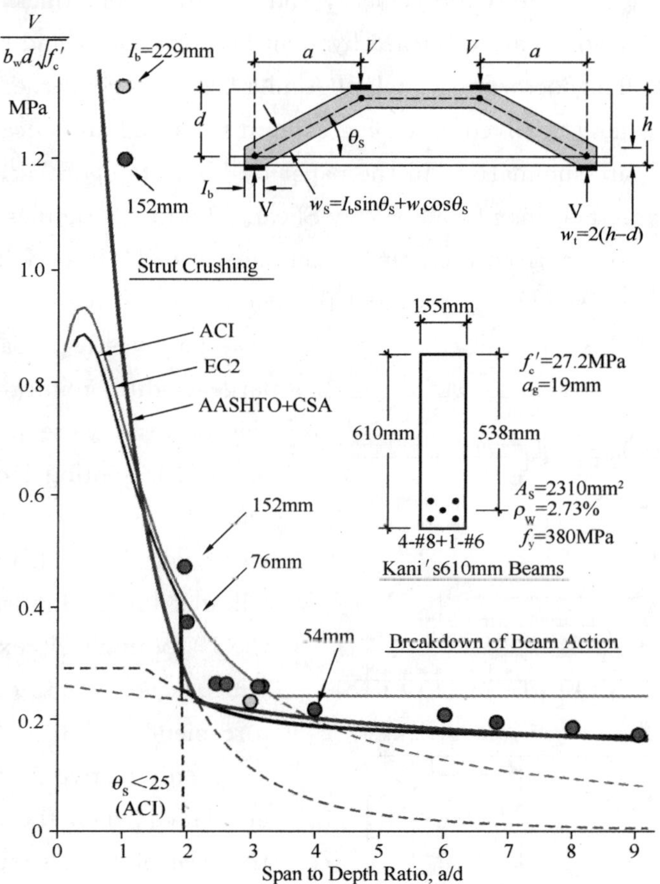


Figure 26　Arch Action versus Beam Action

Figure 26 shows the results of experiments by Kani (1979) in which he investigated the influence on shear strength of changing the shear-span-to-depth ratio (a/d) for a series of beams without shear reinforcement. For larger a/d ratios, the beams failed immediately upon the breakdown of beam action as discussed above. For beams with shorter shear spans, after the breakdown of beam action, the tension in the longitudinal reinforcement became essentially constant from support to support and an arch could form in the concrete enabling the member to carry substantially higher loads. The magnitude of this arch-action can be estimated by using a strut-and-tie model such as that shown in Figure 26. From this figure it can be seen that the CSA procedures lead to accurate estimates of shear strength both for the members governed by breakdown of beam action and for members in which strut crushing is predicted to govern shear capacity. While we have seen that the shear stress at breakdown of beam action is strongly influenced by the absolute size of the member, shorter span members whose capacity is controlled by arch-action are predicted to show no size effect.

While footings such as that shown in Figure 22 and those shown in Figure 27 are usually constructed without shear reinforcement, and are often very thick, they are typically non-slender members, which are governed by strut-and-tie action, and hence will often not exhibit a size effect. The doctoral research of Almila Uzel was concerned with the shear design of large footings and involved extensive experimental and analytical studies. The key aspects of this work are summarized in the paper by Uzel, Podgorniak, Bentz and Collins titled "Design of Large Footings for One-Way Shear." Figure 27 defines the tributary shear length, $L_o$, as the distance from the point of zero shear to the face of the column or wall. The ratio $L_o/d$ is then used as a measure of the slenderness of the footing.

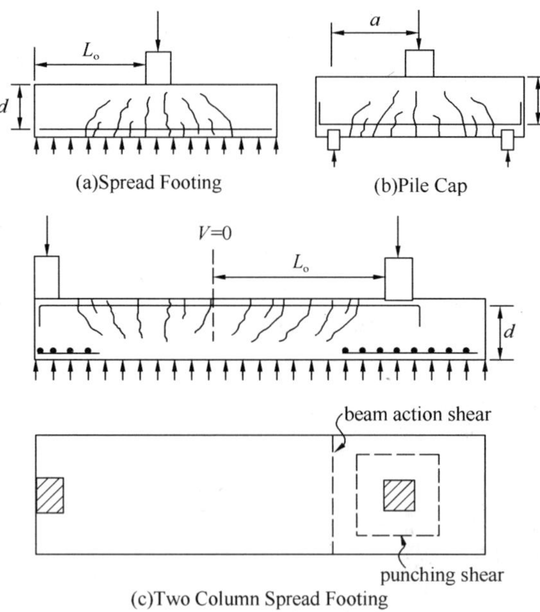


Figure 27 Examples of large footings

To investigate safety concerns with large, lightly reinforced footings, a series of tests were performed on specimens representing footing strips, which had effective depths up to one metre and were typically loaded with uniformly distributed loads. Figure 28 shows two of the 17 footing strip experiments, both of which have 0.76% of longitudinal reinforcement. AF3 had a $L_o/d$ ratio of 4.74, an effective depth, d, of 617 mm, and failed when the shear force $d$ from the face of the "support" reached $2.59\sqrt{f'_c}b_w d$ (psi units). AF11 had a $L_o$/d ratio of 2.00, with an effective depth of 925 mm, and reached flexural yielding when the shear force $d$ from the face reached $4.75\sqrt{f'_c}b_w d$ (psi units). Hence both of these footing strips would have been safe if designed by the traditional ACI procedure, which assumes shear failure at $2\sqrt{f'_c}b_w d$ (psi units).

The observed shear stresses at failure for uniformly loaded members of different

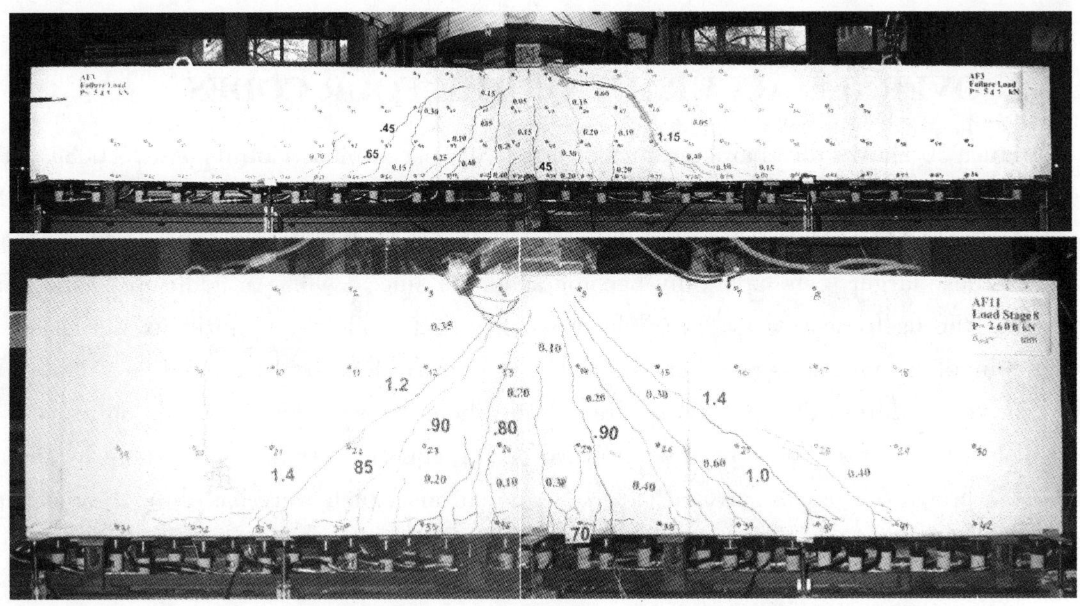

Figure 28　Tests on uniformly loaded footings

depths and different slenderness ratios are shown in Figure 29. It can be seen that while there is a substantial decrease in failure shear stress with increasing depth for the slender tests by Shioya from Japan, where $L_o/d$ equals about six, there is no evidence of a size effect for the specimens where $L_o/d$ equals two. In the paper it is shown that a combination of the simplified MCFT equations to predict the shear at which beam action breaks down, combined with the AASHTO LRFD strut-and-tie model to predict the remaining strength, results in reasonably accurate predictions for the strengths of the 24 footing experiments studied. The paper recommends that the current ACI minimum shear reinforcement exemption for footings be limited to footings where the slenderness ratio, $L_o/d$, does not exceed 2.5.

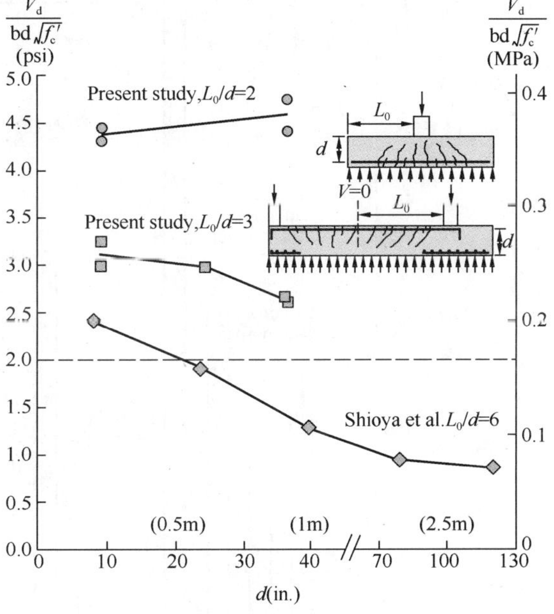


Figure 29　Observed shear stresses at failure for uniformly loaded members of different depths and different slenderness ratios

## 2.10 OVERALL ACCURACY OF THE FOUR CODES

Figure 30 shows the characteristics of a recently assembled (Collins et al. 2008) database of 1849 reported shear tests on members not containing shear reinforcement. It can be seen that only 144 of these specimens have depths exceeding 550 mm and that results from most of these larger tests have only become available since 1996. An additional aspect influencing the usefulness of available shear tests is that while the majority of actual shear designs involve continuous members subjected to uniform loads only about 1% of laboratory shear tests address this situation. Thus while there are limitations on this dataset, it is still of interest to compare how the empirical design equations that were derived by fitting to such a body of data compare with the CSA equations which were based on the modified compression field theory.

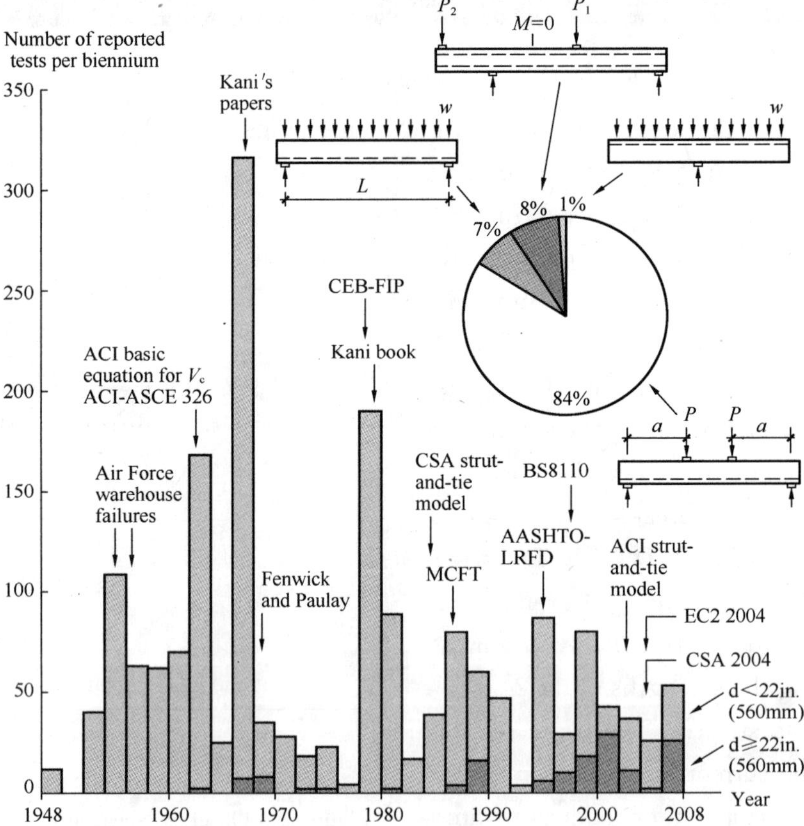


Figure 30 Shear tests of members without shear reinforcement

Figure 31 shows the values of the ratios of the predicted failure shear stresses to the observed failure shear stresses of the large database of members without links shown in Figure 30. The predictions were made using both the sectional analysis method and the strut-and-tie method. It can be seen that the CSA provisions provide the lowest scatter.

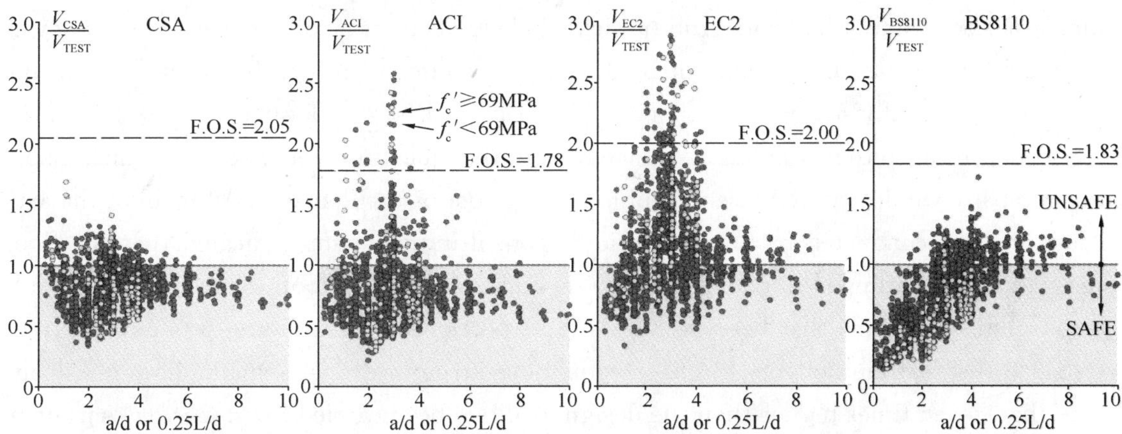


Figure 31　Comparative accuracy of four codes in predicting shear strength of members without shear reinforcement links. Total factor of safety listed for case where live load is 50% of dead load

The dashed horizontal lines indicate the total factor of safety for shear failure of such members used in the different codes for a representative ratio of live load to dead load. Experimental points lying above these lines indicate that failure at service loads is more probable for this set of parameters. It is a troubling aspect of both the ACI code and the EC2 code that so many points are above these lines. Rather than using the a strut-and-tie model for members with short shear spans, BS8110 uses an empirical enhancement factor, and in this region gives very conservative results.

## 2.11 EXAMPLES OF EVALUATION OF SHEAR SAFETY OF EXISTING STRUCTURES

Structural engineers are responsible for making sure that buildings stand up and if buildings are well designed and well constructed, the chance of structural failure is so small that the vast majority of structural engineers will never have one of their buildings fall down or even partially collapse. While extremely rare, such structural failures are an opportunity to advance the art of structural engineering. This is because laboratory experiments are typically performed on only a single member or perhaps a small sub-assembly of a few parts of the structure. Are the models developed from such small component tests capable of predicting the response of the total structure? Lessons learned from failure investigations have strongly influenced the shear research program at Toronto.

An example of such a failure investigation is that of the collapse of the Kimberley-Clark Building in Niagara Falls Ontario, see Figure 32. This study was summarized by Vecchio and Collins in an ACI Concrete International paper called "Investigating the Collapse of a Warehouse" (1990). The third floor of this building, which had been designed for a live load of 125 pounds per square foot, was subjected to a live load of nearly 900 pounds per square foot by drums full of nickel pellets at the time of collapse. On the day of the collapse there was also a paper dust explosion and a major fire on the lower floors of the building. The legal issue was whether the explosion caused the floor to collapse or did the floor collapse cause the explosion. The answer to this question would resolve which insurance company would pay the twenty million dollar cost of the damage. A "state of the art" failure analysis conducted by an international consulting company indicated that the floor must fail at a superimposed load of about 600 pounds per square foot. Shortly after the collapse Collins suggested that because only a local portion of the building was extremely overloaded the restraint effects of the surrounding structure must explain how live loads more than seven times higher than the design load had been carried for some weeks prior to collapse, see Figure 33. The non-linear analysis conducted by Vecchio demonstrated that this was indeed possible and that it was therefore feasible that the paper dust explosion triggered the final collapse, see Figure 34. This investigation showed how non-linear analysis can be a useful tool in understanding the behaviour of reinforced concrete structures subjected to extreme conditions. The key lesson is that flexural failures of reinforced concrete members involve substantial elongations, and if these elongations are restrained by the surrounding structure, significant compression forces can be induced in the members, which in turn should substantially increase their strength.

Figure 32 Portion of collapsed floor of warehouse

On August 23, 1991 the concrete base structure for Sleipner A, the twelfth Condeep offshore platform, was being lowered into a Norwegian fjord in preparation for deck mating when a concrete wall suddenly failed allowing water to flow uncontrollably into the structure. As the platform sank the increasing water pressure imploded the platform, leaving just a pile of rubble on the bottom of the fjord. As stated on the cover of a Norwegian engi-

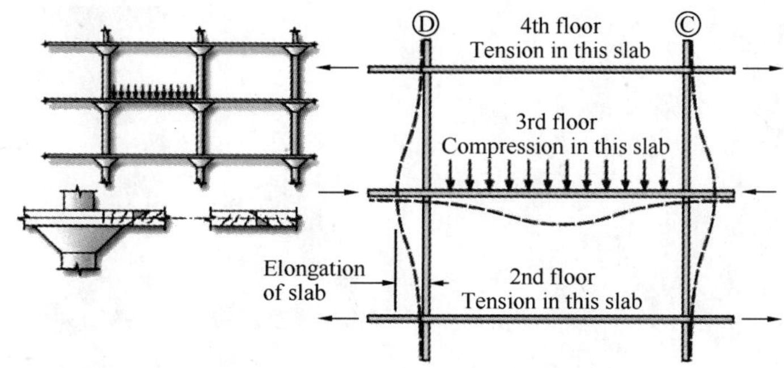


Figure 33　Membrane action in slab

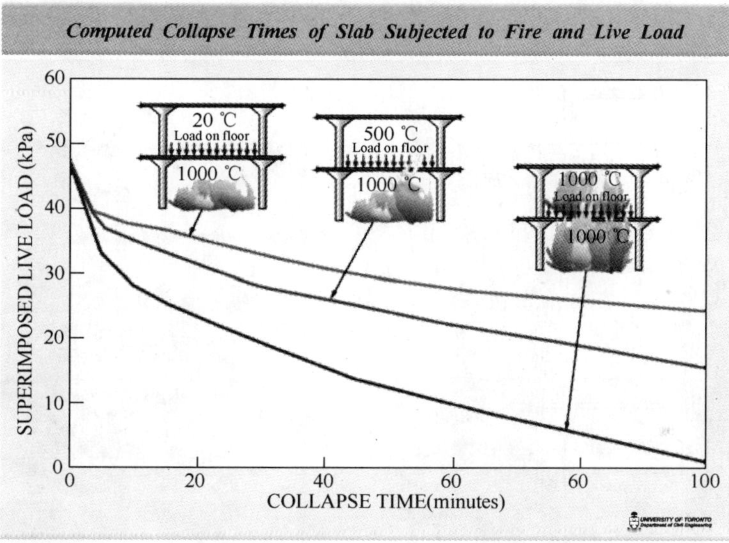


Figure 34　Predicted collapse times to slabs subject to fire and live load

neering journal, Sleipner had gone from an engineered cathedral to concrete scrap, see Figure 35. This very expensive shear failure initiated a period of intense activity for the Toronto shear group. First the cause of the failure needed to be identified and new designs for a replacement structure made. Then other concrete platforms, such as Draugen and Troll, still under construction, had to be checked in every detail to avoid similar problems.

In the Sleipner tricell wall which failed, there was no shear reinforcement because the global analysis and sectional analysis procedures used in the design indicated that such reinforcement was not required. The elastic finite element based global analysis had seriously underestimated the magnitude of the shear while the sectional analysis procedures had seriously overestimated the beneficial effects of axial compression on shear capacity. The shear procedures used to design the wall were those of the 1977 Norwegian concrete code which had been influenced by the 1971 ACI building code. Figure 36 compares the ACI and AASHTO LRFD "hand calculated" shear force - axial load interaction diagrams for the Sleipner wall section which failed. When the platform was lowered into the fjord, the shear

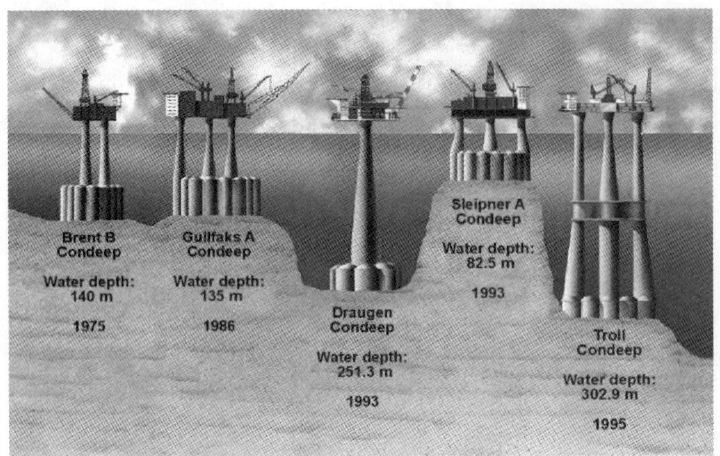

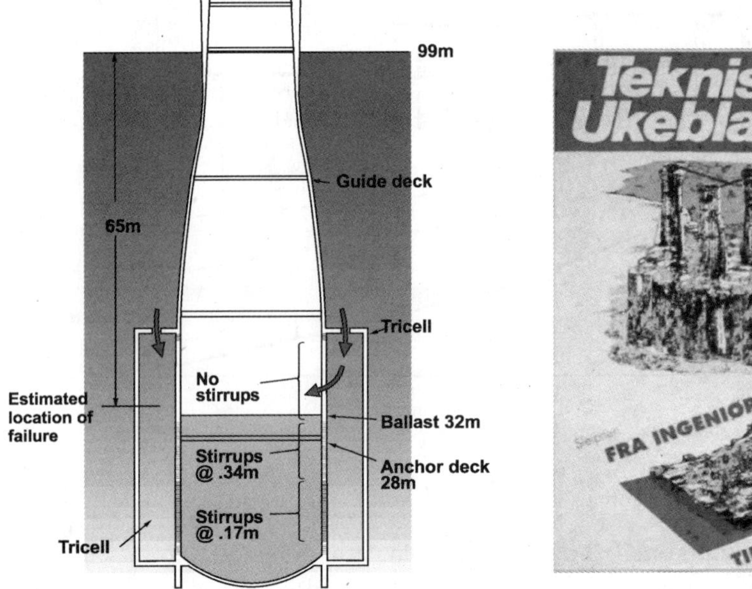


Figure 35 Condeep offshore oil platforms

on the wall increased and the axial compression also increased. The interaction diagram predicted by AASHTO LRFD indicates that failure would occur at a water head of 44 m while the ACI prediction is nearly three times higher at 120m. A second MCFT based prediction shown on the figure is that from program SPARCS (Selby's Program for the Analysis of Reinforced Concrete Solids) a 3D version of program VecTor2 (Vecchio and Selby 1991). This more refined prediction indicates failure at a water head of 62m. The Sleipner wall failed at a water head of about 65m.

A series of SPARCS analyses were conducted in October of 1991 to investigate the consequences of changing the details of the critical region of the Sleipner wall. The analyses demonstrated that the wall would not have failed during deck mating if stirrups had been provided or if the T-headed bars across the throat of the joint had been 0. 5 m longer.

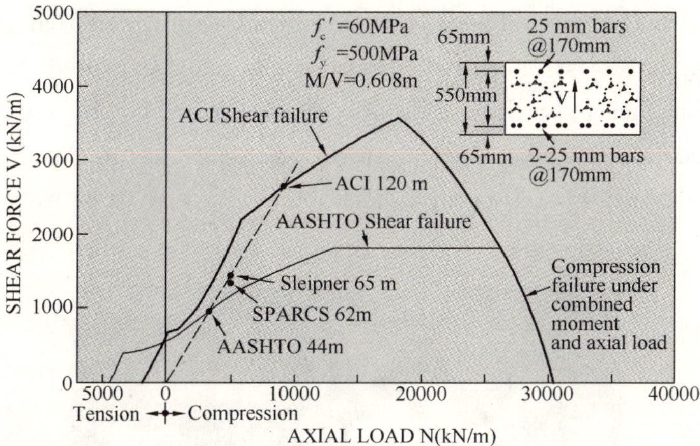


Figure 36　Shear force-axial load interaction diagram for tricell wall of Sleipner

Figure 37 summarizes the results of these analyses and played an important role in the legal proceedings which followed the collapse of the structure.

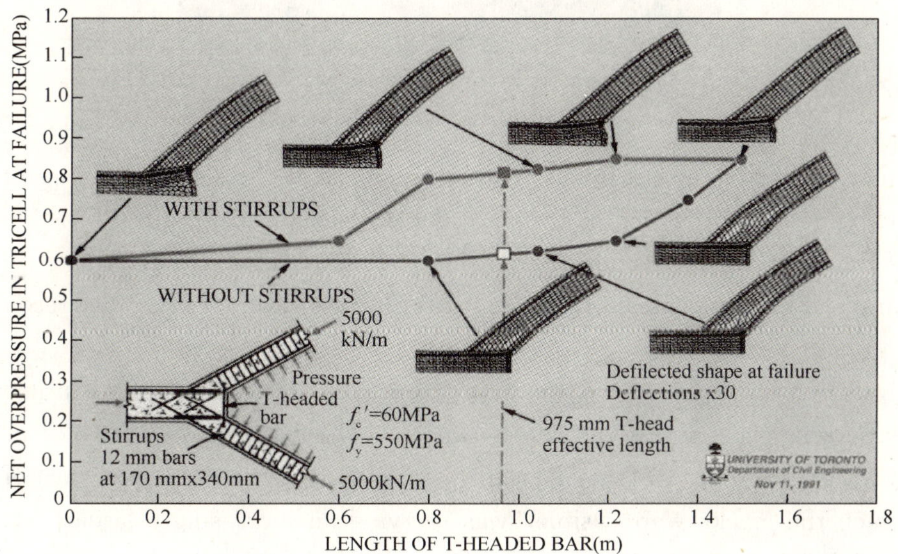


Figure 37　Influence of stirrups and length of tee-headed bar on failure pressure and failure mode of the tricell

After the failure of Sleipner it was clear that there were major problems with the existing design calculations and software. However the financial penalties involved were such that a replacement structure needed to be built in the shortest possible time. A decision was made to proceed with the design using the pre-computer, slide-rule era techniques that had been used for the first Condeep platforms designed 20 years previously. By the time the new software and computer results were available, all of the structure had been designed and most of it had been built. These computer results indicated that one portion of the built structure may have a strength deficiency for the loads expected during deck mating. Analytical and experimental work, see Figure 38, performed by the Toronto group demonstrated that this portion of the rebuilt platform was safe and on April 29th 1993 the new Condeep was successfully mated with the deck and was ready to be towed to sea.

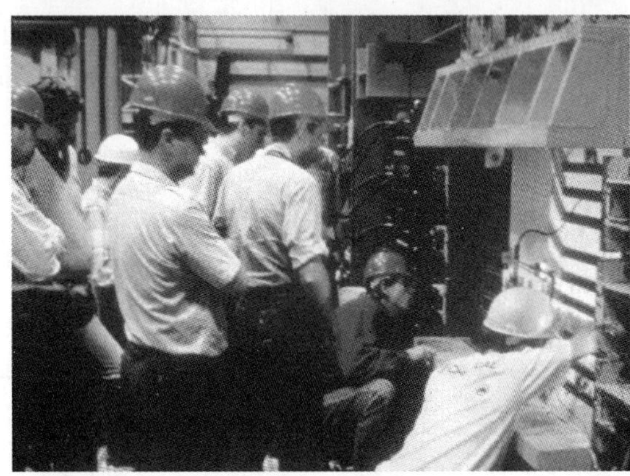

Figure 38 Engineers from the Sleipner project team test a specimen representing a portion of the rebuilt platform

Another major assignment involved the assessment of the safety of a 14 storey apartment building in Toronto which had wide diagonal cracks in major girders visible from the outside of the building, see Figure 39. Again the first issue was to determine the cause of the cracking and then to evaluate the safety of the structure and if necessary recommend repairs.

Analysis of the structure and the supporting soil, revealed that the prime cause of these shear cracks was the differential settlement of the mat foundation which in turn was caused by a significant layer of clay beneath the building. The "outrigger" part of the girders, in which the cracks were visible, was carrying an increasingly large portion of the building weight. It was calculated that under high wind loads these outrigger girders had an unacceptably high risk of failing. A detailed VecTor2 analysis of the 1. 2 m deep transfer girder system at the second floor level predicted that an even more dangerous situation existed in parts of the girders that were not visible, see Figure 40. When the ceiling was removed and these portions examined, wide shear cracks were discovered in the locations and in the orientations predicted. A temporary support system was immediately provided until more permanent repairs could be completed.

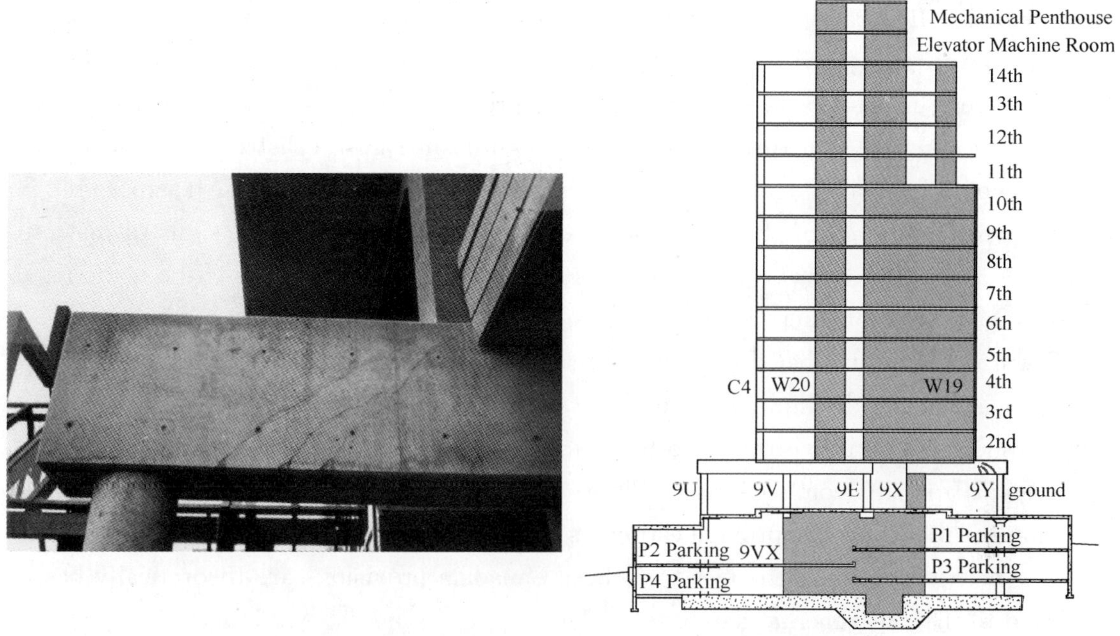


Figure 39 Cracks in condominium tower in Toronto

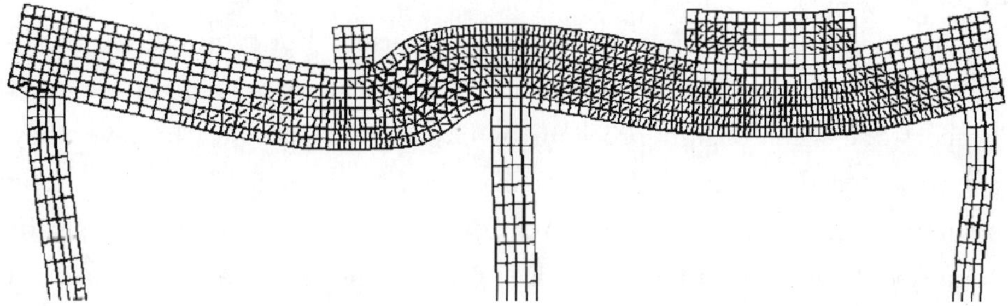

Figure 40 Predicted deformations magnified by 100 and crack patterns for interior portion of second floor beams of high-rise apartment building at load factor 1.20

## 2.12 CONCLUDING REMARKS

The recent collapse of the bridge shown in Figure 1 has emphasized the importance of accurate shear design provisions both for the design of new structures and for the evaluation of older structures, many of which were designed using shear provisions which are now known to be rather unconservative. How could the shear design provisions of 40 years ago been so inadequate and if they were, why was the problem not identified more quickly? The basic problem was that these provisions were developed at a time when there was no adequate theory for the shear strength of reinforced concrete members and the available laboratory experiments were nearly all conducted on rather small specimens with a rather narrow range of parameters. Because predicting shear strength is a more complex problem than predicting flexural strength and is influenced by many more parameters, developing accurate shear design provisions using only the results from a limited range of shear experiments is a very difficult task. If flexural design provisions were seriously unconservative, the resulting concrete structures would display significant signs of distress and, hence, the problem would be quickly rectified. Unfortunately, for the case of shear, it is possible for a member to be at 99% of the failure load and yet show no significant signs of distress. Thus unconservative shear design provisions can be used for many years without the resulting very low factor of safety being discovered.

Given the large number of existing concrete structures designed using unconservative shear provisions and the limited resources available to increase public safety, it is essential that the shear design provisions used to evaluate the structures be as accurate as possible. It is evident from the comparisons in Figure 31 that of the four codes studied, the shear provisions of the CSA code provide estimates with the least scatter. This low scatter is despite, or perhaps because, the fact that these Canadian provisions are theoretically-based, not fitted to the experimental database.

It should be appreciated that one of the advantages of shear design provisions that are based on a theoretical model is that for important evaluations, more refined analyses such as those described in Figs. 34, 36, 37 and 40 may be used. For typical designs, a structural engineer may be concerned that based on the example calculations given above, the theoretically based CSA shear design procedures may be too complex for everyday use. It should be appreciated that while iteration is required to evaluate the shear strength of an existing structure, no iteration is necessary for design. The longitudinal strain at mid-depth, $\varepsilon_x$, can be directly calculated from Eq. 6 using the factored design loads. Based on the assumption that the longitudinal reinforcement will be designed to be just less than the yield stress at factored loads, the CSA code also permits the use of a simplified method for evaluating $\varepsilon_x$ wherein $\varepsilon_x$ is conservatively taken as 0.43 times the yield strain of the longitudinal reinforcement.

One of the key advantages of a theoretically-based shear design procedure is that it is capable of making accurate predictions for the impact of changing different design parameters. Thus when faced with the design of a thick slab such as that shown in Figure 41, engineers may question whether the enhancements in shear behaviour that will result from adding minimum shear reinforcement will justify the cost involved. If they use traditional, empirical design equations, they may conclude that the strength enhancements do not justify the cost. As shown in Figure 41, however, the addition of minimum shear reinforcement to the thick slab strip shown has increased its shear strength by a factor of about three and has increased the energy required to cause collapse by a factor of about 30. The structure without shear reinforcement gave no warning of failure and failed prior to yield of the longitudinal reinforcement. With minimum shear reinforcement, yielding of the longitudinal reinforcement governed the failure load and caused significant deformations and cracking prior to failure. The Canadian bridge in Figure 1 which collapsed is similar to the member without shear reinforcement. Had the current Canadian shear provisions been in force when the bridge was designed, the slab would have required minimum shear reinforcement and the failure would have been prevented.

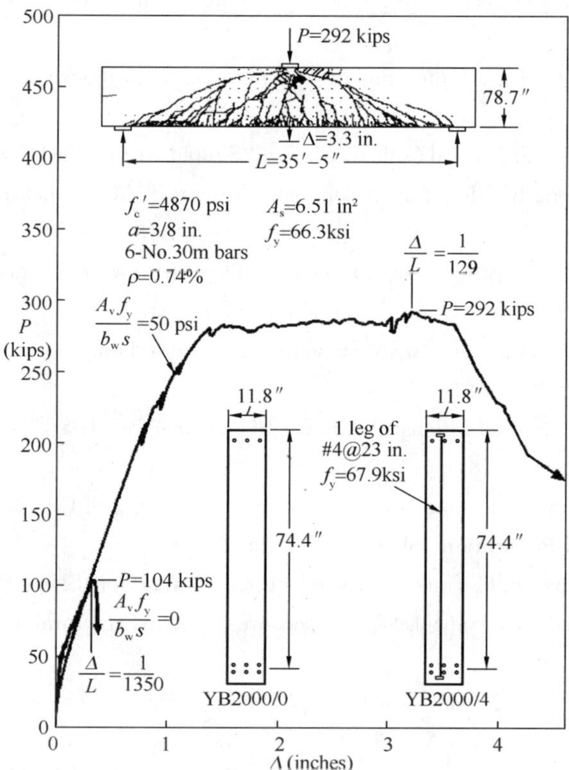


Figure 41　Influence of minimum shear reinforcement on member strength and ductility


## ACKNOWLEDGEMENTS

The authors would like to express their gratitude to the Natural Sciences and Engineering Research Council of Canada for a series of grants that have made possible the long-term research projects on the shear design of reinforced concrete at the University of Toronto


## REFERENCES


[1] AASHTO "LRFD Bridge Design Specifications and Commentary," Third Ed., American Association of State Highway Transportation Officials, Washington, 2004, 1264 pp.

[2] ACI Committee 318, "Building Code Requirements for Reinforced Concrete (ACI 318-08) and Com-

mentary (ACI 318 R-08)," American Concrete Institute, Detroit, 2008, 465 pp.

[3] Bentz, E. C., "Sectional analysis of reinforced concrete members." Ph. D. Thesis, Department of Civil Engineering, University of Toronto, 2000, 198 pp.

[4] Bentz, E. C., and Collins, M. P., "Development of the 2004 CSA A23. 3 Shear Provisions for Reinforced Concrete," Canadian Journal of Civil Engineering, V. 33, No. 5, May 2006, pp. 521-534.

[5] Bentz, E. C., http://www. ecf. utoronto. ca/~bentz/m2k. htm, Membrane-2000 webpage last accessed 2010/08/13.

[6] Bentz, E. C., http://www. ecf. utoronto. ca/~bentz/r2k. htm, Response-2000 webpage last accessed 2010/08/13.

[7] Bentz, E. C., Vecchio, F. J., and Collins, M. P., "Simplified Modified Compression Field Theory for Calculating Shear Strength of Reinforced Concrete Elements," ACI Structural Journal, Vol. 103, No. 4, July 2006, pp. 614-624.

[8] BSI, "BS 8110-1:1997 Structural Use of Concrete - Part 1: Code of Practice for Design and Construction," 1997, British Standards Institute, London, 159 pp.

[9] Canadian Standards Association, "CSA Standard A23. 3-04: Design of Concrete Structures." CSA, Mississauga, Ontario, 2004, 214 pp.

[10] Collins, M. P., "Towards a rational theory for RC members in shear," Journal of the Structural Division, ASCE, 104(4): 1978, pp. 649-666.

[11] Collins, M. P., Bentz, E. C. and Kim, Y. J., "Shear Strength of Circular Reinforced Concrete Columns", ACI Special Publication, SP-197, 2002, pp. 45-85.

[12] Collins, M. P., Bentz, E. C., and Sherwood, E. G., "Where is Shear Reinforcement Required? A Review of Research Results and Design Procedures," ACI Structural Journal, Vol. 105, No. 5, Sept-Oct 2008, pp. 590-600.

[13] Collins, M. P., Mitchell D., and Bentz E. C., "Shear Design of Concrete Structures," The Structural Engineer, Vol. 86, No. 10, May 2008, pp. 32-39.

[14] Collins, M. P., Vecchio, F. J., Selby, R. G. and Gupta, P. R., "The Failure of an Offshore Platform," ACI Concrete International, Vol. 19, No. 8, August 1997, pp. 29-35.

[15] European Committee for Standardization, CEN, EN 1992-1-1:2004 Eurocode 2: Design of Concrete Structures- Part 1-1: General rules and rules for buildings, Brussels, Belgium, 2004, 225 pp.

[16] FIB, "Model Code 2010, First Complete Draft, Volume 2: fib Bulletin 56,", fib, Lausanne, 2010, 288.

[17] Higgins, C., Miller, T. H., Rosowsky, D. V., Yim, S. C., Potisuk, T., Daniels, T. K., Nicholas, B. S., Robelo, M. J., Lee, A. Y., and Forrest, R. W., "Assessment Methodology for Diagonally Cracked Reinforced Concrete Deck Girders", Final Report for Oregon Department of Transportation and the Federal Highway Administration, Final Report SPR 350, SR 500-091, October 2004, 340 pages plus appendices.

[18] Hooke, R., "Lectures De Potentia Restitutiva, or of Spring-Explaining the Power of Springing Bodies", Printed for John Martyn, Printer to The Royal Society at the Bell in St. Paul's Church-Yard, 1678, 24 pp.

[19] Kani, M. W., Huggins, M. W., and Wittkopp, R. R., "Kani on Shear in Reinforced Concrete", Department of Civil Engineering, University of Toronto, Toronto, 1979, 225 pp.

[20] Katsaga, T., Sherwood, E. G., Collins, M. P., and Young, R. P., "Acoustic Emission Imaging of

Shear Failure in Large Reinforced Concrete Structures", International Journal of Fracture, Springer, Netherlands, Vol. 148, No. 1, Nov. 2007, pp. 29-45.

[21] Kowalsky, M. J., Priestley, M. J. N., "Improved analytical model for shear strength of circular reinforced concrete columns in seismic regions," ACI Structural Journal, Vol. 97, No. 3, May-June 2000, pp. 388-396.

[22] Lubell, A., Sherwood, T., Bentz, E. C., and Collins, M. P., "Safe Shear Design of Large, Wide Beams," Concrete International, V. 26, No. 1, Jan. 2004, pp. 66-78.

[23] Mitchell, D. and Collins, M. P., "Diagonal Compression Field Theory - A Rational Model for Structural Concrete in Pure Torsion," Journal of the American Concrete Institute, Vol. 71, No. 8, August 1974, pp. 396-408.

[24] Mörsch, E., "Der Eisenbetonbau (Reinforced Concrete Construction)," Verlag von Konrad Witwer, Stuttgart, Germany, 1907, 368 pp.

[25] Pang, X. -B., Hsu, T. T. C., "Behavior of reinforced concrete membrane elements in shear," ACI Structural Journal, Vol. 92, No. 6, Nov. 1995, pp. 665-679.

[26] Ritter, W., 1899, "Die bauweise hennebique," Schweizerische Bauzeitung, V. 33, No. 7, pp. 59-61. Shioya, T. "Shear Properties of Large Reinforced Concrete Member," Special Report of Institute of Technology, Shimizu Corporation, No. 25, 1989, 198 pp. (in Japanese).

[27] Uzel, A., Podgorniak, B., Bentz, E. C. and Collins, M. P., "Design of Large Footings for One-Way Shear", Accepted for publication by ACI Structural Journal.

[28] Vecchio, F. J. "Nonlinear Finite Element Analysis of Reinforced Concrete Membranes," ACI Structural Journal, Vol. 86, No. 1, Jan. -Feb. 1989, pp. 26-35.

[29] Vecchio, F. J. and Collins, M. P., "An Investigation of the Collapse of a Warehouse Structure," ACI Concrete International, Vol. 12, No. 3, Mar. 1990, pp. 72-78.

[30] Vecchio, F. J. and Collins, M. P., "The Modified Compression Field Theory for Reinforced Concrete Elements Subjected to Shear," ACI Journal, Vol. 83, No. 2, 1986, pp. 219-231.

[31] Vecchio, F. J., http://www.civ.utoronto.ca/vector, VecTor2 webpage, last accessed 2010/08/13.

[32] Vecchio, F. J., Selby, R. G., "Towards compression field analysis of reinforced concrete solids," ASCE Journal of Structural Engineering, Vol. 117, No. 6, June 1991, pp. 1740-1758.

[33] Wagner, H., "Ebene Blechwandträger mit sehr dünnem Stegblech," (Metal Beams with Very Thin Webs), Zeitschrift für Flugtechnik und Motorluftschiffahr, Vol. 20, Nos. 8 to 21, 1929.

[34] Walraven, J. C., "Fundamental Analysis of Aggregate Interlock," Journal of the Structural Division, ASCE, Vol. 107, No. ST11, Nov. 1981, pp 2245-2270.

[35] Xie, L., "The influence of axial load and prestress on the shear strength of web-shear critical reinforced concrete elements," PhD thesis, Department of Civil Engineering, University of Toronto, 2009, 344 pp.

# 第 3 章 Chapter 3

# 混凝土材料:高强度与耐久性

# CONCRETE MATERIALS: HIGH-TECH AND SUSTAINABLE

Hans W. Reinhardt
[1]Department of Construction Materials, University of Stuttgart, Germany.

**Abstract**: The paper surveys the development of modern concretes with respect to their high-tech properties and their sustainability. The aim of modern concrete technology is to increase workability, strength, and ductility of the material. Reviewed are self-compacting concrete (SCC), high-performance (HPC) and ultra-high performance (UHPC) concretes, fiber cement composites (HPFRCC and ECC), textile reinforced concrete (TRC), and concrete with superabsorbent polymers. All these types of concrete are only possible with a suitable grading of particles from the centimeter scale to the nanometer scale and with suitable additions and admixtures. The second aspect is sustainability which, in technical terms, is mostly reduced to ecology, i. e. consumption of raw material and energy, which is demonstrated at the cement and concrete production. Finally, demountable concrete structures are regarded as a challenge for the future.
**Keywords**: High-performance concrete, ultra-high performance concrete, fiber cement composites, self-compacting concrete, textile reinforced concrete, super absorbent polymers, sustainable structures

## 3.1 INTRODUCTION AND DEFINITIONS

Concrete is the most widely used construction material of the globe. The quantity amounts to about 15 billion tons per year. It is certainly true that not every concrete has high-performance properties but it is also true that concrete can be produced with high-performance properties and there are ambitious applications nowadays which show this clearly. What means high-tech or high performance? It can refer to various properties, i. e. to strength, to ductility, to workability, to resistance against frost, chemicals, abrasion, fire, to economy, to sustainability. One of the properties must be excellent and much superior to the average. One speaks about high-strength concrete, high-ductility concrete, high-durability concrete, etc. In the list of properties, economy and sustainability belong to different categories than the others which are technical. Sustainability has environmental and technical aspects but also socio-cultural ones which will not be discussed here. How to gain high-performance and sustainable concretes will be considered in the next chapters.

## 3.2 ACHIEVING CONCRETE WITH HIGH-PERFORMANCE PROPERTIES

### 3.2.1 Composition of concrete

Classically speaking concrete is a compound of cement, water, and aggregates. Cement and water forms the cement paste which hydrates. The water is bound chemically in the calcium-silica hydrates (C-S-H) and also physically to the large specific surface of about $200m^2/g$. The C-S-H contains capillary pores of 5 to 100 nm size and gel pores which have the size of few nanometers. The gel pores are inherent to the hydrate while the capillary pores develop due a surplus of water which can evaporate later. Near the interface of cement paste and aggregate there exists more capillary pores than in the bulk material which is due to the precipitation of calcium hydroxide. Calcium hydroxide is another product of the hydration process. To make concrete stronger one has to reduce the capillary pores as much as possible.

Modern concrete technology treats concrete as a five material compound which includes also admixtures and additions. Admixtures are chemical products and are added in very small quantities. They can improve the workability, the frost resistance, and other properties of fresh and hardened concrete. Additions are mainly mineral products. They can react with cement and water to form also C-S-H or they are fillers which do not react. Additions are added in larger quantities. The next chapters will discuss the use and advantage of admixtures and additions in more detail.

### 3.2.2 Use of admixtures and additions

1. High-strength concrete

The key to high-strength concrete (HSC) is the densification of the structure. It starts with a low water-cement ratio. Fig. 1 shows the distribution of the components of hydrated Portland cement paste as function of the water-cement ratio (w/c) in sealed condition.

One can see that cement can completely hydrate with a $w/c$ equal to 0.42. At this point, the volume of capillary pores amounts to about 8 %. If $w/c<0.42$ the volume of capillary pores is reduced, if $w/c>0.42$ the volume of capillary pores is increased. To produce high-strength concrete the $w/c$ should be in the order of 0.25 to 0.30.

Even lower w/c down to 0.18 is required for an ultra high-strength concrete (UHSC). In these concretes, a maximum packing density of all components has to be achieved. Silica fume is used as an addition to cement. Silica fume has a grain size in the order of 0.1μm compared to 5—20μm of cement, i.e. it fills the gap between the cement grains. A second advantage is that silica fume consumes a part of the weak calcium hydroxide during hydration. Thus, the structure of UHSC is much denser than that of HSC. Fig. 2 illustrates the situation. This densification is especially important at the interface of aggregates.

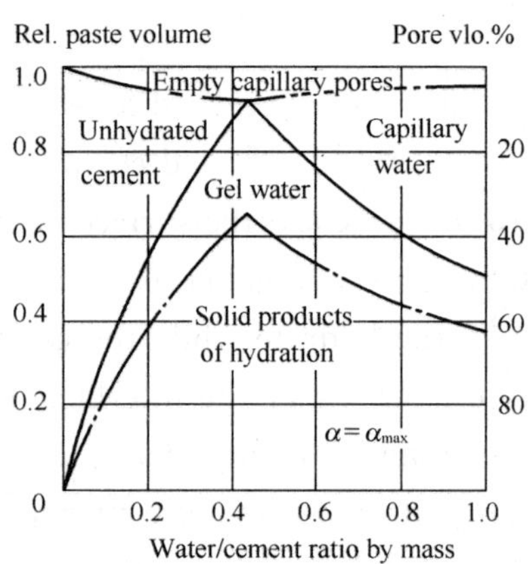


Figure 1 Components of hydrated cement paste

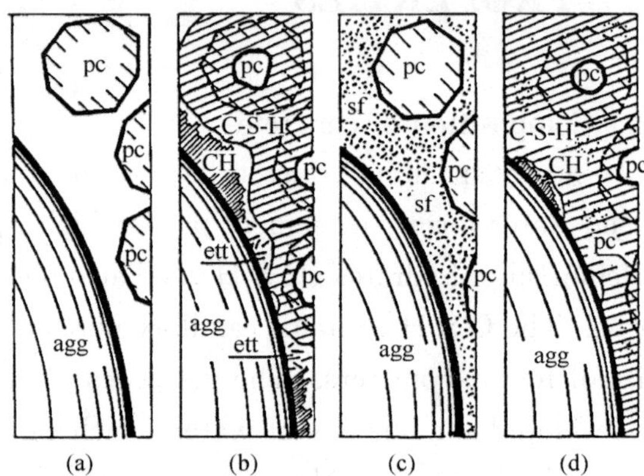


Figure 2 Cement paste in the interface to aggreggate (agg), after Bentur,
(a) fresh cement paste (pc), (b) hardened cement paste (CH =calcium hydroxide, ett=ettringite), (c) fresh PC with silica fume (sf), (d) hardened PC with SF

The added amount of silica fume is in the order of 10 % of the mass of cement in the case of HSC, and up to 30 % in the case of UHSC. It should be noted that the aggregates themselves have to be graded very well so that packing of the grains is optimized.

HSC and UHSC can only be worked in combination with a superplasticizer (or high water reducing agent). Such an admixture on the basis of polycarboxylic ether (PCE) maintains the electrostatic charge on the cement particles and prevents flocculation. The quantity which is added to cement is $<$ 80g per kg of cement.

The result of the dense packing is a compressive strength of HSC up to 140 MPa and of UHSC up to 250 MPa (in the case of teat treatment up to 400MPa).

The high strength can be attributed to the reduction of capillary pores and to the size of the pores. Fig. 3 shows the differential distribution of pores, measured with mercury intrusion, of normal strength concrete (NC), of high performance concrete (HPC), of UHPC, and of a teat treated UHPC (RPC) with the compressive strength indicated by the numbers in the brackets. It can be observed that the peak of the distribution curves shift to the left for higher strength, i.e. the pores get smaller, and that the area under the curves becomes smaller which is extreme for the UHPC and the RPC. Together with these observations one can also predict that water permeability and gas permeability are greatly reduced, and that durability is enhanced[1].

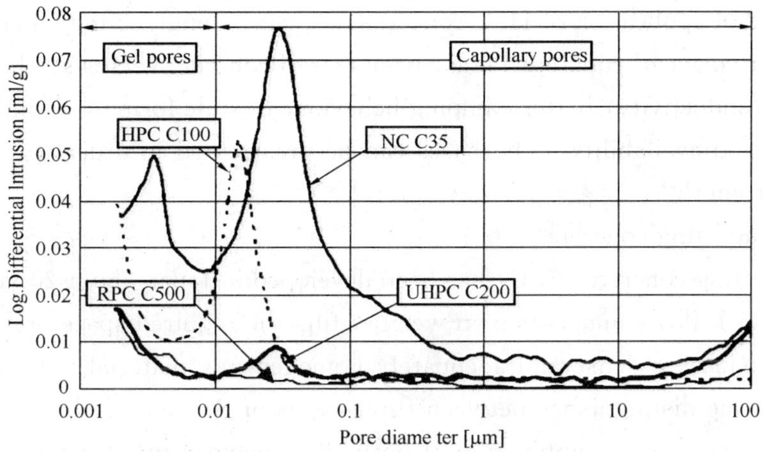


Figure 3　Differential pore size distribution for various concretes [2]

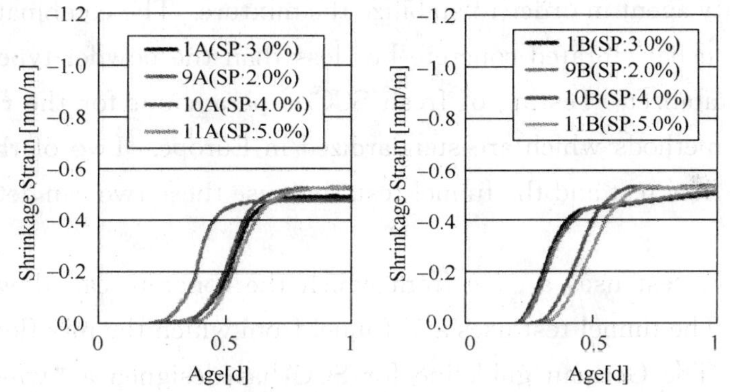


Figure 4　Development of autogenous shrinkage during the first day of hydration with variation of content of superplasticizer, 800kg/m$^3$ cement, $w/c=0.23$[3]

New developments are due to special cements and appropriate superplasticizers[2]. One property should be mentioned which has to be considered in structural design, that is autogenous shrinkage. Autogenous shrinkage develops during the first days of hydration and is due to chemical shrinkage and self desiccation. Chemical shrinkage is inherent to the hydration process because the constituent materials of cement paste occupy a larger space than the hydration products. Self desiccation occurs in those pastes which have a lower water-cement ratio than 0.42. Since UHPC is produced with very low water-cement ratio self desiccation is important. Fig. 4 shows the development of autogenous shrinkage during the first day of hydration with variation of content of superplasticizer.

It can be observed that the autogenous shrinkage amounts to roughly $0.5 \cdot 10^{-3}$. After continuation of the experiments the shrinkage reached a value of about $0.7 \cdot 10^{-3}$. Especially in those structures which have fixed ends the development of stresses due to imposed deformation would inevitably cause tensile cracks.

A new field of application of HSC concerns mechanical engineering where bases of mechanical tools are made of concrete. The advantages of concrete versus steel or cast iron are lower thermal conductivity, better damping behavior, flexible forming at ambient temperature, economy and availability. The bases can be prestressed in order to compensate the lower tensile strength[4].

2. Self-compacting concrete

Self-compacting concrete (SCC) has been developed in Japan about 20 years ago [5]. It is a concrete which flows under its own weight, fills the required space or formwork completely, and produces a dense and adequately homogeneous material without the need of compaction[6]. One distinguishes between three types of SCC, i. e. the powder type, the viscosity-agent type, and a combination of both. The powder type uses a large amount of powder in the mix such that the aggregate grains can float in the paste. The viscosity-agent type uses a viscosity agent in order to stabilize the mixture. The combination type contains more powder than usual vibrated concrete but less than the powder-type SCC and used a small amount of stabilizer. Testing of fresh SCC is paramount for the right consistency. There are several methods which are standardized in Europe. Two of them will be mentioned, the slump-flow test and the funnel test, because these two can determine the suitability of a mix.

The slump-flow test uses a cone from which the concrete can flow out and form a spread on a table. The funnel test uses a V-funnel from which the mix flows out. The flow time is measured. The German guideline for SCC has designed a "window" in a graph which is supposed to determine the suitability of a mix[7]. Fig. 5 shows the window consisting of the V-funnel time and the slump-flow. Concrete mixes outside the window show stagnation, i. e. they do not flow, or air inclusions, i. e. the self compaction is not sufficient, or sedimentation, i. e. the mixes segregate such that the heavier aggregate grains sink in the formwork down.

Fig. 5 is the example from the guideline, however, every producer has to construct his own window for his type of SCC. Once, he has established his window he can assume that all concretes inside it would fulfill the workability requirements.

A typical powder-type SCC contains in $dm^3$ per $m^3$ of concrete: 110 cement, 120 filler, 160 water, 10 air, and the remainder is the aggregate up to 16 mm grain size. The filler can be flyash, limestone filler, or ground granulated furnace slag. The self-compacting workability can only be reached with superplasticizer the amount of which has to be determined in tests.

SCC can be designed as normal strength or as high strength concrete. The properties

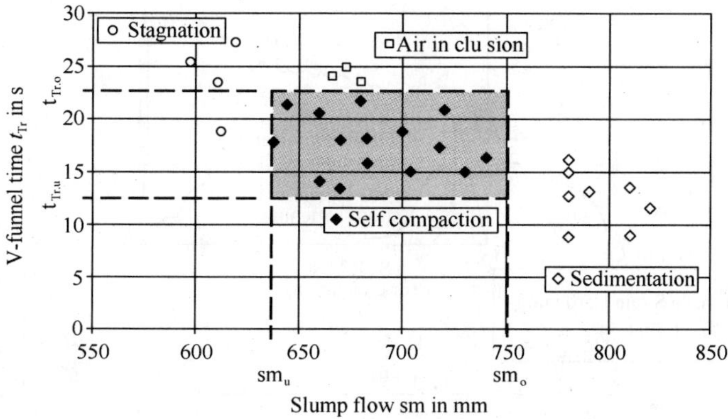


Figure 5 "Window" for SCC suitability[7]

of SCC and traditional vibrated concrete in the hardened state are almost the same so that structures can be designed in the same way.

3. Fiber concrete

(1) Classification

Plain concrete is a brittle material with low tensile strength and very low tensile strain capacity. To increase the ductility fibers are being used. Conventional fiber concrete is appropriate for slabs on grade, for pavements, for foundations, for basement walls, i. e. for those structures where the risk of failure is low or where the consequences of a failure are limited. Fiber concrete can also be used in prestressed structures as shear reinforcement instead of reinforcing bars.

However, there are developments towards high-performance fiber reinforced concrete composites (HPFRCC) which have greater tensile strength and larger ductility. The various fiber reinforced concretes (FRC) can be classified according to Fig. 6. The lower diagram shows the deflection behavior of tensile strain softening FRC with the typical softening curve in deflection for low fiber contents and the deflection hardening curve for higher fiber contents.

But these fiber contents are still lower than the critical fiber content in tension $V_{fcri,\ tension}$. The upper diagram shows tensile stress-strain curves for strain softening and strain hardening. Strain hardening occurs if the fiber content is larger than $V_{fcri,\ tension}$. Strain hardening is always accompanied with multiple cracking. It is observed that a strain-hardening material is considered mechanically more performing than a strain-softening one. Deflection-hardening materials are useful in applications where bending prevails, while deflection-softening composites cover the range which is stated above.

The strain hardening composites are not equivalent. In order to promote the application of HPFRCC there is a need for a classification. Fig. 7 shows a possible classification for strain-hardening composites proposed in [8]. The classification uses only two quantities, the elastic modulus and the tensile stress (or strength) at a tensile strain of 0.5 %.

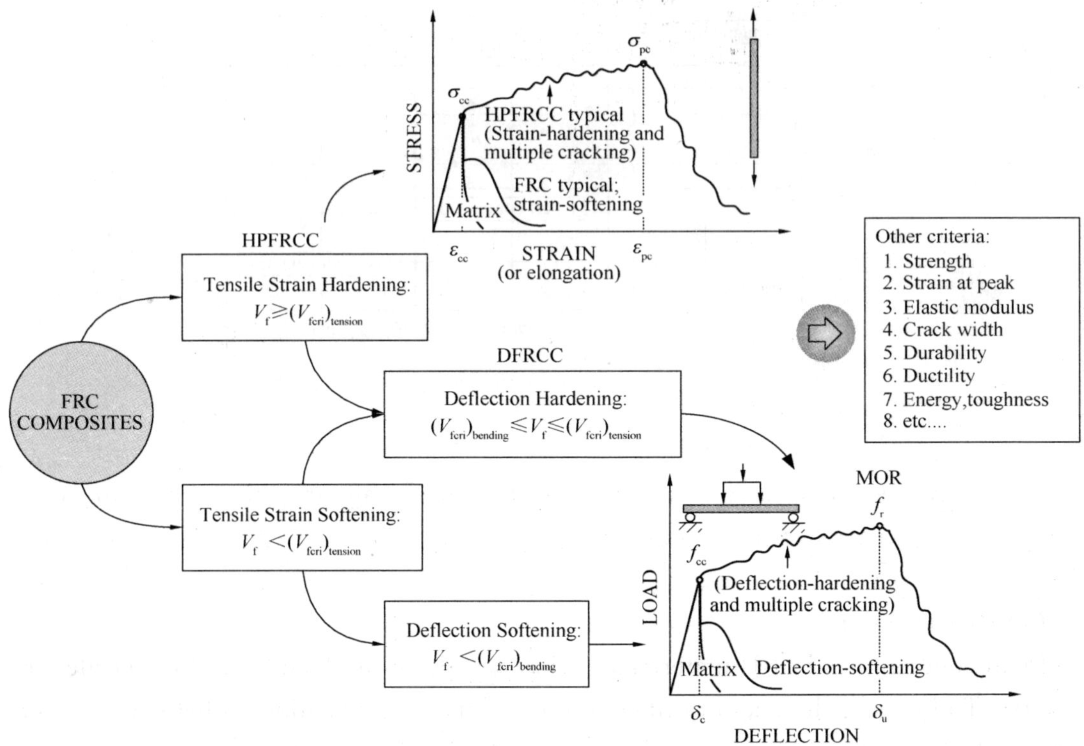


Figure 6 Classification of FRC composites based on their tensile stress-strain response[8]

The elastic modulus is lower than measured in many experiments but it accounts for creep effects during long-term loading. The strain of 0.5% at peak stress is more than twice the yield strain of reinforcing bars. This guarantees that strain-hardening FRC will contribute to crack control in service as well as ductility, energy absorption, and ultimate strength of conventional reinforced elements.

For each class of composites, two curves are shown in Fig. 7, one in full and one dotted, to illustrate different examples of curves in the same class. A class T-10 would guarantee a post-cracking tensile strength of 10 MPa. For all classes defined in Fig. 7, there is the same required minimum level of strain at peak stress, and the same required minimum elastic modulus. A direct tensile test should verify such constraints. The elastic modulus represents the tangent modulus at the origin. The first cracking strain is considered to be equal to the ultimate strain of the matrix which is about 0.0002. To account for statistical variability, 90 % of experimental data should have a peak stress of 10 to 15 MPa in order to satisfy the class T-10. It is supposed that such a classification is sufficient for most preliminary design situations.

(2) UHPC with fibers (UHPFRC)

Plain UHPC is a linear elastic brittle material. It changes the behavior when fibers are added according to the schematic of Fig. 8. The critical fiber content is the fiber content

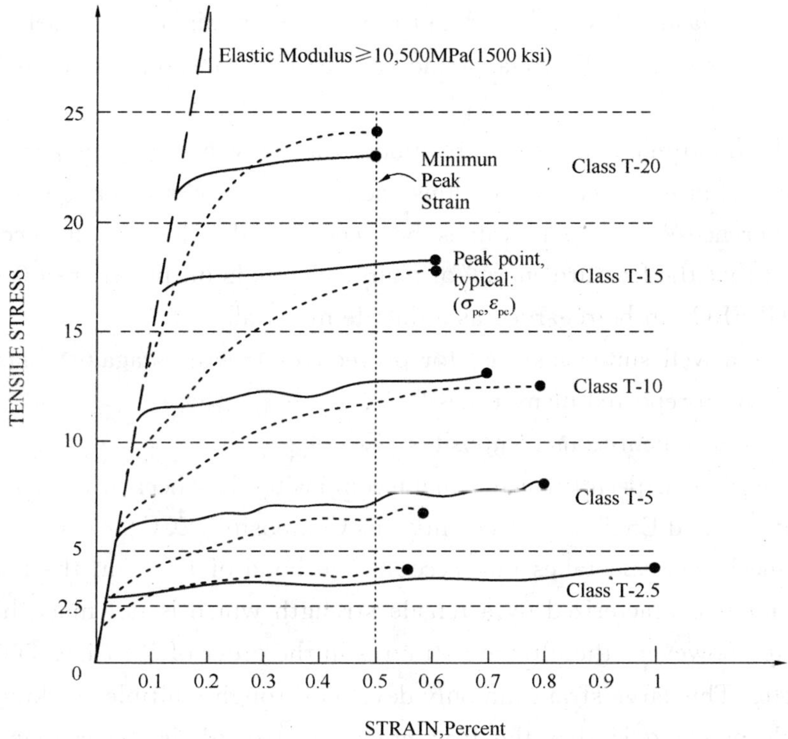


Figure 7 Definition of different classes of strain-hardening composites[8]

which equilibrates the stress at the peak load. Under-critical fiber content leads to strain-softening, over-critical fiber content leads to hardening.

The diagram is valid for compressive and for tensile loading, of course, the scale on the vertical axis would be about ten times larger for compressive loading. Fig. 9 shows test results of a UHPFC (DUCTAL®) with 2 Vol. -% steel fibers in tension[9].

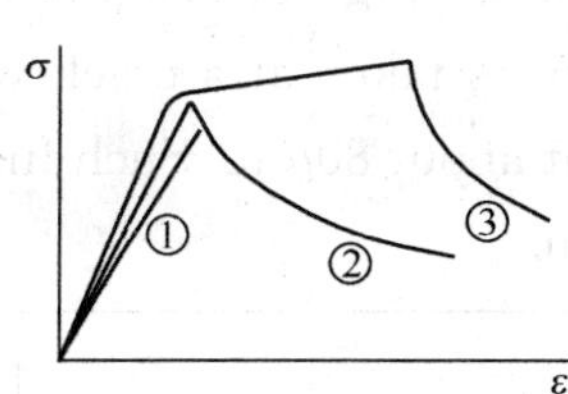


Figure 8 Schematic stress-strain curves of UHPC.
1) plain, 2) with under-critical fiber content, 3) with over-critical fiber content

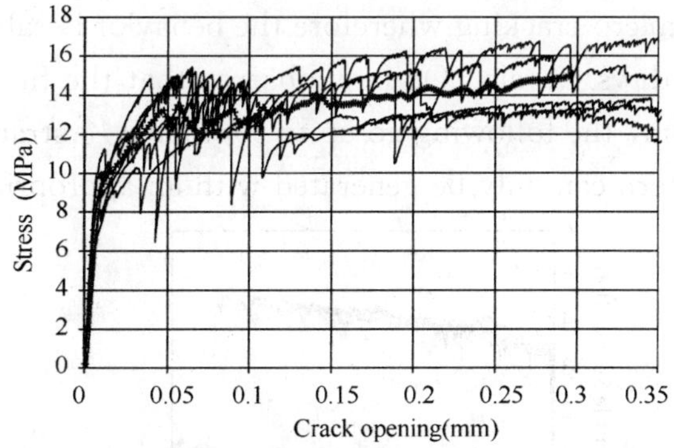


Figure 9 Stress-crack opening curve of DUCTAL®[9]

The lines start with a linear branch up to about 8 MPa. Then the non-linear part begins, and the curve rests almost stable up to a crack width of 0. 35 mm. The behavior can be characterized as elastic-plastic. The data from [9]are: strain at first crack 0. 00019, tensile strength of matrix 10. 6 MPa, elastic modulus 58 GPa, mean peak stress 15 MPa. Applying the classification of Fig. 7 would lead to T-10.

The specific fracture energy is the area under the stress crack-opening curve. This value is necessary for finite-element calculations and also for the characterization of the material. Various references[10~12] report values between 6 and 30kJ/$m^2$. For comparison, it should be noted that the fracture energy of plain concrete is in the order of 0. 1kJ/$m^2$. This shows that UHPFRC can be regarded as a ductile material.

UHPFRC is a well suited material for protective structures against blast and impact loads which has been reported many times[13, 14].

(3) Engineered cementitious composites (ECC)

A new type of high performance cement composites has been invented at the end of last century, it is called Engineered Cementitious Composite (ECC)[15]. It is derived on the basis of micromechanics and takes into account the bond of fibers in the matrix[16]. The material behavior is characterized by a tensile strength which is not much higher than of normal concrete, however, the ultimate strain is in the order of 5%, i. e. 500 times larger than of concrete. This large strain can only develop through multiple cracking. The condition of multiple cracking is that the first cracking strength is lower than the bridging strength of the fibers.

Fig. 10 shows typical stress-strain curves of ECC with 2 vol. -% PVA fibers[17]. The first cracking strength amounts to 3. 5 MPa. After the linear elastic behavior, it follows a steady strain-hardening response with a peak stress of 4. 5MPa and an ultimate strain of 5. 5%. The crack pattern is so fine that the cracks cannot be observed with the naked eye up to a strain of about 1%.

The material shows the typical feature of plasticity, however, the plasticity is caused by micro-cracking wherefore the behavior is called pseudo-plastic. Fig. 11 is a plot of crack width vs. strain. One can observe that the first measurement is taken at a crack width of 20μm, the following go up to 60μm at 1% strain and finish at about 80μm. Such fine crack pattern can only be generated with an appropriate mix design.

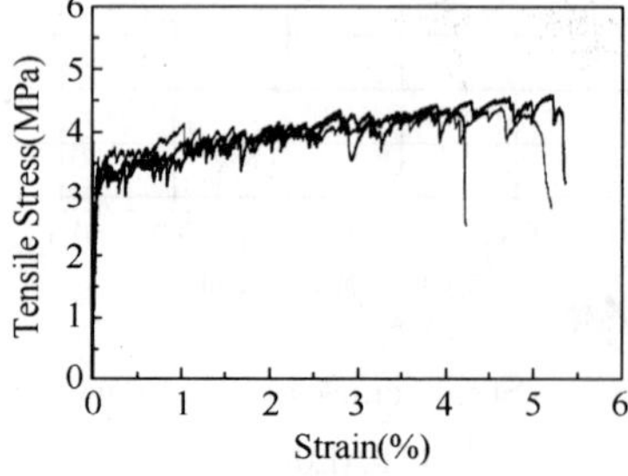


Figure 10 Uniaxial tensile stress-strain curves of an ECC with 2 vol. -% PVA fibers[17]

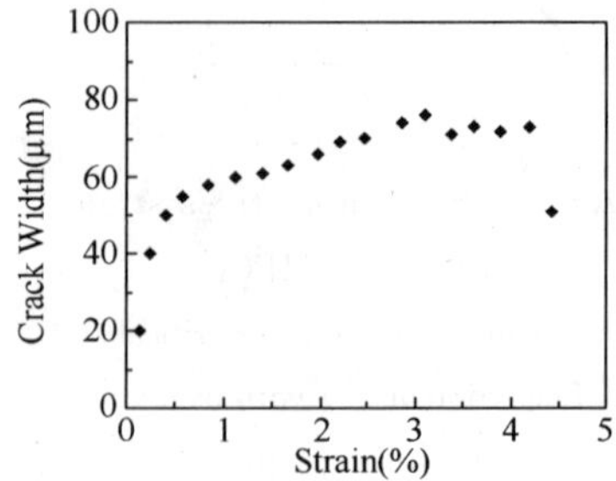


Figure 11 Crack width vs. strain of an ECC with 2 vol. -% PVA fibers[17]

The following Table 1 shows an example of two concrete compositions. The binder content (cement plus fly ash) is rather high with 1080kg/m$^3$ and 1102kg/m$^3$ resp. , but it is necessary in order to receive a matrix-rich mix which can lead to fine multiple cracking.

**Composition of ultra-ductile concrete (ECC)[18]** **Table 1**

| Concrete | Cement [kg/m$^3$] | Fly ash [kg/m$^3$] | Quartz sand [kg/m$^3$] | LWA sand [kg/m$^3$] | Water [kg/m$^3$] | SP1) [kg/m$^3$] | VA2) [kg/m$^3$] | PVA fibre [kg/m$^3$] |
|---|---|---|---|---|---|---|---|---|
| Ⅰ | 320 | 750 | 535 | — | 335 | 16. 1 | 3. 2 | 29. 3 |
| Ⅱ | 551 | 551 | — | 186 | 344 | 16. 5 | 3. 3 | 26. 0 |

1) SP=superplasticizer, 2) VA=viscosity agent

Fine quartz sand is used in the first mix while a light-weight sand (expanded glass) is used in the second. The water-binder ration amounts to 0. 31 in both cases while the water-cement ratio is 1. 04 and 0. 62 resp. which leads to a compressive strength of about 30 MPa. If it is assumed that the fly ash also reacts with cement and water with a replacement factor of 0. 4 then the effective water-cement ratio computes to 0. 54 and 0. 45 resp. The first concrete had 2. 2 vol. -% polyvinyl alcohol fibers (PVA) while the second had only 2. 0 vol. -%.

(4) Textile reinforced concrete

Textile reinforced concrete (TRC) can be regarded as a special class of fiber concrete. While short fibers are used in conventional fiber concrete TRC uses long strands in the form of woven or knitted fabrics. Fig. 12 shows a few examples. The fabrics can be flat or three-dimensional since the weaving technology is very advanced.

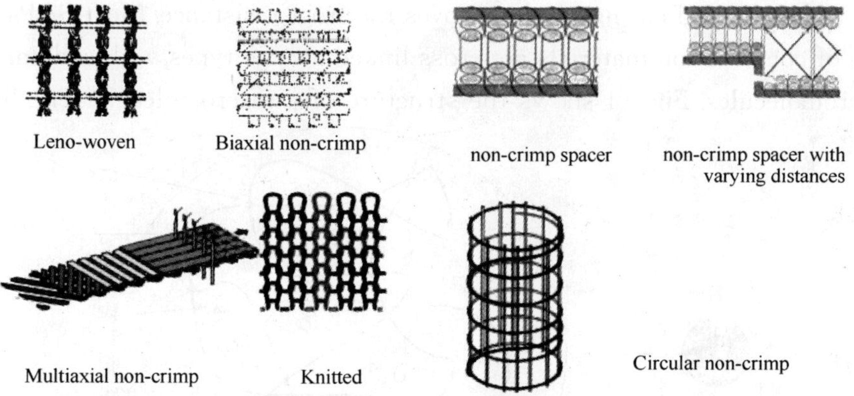


Figure 12　Some semi-finished textile reinforcement products for the application in concrete[19]

The material used is either alkali-resistant glass or polymers. The tensile strength of glass is about 1. 500 MPa, of aramide is 2. 000 MPa, and of carbon is about 3. 000 MPa.

However, in the woven state, the tensile strength is reduced to one half or two thirds due to the interaction at the intersection points. The matrix is made of cement, fly ash, silica-fume, and fine quartz sand up to 1.2 mm grain size. The water-binder ratio is in the order of 0.30 which makes that the compressive strength goes up to 100 MPa. Fig. 13 shows an example of load-deflection curves of a bending experiment. The plot starts with the linear elastic behavior until the first cracking load.

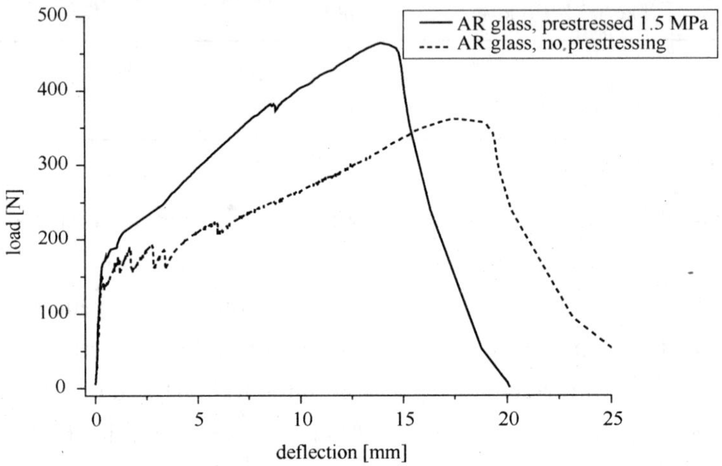


Figure 13 Load-deflection curves of an AR-glass reinforced plate under cyclic load[20]

After this point, several cracks occur which can observed from the saw-tooth shaped line. Although the reinforcing material as such is brittle the plate behaves quasi-plastic.

(5) Concrete with superabsorbent polymers

Superabsorbent polymers (SAP) are a class of additions which influences the workability, provides internal curing, and improves the frost resistance. Most SAPs used in the production of construction materials are cross-linked acrylic types with sodium ions bound to the macromolecules. Fig. 14 shows the structure of a macromolecule. The linear chains

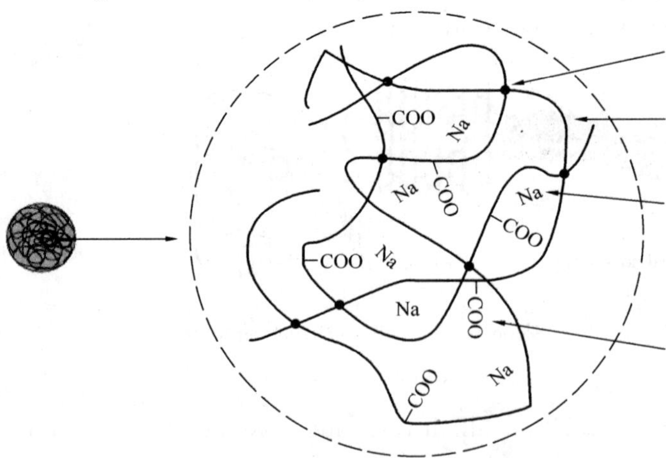


Figure 14 Structure of superabsorbent polymer

are cross-linked which makes the molecule insoluble in water.

Functional side groups of polymer chains attract water molecules. Dissociated sodium ions are trapped inside the polymer network. A diffusion equilibrium with the surrounding fluid cannot be reached wherefore an osmotic pressure build up which sucks water into the interior. Polymers with ionic groups are able to store a large quantity when the ions are forced closely together and build up high osmotic pressure inside. The osmotic pressure depends on the concentration of the surrounding fluid and therefore the water absorption depends also on the concentration. The SAP are able to absorb up to 1. 500 g of water per gram of SAP. However, tests with concrete have shown that they can absorb only 24 g pore water/g SAP[21].

When mixed in cement paste SAP stores some water which can be released during hydration. The example of Fig. 15 shows the situation for a cement paste with the ratio of 0. 42 of total water to cement: 0. 43 parts are cement, 0. 02 is SAP, 0. 08 is water stored in SAP, and 0. 47 is water. After the cement has hydrated completely there are 0. 65 parts cement gel, 0. 25 gel pores and 0. 10 SAP plus air voids.

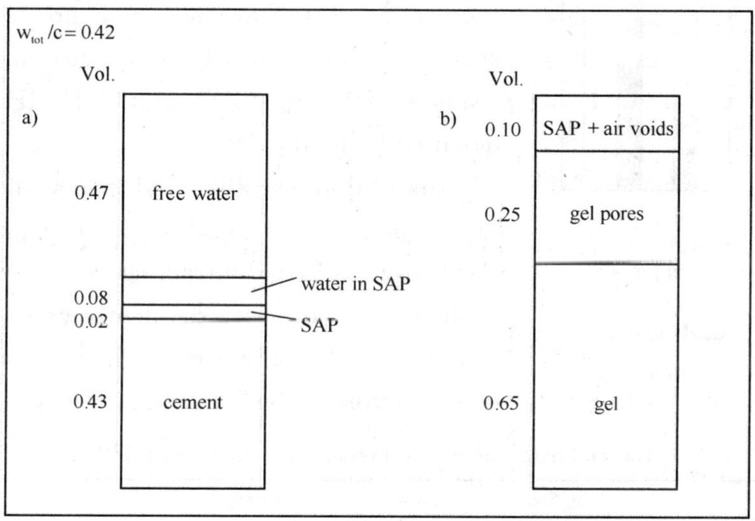


Figure 15 Volume partition between the components of cement paste, a) fresh state, b) hardened state

There are two counteracting effects of SAP in concrete. One is a positive effect which is the internal curing without supplying water from outside, the other is a negative effect which is the formation of pores with simultaneous reduction of strength. However the pores have also a positive effect on frost resistance of concrete. Tests with various contents of SAP have resulted in Fig. 16 which shows a reduction of compressive and flexural strength and Young's modulus.

The graph is normalized with respect to concrete without SAP. The water-cement ratio

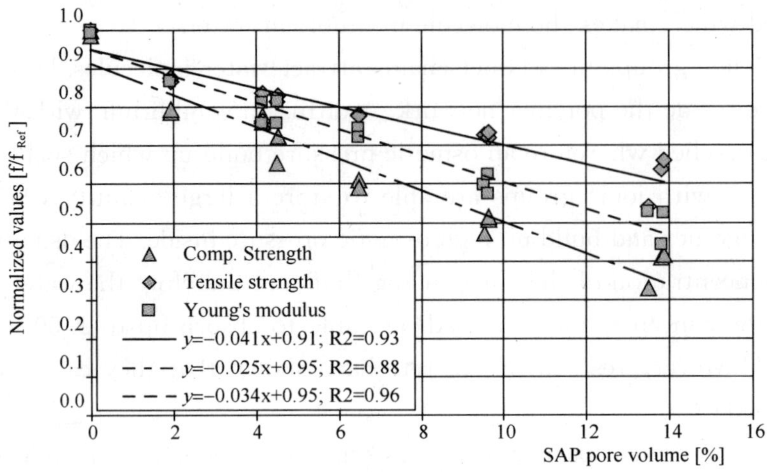


Figure 16 Reduction of mechanical properties due to addition of SAP[22]

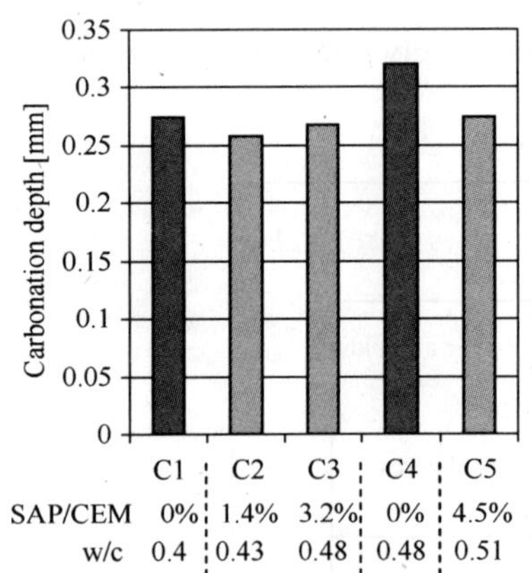


Figure 17 Carbonation depth after 14 months

was in the order of 0. 50. When SAP is used the SAP pore volume should be limited to a few percent in order not to impair the mechanical properties too much. On the other hand, tests have shown a greater resistance against carbonation which is depicted in Fig. 17. The prisms were stored at 20°C and 65 % RH. The reduction amounted to about 15%.

It is well known that well graded and well distributed air pores have a positive effect on frost resistance. SAP is used as an air-entraining measure[23]. The SAP of this investigation was a covalently crosslinked acrylamide/acrylic acid copolymer[24]. The concrete had the composition according to Table 2.

**Material data and composition of 1 m³ concrete used[24]** **Table 2**

| Material | Size [mm] | Absorption [%] | Density [kg/m³] | Mass [kg] | Volume [m³] |
|---|---|---|---|---|---|
| Cement CEM I 52. 5 | — | — | 3150 | 390 | 0. 124 |
| Tap water | — | — | 1000 | 164 | 0. 164 |
| Entrapped air (estimate) | — | — | — | — | 0. 015 |
| Sea sand | 0—4 | 0. 5 | 2601 | 774 | 0. 297 |
| Sea gravel | 4—8 | 0. 6 | 2642 | 333 | 0. 126 |
| Sea gravel | 8—16 | 0. 5 | 2637 | 722 | 0. 274 |

The dry SAP particles with a density of approximately 1500 kg/m³ were divided into the fractions 38 to 63μm and 90 to 125μm by sieving. During mixing of the concrete SAP absorbs approximately 12. 5 g water per g dry SAP. The dry particles swell by water absorption and the two fractions of SAP got average diameters of 150 and 300 μm respectively. The total volume of SAP pores were adjusted to 1, 2, 3, 5 and 10 % with 150μm SAP and 5, 10, 15, 25 and 35 % with 300μm SAP relative to the volume of cement paste.

The freeze-thaw tests were carried out acc. to the European standard CEN/TS 12390—9[25] where temperature cycles of −20 to +20℃ during 24 hours apply. The test is continued up to the end of the 56th cycle.

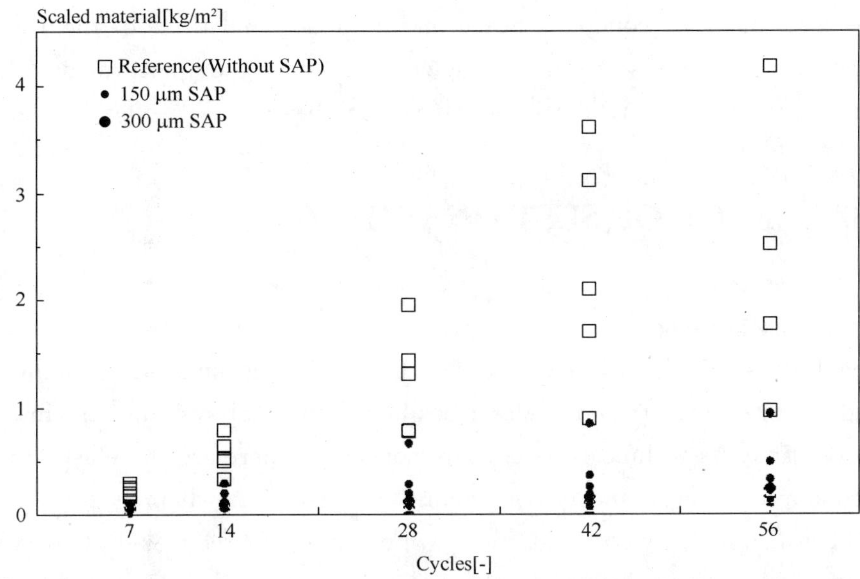


Figure 18　Amount of scaled material as function of number of cycles[24]

Figure 19 shows the mean values of scaled material after 56 cycles as function of pore volume and spacing factor. The spacing factor was estimated to be in the range of 0. 15 to

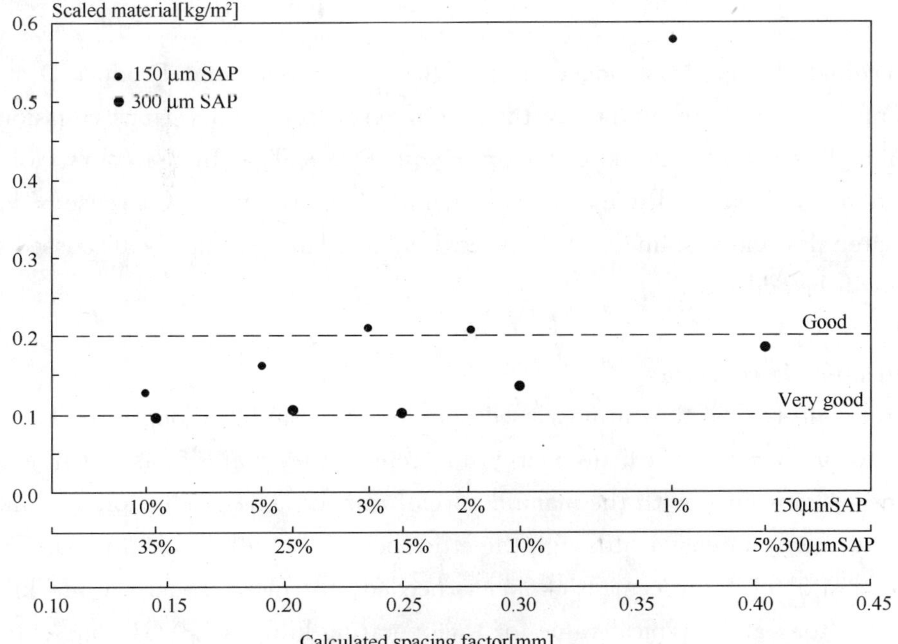


Figure 19　Amount of scaling after 56 cycles as function of pore volume and spacing factor[24]

0.40 mm depending on the number of SAP pores.

It can be seen that scaling is reduced with higher amount of SAP pores. The Swedish standard SS 137244[26] defines how to classify concrete after the frost test. "Good" performance is achieved when scaling amounts to 0.2 $kg/m^2$ and "very good" when it is only 0.1 $kg/m^2$. Most SAP modified concretes range under "good" and three of them under "very good". All reference concretes are outside the good and very good performance. As the pore size is concerned the larger SAP particles behaved the best.

## 3.3 CHALLENGE OF SUSTAINABILITY

### 3.3.1 General background

Sustainability is a term with many aspects. Generally speaking sustainability is the capacity to endure. In ecology the word describes how biological systems remain diverse and productive over time. As technical systems are concerned there are three aspects which are relevant: economy, society, and environment. The main issue here is the environment. There was a German forestry commissioner[27] who has postulated in 1713 that only as many trees can be cut as new trees replace them. That means in a wider sense that a system should renew itself. This will not be possible in the construction industry but one should strive a minimum of consumption of resources and a maximum of reuse of materials. Additionally, one should save primary energy as much and the ecological impact should be as little as possible.

The ecological impact is judged in so-called "Environmental Product Declarations" which count the energy consumption, the use of resources, and various emissions during production and use which are stated in the Kyoto Protocol[28]. In the contact of the construction industry there are three aspects which deserve attention, i.e. reuse of land, sustainable materials, and sustainable structures. Reuse of land will not be discussed while the other two will be addressed.

### 3.3.2 Sustainable concrete

Concrete will remain the outmost used building material for a long time. Therefore one should try to produce it with little energy and reuse other material as much as possible. Most energy is consumed with the manufacture of cement. The production of clinker in the rotary kiln is energy-intensive although the efficiency of the kiln has increased. Thermodynamically about 3.0 GJ/ton cement clinker is needed, the most modern rotary kilns can achieve 3.6 GJ/ton, and a typical value for the operating kilns is 5.0 GJ/ton[29]. In order to save primary energy one can replace fossil fuel by low-cost secondary fuel, e.g. car tires, bone meal, sewage sludge, paper sludge etc. Fig. 20 shows an example from the German cement industry. The largest part is covered by industrial waste, the second are the ground car tires, the third is bone meal, finally municipal waste, and waste oil. The sum of secondary fuel covers about 50 % of the total energy consumption.

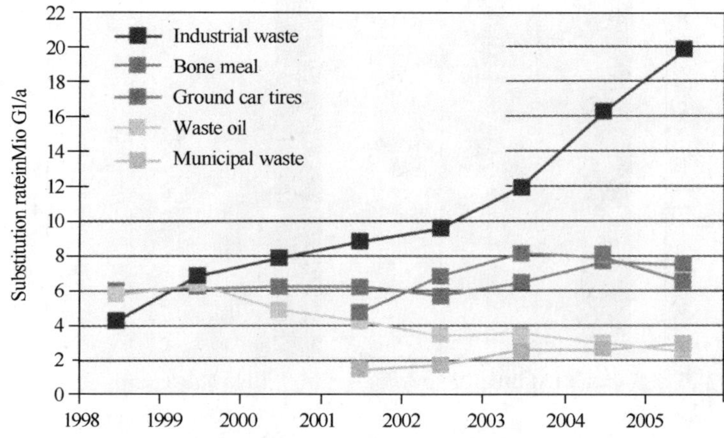


Figure 20　Secondary fuel for the production of cement[30]

Savings are also possible on the raw materials side. When they have puzzolanic or hydraulic properties they can partly replace the clinker. Fig. 21 shows the A-S-C ternary phase diagram of various materials. Blast furnace slag (BFS) is the nearest to Portland cement, before fly ash and natural pozzolans.

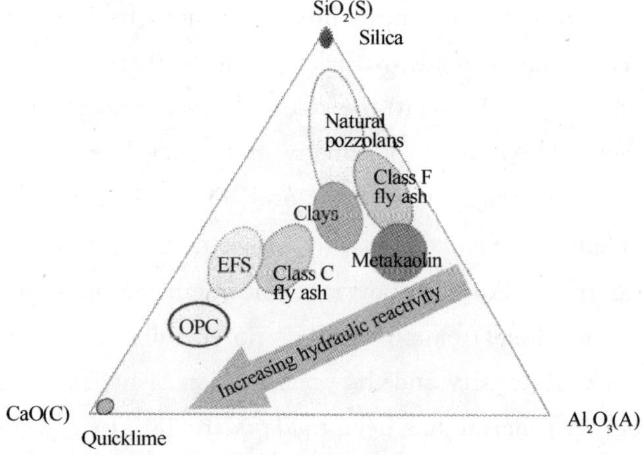


Figure 21　Ternary phase diagram of various binders and by-products[29]

In Germany, about 60 % of the total cement production refers to ordinary Portland cement (OPC), die other 40 % are Portland composite cements and blast furnace slag cement.

The production of blended cements has the great advantage of saving energy and reducing the emission of carbon dioxide. Fig. 22 gives an overview. CEM II/B-S contains up to 35 % blast furnace slag, CEM III/A contains up to 65 % slag.

The more slag the cement has the lower is the energy consumption compared to OPC (first column). The other columns refer to the ecological indicators. These are also much lower than for OPC.

Concrete consists of five components, i. e. cement, aggregate, water, additions, and admixtures. The amount of cement is the most important component with relation to ecolo-

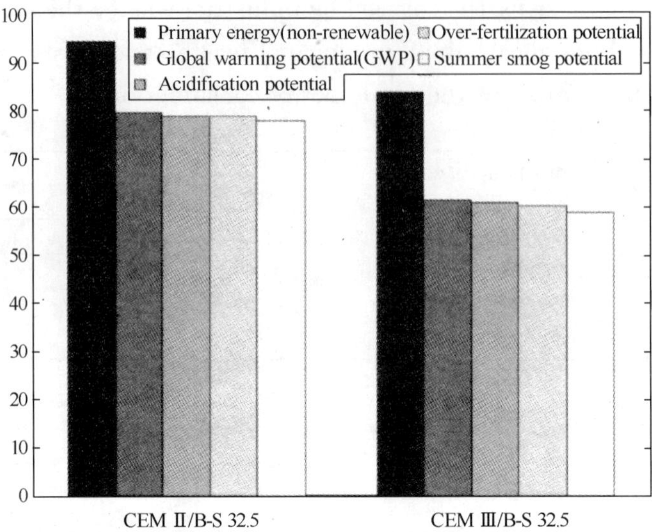


Figure 22 Overview of ecological indicators of two cements in comparison to OPC[31]

gy. It has been shown how the ecological "foot print" of cement can be reduced. The foot print of concrete can mainly be reduced by reducing the cement content. Conventional concretes can be made with less cement than is usual today. Fly ash is a pozzolanic material which contributes to the strength development and makes concrete denser. The replacement factor (k-value) in Germany and other European countries is 0. 40. Many mixes are produced with 240 kg cement and 80 kg fly ash per $m^3$ concrete. For moderate compressive strength, even cement contents can be lowered to 120 kg/$m^3$ with a fly ash content of 200 kg/$m^3$[32]. An even better replacement factor is valid for blast furnace slag, about 0. 80. Although limestone powder has no chemical reactivity there is a positive influence due to the packing density and the precipitation of hydration products on the large surface of the powder. A concrete has been made with 200 kg/$m^3$ cement content and 390 kg/$m^3$ marble powder. It had self-compacting properties and a medium strength[33].

### 3. 3. 3 Sustainable structures

Sustainability can finally be judged only on the structure, e. g. a building can be made of timber, concrete, or steel. It is not fair to compare a ton of steel with a ton of concrete. One has to compare a wall, a ceiling, a façade etc. on the basis of the same capability and performance. However, for all buildings irrespective of the material used it is paramount for good sustainability that the live cycle of the structure is long. Buildings should be designed such that the ground plan is flexible, i. e. apartments should be changeable to offices and the other way round[34]. There should be no walls on the ground plan, it is better to work with columns. The installations should remain accessible for repair and replacement.

The same is true for the façade. Fig. 23 shows an example of a floor slab with consists of a lower bearing slab with ribs on top which allow the installation of all types of piping, electricity, communication devices etc. Since the cover slabs are demountable the inner space of the floor is always accessible.

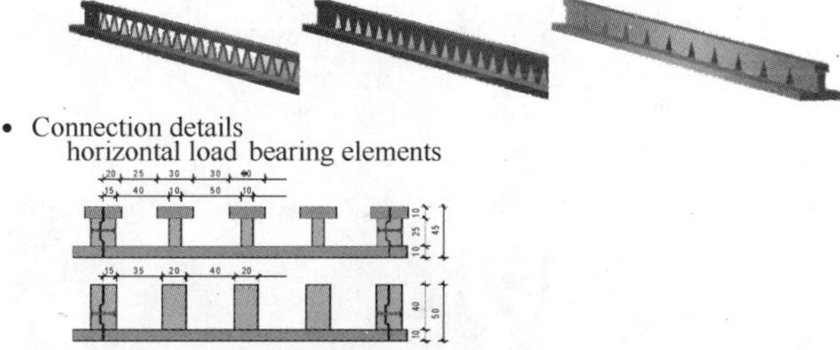



Figure 23　Flexible floor slab[34]

A step further concerns the total demountability of concrete structures[35]. The demountability is feasible with prefabricated structures with steel connectors. There are already precast systems available[36] which are being used in school buildings, hospitals, university buildings, and other offices. They consist of very accurately fabricated columns, floor slabs, walls, and stairs. They can be mounted in a modular system, i. e. the ground plan is very flexible. Fig. 24 shows an example.

Efforts should be made to develop the idea of demountability further because this gives the opportunity to demount a structure and to use parts of it again. This would save energy and raw material consumption which is a great contribution to sustainability. When this is combined with ecological concrete and with high performance concrete with smaller dimensions the aim of high tech and sustainable construction is reached.

Figure 24　Demountable system CD 20[36]

## 3.4 STRUCTURES

Only two examples of the use of UHPC are shown, a bridge which has been completed recently and a tower of an offshore wind energy plant which is in the design stadium. The "Gärtnerplatz" bridge[37] is a 132 m long composite bridge with a steel truss girder and precast UHPC slab. Fig. 25 shows the bridge after completion, Fig. 26 gives a detail of the pedestrian bridge. The upper chors of the truss consists of UHPC with a compressive strength of >165 MPa while the lower chord is made of steel. The slab elements are glued to the upper chords.

Figure 25 Gärtnerplatz bridge after completion

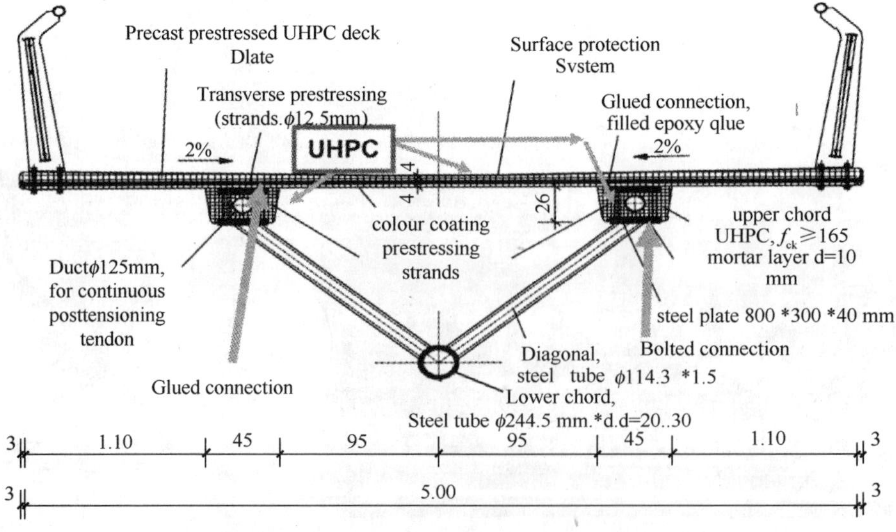


Figure 26 Cross-section of Gärtnerplatz bridge[37]

The other example refers to a tower of an offshore wind energy plant. The tower consists of a concrete foundation, a cast in place lower segment of 65 m height, and an upper part made of prefabricated UHPC rings which are post-tensioned together by exterior prestressing cables. The total tower height is 135 m and the water depth amounts to 40 m. It is designed for the installation in the near-coast area of the North Sea.

## 3.5 CONCLUSIONS

There is a great variety of high-tech concrete materials. The aim of modern concrete technology is to increase workability, strength, and ductility of the material. Self-compacting concrete flows under its own eight and fills the required space without vibration. High-performance (HPC) and ultra-high performance (UHPC) concretes aim at very high strength up to 400 MPa. New types of fiber concrete composites exhibit an ultimate strain of several percent at moderate strength on the one hand, whereas there are concretes on the other hand which show high strength and large ductility at the same time. Textiles are used as a special type of fibers. Textile reinforced concrete is mainly designed for thin products like plates and shells. Superabsorbent polymers can enhance the durability of concrete. All these types of concrete are only possible with a suitable grading of particles from the centimeter scale to the nanometer scale which results in a dense packing. The second prerequisite is the availability of suitable additions and admixtures. Without those, a good workability is not possible.

The second aspect is sustainability which, in technical terms, is mostly reduced to ecology, i. e. consumption of raw material and energy. Cement can be produced with less energy and less raw material by use of waste products, concrete can be made with a smaller amount of cement by use of industrial by-products. Finally, demountable concrete structures save raw material and energy by reuse.

## REFERENCES

[1] Roux, N.: Durabilité des bétons de poudres réactives. Centre d'assistance technique et de documentation, St-Rémy-lès-Chevreuse, 08/97, pp 46-58.

[2] Fehling, E., Schmidt, Stürwald, S.: Ultra High Performance Concrete (UHPC), No. 10, Structural Materials and Engineering Series, Kassel University Press, 2008.

[3] Eppers, S., Müller, C.: Autogenous shrinkage strain of ultra-high-performance concrete (UHPC). In: Fehling, E., Schmidt, Stürwald, S. Ultra High Performance Concrete (UHPC), No. 10, Structural Materials and Engineering Series, Kassel University Press, 2008, pp. 433-441.

[4] Bögl Reitz, HIPERCON®, Asslar, Germany, 2010.

[5] Okamura, H., Ozawa, K. Self-compactable high-performance concrete in Japan. In P. Zia (Ed.) High-performance concrete. SP-159, ACI, Farmington Hills 1996, pp 31-44.

[6] De Schutter, G., Bartos, P. J. M, Domone, P., Gibbs, J.: Self-compacting concrete. Whittles

Publ., Dunbeath 2008.

[7] Guideline for self-compacting concrete (in German), DAfStb, Berlin 2003.

[8] Naaman, A. E., Reinhardt, H. W.: Proposed classification of HPFRC composites based on their tensile response. Materials and Structures 39 (2006), No. 5, pp 547-555.

[9] Chanvillard, G., Rigaud, S.: Complete characterization of tensile properties of DUCTAL® UHPFRC according to the French recommendations. In Naaman, A. E., Reinhardt, H. W. (Eds.) "High Performance Fiber Reinforced Cement Composites (HPFRCC 4)", RILEM Pro 30, RILEM SARL, Bagneux 2003, pp 21-34.

[10] Karihaloo, B. L., de Vriese, K.: Short-fibre reinforced reactive powder concrete. In Reinhardt, H. W., Naaman A. E. (Eds.) "Third International Workshop on High Performance Fiber Reinforced Cement Composites (HPFRCC 3)", RILEM Publications, Cachan 1999 (RILEM proceedings, PRO 6), S. 53-63.

[11] Sun, W., Liu. S., Lai, J.: Study on the properties and mechanism of ultra-high performance ecological reactive powder concrete (ECO-RPC). In Naaman, A. E., Reinhardt, H. W. (Eds.) "High Performance Fiber Reinforced Cement Composites (HPFRCC 4)", RILEM Pro 30, RILEM SARL, Bagneux 2003, pp 409-417.

[12] Orange, G., Acker, P., Vernet, C.: A new generation of UHP concrete: DUCTAL®. In Reinhardt, H. W., Naaman A. E. (Eds.) "Third International Workshop on High Performance Fiber Reinforced Cement Composites (HPFRCC 3)", RILEM Publications, Cachan 1999 (RILEM proceedings, PRO 6), pp 101-111.

[13] Sun Wei, Lai Jianzhong, Rong Zhidan, Zhang Yunsheng, Zhang Yamei: Dynamic mechanical behaviour of ultra-high performance cementitious composites under repeated impact. In H. W. Reinhardt and A. E. Naaman (Eds.) "High Performance Fiber Reinforced Cement Composites (HPFRCC5)", RILEM PRO 53, 2007, pp 471-479.

[14] Rebentrost, M., Wight, G.: Behaviour and resistance of Ultra High Performance Concrete to blast effects. In: Fehling, E., Schmidt, Stürwald, S. Ultra High Performance Concrete (UHPC), No. 10, Structural Materials and Engineering Series, Kassel University Press, 2008, pp. 735-742.

[15] Li, V. C.: Engineered cementitious composites - Tailored composites through micromechanical modeling. In: Fiber reinforced concrete: Present and the Future, Montreal, Can. Soc. for Civil Eng., 1998, pp. 64-97.

[16] Li, V. C., Leung, C. K. Y.: Steady-state and multiple cracking of short random fiber composites, ASCE J. Engineering Mechanics 118 (1992), No. 11, pp. 2246-2264.

[17] Li, V. C.: Reflections on the research and development of engineered cementitious composites (ECC), Proc. JCI Intern. Workshop on Ductile Fiber Reinforced Cementitious Composites (DFRCC), Japan Concrete institute, Tokyo 2002, pp. 1-21.

[18] Mechtcherine, V., Schulze, J.: Ultra-ductile concrete - Material design and testing, In: V. Mechtcherine "Ultra-ductile concrete with short fibres", ibidem Stuttgart, 2005, pp. 11-35.

[19] Brameshuber, W.: Textile reinforced concrete. State-of-the-art report RILEM 61, Bagneux 2006.

[20] Reinhardt, H. W., Krueger, M., Grosse, C. U.: Thin plates prestressed with textile reinforcement. In: P. Balaguru, A. E. Naaman, J. Weiss, Concrete: Material science to application, ACI SP 206-22, Farmington Hills, 2002, pp. 355-372.

[21] Reinhardt, H. W., Assmann, A.: Enhanced durability of concrete by superabsorbent polymers. In:

A. M. Brandt, J. Olek, I. H. Marshall (Eds.) "Brittle Matrix Composites 9", 2009, pp 291-300.

[22] Reinhardt, H. W., Assmann, A., Mönnig, S.: Superabsorbent polymers (SAPs) - an admixture to increase the durability of concrete. In: Wei Sun, K. van Breugel, C. Miao, G. Ye, H. Chen (Eds.) "Microstructure related durability of cementitious composites", RILEM Pro 61, Vol. 1, pp 313-322.

[23] Laustsen, S., Hasholt, M. T., Jensen, O. M.: A new technology for air-entrainment of concrete. RILEM Publ. Pro 61, SARL, Bagneux 2008, pp. 1223-1230.

[24] Laustsen, S., Hasholt, M. T., Jensen, O. M.: A new technology for air-entrainment of concrete. In Wei Sun, K. van Breugel, Changwen Miao, Guang Ye and Huisu Chen (Eds.) Microstructure Related Durability of Cementitious Composites, RILEM Publ. SARL, Bagneux 2008, pp. 1223-1230.

[25] CEN/TS 12390-9 Testing hardened concrete - Part 9: Freeze-thaw resistance, August 2006.

[26] SS 137244, Concrete testing - Hardened concrete - Scaling and freezing. Stockholm: Swedish Standards Institute, 2005, 10 pages.

[27] Carlowitz, Hans Carl von: Sylvicultura oeconomica. Leipzig 1713.

[28] Kyoto Protocol of the UN, 1997.

[29] Brouwers, H. J. H.: Recipes for porous building materials. TU Eindhoven2010.

[30] Association of German Cement Factories (VDZ). Activity Report 2005-2007.

[31] Hauer, B. et al.: Potential of the use of secondary materials in concrete construction (in German), DAfStb, No. 572, Berlin 2007, pp. 131-222.

[32] Malhotra, V. K., Mehta, P. K.: High-performance, high-volume fly ash concrete: Materials, mixture proportioning, properties, construction practice, and case histories, Ottawa 2002.

[33] Hunger, M.: An integral design concept for ecological self-compacting concrete (Bowstenen 146), PhD thesis, Eindhoven 2010.

[34] Sustainable construction with concrete (in German), DAfStb No. 572, Berlin 2007.

[35] Weiss, G., Reinhardt, H. W.: Demountable construction using concrete and demountable ceiling plates. Concrete Plant International 10 (2007), No. 6, pp. 170-176.

[36] Demountable building systems in concrete (in Dutch), Rijksgebouwendienst, The Hague 1996.

[37] Fehling, E., Bunje, K., Schmidt, M., Schreiber, W.: The "Gärtnerplatzbrücke", Design of first hybrid UHPC-steel bridge across the river Fulda in Kassel Germany, In: Fehling, E., Schmidt, Stürwald, S. Ultra High Performance Concrete (UHPC), No. 10, Structural Materials and Engineering Series, Kassel University Press, 2008, pp. 581-588.

# 第 4 章 Chapter 4

# 混凝土性能的多尺度模拟—融合材料与结构力学

# MULTI-SCALE MODELING OF CONCRETE PERFORMANCE-INTEGRATED MATERIAL AND STRUCTURAL MECHANICS

**Koichi MAEKAWA[1], Tetsuya ISHIDA[2], Toshiharu KISHI[3]**

[1] Professor, University of Tokyo, Japan Email: maekawa@concrete. t. u-tokyo. ac. jp

[2] Assistant Professor, University of Tokyo, Japan

[3] Associate Professor, University of Tokyo, Japan

**Abstract:** Multi-scale modeling of structural concrete performance is presented as a systematic knowledge base of coupled cementitious composites and structural mechanics. An object-oriented computational scheme is proposed for life-span simulation of reinforced concrete. Conservation of moisture, carbon dioxide, oxygen, chloride, calcium and momentum is solved with hydration, carbonation, corrosion, ion dissolution, damage evolution and their thermodynamic/mechanical equilibrium. The holistic system is verified by the reality.

**Keywords:** Multi-scale modeling, Serviceability, Durability, Thermodynamics, Structural mechanics

## 4.1 INTRODUCTION

For sustainable development of gigantic urban universe where nature and people reside together, infrastructures are required to retain their performances over the long term. In order to construct durable and reliable social backbones, functionality, safety and their life shall be estimated on the basis of scientific and engineering knowledge. For present deteriorated infrastructures, a rational maintenance plan should be implemented in accordance with current situations. It is indispensable to grasp entire structural performances under predictable ambient and load conditions during the designated service life. The objective of this research is to develop multi-scale modeling of structural concrete performance and to realize a so-called lifespan simulator capable of predicting overall structural behaviors as well as constituent materials.

Figure 1 shows a schematic representation of the targeted multi-scale modeling of material science and structural mechanics. Macroscopic characters of concrete composites are rooted in micro-pore structures and thermodynamic states having much to do with durability. In turn, chemo-physical states of substances in nano～micro-pores are greatly associated with structural mechanics with damaging induced by loads and weather actions. For considering the practical use of knowledge, two sub-fields of structural mechanics and material science have been investigated with a couple of interacting engineering factors with a macroscopic feature of composites like uniaxial strength. Currently, much valuable knowledge is in our hands. In view of academia, the authors have recognized that it is natural to synthesize plenty of concrete knowledge on a systematic platform like trees with leaves. The knowledge system should not be mere archives or indexers, but quantitative estimation shall be drawn from it. Thus, overlaid multiple linkage is aimed on an info-technological basis.

For practice, multi-scale behavioral simulation of coupled constituent materials and structures is expected to serve engineers in charge of design, planning and policymaking. As a matter of fact, real maintenance issues require comprehension of materials, structures, construction and management. Thus, the computerized scheme of holistic approach is sought by means of explicit mathematical formula originated from the past research.

This paper will propose one of the possible means to build up a systematic knowledge

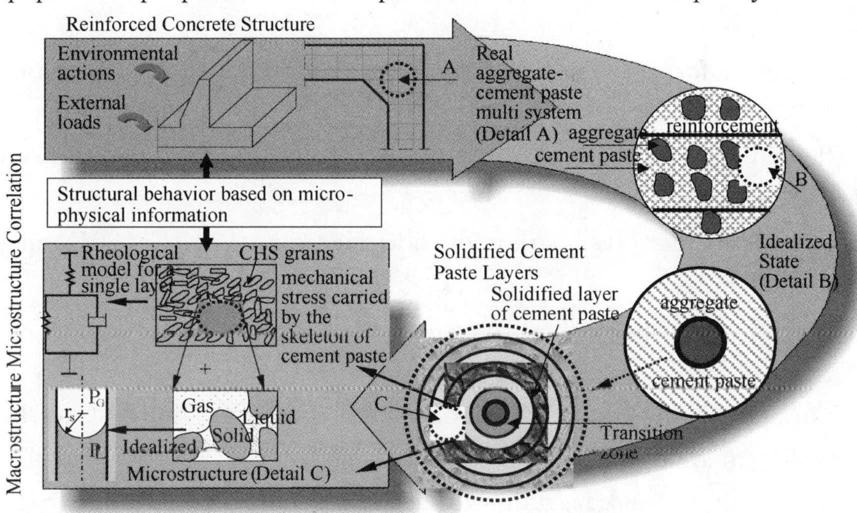


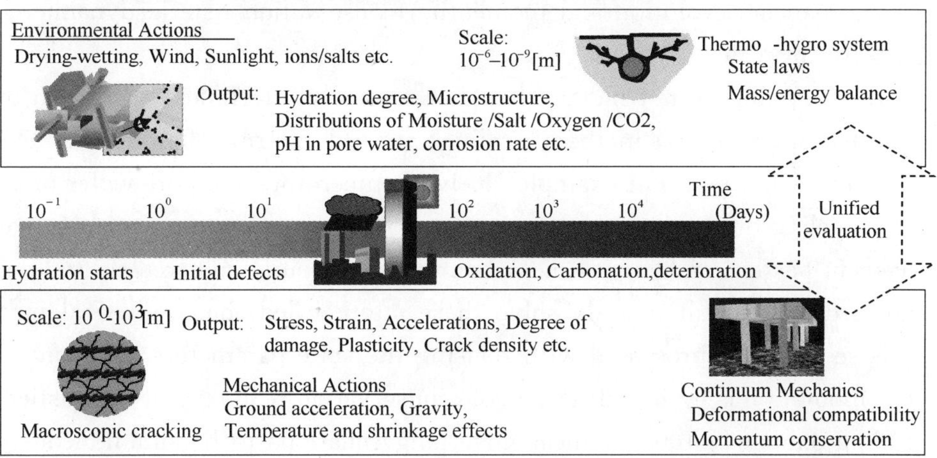


Fig. 1　Multi-scale scheme and lifespan simulation for materials and structures

basis to amalgamate material and structural engineering. As a main discussion point is addressed to the system formation, some models of microphysical chemistry are forced to be somehow simplified. Doubtlessly, system competence significantly relies on its constituent elements. Thus, individual modeling shall be experimentally verified for continuous development as a quantitative knowledge for engineers.

## 4.2 BASIC COMPUTATIONAL SCHEME

In principle, a dispersed object-oriented scheme on open memory space is implemented as shown in Fig. 2. Chemo-physical and mechanical events in 3D extent are individually solved in a particular time step. The authors select the following kinematic chemo-physical and mechanical events (Fig. 2) with different geometrical scales of referential control volume (REV) where governing equations hold as,

(1) Cement heat hydration and thermal conduction (thermal conservation) —$10^{-6}\sim 10^{-4}$ m scale

(2) Pore structure formation and moisture equilibrium/transport (water mass conservation) —$10^{-10}\sim 10^{-6}$ m scale

(3) Free/bound chloride equilibrium and chloride ion transport (chloride ion conservation) —$10^{-8}\sim 10^{-6}$ m scale

(4) Carbonation and dissolved carbon dioxide migration ($CO_2/Ca(OH)_2$ mass conservation) —$10^{-9}\sim 10^{-6}$ m scale

(5) Corrosion of steel and dissolved oxygen transport ($O_2$/proton conservation) —$10^{-9}\sim 10^{-6}$ m scale

(6) Calcium ion leaching from $Ca(OH)_2$ and transport ($Ca^{++}$ ion + $Ca(OH)_2$ conservation)—$10^{-9}\sim 10^{-6}$ m scale

(7) Chrome dissolution and migration (Cr ion mass conservation)—$10^{-9}\sim 10^{-6}$ m scale

(8) Macro-damage evolution and momentum conservation (static/dynamic equilibrium)—$10^{-3}\sim 10^{-0}$ m scale

Here, these events are not independent but mutually interlinked with each other. A complex figure of interaction is mathematically expressed in terms of state parameters commonly shared by each event. For example, Kelvin temperature and pore water pressure are seen in the modeling of cement hydration rate, moisture equilibrium, constitutive law of hardened cement paste, conductivity of carbon dioxide, bound and free chloride equilibrium, etc. In order to simultaneously solve these multi-scaled chemo-physical/mechanics, each event is sequentially processed with revising the state parameters commonly referred to and the computation is cycled till the whole conservation requirement is satisfied at each time step. Approximately 100 nonlinear governing equations to be simultaneously solved and their independent variables of the same numbers are summarized in the Appendix. In other words, concrete performance of micro and macro scales is characterized by these set of variables having multi-overlaid structures of cement composites. Individual chemo-physical events will be discussed subsequently.

Each sub-event is discretized in space by applying Galarkin's method of weighted residual function on finite element scheme. This three-dimensional multi-scale coupled system is built on the thermo-hydro physics coded by *DuCOM* for early aged concrete (Maekawa et al. 1999) and nonlinear mechanics FEM coded by *COM*3 (Maekawa et al. 2003) for seismic performance assessment of reinforced concrete. The reason we select an object-oriented scheme is its stability of convergence and easy extensibility. In fact, soil foundation space, microorganism activity and electric potential field were easily overlaid on this system for further extension and are being investigated.

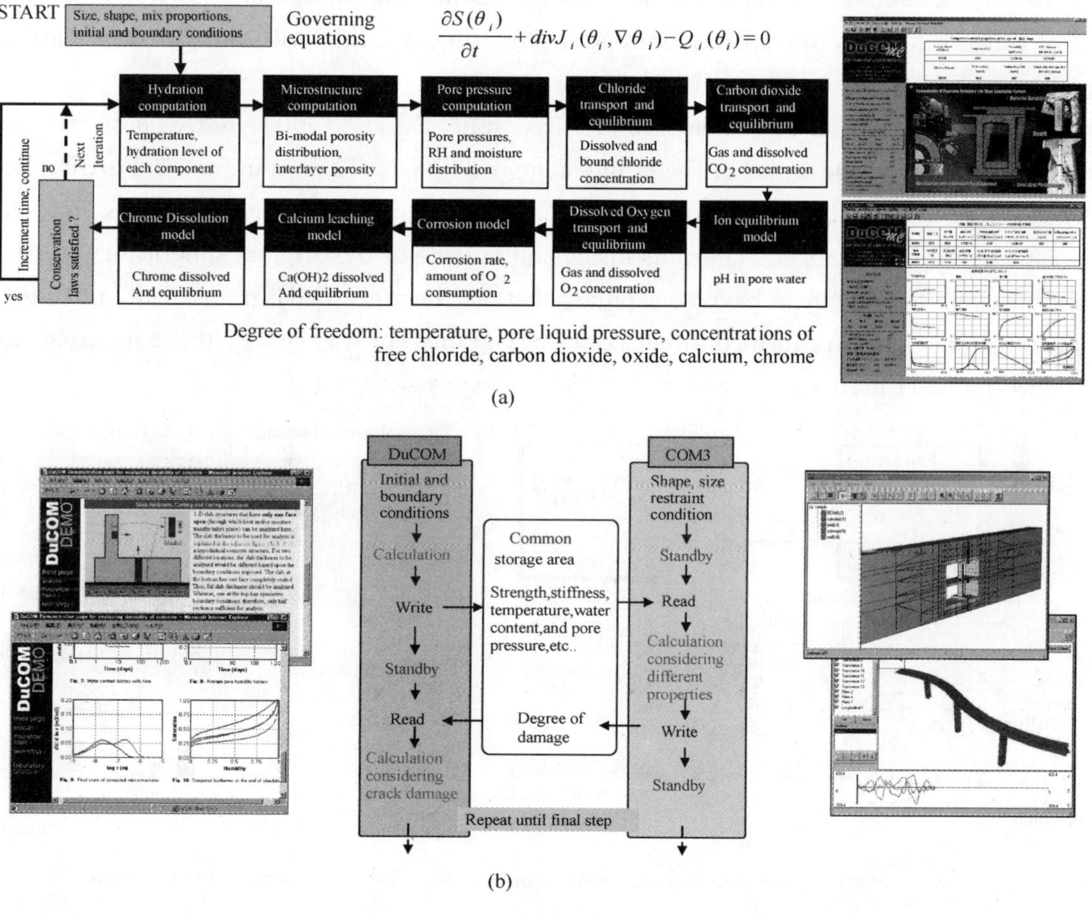


Fig. 2

(a) Sub-structure platform of *Concrete Model for Durability*-DuCOM-(Maekawa et al. 1999, Ishida 1999)

(b) Integration of microphysics and macro-structural analysis

(Maekawa et al. 1999, Ishida and Maekawa 2000, Maekawa et al. 2003)

## 4.3 MULTI-SCALE MODELING

### 4.3.1 Pore-Structure Model of Cement Paste—$10^{-10}$～$10^{-6}$m Scale-

Computational modeling of varying micro-pore structures of hardening cement paste media is a central issue of multi-scale analysis. Moisture migration and durability related substances (in this study, calcium, chloride, dissolved $CO_2$ and $O_2$ are treated) and diffusion of gaseous phases are greatly governed by micro-pore structures. Without information of these mass transport and equilibrated thermo-dynamic states, we cannot predict hydration of cement, carbonation and corrosion of steel. Volumetric change caused by drying and stresses invoked are affected by the micro-pore moisture and the solidified skeleton mechanics of concrete. Here, the authors initiate multi-scale modeling with nm-μm scale hardening cement with CSH hydrates (Maekawa et al. 1999).

Suspended particles of cement are idealized spherical and representative unit cell consisting of a particle and water at the initial stage is defined as shown in Fig. 3. For simplicity, radius of individual cement particle is assumed constant equal to average representative. Cement hydration creates CSH grains inside and outside of the original spherical geometry. Inner product is thought to have no capillary distance between CSH grains. But, outside original sphere, gaps are formed in between CSH grains. In this study, this gap space surrounded by CSH grains is defined as capillary pores as shown in Fig. 3.

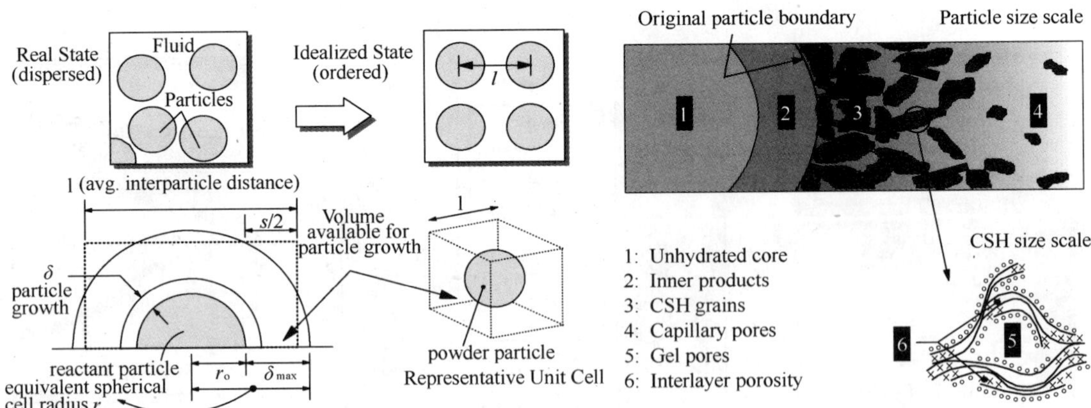


Fig. 3 Statistical modeling of micro-pore geometry and size for hardening cement paste

For statistically identifying micro-pores of nm～μm scales, the authors introduced the following hypothesis and computational assumption according to cement solid science earned in the past several decades.

(G1) The particle size and geometry of CSH grains are constant. (Volume to surface area ratio=19nm for OPC)

(G2) No capillary space is made inside the original sphere and porosity of outer product varies linearly (Fig. 4).

(G3) CSH grains contain interlayer pores and gel pores as shown in Fig. 3.

When we direct our attention to the inside of CSH grains, interlayer pores whose size is similar to that of a water molecule are identified. The authors further assume the following:

(I1) Total porosity of CSH grains is constant (0. 28) and the size of interlayer in CSH grains is assumed 2. 8A.

(I2) The specific surface area of interlayer is $510m^2/g$ for OPC and that of gel pores is $40m^2/g$ as constant.

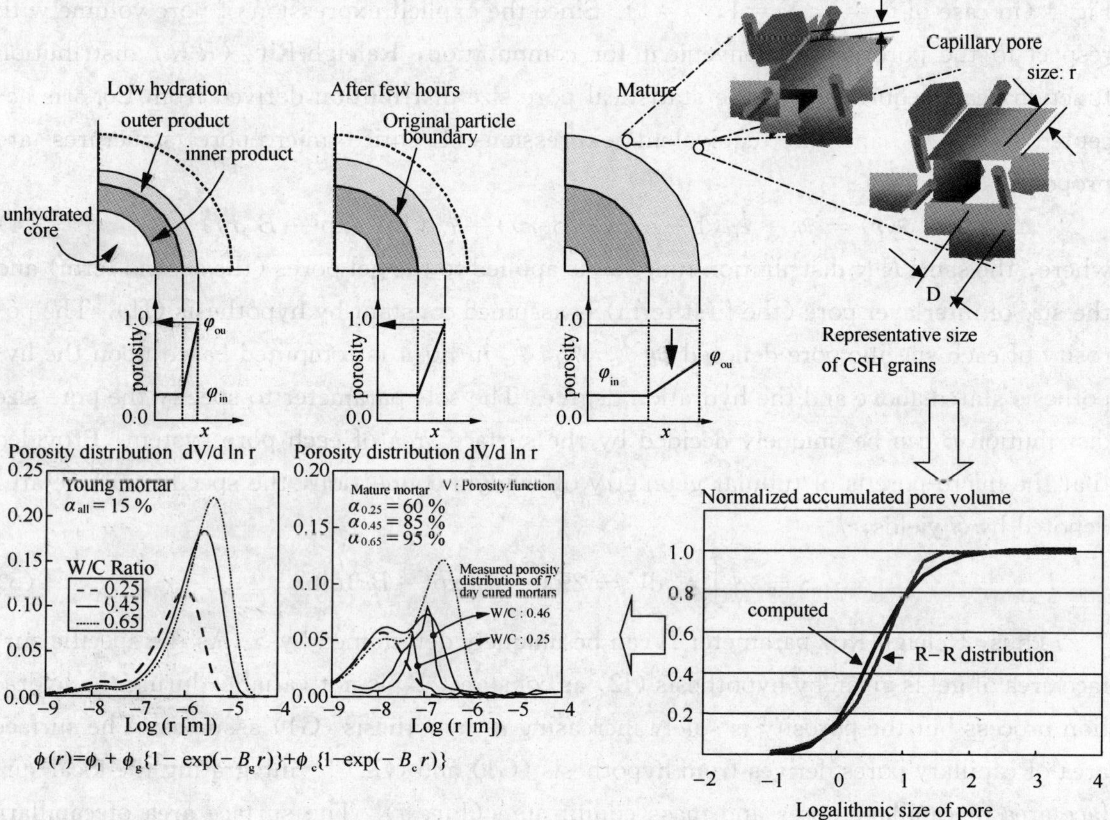


Fig. 4　Micro-pore development and statistical expression

From hypothesis G1 and G2, the capillary pore size distribution can derive (Fig. 4). As capillary porosity at location "$x$" from the original boundary of the cement particle is proportional to the gap size "$r$" (capillary pore size) between CSH grains and assumed to linearly vary in space, we have,

$$\phi_c(x)=\frac{r}{D+r}=\left(\frac{x}{\delta_{max}}\right)\phi_{ou} \tag{1}$$

where, $D$: size of CSH grains, $\delta_{max}$: distance between the extreme radius of unit cell (REV) and the original boundary (Fig. 3), $\phi_{ou}$: porosity at the extreme radius of the unit cell (Fig. 3 and Fig. 4). Accumulated volume of capillary pores formed up to distance x is computed by integrating porosity with respect to location as,

$$Vx = \int_0^x \phi_c(x) 4\pi (x + x_o)^2 \mathrm{d}x \tag{2}$$

Substitution of Eq. 1 into Eq. 2 yields,

$$Vx = A\left(\frac{1}{4}x^4 + \frac{2}{3}x_o x^3 + \frac{1}{2}x_o^2 x^2\right) \tag{3}$$

Since location $x$ is nonlinearly related to the capillary pore size by Eq. 1, the accumulated volume of capillary pores can be implicitly obtained in terms of the pore size as shown in Fig. 4 (in case of $\delta_m = 1$, $X_o = 1$, $\phi_{ou} = 1$). Since the explicit expression of pore volume with respect to the pore size is convenient for computation, Raleigh-Ritz (*R-R*) distribution function nearly equivalent to the statistical pore size distribution derived from Eq. 3 is accepted (Fig. 4) and the equivalent expression of entire micro-pore structures are proposed as,

$$\phi(r) = \phi_{lr} + \phi_{gl}(1 - \exp(-B_{gl}r)) + \phi_{cp}(1 - \exp(-B_{cp}r)) \tag{4}$$

where, the same *R-R* distribution function is applied to the gel pores (the second term) and the size of interlayer pore (the first term) is assumed constant by hypothesis (I1). The porosity of each specific pore denoted by $\phi_{lr}$, $\phi_{gl}$, $\phi_{cp}$ in Eq. 4 is computed based upon the hypothesis stated above and the hydration degree. The sole parameter to specify the pore size distribution $B$ can be uniquely decided by the surface area of each pore system. Provided that the micro-pore is of tubular geometry of random connection, the specific surface area denoted by $S$ yields,

$$S = 2\phi \int r^{-1} \mathrm{d}V = 2\phi \int_{r_{min}}^{\infty} B\exp(-Br)\,\mathrm{d}\ln r \tag{5}$$

Thus, Raleigh-Ritz parameter $B$ can be uniquely determined by $S$. As the specific surface area of gel is given by hypothesis (I2) as constant, $B_{gl}$ is not variable during the hydration process but the porosity is solely increasing as hypothesis (G1) assumes. The surface area of capillary pores derives from hypothesis (G1) and (G2) by integrating the local surface area of capillary pores and mass equilibrium (Fig. 5). The surface area of capillary pores drastically varies in accordance with cement hydration and its mean size is decreasing. The detailed derivation of porosity and surface area of each pore system from hydration degree is presented in reference 5 and the governing equations are summarized in Appendix.

The Multi-component cement hydration model by Kishi and Maekawa (1996, 1997) is used in this study. As cement consists of several minerals, the averaged activation energy is not constant in nature and the hydration pattern looks complex. In considering the engineering practice for different sorts of Portland cement and mixed cement with pozzolans, the authors computationally define referential hydration (heat) rate and activation energy of each mineral compound as shown in Fig. 6. Although each mineral activation is constant, apparently varying activation energy of cement, which is extracted from adiabatic temperature rise tests, can be simulated in terms of hydration degree. The effect of free water on hydration rate is modeled by hard shell concept of hydrated cluster. The detailed formulation is also lined up in Appendix.

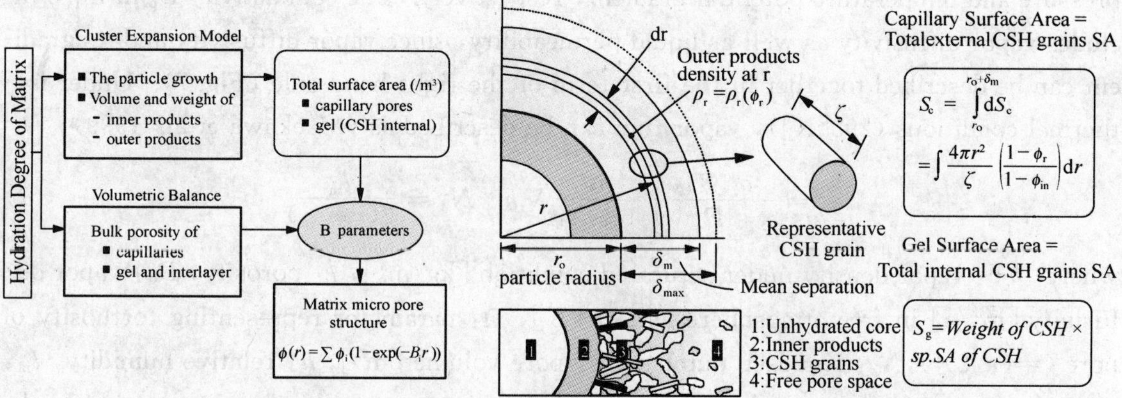


Fig. 5　Computation of varying surface area of hydrates and capillary pores

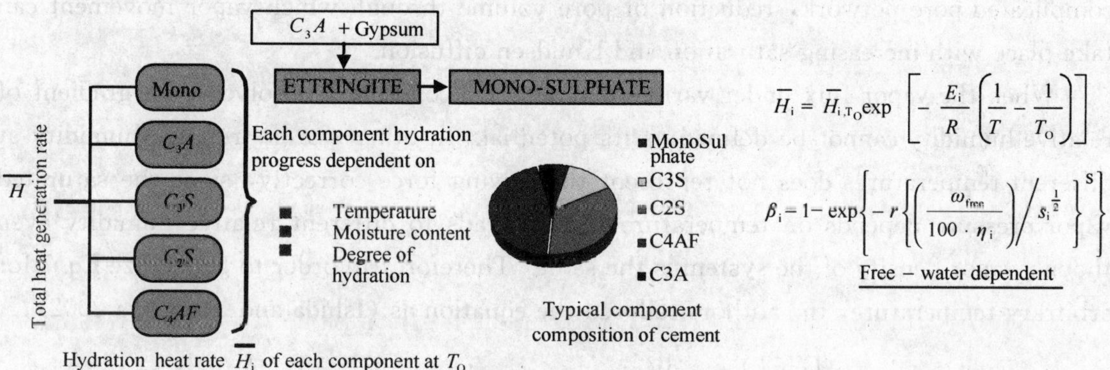


Fig. 6　Multi-mineral component model of cement hydration and heat generation

## 4.3.2　Moisture Transport and Equilibrium $-10^{-10}\sim10^{-6}$m Scale-

Moisture mass balance must be strictly solved in both vapor and condensed water. The conservation equation is expressed with capacity, conductivity and sink terms on the referential volume as shown in Fig. 7 (Maekawa et al. 1999). Pore pressure of condensed water is selected as a chief variable so that both saturated and unsaturated states can be covered with perfect consistency. Another key issue here is that material characteristic parameters are variable with respect to micro-pore development.

Knudsen diffusion theory was applied for vapor transport and conductivity component is formulated based on vacant micro-pores where vapor can move as shown in Fig. 7. For liquid transport, random pore model described by percolation threshold was applied and its conductivity is computed based on the micro-pore distribution occupied by condensed water. Moisture transport in a porous body would be driven by temperature, pore pressure, and vapor pressure difference. In this study, total flux of moisture $J$ is formulated as,

$$J = -(D_p \nabla P_l + D_T \nabla T) \tag{6}$$

where, $D_p$ and $D_T$ represent the macroscopic moisture conductivities with respect to pore pressure and temperature potential gradient, respectively. The conductivity $D_p$ in Eq. 6 includes vapor diffusivity as well as liquid permeability, since vapor diffusivity and its gradient can be described together in the first term of the right hand side (Fig. 7). Under isothermal conditions (293[K]), vapor flux can be described as (Maekawa et al. 1999),

$$q_v = -\frac{\rho_v^{sat}\phi D_0}{\Omega}\int_{r_s}^{\infty}\frac{dV}{1+N_k}\nabla h \quad N_k = \frac{l_m}{2(r-t_a)} \tag{7}$$

where, $\rho_v^{sat}$: vapor density under saturated condition [kg/m$^3$], $\phi$: porosity, $D_0$: vapor diffusivity [m$^2$/s] in free atmosphere at 293 [K], $\Omega$: parameter representing tortuosity of pore $(=(\pi/2)^2)$, $N_k$: Knudsen number, $V$: pore volume [m$^3$], $h$: relative humidity, $l_m$: mean free path of a water molecule [m], and $t_a$: thickness of adsorbed layer [m]. In the model, factors reducing the apparent diffusivity of vapor are taken into account, such as complicated pore network, reduction of pore volume through which vapor movement can take place with increasing saturation and Knudsen diffusion.

When the vapor flux under various temperature conditions is solved, the gradient of relative humidity cannot be defined as its potential. In other words, relative humidity at different temperatures does not represent the driving force correctly, since the saturated vapor pressure depends on temperature, which leads to different relative humidity even though vapor density of the system is the same. Therefore, in order to generalize Eq. 7 for arbitrary temperature, the authors tailored the equation as (Ishida and Maekawa 2002),

$$q_v = -\frac{\phi D_0(T)}{\Omega}\int_{r_c}^{\infty}\frac{dV}{1+N_k}\nabla\rho_v = -D_v\nabla\rho_v \quad N_k = \frac{l_m}{2(r-t_a)} \tag{8}$$

The vapor flux described by Eq. 8 is driven by the gradient of absolute vapor density $\rho_v$ of the system, instead of relative humidity. Here, absolute vapor density $\rho_v$ corresponds to the product of relative humidity $h$ and $\rho_v^{sat}$ in Eq. 7. It has to be noted that vapor diffusivity has a strong dependence on temperature of the system. From a thermodynamic point of view, mass diffusivity under different temperature conditions can be derived by,

$$\frac{D_0(T_1)}{D_0(T_2)} = \left(\frac{T_1}{T_2}\right)^{3/2}\left(\frac{\Omega_{D,T_2}}{\Omega_{D,T_1}}\right) \tag{9}$$

where, $\Omega_{D,T_1}$ and $\Omega_{D,T_2}$ are collision integrals for molecular diffusion at temperature $T_1$ and $T_2$, respectively, which are functions of Boltzmann constant, temperature, and so on (Welty et al. 1969).

Next, the flux of liquid water $q_l$ can be described as (Maekawa et al. 1999),

$$q_l = -\frac{\rho_l \phi^2}{50\eta}\left(\int_0^{r_c} r\mathrm{d}V\right)^2 \nabla P_l = -K_l \nabla P_l \tag{10}$$

where, $\eta$ is viscosity of fluid. It has been reported that the actual viscosity in the cementitious microstructure is far from the one of bulk water under ideal conditions. To account such a phenomenon, Chaube et al. proposed the following model based on the thermodynamics as (Maekawa et al. 1999),

$$\eta = \eta_i \exp\left(\frac{G_e}{RT}\right) \tag{11}$$

where, $\eta_i$: referential viscosity under ideal conditions and $G_e$: free energy of activation of flow in excess of that required for ideal flow conditions. Considering a pore-water and microstructure interaction, Gibbs free energy $G_e$ has been formulated as a function of characteristics of pore structure and moisture history. In the proposed model, the different value of viscosity $\eta_i$ is given according to temperature in a pore, whereas the authors use the same formulation to evaluate the Gibbs energy since the effect of temperature is implicitly considered by the original formula. Then, we have (Ishida and Maekawa 2002),

$$\begin{aligned} J &= -(D_p \nabla P_l + D_T \nabla T) \\ &= -\left\{D_v\left(\frac{\partial \rho_v}{\partial P_l}\nabla P_l + \frac{\partial \rho_v}{\partial T}\nabla T\right) + K_l \nabla P_l + K_T \nabla T\right\} \\ &= -\left(D_v \frac{\partial \rho_v}{\partial P_l} + K_l\right)\nabla P_l - \left(D_v \frac{\partial \rho_v}{\partial T} + K_T\right)\nabla T \end{aligned} \tag{12}$$

In general, mass transport due to temperature gradient is known as the Soret effect or thermal diffusion that is represented by the last term of the right hand side, i. e., $K_T \cdot \mathrm{grad}(T)$. However, contribution of thermal diffusion to the total flux has not been cleared in concrete engineering. Moreover, this phenomenon normally plays a minor role in diffusion compared with moisture transfer driven by pore pressure and vapor pressure gradient

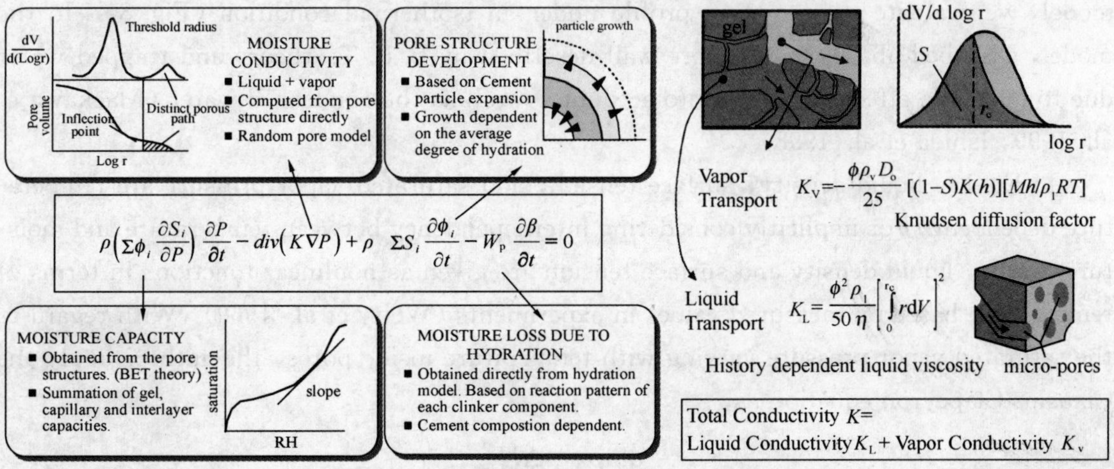


Fig. 7 Moisture conservation and transport based on statistical structural model of micro-pores

(Welty et al. 1969). Then, as the first approximation, the thermal diffusion is neglected in this study ($K_T \approx 0$).

The moisture capacity term enumerates moisture content with regard to the pore pressure. Quasi-thermo equilibrium is assumed at any time in nm~μm sized pores residing in REV. Vapor partial pressure can be obtained by equating Gibbs free energy of vapor to that of condensed water as,

$$P_l = \frac{\rho_l RT}{M_w} \ln \frac{p_{vap}}{p^*} \tag{13}$$

where, $R$: gas constant [J/mol. K], $p^*$: saturated vapor pressure [Pa], $M_w$: molecular mass of water [kg/mol] and $\rho_l$: density of liquid water [kg/m$^3$]. Under isothermal conditions, $p_{vap}/p^*$ corresponds to relative humidity $h$ inside pores.

By considering local thermodynamic and interface equilibrium, vapor and liquid interfaces would be formed in the pore structure due to pressure gradients caused by capillarity. When the interface is a part of an ideal spherical surface, the relation can be described by the Laplace equation as,

$$P_l = \frac{2\gamma}{r_s} \tag{14}$$

where, $\gamma$: surface tension of liquid water [N/m] and $r_s$: the radius of the pore in which the interface is created [m]. Based on the thermodynamic conditions represented by Eq. 13 and Eq. 14, a certain group of pores whose radii is smaller than the specific radius $r_s$ at which a liquid-vapor interface forms are completely filled with water, whereas larger pores remain empty or partially saturated.

If the porosity distribution of micro-pore structures is known, equations 13 and 14 allow us to obtain the amount of condensed water at a given relative humidity or absolute vapor pressure. By combining the above theory with the micro-pore structure distribution model, we evaluate the moisture profile under an isothermal condition (Fig. 8). In the model, adsorbed phases on the pore wall described by B. E. T. theory and trapped water due to inkbottle effect are taken into account as well as the condensed water (Maekawa et al. 1999, Ishida et al. 1998).

In Eq. 13, liquid density, surface tension, and saturated vapor pressure are temperature-dependent. For implicitly considering interdependency between temperature and moisture profile, liquid density and surface tension are given as nonlinear functions in terms of temperature based on measured values in experiments (Welty et al. 1969). With regard to the saturated vapor pressure varying with temperature inside pores, the authors apply the Clausius-Clapeyron equation as,

$$\frac{\mathrm{d}\ln p}{\mathrm{d}T} = \frac{\Delta H_{vap}}{RT^2} \tag{15}$$

where, $\Delta H_{vap}$ is the heat of vaporization [kJ/mol]. In this study, $\Delta H_{vap}$ is assumed to be constant (40.7 [kJ/mol]), since temperature variation is limited to the range from 273[K] to 373[K] in this study. By assuming the constant $\Delta H_{vap}$, integral of Eq. 15 yields,

$$p_{vap}=p^{*}\exp\left(\frac{P_l M_w}{\rho_l RT}\right)=p_0\exp\left\{-\left(\frac{\Delta H_{vap}}{R}\right)\left(\frac{1}{T}-\frac{1}{T_0}\right)\right\}\exp\left(\frac{P_l M_w}{\rho_l RT}\right) \tag{16}$$

where, $p_0$ and $T_0$ are the referential pressure and temperature, respectively.

From the above discussions, the relationship of pore pressure, saturated vapor pressure and absolute vapor pressure is specified. By combining the pore structure model with these formulations, moisture in terms of both vapor and liquid inside gel and capillary pores can be related to pore pressure.

Thermo-dynamics is hardly applied to behaviors of moisture inherent in interlayer pores with the angstrom scale but individual molecular dynamics is the matter. Here, empirically observed isotherm as shown in Fig. 9 is used for evaluating total moisture capacity. Temperature is thought to be a key parameter especially for low water to cement ratio concrete having a great deal of interlayer pores. Currently, general modeling is under investigation.

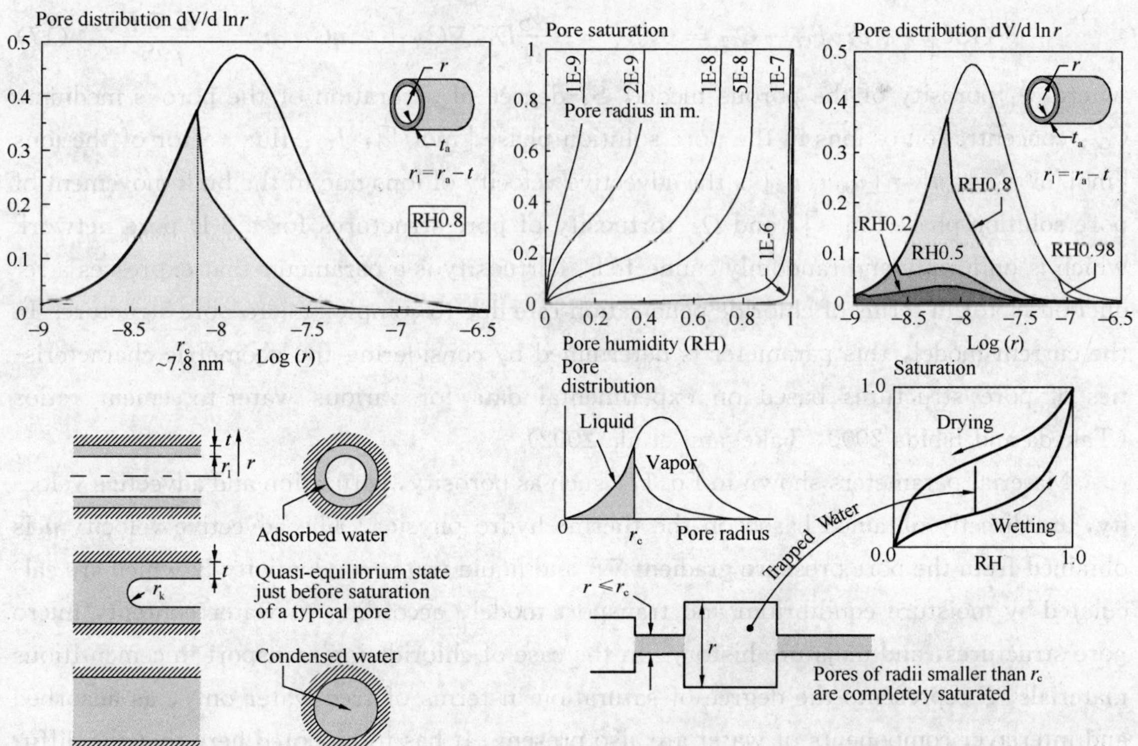


Fig. 8　Thermo-dynamic moisture equilibrium of condensed and adsorbed water in micro-gel and capillary pores

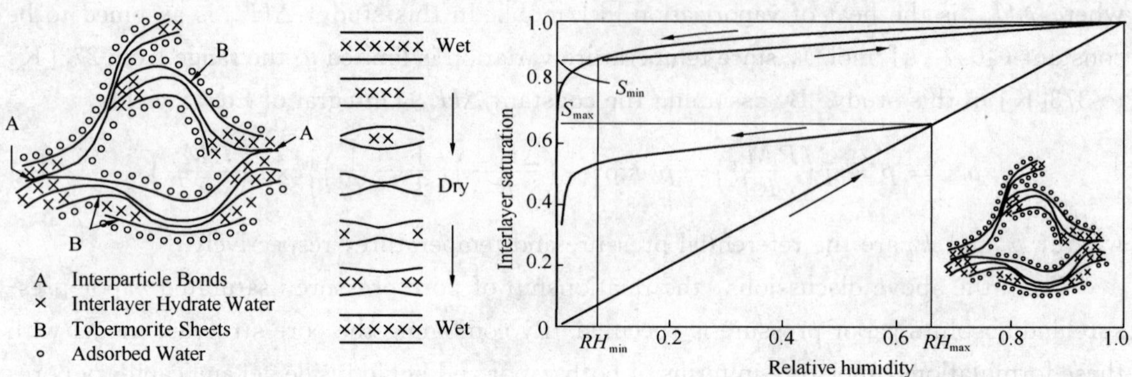


Fig. 9 Thermo-dynamic moisture equilibrium in micro-interlayer pores

### 4.3.3 Chloride Ion Transport and Equilibrium $-10^{-8}\sim10^{-6}$ m Scale-

Chloride transport in cementitious materials under usual conditions is an advective-diffusive phenomenon. In modeling, the advective transport due to bulk movement of pore solution phase is considered, as well as ionic diffusion due to concentration gradients. Mass balance for free (movable) chlorides can be expressed as (Ishida 1999a, Takeda and Ishida 2000) (Fig. 10),

$$\frac{\partial}{\partial t}(\phi SC_{cl})+divJ_{cl}-Q_{cl}=0,J_{cl}=-\frac{\phi S}{\Omega}D_{cl}\ \nabla C_{cl}+\phi S\boldsymbol{u}C_{cl},\boldsymbol{u}=-\frac{K\ \nabla P}{\rho\phi S} \qquad (17)$$

where, $\phi$: porosity of the porous media, $S$: degree of saturation of the porous medium, $C_{cl}$: concentration of ions in the pore solution phase [mol/l], $J_{cl}$: flux vector of the ions [mol/m$^2$ · s], $\boldsymbol{u}^{T}=[u_x\ u_y\ u_z]$ is the advective velocity of ions due to the bulk movement of pore solution phase [m/s], and $\Omega$: tortuosity of pore-structures for a 3-D pore network which is uniformly and randomly connected. Tortuosity is a parameter that expresses a reduction factor in terms of chloride penetration rate due to complex micro-pore structure. In the current model, this parameter is determined by considering the geometric characteristics of pore structures based on experimental data for various water-to-cement ratios (Takeda and Ishida 2000, Takegami et al. 2002).

Material parameters shown in Eq. 17, such as porosity, saturation and advective velocity, are directly obtained based on the thermo-hydro physics. The advective velocity $\boldsymbol{u}$ is obtained from the pore pressure gradient $\nabla P$ and liquid water conductivity $K$ which are calculated by moisture equilibrium and transport model, according to water content, micro pore structures, and moisture history. In the case of chloride ion transport in cementitious materials, $S$ represents the degree of saturation in terms of free water only, as adsorbed and interlayer components of water are also present. It has to be noted here that the diffusion coefficient $D_{cl}$ in the pore solution may be a function of ion concentration, since ionic interaction effects will be significant in the fine microstructures at increased concentrations, thereby reducing the apparent diffusive movement driven by the ion concentration gradient (Gjørv and Sakai 1995).

This mechanism, however, is not clearly understood, so we neglect the dependence of ionic concentration on the diffusion process in the model. From several numerical sensitivity analyses, a constant value of $3.0\times10^{-11}[m^2/s]$ is adopted for $D_{Cl}$. From the above discussions, the total diffusivity of concrete is described as the product of $D_{Cl}$ and $\phi S/\Omega$, which depends on the achieved micro-structures (the initial mix and curing condition dependent variables) and moisture history.

As well-known, chlorides in cementitious materials have free and bound components. The bound components exist in the form of chloro-aluminates and adsorbed phases on the pore walls, making them unavailable for free transport. In this study, the relationship between free and bound components of chlorides is expressed by the equilibrium model proposed by Maruya et al. (1998), Takeda and Ishida (2000), and Takegami et al. (2002) as shown in Fig. 10. Here, the bound chlorides are classified into two phases: adsorbed and chemically combined components. Through these studies, it can be assumed that the amount of combined phases is approximately constant in terms of weight percent of hydrated gel products, whereas that of the adsorbed phase is strongly dependent on the constituent powder materials. For example, in the case of BFS, the amount of adsorbed phases becomes larger compared to the case of OPC, which leads to a higher binding capacity of BFS concrete and mortar. It means that the adsorbed component plays a major role in the chloride binding capacity of concrete.

By assuming local equilibrium conditions based on this relationship, the rate of binding or the change of free chloride to bound chloride per unit volume $Q_{Cl}$ can be obtained. From the above discussions and formulations, the distribution of bounded and free chloride ions can be obtained without any empirical formula and/or intentional fittings, once mix proportions, powder materials, curing and environmental conditions are applied to the analytical system.

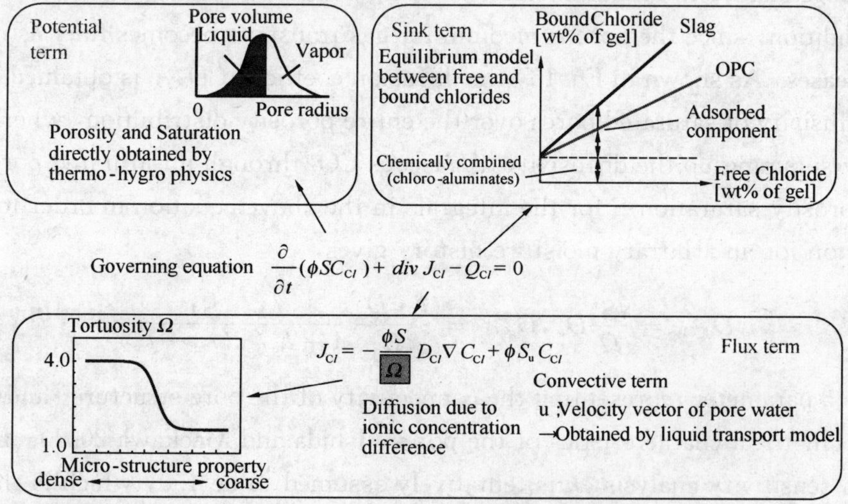


Fig. 10　Governing equation and constitutive models for chloride ion

### 4.3.4 $CO_2$ Transport, Equilibrium and Carbonation $-10^{-9} \sim 10^{-6}$m Scale -

For simulating carbonation in concrete, equilibrium of gaseous and dissolved carbon dioxide, their transport, ionic equilibriums and carbonation reaction process are formulated on the basis of thermodynamics and chemical equilibrium theory by Ishida and Maekawa (2001). Mass balance condition for dissolved and gaseous carbon dioxide in porous medium can be expressed as,

$$\frac{\partial}{\partial t}\{\phi[(1-S)\cdot\rho_{gCO_2}+S\cdot\rho_{dCO_2}]\}+divJ_{CO_2}-Q_{CO_2}=0 \tag{18}$$

where, $\rho_{gCO_2}$: density of $CO_2$ gas [kg/m$^3$], $\rho_{dCO_2}$: density of dissolved $CO_2$ in pore water [kg/m$^3$], $J_{CO_2}$: total flux of dissolved and gaseous $CO_2$ [kg/m$^2\cdot$s]. The local equilibrium of gaseous and dissolved carbon dioxide is represented by Henry's law, which states the relation of gas solubility in pore water and the partial gas pressure. $CO_2$ transport is considered in both phases of dissolved and gaseous carbon dioxide. By considering the effect of Knudsen diffusion, tortuosity and connectivity of pores on the diffusivity, Fick's first law of diffusion yields the flux of $CO_2$ as,

$$J_{CO_2}=-(D_{dCO_2}\nabla\rho_{dCO_2}+D_{gCO_2}\nabla\rho_{gCO_2}),D_{dCO_2}=\frac{\phi D_0^d}{\Omega}\int_0^{r_c}dV\,D_{gCO_2}=\frac{\phi\cdot D_0^g}{\Omega}\int_{r_c}^{\infty}\frac{dV}{1+N_k} \tag{19}$$

where, $D_{gCO_2}$: diffusion coefficient of gaseous $CO_2$ in porous medium[m$^2$/s], $D_{dCO_2}$: diffusion coefficient of dissolved $CO_2$ in porous medium[m$^2$/s], $D_0^g$: diffusivity of $CO_2$ gas in a free atmosphere[m$^2$/s], $D_0^d$: diffusivity of dissolved $CO_2$ in pore water [m$^2$/s], $V$: pore volume, $r_c$: pore radius in which the equilibrated interface of liquid and vapor is created, $N_k$: Knudsen number, which is the ratio of the mean free path length of a molecule of $CO_2$ gas to the pore diameter. Knudsen effect on the gaseous $CO_2$ transport is not negligible in low RH condition, since the porous medium for gas transport becomes finer as relative humidity decreases. As shown in Eq. 19, the diffusion coefficient $D_{dCO_2}$ is obtained by integrating the diffusivity of saturated pores over the entire porosity distribution, whereas $D_{gCO_2}$ is obtained by summing up the diffusivity of gaseous $CO_2$ through unsaturated pores. Substitution of porosity saturation $S$ for the integrals in the above equation in order to generalize the expression for an arbitrary moisture history gives,

$$D_{dCO_2}=\frac{\phi S^n}{\Omega}D_0^d,D_{gCO_2}=\frac{\phi\cdot D_0^g}{\Omega}\frac{(1-S)^n}{1+l_m/2(r_m-t_m)} \tag{20}$$

where, $n$ is a parameter representing the connectivity of the pore structure, and might vary with the geometrical characteristics of the pores (Ishida and Maekawa 2001). In the model, through sensitivity analysis, $n$ is tentatively assumed to be 4.0, which is the most appropriate value for expressing the reduction of $CO_2$ diffusivity with the decrease of relative humidity (Fig. 11). In Eq. 19, the integral of the Knudsen number is simplified so that it can be easily put into practical computational use; $r_m$ is the average radius of unsaturated pores, and $t_m$ is the thickness of adsorbed water layer in the pore whose radius is $r_m$.

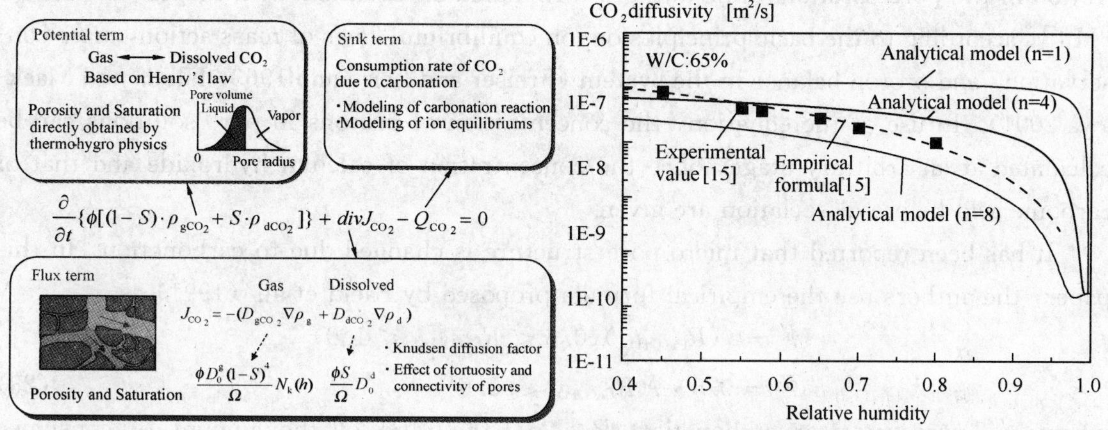


Fig. 11 Relationship between $CO_2$ diffusivity and relative humidity

$Q_{CO_2}$ in Eq. 18 and Eq. 21 is a sink term that represents the rate of $CO_2$ consumption due to carbonation [$kg/m^3 \cdot s$]. The rate of $CO_2$ consumption can be expressed by the following differential equation, assuming that reaction is of the first order with respect to $Ca^{2+}$ and $CO_3^{2-}$ concentrations as,

$$Ca^{2+} + CO_3^{2-} \longrightarrow CaCO_3,$$

$$Q_{CO_2} = \frac{\partial(C_{CaCO_3})}{\partial t} = k[Ca^{2+}][CO_3^{2-}] \quad (21)$$

$$H_2O \rightleftharpoons H^+ + OH^-$$

$$H_2CO_3 \rightleftharpoons H^+ + HCO_3^- \rightleftharpoons 2H^+ + CO_3^{2-}$$

$$Ca(OH)_2 \rightleftharpoons Ca^{2+} + 2OH^-$$

$$CaCO_3 \rightleftharpoons Ca^{2+} + CO_3^{2-} \quad (22)$$

where, $C_{CaCO_3}$: concentration of calcium carbonate, $k$ is a reaction rate coefficient. In this paper, a unique coefficient is applied ($k = 2.08$ [l/mol. sec]), although the reaction rate coefficient involves temperature dependency.

In order to calculate the rate of reaction with Eq. 21, it is necessary to obtain the concentration of calcium ion and carbonic acid in the pore water at an arbitrary stage. In this study, we consider the ion equilibriums; dissociation of water and carbonic acid, and dissolution and dissociation of calcium hydroxide and calcium carbonate as above. Here, the presence of chlorides is not considered, although chloride ions are likely to affect the above equilibrium conditions. The formulation including chlorides remains a subject for future study.

As shown in Eq. 22, carbonation is an acid-base reaction, in which cations and anions act as a Br? nsted acid and base, respectively. Furthermore, the solubility of precipitations is dependent on the pH of the pore solutions. Therefore, for calculating the ionic concentration in the pore solutions, the authors formulated an equation with respect to protons $[H^+]$, according to the basic principles on ion equilibrium; laws of mass action, mass conservation, and proton balance in the system (Freiser and Fernando 1963, Ishida and Maekawa 2001). In use of the equation, the concentration of protons in pore solutions can be calculated at an arbitrary stage, once the concentration of calcium hydroxide and that of carbonic acid before dissociation are given.

It has been reported that micro-pore structure is changed due to carbonation. In this paper, the authors use the empirical formula proposed by Saeki et al. (1991) as,

$$\phi' = \phi(R_{Ca(OH)_2})(0.6 < R_{Ca(OH)_2} < 1.0)$$
$$\phi' = 0.5 \cdot \phi(R_{Ca(OH)_2} \leqslant 0.6) \tag{23}$$

where, $\phi'$: porosity after carbonation, $R_{Ca(OH)2}$: the ratio of the amount of consumed $Ca(OH)_2$ for the total amount of $Ca(OH)_2$.

### 4.3.5 Oxygen Transport and Micro-cell Based Corrosion Model $-10^{-9} \sim 10^{-6}$m Scale-

In this section, a general scheme of a micro-cell corrosion model is introduced based on thermodynamic electro-chemistry (Ishida 1999a). Corrosion is assumed to occur uniformly over the surface areas of reinforcing bars in a referential finite volume, whereas formation of pits due to localized attack of chlorides and corrosion with macro cell remains for future study. Fig. 12 shows the flow of the corrosion computation. When we consider the micro-cell based corrosion, it can be assumed that the anode area is equal to that of the cathode and they are not separated from each other. Then, electrical conductivity of concrete, which governs the macroscopic transfer of ions in pore water, is not explicitly treated.

First of all, electric potential of corrosion cell is obtained from the ambient temperature, pH in pore solution and partial pressure of oxide, which are calculated by other subroutine in the system. The potential of half-cell can be expressed with the Nernst equation as (West 1986),

$$Fe(s) \longrightarrow Fe^{2+}(aq) + e(pt)$$
$$E_{Fe} = E_{Fe}^{\ominus} + (RT/z_{Fe}F)\ln h_{Fe^{2+}}$$
$$O_2(g) + 2H_2O(l) + 4e(Pt) = 4OH^-(aq)$$
$$E_{O_2} = E_{O_2}^{\ominus} + (RT/z_{O_2}F)\ln(P_{O_2}/P^{\ominus}) - 0.06pH \tag{24}$$

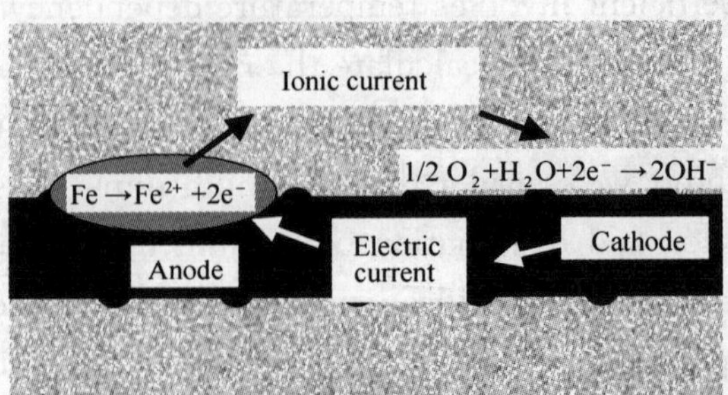

where, $E_{Fe}$: standard cell potential of Fe, anode (V, SHE), $E_{O_2}$: standard cell potential of $O_2$, cathode (V, SHE), $E^{\ominus}_{Fe}$: standard cell potential of Fe at 25℃(=−0.44V, SHE), $E^{\ominus}_{O_2}$: standard cell potential of $O_2$ at 25℃ (=0.40V, SHE), $z_{Fe}$: the number of charge of Fe ions (=2), $z_{O_2}$: the number of charge of $O_2$ (=2), $P^{\ominus}$: atmospheric pressure. By assuming an ideal condition, we neglect the solution of other ions in pore water on the half-cell potentials. Further study is intended in the multi-scale scheme.

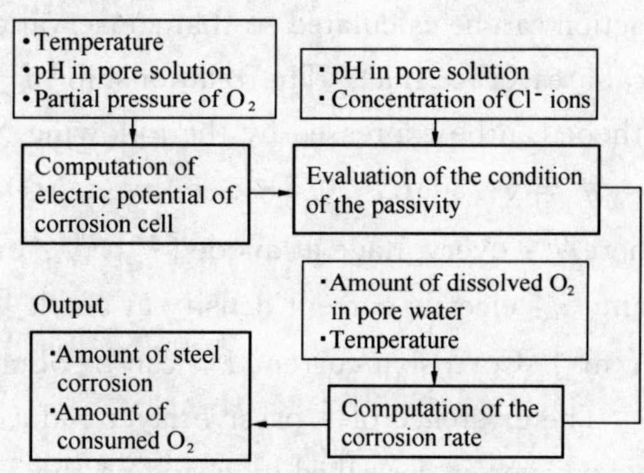


Fig. 12　Overall scheme of corrosion computation

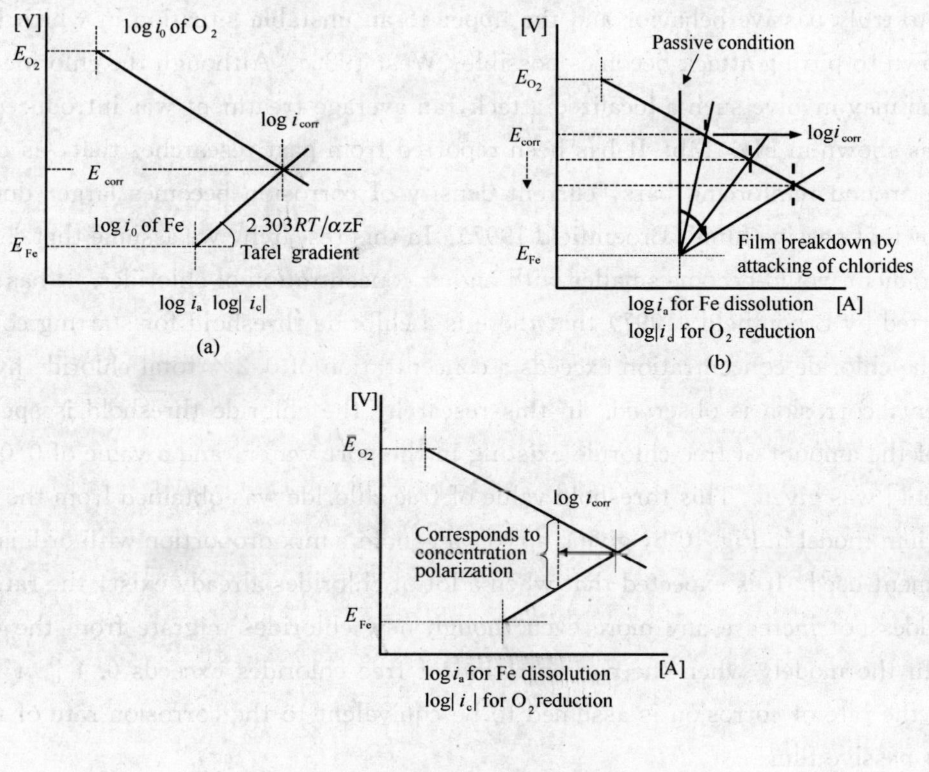


Fig. 13

(a) Relationship of electric current and voltage for anode and cathode;

(b) Rate of corrosion accelerated by chloride migration; (c) Rate of corrosion under $O_2$ diffusion control

Next, based on thermo-dynamic conditions, the state of passive layers is evaluated by the Pourbaix diagram, where steel corrodes, areas where protective oxides from, and an area of immunity to corrosion depending on pH and the potential of the steel. From the electric potential and the formation of passive layers, electric current that involves chemical reaction can be calculated so that conservation law of electric charge should be satisfied in a local area (Fig. 13a). The relationship of electric current and voltage for the anode and cathode can be expressed by the following Nernst equation as,

$$\eta^a = (2.303RT/0.5 \cdot z_{Fe}F)\log(i_a/i_0), \eta^c = -(2.303RT/0.5 \cdot z_{O_2}F)\log(i_c/i_0) \quad (25)$$

where, $\eta^a$: overvoltage at anode [V], $\eta^c$: overvoltage at cathode [V], $F$: Faraday's constant, $i_a$: electric current density at anode [A/m²], $i_c$: electric current density at cathode [A/m²]. Corrosion current $I_{corr}$ can be obtained as the point of intersection of two lines.

The existence of a passive layer reduces the corrosion progress. In this model, this phenomenon is described by changing the Tafel gradient (Fig. 13b). When chlorides exist in the system, the passive region hitherto occupied by $Fe_3O_4$ will disappear. In addition, the protective region of $Fe_2O_3$ will be divided into two regions, the lower of which corresponds to truly passive behavior and the upper to an unstable situation in which localized breakdown to pitting attack becomes possible (West 1986). Although the chloride induced corrosion may involve such a localized attack, an average treatment was introduced in this work, as shown in Fig. 13b. It has been reported from past researches that, as chlorides increase around reinforcing bars, current density of corrosion becomes larger due to the breakdown of passive films (Broomfield 1997). In this research, we assume that the anodic Tafel gradient would become smaller with higher concentration of chlorides. It has been also reported by Broomfield (1997) that there is a chloride threshold for starting corrosion. When the chloride concentration exceeds a concentration of 0.2% total chloride by weight of cement, corrosion is observed. In this research, the chloride threshold is specified in terms of the amount of free chloride existing in the pore water, and a value of 0.04 [wt% of cement] was given. This threshold value of free chloride was obtained from the chloride equilibrium model in Fig. 10 by giving a typical concrete mix proportion with ordinary portland cement used. It is expected that when a lot of chlorides already exist, the rate of corrosion does not increase any more even though new chlorides migrate from the environment. In the model, when the concentration of free chlorides exceeds 0.4 [wt% of cement], the rate of corrosion is assumed to be equivalent to the corrosion rate of the steel without passive films.

When the amount of oxygen supplied to the reaction is not enough, corrosion rate would be controlled by diffusion process of oxygen (Fig. 13c). By coupling with an oxygen transport model, this phenomenon can be logically simulated. Current density $i_{corr}$, which is obtained as the intersection of anodic and cathodic polarization curve, corresponds to the corrosion under sufficient availability of oxygen (Fig. 13a). When oxygen supply shortens, the rate of corrosion will be limited by the slow diffusion of oxygen. A limited value of current density $i_L$ can be expressed as,

$$i_L / z_{Fe}F = O_2^{sup} \quad (26)$$

where, $O_2^{sup}$[mol/m$^2$ · s] is the amount of oxygen supplied to the surface of metal, which is obtained by the equilibrium and transport model for oxygen discussed below. In this research, the rate of corrosion under diffusion control of oxygen $i_{corr}$[A/m$^2$] is assumed to be as (Fig. 13c),

$$i_{corr} = i_L \tag{27}$$

The potential difference between the anode and cathode corresponds to the concentration polarization.

Fig. 14 summarizes the formulation of the key governing equation, which is almost the same as that of carbon dioxide (Ishida and Maekawa 2001, Maekawa and Ishida 2002). Finally, using the Faraday's law, electric current of corrosion is converted to the rate of steel corrosion. These models derive from thermodynamic electrochemistry. Further development and improvement are still needed thorough various verification of corrosion phenomena in real concrete structures.

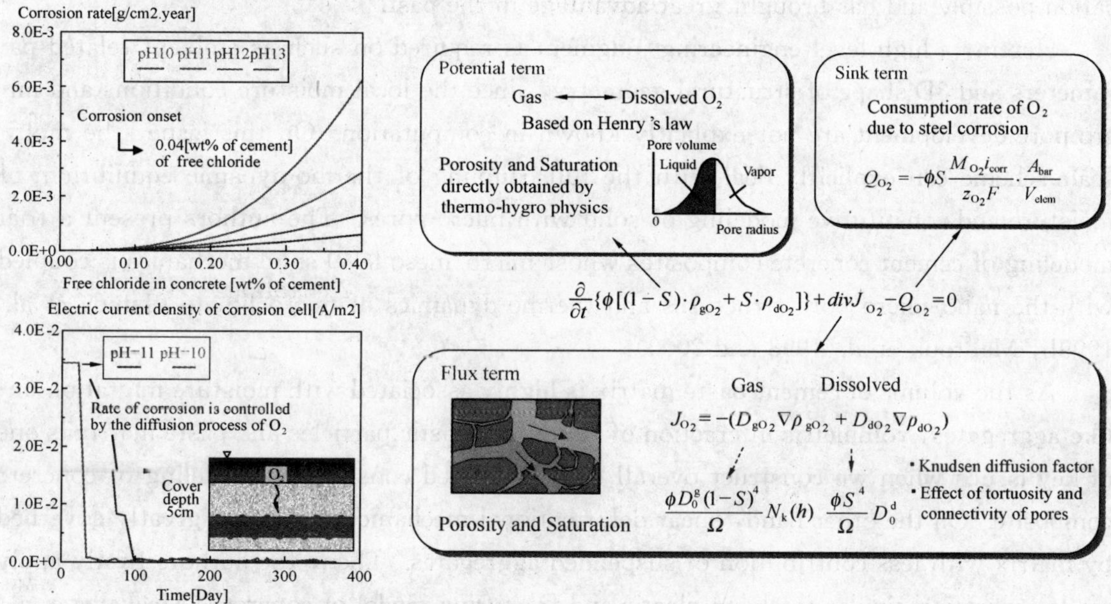


Fig. 14　Oxygen diffusion model linked with micro-corrosion of steel

### 4.3.6　Calcium Ion Transport and Leaching $-10^{-9} \sim 10^{-6}$m Scale-

Calcium is one of the main chemicals and in pore solution, $Ca^{2+}$ is equilibrated with $Ca(OH)_2$ solids and other ionic substances as stated in Eq. 22 and the Appendix in detail. Calcium leaching may change pore structures due to lost $Ca(OH)_2$ and long-term performance of cementitious solids would be influenced especially when exposed to pure water. Similar to chloride conservation in Eq. 22, we have,

$$\frac{\partial}{\partial t}(\phi S C_{ca}) + divJ_{ca} - Q_{ca} = 0, J_{ca} = -\frac{\phi S}{\Omega} D_{ca} \nabla C_{ca} + \phi S u C_{ca}, u = -\frac{K \nabla P}{\rho \phi S} \tag{28}$$

where, $C_{ca}$: calcium ion concentration, $D_{ca}$: diffusivity of Ca ion, $Q_{ca}$: dissolution rate. The saturated calcium ion concentration can be computed by Eq. 22 and solution product constant in Appendix with pH value. Then, $Q_{ca}$ is given as the rate of dissolved ion concentration. In considering much smaller solubility, calcium dissolution from CSH hydrates is not considered in this study. In the same manner, chrome dissolution can be incorporated in the multi-scale platform, too.

### 4.3.7 Mechanics of Young Concrete before Cracking $-10^{-6}\sim10^{-2}$m Scale-

Degree of hydration and moisture in micro-pores greatly influences the solid mechanics of cementitious composites. Constitutive modeling has been proposed with respect to stresses/strains and the micro-pore related solid properties of young concrete composites have been considered in terms of varying elasticity and creep coefficients under specified ambient conditions. This macroscopic expression of micro-pore development makes practical application possible and has brought great advantage in the past.

Herein, a high level engineering judgment is required on such as ambient related parameters and 3D shape of structural geometry, since the local moisture conditions and micro-pore development are not explicitly known in computation. On this issue, the multi-scale scheme can explicitly deal with the full coupling of thermo-dynamic equilibrium of moisture and constitutive modeling of solid with micro-pores. The authors present a trial modeling of cement concrete composites whose micro/meso level solid mechanics is coupled with the nano/micro pore structures and thermo-dynamics of microclimate (Ishida et al. 1999b, Mabrouk et al. 1998 and 2000).

As the volume of cement paste matrix is highly associated with moisture migration unlike aggregates, volumetric interaction of aggregate elastic particles and paste matrix is one of key issues when we construct overall space-averaged constitutive modeling of concrete composite. On the other hand, shear deformational mechanics is thought greatly governed by matrix with less contribution of suspended aggregates. Then, the authors firstly apply mode separation similar to elasto-plastic and fracturing model of concrete (Maekawa et al. 2003) and secondly, we compose them into unified constitutive modeling for structural analysis.

#### 4.3.7.1 Particle dispersion and solidification

Aggregate particles can be modeled as a suspended elastic body by surrounding cement paste as shown in Fig. 15. Local stresses developing in both aggregates and cement paste are non-uniform. Here, let us define the referential volume (REV) whose scale is 1-5cm including several gravels and lots of sands. The volumetric virtual work principle yields equilibrium of stress components and compatibility of strain fields on this REV as,

$$\bar{\sigma}_o = \rho_{ag}\,\bar{\sigma}_{ag} + \rho_{cp}\,\bar{\sigma}_{cp},\ \bar{\varepsilon}_o = \rho_{ag}\,\bar{\varepsilon}_{ag} + \rho_{cp}\,\bar{\varepsilon}_{cp} \tag{29}$$

where, $\bar{\sigma}_o$, $\bar{\sigma}_{ag}$ and $\bar{\sigma}_{cp}$ are the mean volumetric stresses on concrete, aggregate and cement paste, respectively, and $\bar{\varepsilon}_o$, $\bar{\varepsilon}_{ag}$ and $\bar{\varepsilon}_{cp}$ are the mean volumetric strains, $\rho_{ag}$ and $\rho_{cp}$ are the volume fractions of aggregate and cement paste, respectively. As the aggregate phase is assumed elastic, we have,

$$\bar{\varepsilon}_{ag} = \frac{1}{3K_{ag}}\bar{\sigma}_{ag},\ \bar{\varepsilon}_{cp} = f(\bar{\sigma}_{cp}), K_{ag}\text{: volumetric stiffness of aggregate} \tag{30}$$

Here, let us consider the local equilibrium and compatibility between two elements. If the cement paste matrix be a perfect liquid losing resistance to the shear deformation, we have $\bar{\sigma}_{ag} = \bar{\sigma}_{cp}$, where the shear stiffness of cement paste becomes zero. This system corresponds to Maxwell chain idealization. If the shear stiffness, $G_{cp}$, is infinitely large on the contrary, it brings no change of geometrical shape and results in $\bar{\varepsilon}_{ag} = \bar{\varepsilon}_{cp}$. This system corresponds to so called Kelvin chain. As the reality is in between two extremes, the Lagragian method of linear summation is applied as,

$$\left(\frac{\bar{\sigma}_{ag} - \bar{\sigma}_{cp}}{G_{cp}}\right) + (\bar{\varepsilon}_{ag} - \bar{\varepsilon}_{cp}) = 0 \tag{31}$$

Under the deviatoric shear mode of deformation, shear stress is hardly transferred through the contact link network of aggregates except for pre-packed concrete, and rotational resistance of individual particles is no longer expected when cement paste matrix deforms in shear. Thus, the authors assume that the space averaged deviatoric stress and strain of cement paste in concrete coincides with those of overall concrete composite. Then, we have,

$$S_{ij} = f(e_{ij}) \tag{32}$$

where, $S_{ii}$ and $e_{ii}$ are the deviatoric stress and strain tensors for both cement paste and concrete composite. The function $f$ for cement paste is formulated as below.

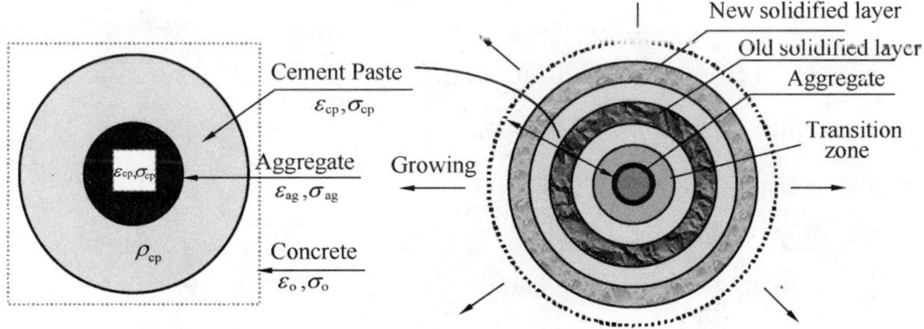


Fig. 15　Two-phase modeling of concrete composite -elastic suspension and nonlinear matrix-

The structural growth of cement paste is rather complex. Upon contact with water, powder particles start to dissolve and reaction products start to form. Due to gradual solidification of hydration products, properties of paste matrix vary with time. Solidification theory (Bazant and Prasannan 1989) proved a potential way of taking this aging effect into consideration.

As shown in Fig. 15 and Fig. 16, growth of cement paste is idealized by forming finite fictitious clusters. When a new cluster is formed onto the already formed assembly of old clusters, strain of the cluster is originated at the time of birth. Aging process is represented by solidification of new cluster afterwards. In order to set a criterion for cluster formation, a volume fractional function is introduced. This function is taken as the hydration ratio at a certain time $t$, $\psi(t)$. It is defined as a ratio of hydrated volume of cement powder, $V(t)$, to total volume of cement that is available for hydration, $V_{cp}$, as,

$$\psi(t) = V(t)/V_{cp} \tag{33}$$

The structure of cement paste at any time is represented using the number of clusters already solidified at that time. According to hydration degree, numbers of clusters, $N$, at a certain time can be determined and when the hydration increment reaches a certain value a new cluster is developed and attached to the older ones. As these clusters share bearing stresses carried by the cement paste, we introduce an infinitesimal stress in each cluster, $S_{cp}$. When a new cluster is formed, these infinitesimal stresses are redistributed among the clusters. It also means that the stress condition in a certain cluster is a function of both current time and the location of this cluster or the time when this cluster was born.

Let $S_{cp}=S_{cp}(t, t')$ denotes the average stress in a general cluster, where $t$ is the current time and $t'$ is the time when this cluster is solidified. Total volumetric stress in cement paste at a certain time is the summation of average stresses in all clusters activated at that time and strain in cement paste is equal to the one induced in each cluster. Solidification concept of cement paste cluster regarding growth of microstructure yields,

$$\bar{\sigma}_{cp}(t) = \int_{t'=0}^{t} S_{cp}(t',t)\mathrm{d}\psi(t'), S_{ij}(t) = \int_{t'=0}^{t} S_{ij}(t',t)\mathrm{d}\psi(t') \tag{34}$$

where, $S_{cp}(t', t)$ is the mean volumetric stress acting on a certain cluster, $t'$ is the time when this cluster formed, $t$ is the current time, $\psi$ is the hydration degree, $S_{ij}(t', t)$ is the deviatoric stress tensor acting on the cluster concerned. When new cluster is formed, specific cluster stress is null. Then, we have $S_{cp}(t, t)= 0$ and $S_{ij}(t, t)=0$.

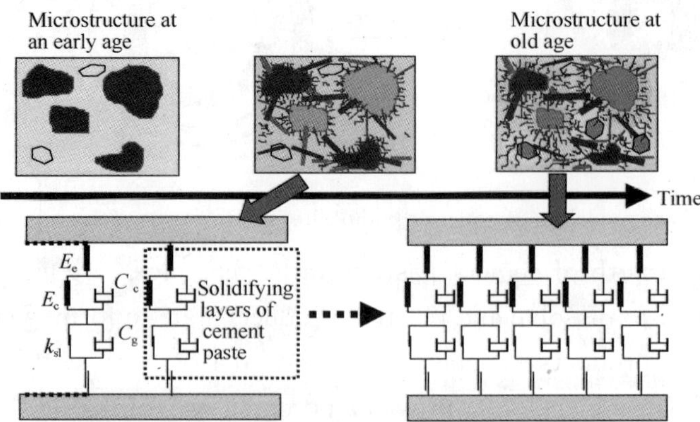


Fig. 16 Aging effect by solidified elasto-plastic assembly with different stress/strain histories

4.3.7.2 Modeling of solidifying cement paste cluster

The solidifying mechanical unit as shown in Fig. 16 is thought to be associated with CSH gel grains. Thus, the time-dependent deformation rooted in moisture in gel and interlayer pores should be taken into account in each individual solidifying component. The assembly of solidifying components corresponds to cement paste. Here, the moisture existing in capillary pores between gel grains has to be taken into account in the model, too. The authors simply assume two fictitious time-dependent deformational components rooted in both interlayer/gel pores and capillary one as shown in Fig. 17 (Mabrouk et al. 1998 and 2000). Moisture transport actually takes place into the capillary and gel pores. Under severe drying, interlayer water starts to be in motion. Thus, moisture transport related to each of these pores categories can be correlated to a certain aspect of the creep behaviors (Glucklich 1959, Neville 1959).

The rate of flow of capillary water is comparatively high and easily reversible. Thus, moisture transport within capillary pores can be assumed as a cause of short-term creep at earlier ages. This creep rate drops steeply with time and is highly related to the hydration process. Moisture transport within gel pores is thought to be slow and prolonged in nature. Kinematics of gel water can be reversible up to a certain limit. It can be assumed as a cause of long-term creep and is responsible for the main part of the unrecoverable creep. Moisture migration within interlayer pores can be assumed to drive creep under severe conditions. This creep is highly irreversible accompanying so called disjoining pressure.

In order to reflect three aspects of creep behaviors, a simple rheological model for each cluster component is assumed as shown in Fig. 17. The same model is adopted for both volumetric and deviatoric components. Here, the volumetric component is discussed in detail. Total strain of each solidifying cluster is decomposed into instantaneous elastic strain $\varepsilon_e$, visco-elastic strain $\varepsilon_c$, visco-plastic strain $\varepsilon_g$ and instantaneous plastic strain $\varepsilon_l$.

$$\varepsilon'_{cp} = \varepsilon_e + \varepsilon_c + \varepsilon_g + \varepsilon_l \tag{35}$$

where, $\varepsilon'_{cp}$ is the volumetric strain in a general layer.

It should be noted here that $\varepsilon'_{cp}$ is the strain induced in an individual cluster after its solidification. As the clusters are assumed to join together, these strains should be common to corresponding volumetric strain in cement paste. However, clusters solidify at zero stress state. Thus, as given by Eq. 36, this strain at a certain time, $t$, is defined as the difference between the strain of cement paste at time, $t$, and the strain of cement paste at the time when the cluster solidified, $t'$.

$$\varepsilon'_{cp}(t) = \varepsilon_{cp}(t) - \varepsilon_{cp}(t') \tag{36}$$

a) Instantaneous elasticity

The elastic spring of solidifying cluster represents the instantaneous deformation and assumed to be perfect elasticity as,

$$S_{cp} = E_e \varepsilon_e \tag{37}$$

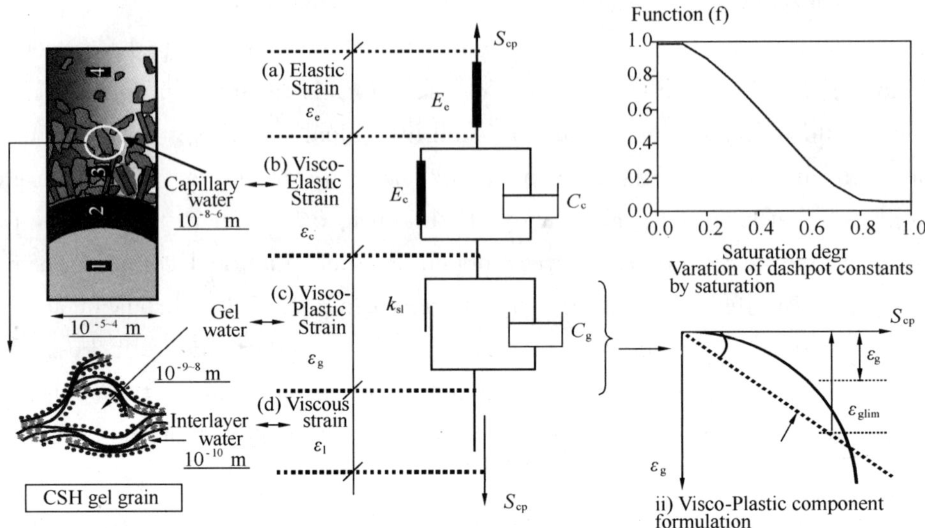


Fig. 17 Modeling of solidifying cluster of cement hydrate

where, $S_{cp}$ is the volumetric stress in a cluster, $E_e$ is the stiffness and $\varepsilon_e$ is the instantaneous elastic strain. The elastic stiffness of the component is calculated so that summation of stiffness over all solidified clusters at a certain time is equal to that of cement paste.

b) Visco-elasticity

The visco-elasticity represents delayed recoverable deformation. This is thought associated with rather larger sizes of micro-pores like capillaries and it macroscopically represents the short-term creep behaviors as,

$$S_{cp} = E_c \cdot \varepsilon_c + C_c \frac{d\varepsilon_c}{dt} \tag{38}$$

where, $E_c$ is the stiffness and assumed $2.0^* E_e$. $\varepsilon_c$ is visco-elastic strain and $C_c$ is defined as a constant of the dashpot fluidity related to moisture kept in capillary pores. The dashpot is related to condensed water motion through capillary pores associated with thermo hydro-physical requirements, and the following assumption is presented as,

$$C_c = a \cdot f(S_{cap}) \cdot \eta \cdot \phi_{cap}^{-1} \tag{39}$$

where, $a$ is a constant, $\phi_{cap}$ is capillary porosity of the paste, $S_{cap}$ is saturation of capillary pores, $\eta$ is average viscosity of condensed water in micro pores. Function, $f$, can be defined as shown in Fig. 17.

c) Visco-Plasticity

The visco-plasticity represents the time-dependent unrecoverable deformation mainly rooted in moisture kinematics in gel and interlayer pores. First, let us consider a simple case where applied stress on a cluster is constant. Plastic strain increases with time and finally may reach convergence denoted by $\varepsilon_{glim}$. Rate of plasticity to convergence is deemed influenced by the moisture residing inside gel pores. Dry empty pores shall bring rapid convergence of plasticity and wet moisture occupying pores may retard plasticity due to seepage.

As a matter of fact, possible converged plasticity is affected by applied stress level. At the same time, moisture in gel pores should be considered, because condensed and adsorbed water occupy micro-pore space. Here, the authors assume the simplest formula under generic stress paths, i. e. , linear viscosity in terms of moisture mass in gel pores, linear converging strain in terms of applied stress, linear rate of plasticity in terms of updated plastic strain as follows.

$$\frac{d\varepsilon_g}{dt} = (\varepsilon_{glim} - \varepsilon_g)/C_g, C_g = d \cdot f(S_{gel}) \cdot \eta \cdot \phi_{gel} \tag{40}$$

$$\begin{gathered} \varepsilon_{glim} = f_1(S_{cp}) \cdot f_2(S_{gel}) \\ f_1 = S_{cp}/E_g, E_g = E_e/4 \\ f_2 = \exp(-a \cdot S_{gel}) \end{gathered} \tag{41}$$

where, $S_{gel}$: saturation of gel pores, $d$: constant, $\phi_{gel}$: gel porosity of paste related to this arbitrary cluster, $\eta$: viscosity of micro pore water.

d) Instantaneous plasticity

Motion of interlayer moisture or some of gel water causes instantaneous volume change since the size of the pores is close to that of water molecule. The instantaneous plasticity related to ambient condition is assumed associated with saturation of interlayer pore as,

$$\Delta\varepsilon_l = z \cdot \Delta S_{int} \tag{42}$$

where, $\varepsilon_l$: moisture related instantaneous strain, $z$: constant, and $S_{int}$: saturation of interlayer. This component is not in motion under usual ambient conditions. But, it may work under very dry situations and/or higher temperatures. Entire set of assumed material functions and constants are summarized in Fig. 18.

4. 3. 7. 3　Coupling of skeleton stresses and pore water pressure

In this section, effect of micro-pore pressure on deformation of cement paste solid is introduced. Pressure drop of condensed water in micro-pores is equilibrated with capillary surface tension acting between liquid and gas phases, and it is assumed as one of causes of shrinkage (Shimomura and Maekawa 1997). Since surface tension developing in micro-pores is isotropic, only the volumetric mode of deformation is targeted. Similar to Biot's theorem of two-phase continuum, volumetric stress of cement paste is idealized as carried by both skeleton solid and pore pressure (Maekawa et al. 1999, Ishida et al. 1999a and 1999b) as,

$$\bar{\sigma}_{cp} = \overline{\sigma'}_{cp} + \beta \cdot \sigma_s \tag{43}$$

where, $\sigma'_{cp}$ is volumetric stress by cement paste skeleton and $\sigma_s$ is pore-water pressure drop. Factor $\beta$ indicates effectiveness to represent the volume fraction where moisture can act. If the whole space is occupied by water continuum, it is unity as Biot's theory reveals.

Pore water pressure can be evaluated by Kelvin's formula and factor b is estimated in

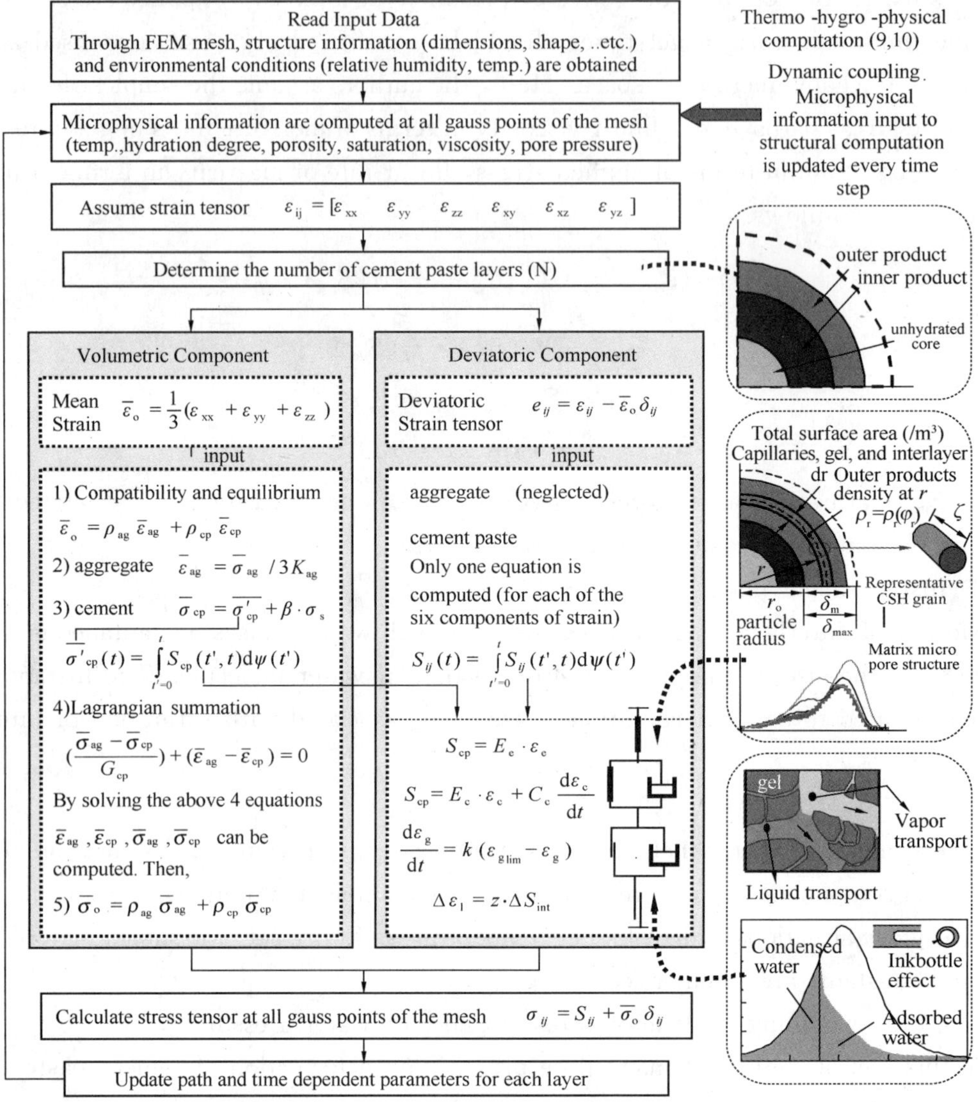


Fig. 18 System dynamics and flow of computation based on particle dispersion and solidification

use of micro-pore water information as,

$$\sigma_s = -\frac{2.\gamma}{r_s} = -\frac{\rho.R.T}{M}\ln h, \ \beta = \frac{\phi_{cap} \cdot S_{cap} + \phi_{gel} \cdot S_{gel}}{\phi_{cap} + \phi_{gel}} \tag{44}$$

where, $\gamma$: surface tension of condensed water, $r_s$: pore radius at which the interface is created, $R$: universal gas constant, $T$: absolute temperature of vapor-liquid system, $M$: molecular mass of the water, $h$: relative humidity i. e. the ratio of the vapor pressure to the saturated vapor pressure.

This coupling is a key to integrate thermo-hydro dynamics and deformational solid mechanics of concrete composite, and effect of moisture on deformation is inherently taken into account. Moisture loss may occur due to cement hydration and/or moisture migration invoked by drying climates. The volumetric change caused by both can be systematically included in the constitutive modeling of multi-scale. Thus, there is no need to assume specific drying/autogenous shrinkage, and basic/drying creep. Mathematical formula based on the basic hypothesis -particle dispersion of aggregates and solidification of cement-are summarized in Fig. 18.

4.3.7.4　Simulation and verification

In this section, the framework involving both the computation of thermo-hydro-physical information and the simulation of the structure behavior is presented. Fig. 19 shows a summary of the model parameters and their computation. Creep and shrinkage are targets of experimental verification and comparison with available experimental data is conducted.

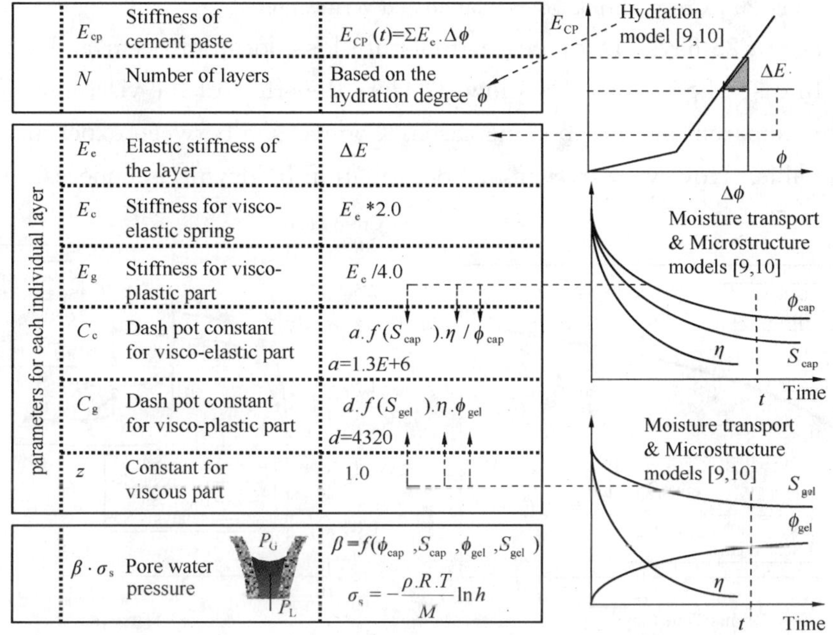


Fig. 19　Model parameters and their computations

Firstly, two pure cement paste specimens (S1, S2) of young age are studied as shown in Fig. 20. Here, aging process represented by solidification of clusters, creep at young age as represented by the visco-elastic part and coupling with pore water pressure for computation of shrinkage are the main parts affecting the results. From the figure, good agreement can be obtained between analysis and experiment in case of drying shrinkage, S1, while the results are rather overestimated in the case of loaded specimen, S2. Concrete based verification follows the verification as S3 in Fig. 21. Reasonable agreement is seen. Here, coupling between the skeleton stress of cement paste and the pore water pressure is a main concern of verification.

In Fig. 22, two concrete cylinder specimens, S4 and S5 (Neville 1959), are studied in terms of normalized creep ratio per unit stress. Case S5 shows creep under sealed condi-

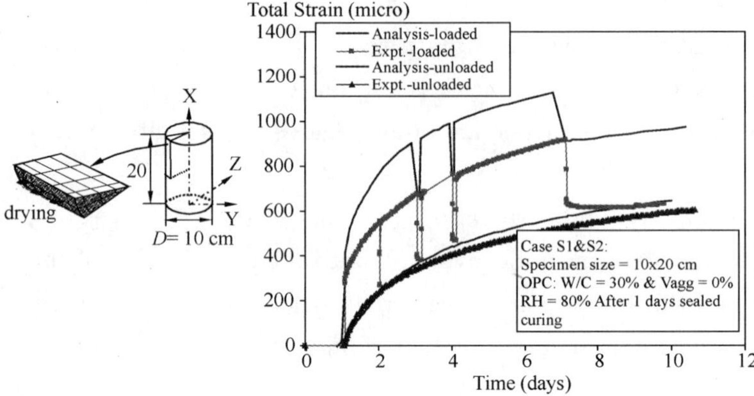


Fig. 20 Creep/shrinkage simulation and verification

tions at the age of 28 days. In this case, moisture loss does not occur and aging is almost completed. In case of S4, drying is allowed. Here the effect of the visco-plastic component becomes more apparent and creep is increased. Comparison between experiment and analysis are reasonable. However, irreversible deformation by drying is underestimated.

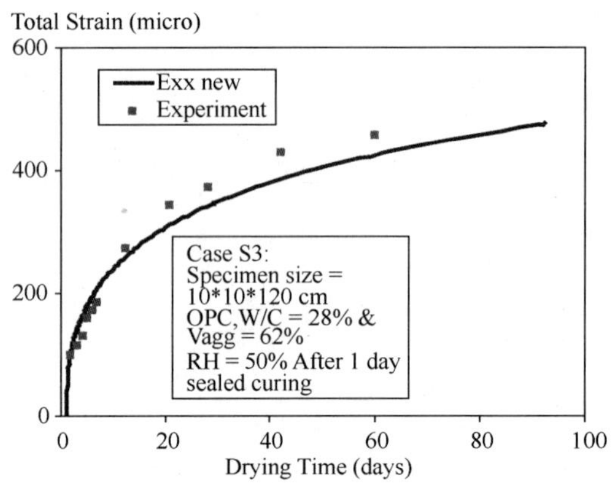


Fig. 21 Free shrinkage simulation and verification

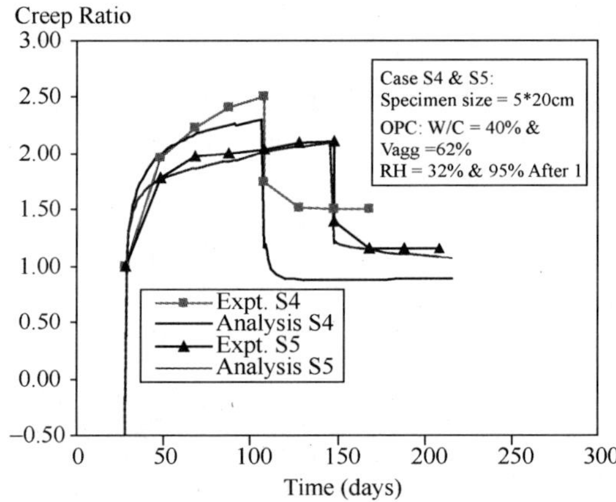


Fig. 22 Basic and drying creep simulation

In addition, two cement paste beams, S6 and S7 (Glucklich 1959), were studied as shown in Fig. 23. These cases represent a simple verification on the structure level. Once again reasonable results are obtained. It can also be shown that the loading-unloading pattern in the case of S7 can be well represented. This part is mainly linked to the slider and dashpot of the visco-plastic component.

Combination of autogenous and drying shrinkage is studied as below. Drying accerlates dessication insde micro-pores together with cement hydration. At the same time, moisture loss close to surfaces may retard hydration at early age accompanying corser solids and eacy loss of moisture. Computation seems acceptable as shown in Fig. 24. As the micro-climate is automatically computed in this holistic approach, relative humidity in pores can be shown, too. Self-dessication of W/C=30% is more severe.

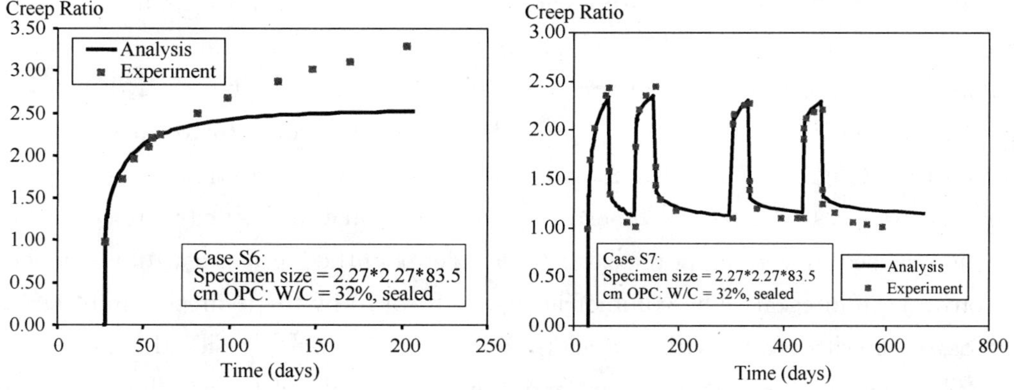


Fig. 23　Simulation of time-dependent deflection of beams under sealed condition

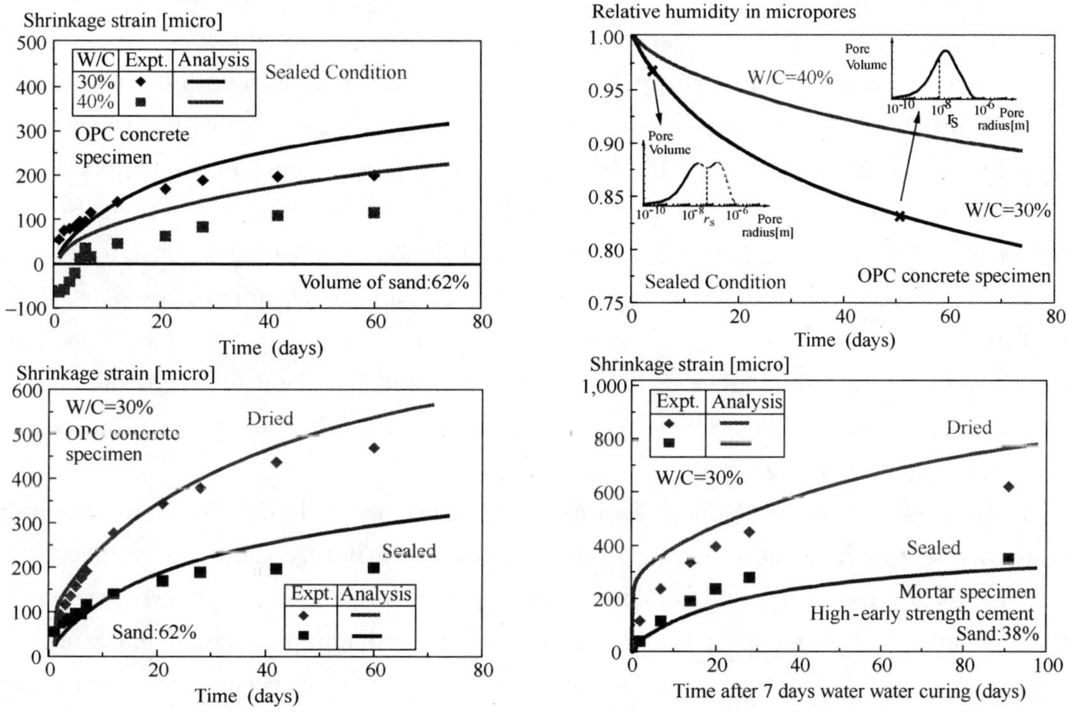


Fig. 24　Volume change simulation under coupled autogenous and drying shrinkage and internal states

### 4.3.8　Model of Macro-Cracks in Concrete $-10^{-3}\sim10^{-0}$m Scale -

For simulating structural behaviors expressed by displacement, deformation, stresses and macro-defects of materials in view of continuum plasticity, fracturing and cracking, well established continuum mechanics is available in multi-scale modeling. Compatibility condition, equilibrium and constitutive modeling of material mechanics are the basis and spatial averaging of overall defects in control volume of meso-scale finite element is incorporated into the constitutive model of quasi-continuum. Here, the size of referential volume is $10^{-3} \sim 10^{-1}$m order. The authors adopted a 3D finite element computer code of non-linear structural dynamics (Maekawa et al. 2003).

This frame of structural mechanics has an inter-link with thermo-hydro physics in terms of mechanical performances of materials through the constitutive modeling in both space and time (Fig. 2). The instantaneous stiffness, short-term strengths of concrete in tension and compression, free volumetric contraction rooted in coupled water loss and self-desiccation caused by varying pore sizes are considered in the creep constitutive modeling of liner convolution integral. The volumetric change invoked by the hydration in progress and water loss is physically tied with surface tension force developing inside the micro-capillary pores. The micro-pore size distribution and moisture balance of thermo-dynamic equilibrium are given from the code *DuCOM* at each time step as discussed in previous chapters.

Cracking is the most important damage index associated with mass transport inside the targeted structures. Cracks are assumed normal to the maximum principal stress direction in 3D extent when the tensile principal stress exceeds the tensile strength of concrete. After crack initiation, tension softening on progressive crack planes is taken into account in the form of fracture mechanics. In the reinforced concrete zone, in which bond stress transfer is expected being effective, tension stiffness model is brought together. Since the external load level, with which the environmental action be coupled in design, is rather lower than ultimate limit states, compression induced damage accompanying dispersed micro cracking is disregarded. Fig. 25 shows the overall frame of constitutive modeling of macro-scale REV including cracks. This macroscopic modeling, which can be directly used for behavioral simulation of entire structures, has the mutual link with nanometer-scale structures as shown in Fig. 1.

In the case where mechanical actions on structures are dominant, multi-directional cracks intersect orthogonally, because the principal stress hardly rotates in nature. In the case where both mechanical and ambient weather actions are applied to concrete structures, however, principal stress axes can drastically rotate and non-orthogonal cracking easily takes place in the REV domain. As highly nonlinear interaction between multi-directional cracking is witnessed, the rotating crack model or single directional smeared crack approach is hardly used for structural damage analysis under coupled loads and ambient conditions.

In this study, multi-directional fixed crack modeling of the RC domain is applied. This modeling was originally developed for reversed cyclic finite element analysis for three-dimensional RC under arbitral ground motions. As a matter of fact, multi-directional forces may cause non-orthogonal cracks in nature. Here, the active crack method is utilized for simplifying numerical processes with reasonable accuracy for practice, and currently used for seismic performance assessment of underground RC and energy facilities.

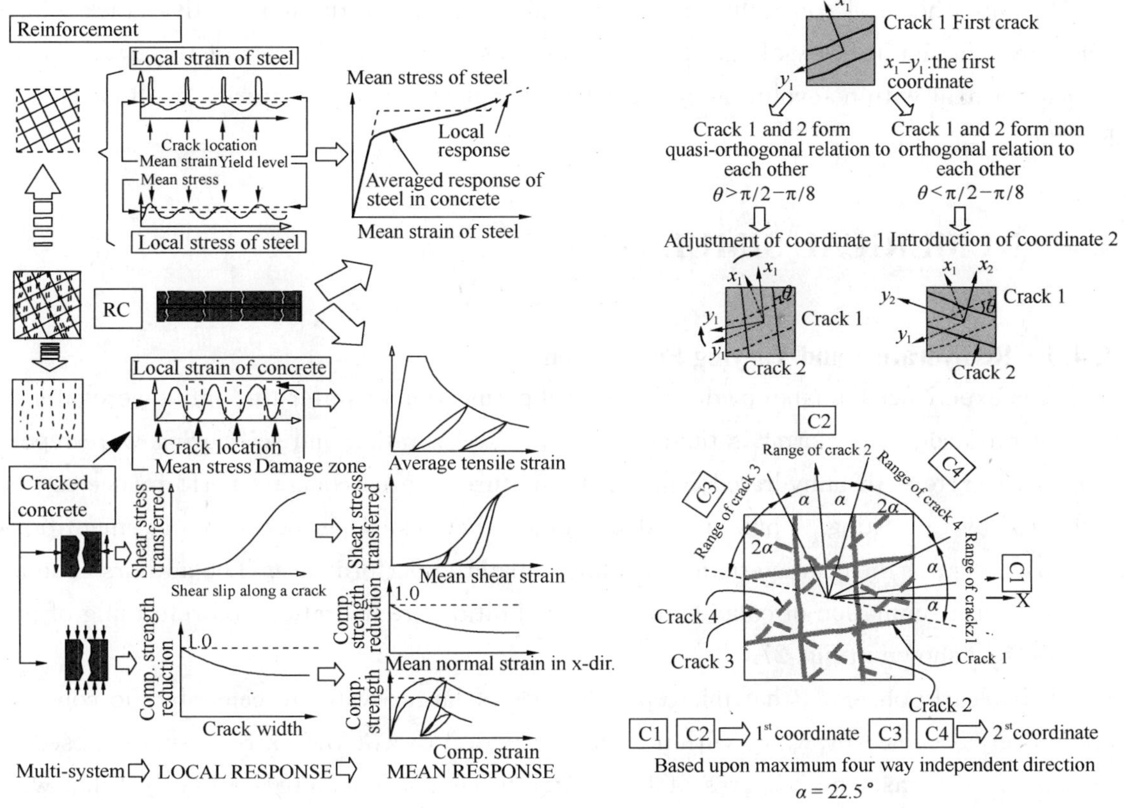


Fig. 25　Macroscopic modeling of reinforced concrete with cracking: REV size is about $10^{-3}\sim10^{-0}$m

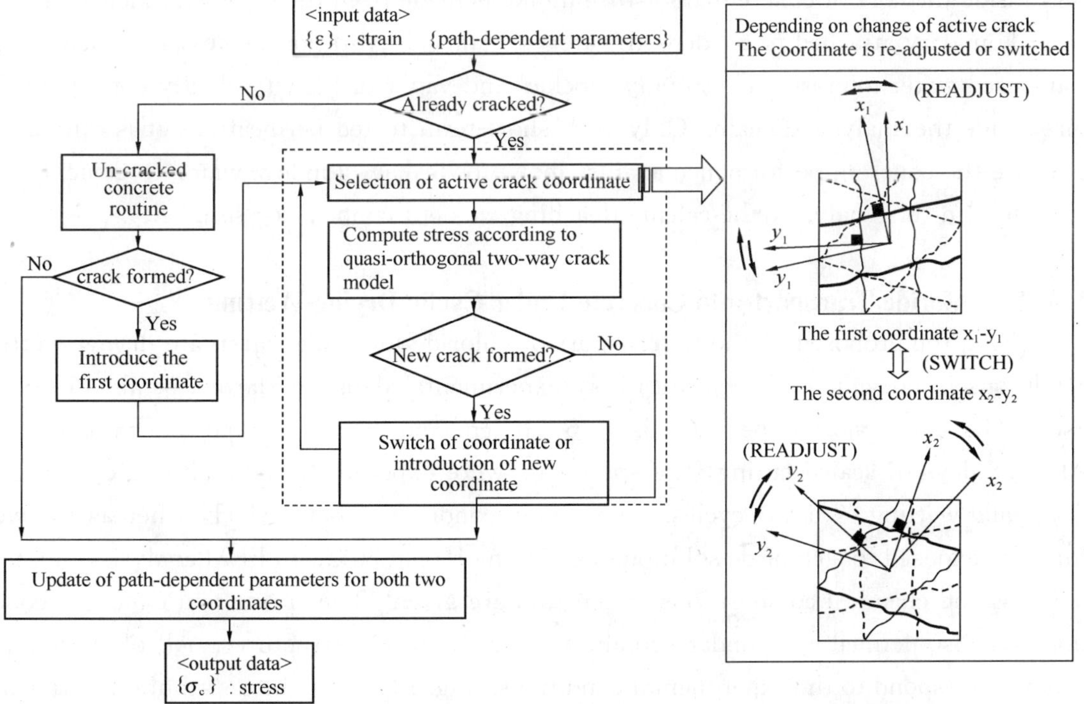


Fig. 26　Active and dormant crack scheme for multi-directional cracking model

In this scheme of controlling multi-directional cracking, the most active crack which dominates the nonlinearity of the whole domain is selected and the rest of the cracks is assumed dormant with no evolution of shear transfer plasticity and fracturing as illustrated in Fig. 26.

## 4.4 NUMERICAL SIMULATIONS

### 4.4.1 Re-hydration and Varying Permeation

It is experienced in laboratories that tested permeation of water through lower water to cement ratio concrete/mortar is time-dependent. Re-hydration and self-curing were reported for concrete with unhydrated cement. Prematurely cured concrete performance can be recovered by re-wetting. Thus, it will be agreed that tested permeability of cementitious composites is not the characteristic constant of material performance. The authors conducted long-term simulation of coupled moisture migration, re-hydration and reforming of micro-pores as shown in Fig. 27.

It is clearly observed that micro-pore spaces of larger water to cement ratio concrete are entirely occupied by water within a shorter period except the extreme end exposed to dry air. In contrast, micro-pores of low water to cement concrete are hardly filled with moisture and it takes many years for stability in computation. As severe self-desiccation occurs inside the core of concrete, moisture intake is made from the drying boundary. At the other boundary attached to condensed water, re-started hydration makes pores denser and condensed water migration is strongly blocked and vapor diffusivity chiefly conveys moisture inside the analysis domain. Only with short-term tested permeation, it is difficult to evaluate the concrete performance against moisture leakage for low water to cement ratio concrete. In this analysis, the calcium leaching was not coupled for simplicity.

### 4.4.2 Chloride Transport into Concrete Under Cyclic Drying-Wetting

Using the proposed method, transport of chloride ion under alternate drying wetting conditions is presented. For verification, experimental data by Maruya et al. (1998) is used. The size of mortar specimens is 5×5×10[cm]and the water to powder ratio is 50%. After 28 days of sealed curing, the specimens were exposed to cyclic alternate drying (7 days) and wetting (7 days) cycles. The drying condition was 60%RH, whereas the wetting was exposed to a chloride solution of 0.51 [mol/l] at 20℃. In FEM analysis, mix proportions and chemical compositions of cements are given. The curing and exposure conditions are also defined as boundary conditions for the target structures. All of these input values correspond to the experimental conditions. Fig. 28 shows the distribution of free and bound chlorides from the surface of exposure. Two cases are compared; one considering

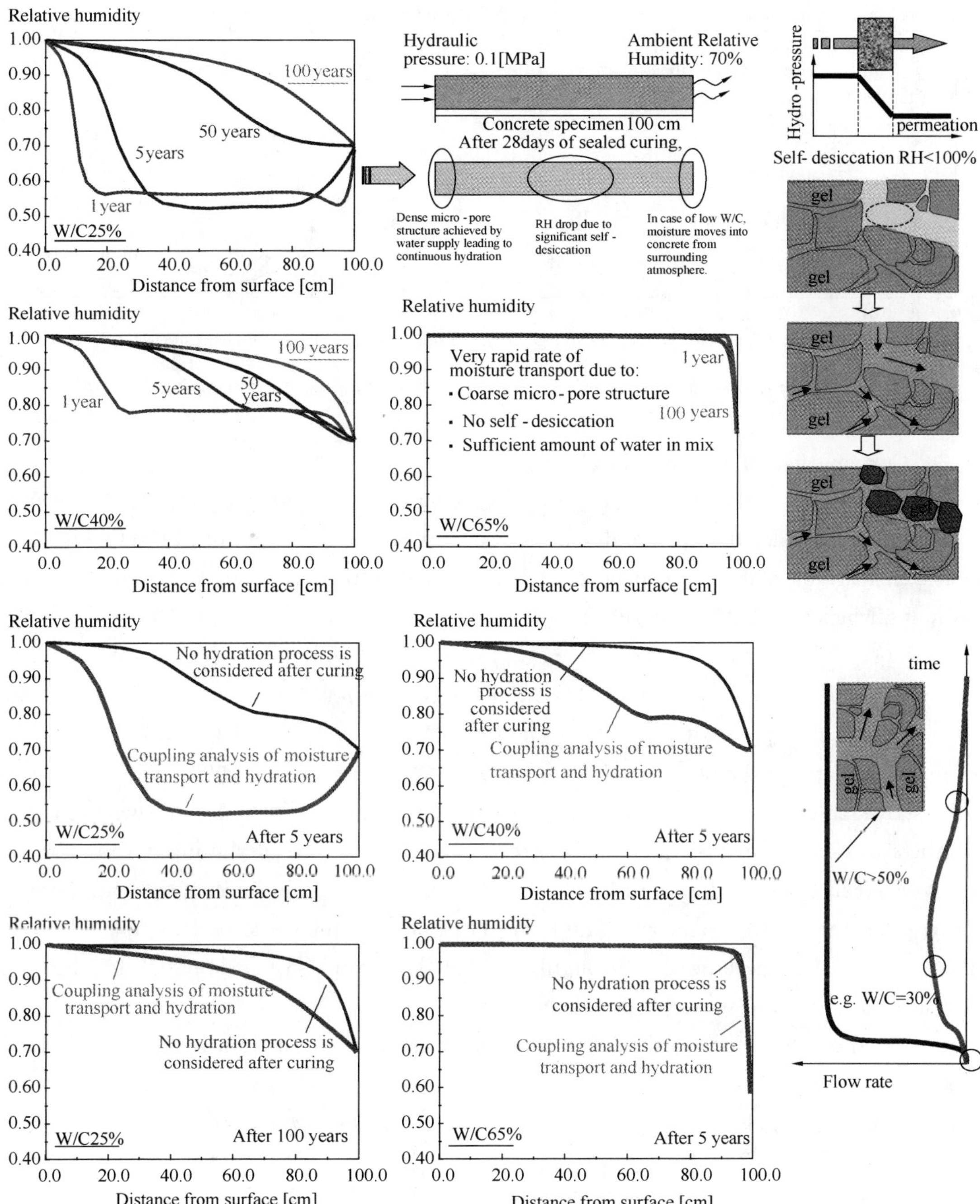


Fig. 27 Coupled moisture migration and transient evolution of micro-pore structures for low W/C

only diffusive movement and the other including the advective transport due to the bulk movement of pore water as well as the diffusion process. The distribution of bound and free chlorides can be reasonably simulated with advective transport due to the rapid suction of pore water under the wetting phase.

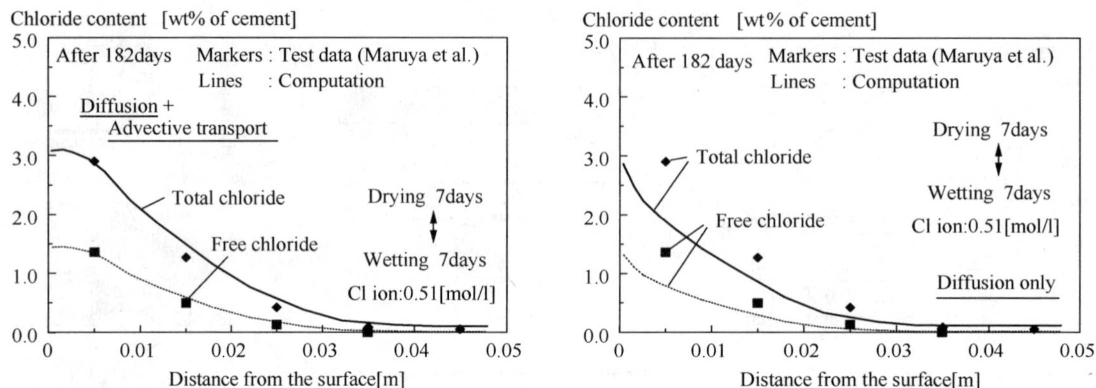


Fig. 28 Chloride content profile in concrete exposed to cyclic drying wetting and drying

### 4.4.3 Carbonation Phenomena in Concrete

In this section, computations were performed to predict the progress of carbonation for different $CO_2$ concentrations and water to cement ratio. The amount of $Ca(OH)_2$ existing in cementitious materials can be obtained by the multi-component hydration model as (Kishi and Maekawa, 1996 and 1997),

$$\begin{aligned} &2C_3S+6H \rightarrow C_3S_2H_3+3Ca(OH)_2 \\ &2C_2S+4H \rightarrow C_3S_2H_3+Ca(OH)_2 \\ &C_4AF+2Ca(OH)_2+10H \rightarrow C_3AH_6 \end{aligned} \tag{45}$$

For verification, the experimental data done by Uomoto et al. (1993) were used. Figure 29 shows the comparison of analytical results and empirical formula that was regressed with the *square root t equation*. Similar to the previous case, all of the input values in the analysis corresponded to the experimental conditions. Analytical results show the relationship between the depth of concrete in which pH in pore water becomes less than 10.0 and exposed time. The simulations can roughly predict the progress of carbonation for different $CO_2$ concentrations and water to powder ratios.

Fig. 30 shows the distribution of pH in pore water, $CO_2$, calcium hydroxide, and cal-

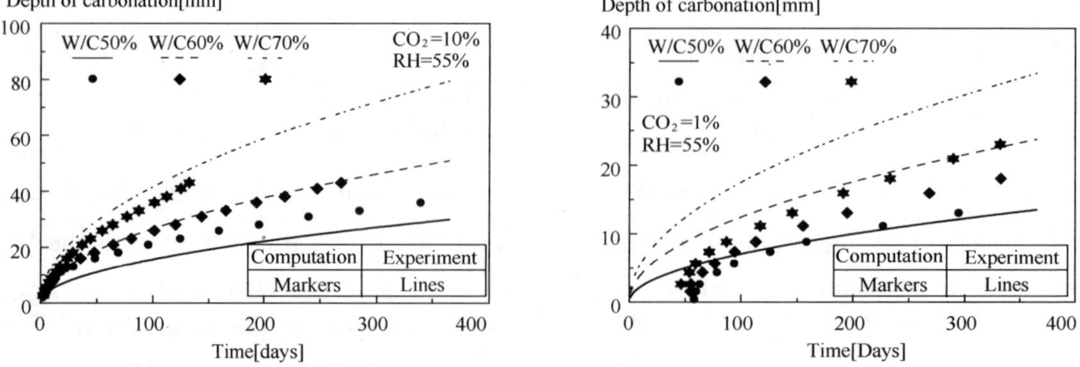


Fig. 29 Carbonation phenomena for different $CO_2$ concentrations and water-to-cement ratio

cium carbonate inside concrete, exposed to the $CO_2$ concentration of 3%. Two different water to powder ratio, W/C=25% and 50%, were analyzed. It can be shown that higher resistance for the carbonic acid action is achieved in the case of a low W/C.

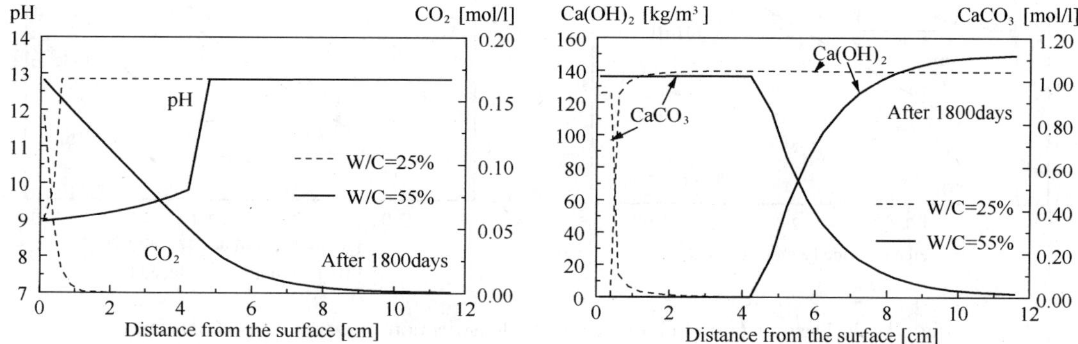


Fig. 30　Distribution of pH, calcium hydroxide, and calcium carbonate under the action of carbonic acid

### 4.4.4 Coupled Carbonation and Chloride Induced Corrosion

Attention is directed to steel corrosion in concrete due to the simultaneous attack of chloride ions and carbon dioxide. Provided that concrete is totally submerged, corrosion rate is small to a great extent due to the much lower supply of oxygen. Thus, drying is also coupled. Interaction of chloride and $CO_2$ sub-phases is modeled such that bound versus free chloride equilibrium is influenced by pH of pore solution. The authors assume pH sensitivity as shown in Fig. 31 from some literature and densification of micro-pores caused by carbonation is simply expressed by the reduction factor of porosity. This is to express the replacement of $Ca(OH)_2$ and calcite.

Figure 32 solely shows simulation. The carbonated zone close to the surface shows evidence of smaller total chloride due to released bound chloride by carbonation and loss of moisture by drying, but dissolved chloride ion is concentrated in the pore solution. Around the carbonation front, a couple of peaks of total chloride are formed. Here, the free chloride is released from formerly bound chloride. Since the carbonation brings densification of pores, chloride ion penetration is also held back and twin peaks can be formed just close to the carbonation front. In the future, detailed discussion will be beneficial in this regard.

For life-span assessment, concrete members with different water to powder ratios, W/C=40, 50, 60%, with only one face exposed to the environment were considered. The stage where concrete cracking occurs was defined as a limit state with respect to the steel corrosion. The progressive period until the initiation of longitudinal cracking was estimated by the equation proposed by Yokozeki et al. (1997), which is a function of cover depth. Fig. 33 shows relationships between cover depth and structural age until cracking due to corrosion obtained by the proposed thermo-hydro system. It can be seen that the concrete closer to the exposure surface will show an early sign of corrosion induced cracking, and

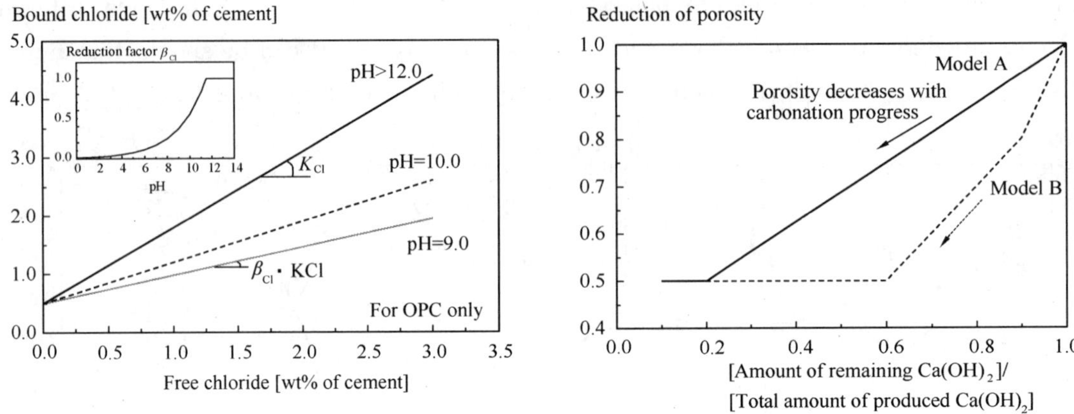


Fig. 31 Release of bound chloride and re-densification of pores by carbonation

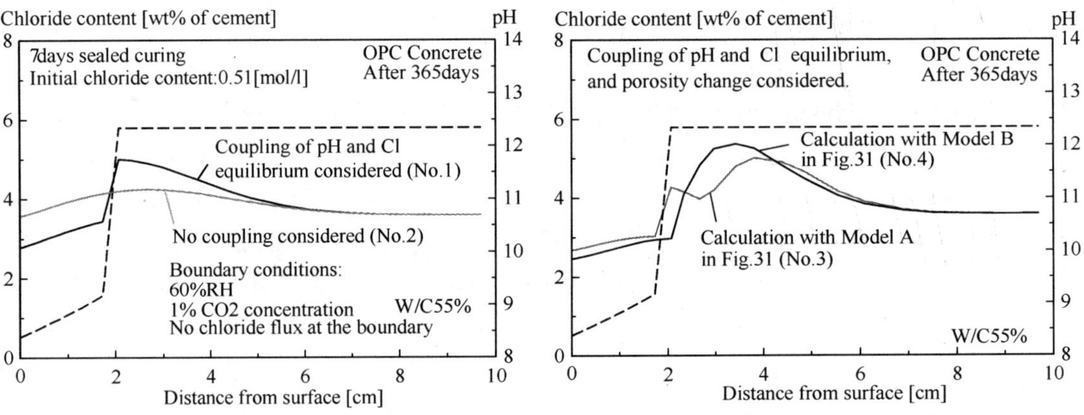


Fig. 32 Computation of total chloride contents under coupled drying, carbonation and chloride penetration

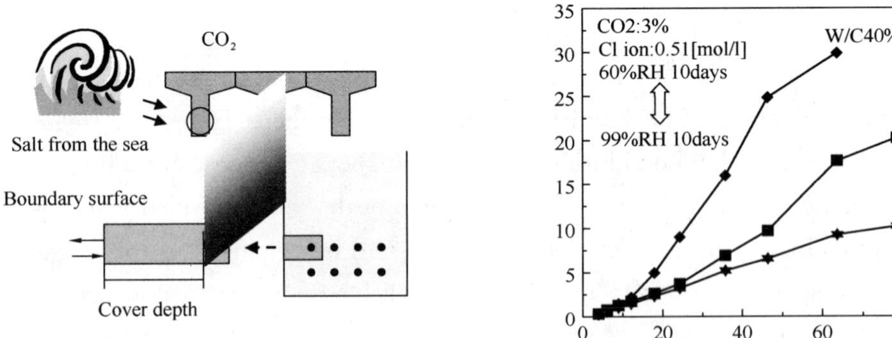


Fig. 33 Time till first sings of cracking due to corrosion for concrete exposed to $CO_2$ gas and salty water

low W/C concrete has the higher resistance against corrosion.

### 4.4.5 Moisture Distribution in Cracked Concrete

In this section, in order to show the possibility of a unification of structure and durability design, a simple simulation was conducted by using the proposed parallel computational system. It has been reported that there should be a close relationship between moisture conductivity and the damage level of cracked concrete; that is, moisture conductivity should be dependent on the crack width or the continuity of each crack. The proposed system, in which information is shared between the thermo-hydro and structural mechanics processes, is able to describe this behavior quantitatively by considering the inter-relationship between moisture conductivity and cracking properties. For representing the acceleration of drying out due to cracking, the following model proposed by Shimomura (1998) was used in this analysis.

$$J_w = \begin{cases} J_V + J_L & before\ cracking \\ J_V + J_L + J_V^{cr} + J_L^{cr} & after\ cracking \end{cases} \tag{46}$$

where, $J_w$ is the total mass flux of water in concrete, $J_V$ and $J_L$ are the mass flux of vapor and liquid in non-damaged concrete respectively, and $J_V^{cr}$ and $J_L^{cr}$ are the mass flux of vapor and liquid water through cracks. In this simulation, only $J_V^{cr}$ is taken into account for the first approximation, since diffusion of vapor would be predominant when concrete are exposed to drying conditions. From the experimental study done by Nishi et al. (1999), it has been confirmed that the flux $J_V^{cr}$ can be expressed as,

$$J_V^{cr} = -\bar{\varepsilon}\rho_V D_a \nabla h \tag{47}$$

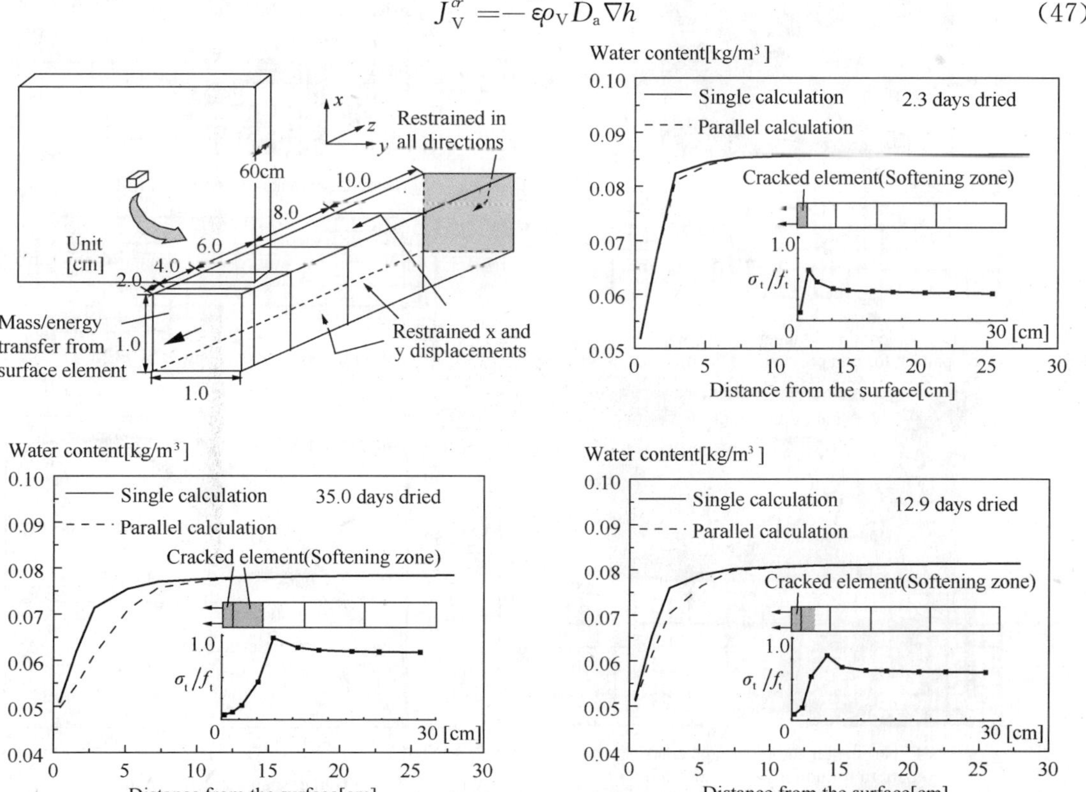


Fig. 34 Moisture and internal stress distribution in concrete exposed to drying condition

where, $\bar{\varepsilon}$: average strain of cracked concrete, which can be computed by COM3, $\rho_V$: density of vapor, $D_a$: vapor diffusivity in free atmosphere, $h$: relative humidity.

The target structure in this analysis was a concrete slab, which has 30% water to powder ratio using medium heat cement. Mesh layout and the restraint conditions used in this analysis are shown in Fig. 34. The volume of aggregate was 70%. After 3 days of sealed curing, the specimen was exposed to 50%RH. Figure 34 shows the cracked elements, the distribution of moisture, and normalized tensile stress at each point from the boundary surface exposed to drying condition. Moisture distribution calculated without stress analysis is also shown in Fig. 34. As shown in the results, the crack occurs from the element near the surface, and the crack progresses internally with the progress of drying. It is also shown that the amount of moisture loss becomes large due to cracking.

### 4.4.6 Coupling of Shear and Axial Forces Invoked by Volume Change

When thermally induced cracks penetrate through a whole section, a diagonal shear crack may be arrested by pre-existing damage induced by heat and/or drying as shown in Fig. 35. The stiffness of the member is certainly reduced by the crack penetrating the entire section, but the shear capacity is, on the contrary, elevated since the shear crack propagation is retarded or has ceased. Non-orthogonal crack-to-crack interaction can be well simu-

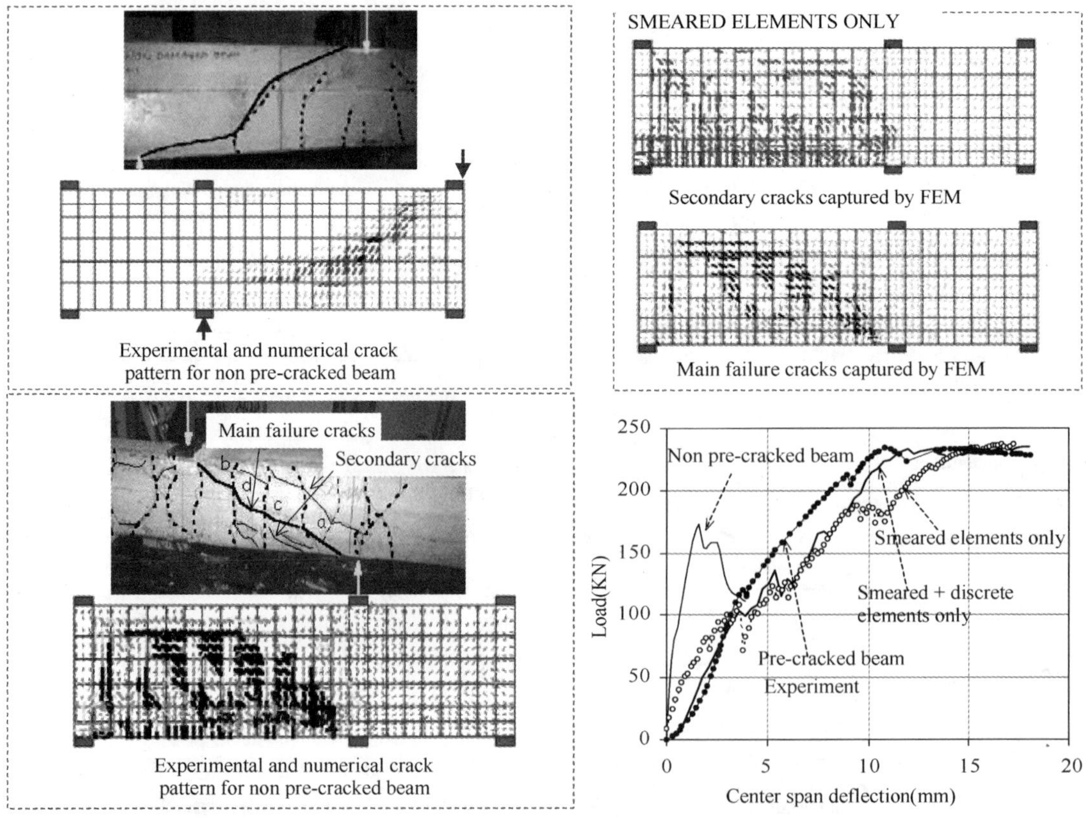


Fig. 35 Interaction of non-orthogonal cracking and safety performance simulation

lated and the localization pattern of force-induced cracks is greatly affected by the damage level of pre-cracks in terms of average strain as shown in Fig. 36. This is attributed to shear transfer performance of concrete crack planes. Here, the contact density model for concrete rough cracks is applied. The shear transfer which causes principal stress rotation is one of crucial mechanics for coupled force-environmental actions.

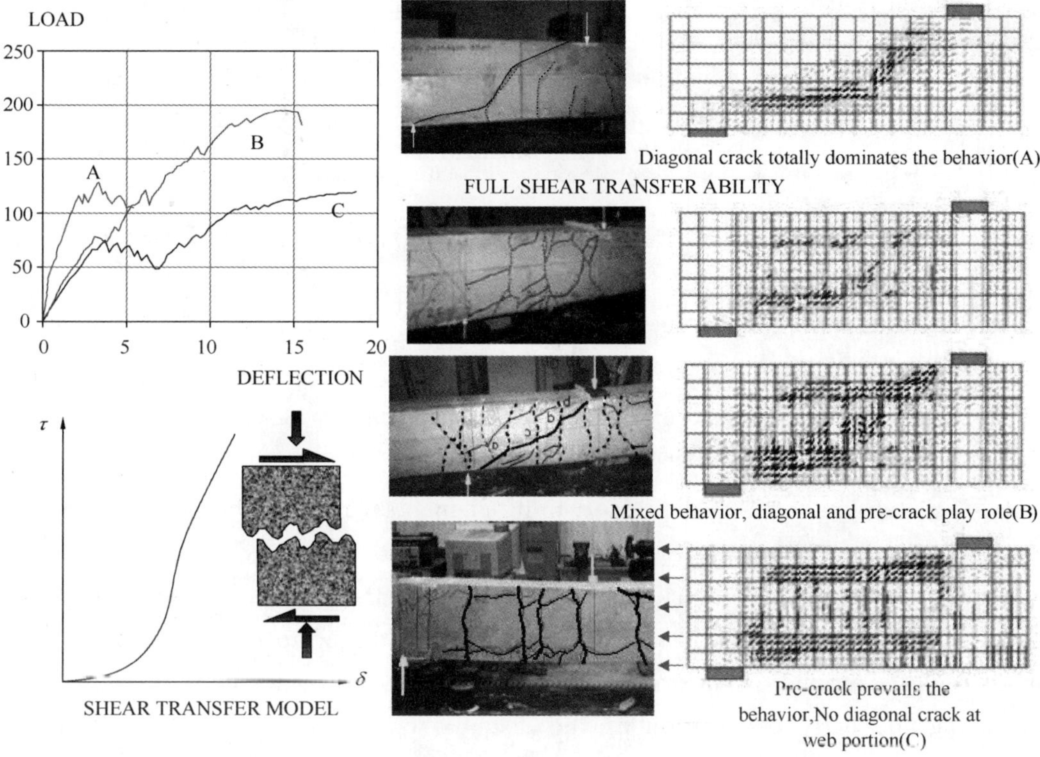


Fig. 36 Effect of pre-induced crack width on the shear capacity

### 4.4.7 Corrosion-Induced Cracking and Structural Capacity

Figure 37(a) shows the chloride ion penetration into the beam subjected to flexure. The bending cracks are distributed around the lower extreme fiber and the averaged tensile strain is transferred to a DuCOM sub-system. As the chloride ion diffusivity is formulated based upon both pore-structural characters (micro-information) and crack strain (macro-information) dependent on applied loads, the deep penetration is seen at the center span of the beam.

The cracking pattern of an RC beam subjected to shear and flexure is illustrated in Fig. 37(b). As the shear capacity is less than the flexural one, a localized shear crack band can be seen after failure. Firstly, partial corrosion is numerically reproduced by concentrated chloride with longitudinal cracking and followed by load. Here, the stiffness of corroded substance treated as perfect elasticity is assumed 7GPa and its volume expansion rate is set

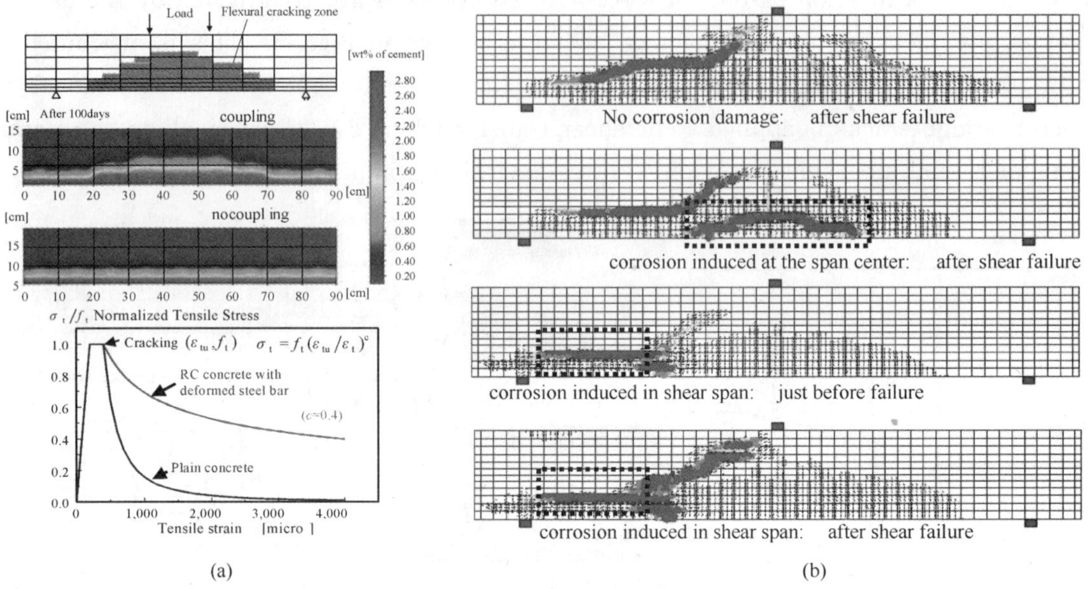


(a) (b)

Fig. 37 Corrosion induced cracking and remaining structural performance assessment

(a) Crack-diffusion interaction; (b) Shear and corrosion crack interaction

3 times. The tension-stiffness defined in finite elements with corroded steel is reduced to plain concrete softening since the bond is thought deteriorated.

If the corrosion induced cracking is located around the center span, no interaction of shear and flexural cracks is seen and shear capacity of the beam is computed unchanged. When it is placed around the shear span close to the support, diagonal shear crack joins this pre-cracking and early penetration of the diagonal crack is computed. Non-orthogonal cracking is the key of this analysis.

### 4.4.8 Time-Dependent Deflection of Beams under Drying

Two beams as shown in Fig. 38 are discussed concerning the load-drying combination. Under pure drying with no external forces, no deflection takes place in nature as no curvature is developed. Under the sealed condition without moisture loss, creep deflection is caused by concrete creep in compression and tension. When two actions are simultaneously combined, time-dependency of deflection is significantly stepped up as shown in Fig. 38. This apparent nonlinearity is known as the Pikket effect and practically, the creep function is specified differently from basic creep. In multi-scale analysis, there is no need to change material functions under different ambient conditions.

The drying effect is explicitly considered as a moist boundary condition, and the pore-water pressure is computed in DuCOM. The self-equilibrated stress and deformation concurrently solved space by space and creep properties of skeleton stress and strain are evaluated with micro-damage if any.

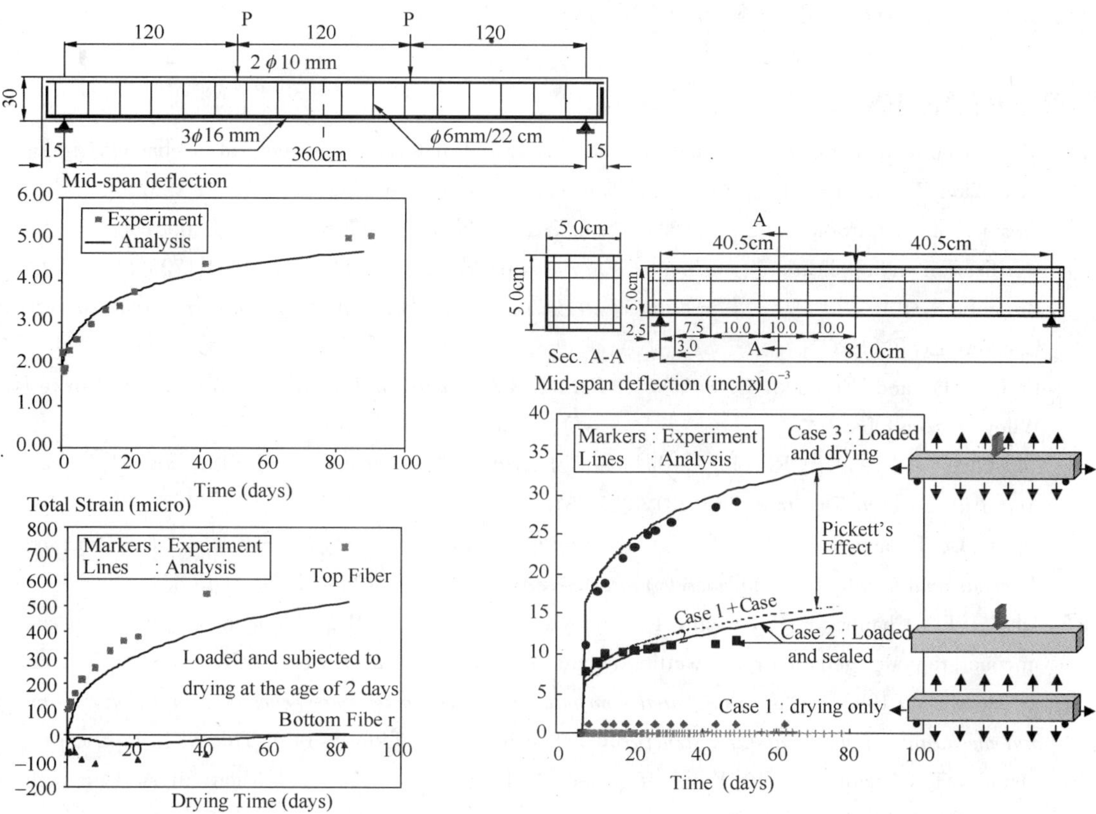


Fig. 38　Coupled sustained load and drying onto beams subjected to flexure

## 4.5　CONCLUSIONS

Chemo-physical and mechanical modeling of concrete with greatly different scales of geometry was presented, and synthesized on a unified computational platform from which quantitative assessment of structural concrete performances may derive. Experimental verifications show its possibility as a holistic approach, while individual modeling of chemical-physical events is required to be enhanced in the future with continuous effort. Currently granted is a great deal of knowledge earned by the past development. At the same time, we face a difficulty to quantitatively extract consequential figures from them. The authors expect that the systematic framework on the knowledge-based technology will be extended efficiently and can be steadily taken over by engineers in charge.

The authors express their sincere gratitude to Prof. Hajime Okamura of Kochi University of Technology for his continuing supervision and Dr. Tsuyoshi Maruya of Taisei Corporation for valuable discussion and suggestions. Cooperation of Dr. Soltani Mohamad, Dr. Rasha Mabrouk and Mr. G. Zhu, research fellows (1999-2002) of the University of Tokyo, is deeply appreciated. This study was financially supported by JSPS Grant-in-aid for scientific research 14205065 and 14655160.

# REFERENCES

[1] Multi-scale modeling of concrete performance-Integrated material and structural mechanics, Maekawa, K., Ishida, T. and kishi, T., Journal of Advanced Concrete Technology, 1(2) pp. 91-126, 2003.

[2] Bazant Z. P. and Prasannan S., (1989). "Solidification theory for concrete creep, I. Formulation, II.

[3] Verification and application" *Journal of Engineering Mechanics*, 115 (8), 1691-1725.

[4] Broomfield, J. P., (1997). "*Corrosion of steel in concrete Understanding, investigation and repair*" London: E&FN SPON.

[5] Freiser, H. and Fernando, Q., (1963). "*Ionic Equilibria in Analytical Chemistry*" New York: John Wiley & Sons, Inc.

[6] Glucklich J., (1959). "Rheological behavior of hardened cement paste under low stress" *Journal of American Concrete Institute*, 56 (23), 327-337.

[7] Gjørv, O. E. and Sakai, K., (1995) "Testing of chloride diffusivity for concrete", *In: Proc. of the International Conference on Concrete under Severe Conditions*, CONSEC95, 645-654.

[8] Ishida, T., Chaube, R. P., Kishi, T., and Maekawa, K., (1998). "Modeling of pore water content in concrete under generic drying wetting conditions" *Concrete Library of JSCE*, 31, 275-287.

[9] Ishida, T., (1999a). "*An integrated computational system of mass/energy generation, transport and mechanics of materials and structures*", Thesis (PhD). University of Tokyo (in Japanese).

[10] Ishida, T., Chaube, R. P., Kishi, T., and Maekawa, K., (1999b). "Micro-physical approach to coupled autogenous and drying shrinkage of concrete" *Concrete Library of JSCE*, 33, 71-81.

[11] Ishida, T. and Maekawa, K., (2000) "An integrated computational system for mass/energy generation, transport, and mechanics of materials and structures" *Concrete Library of JSCE*, 36, 129-144.

[12] Ishida, T. and Maekawa, K., (2001) "Modeling of pH profile in pore water based on mass transport and chemical equilibrium theory" *Concrete Library of JSCE*, 37, 131-146.

[13] Ishida, T. and Maekawa, K., (2002) "Solidified cementitious material-structure model with coupled heat and moisture transport under arbitrary ambient conditions", *Proceedings of fib congress* 2002, Osaka.

[14] Kishi, T. and Maekawa, K., (1996). "Multi-component model for hydration heating of Portland cement" *Concrete Library of JSCE*, 28, 97-115.

[15] Kishi, T. and Maekawa, K., (1997). "Multi-component model for hydration heating of blended cement with blast furnace slag and fly ash" *Concrete Library of JSCE*, 30, 125-139.

[16] Mabrouk, R., Ishida, T. and Maekawa, K., (1998) "Solidification model of hardening concrete composite for predicting autogenous and drying shrinkage", *In*: E. Tazawa Ed. *Autogenous Shrinkage of Concrete*. London: E&FN SPON, 309-318.

[17] Mabrouk, R., Ishida, T., and Maekawa, K., (2000). "A unified solidification model of hardening concrete composite at an early age" *Proceedings of JCI*, 22 (2), 661-666.

[18] Maekawa, K., Chaube, R. P., and Kishi, T., (1999). "*Modelling of Concrete Performance*" London: E&FN SPON.

[19] Maekawa, K., Pimanmas, A. and Okamura, H., (2003). "*Nonlinear Mechanics of Reinforced*

*Concrete*"London: Spon Press.

[20] Maekawa, K. and Ishida, T., (2002) "Modeling of structural performances under coupled environmental and weather actions"*Materials and Structures*, 35, 591-602.

[21] Maruya, T., Tangtermsirikul, S. and Matsuoka, Y., (1998). "Modeling of chloride ion movement in the surface layer of hardened concrete" *Concrete Library of JSCE*, 32, 69-84.

[22] Neville AM., (1959) "Creep recovery of mortars made with different cements" *Journal of American Concrete Institute*, 56 (13), 167-174.

[23] Nishi, T., Shimomura, T. and Sato, H., (1999). "Modeling of diffusion of vapor within cracked concrete" *Proceedings of JCI*, 21 (2), 859-864 (In Japanese).

[24] Pickett, G., (1956). "Effect of aggregate on shrinkage of concrete and hypothesis concerning shrinkage", *Journal of ACI*, 52, 581-590.

[25] Shimomura, T. and Maekawa, K., (1997) "Analysis of drying shrinkage behaviour of concrete using a micromechanical model based on the micropore structure of concrete" *Magazine of Concrete Research*, 49, No. 181, 303-322.

[26] Shimomura, T., (1998). "Modelling of initial defect of concrete due to drying shrinkage", *Concrete Under Severe Conditions* 2, *CONSEC* 98, 3, (1998), 2074-2083.

[27] Saeki, T., Ohga, H., and Nagataki, S., (1991). "Mechanism of carbonation and prediction of carbonation process of concrete" *Concrete Library of JSCE*, 17, 23-36.

[28] Takeda, H. and Ishida, T., (2000). "Chloride equilibrium between chemically combined and adsorbed components in cementitious materials" *Proceedings of JCI*, 22 (2), 133-138.

[29] Takegami, H., Ishida, T. and Maekawa, K., (2002) "Generalized model for chloride ion transport and equilibrium in blast furnace slag concrete" *Proceedings of JCI*, 24 (1), 633-638.

[30] Uomoto, T. and Takada, Y., (1993). "Factors affecting concrete carbonation ratio" *Concrete Library of JSCE*, 21, 31-44.

[31] Welty, J. R., Wicks, C. E., and Wilson, R. E., (1969). "*Fundamentals of momentum, heat, and mass transfer*" New York: John Wiley & Sons, Inc.

[32] West, J. M., (1986). "*Basic Corrosion and Oxidation*" 2nd Ed. New York: John Wiley & Sons, 1986.

[33] Yokozeki, K., Motohashi, K., Okada, K. and Tsutsumi, T., (1997)"A rational model to predict the service life of RC structures in marine environment", *Forth CANMET/ACI International Conference on Durability of Concrete*, SP170-40, 777-798.

# APPENDIX

Conservation laws (mass, energy, and momentum conservation)

Conservation of mass and energy

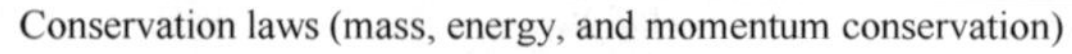

$$\alpha \frac{\partial X}{\partial t} + divJ_X - Q_X = 0$$

Moisture: $\left[\phi_{tot} S_{tot} \frac{\partial \rho_l}{\partial P_l} + \rho_l \frac{\partial(\phi_{tot} S_{tot})}{\partial P_l}\right] \frac{\partial P_l}{\partial t}$

Chloride ions: $\frac{\partial}{\partial t}\left[(\phi_{gl} + \phi_{cp}) S_{cg} C_{Cl}\right]$ Heat energy: $\rho c \frac{\partial T}{\partial t}$

Carbon dioxide: $\frac{\partial}{\partial t}\left\{\phi_{cg}\left[(1 - S_{cg}) \cdot \rho_{gCO_2} + S_{cg} \cdot \rho_{dCO_2}\right]\right\}$

Oxygen: $\frac{\partial}{\partial t}\left\{\phi_{cg}\left[(1 - S_{cg}) \cdot \rho_{gO_2} + S_{cg} \cdot \rho_{dO_2}\right]\right\}$

Balance equation

$$\frac{\partial \sigma_{ij}}{\partial x_j} + X_i = 0$$

**Mass, energy, and momentum transport terms**

Moisture

$J_w = -D_w \nabla P_l \quad D_w = K_l + K_v$

$K_l = \frac{\rho_l \phi_{tot}^2}{50\eta}\left(\int_0^{r_c} r dV\right)^2 \quad K_v = \frac{\rho_v \phi_{tot} D_0}{\Omega} \int_{r_{c1}}^{\infty} \frac{dV}{1+N_k} \quad N_k = \frac{l_m}{2(r - t_a)}$

$\eta = \eta_i \exp(G_e / RT) \quad G_e = G_{max} H_d \quad \ddot{H}_d + \left(\frac{1+\ddot{\eta}_e}{\eta_e}\right) H_d = \frac{RH}{\eta_e}$

$\eta_e = a(1 + bH_d^c) \quad a = \left[1.59\left(\frac{\phi_{fr}}{\phi_{cp} + \phi_{gl}}\right) + 0.7\right]^5 \quad b = 2.5a \quad c = 2.0$

After cracking

$D_w = 10 \cdot (K_l + K_v)$, $or = 50 \cdot (K_l + K_v)$

Chloride ion

$J_{Cl} = (\phi_{gl} + \phi_{cp}) S_{cg} \cdot \left(-\frac{D_{Cl}}{\Omega(B_{cp})} \nabla C_{Cl} + u_w C_{Cl}\right)$

$u_w = J_l / \rho_l = -(K_l \nabla P_l) / \rho_l$

Heat energy

$J_H = -K_H \nabla T$

Carbon dioxide

$J_{CO_2} = -(D_{gCO_2} \cdot K_{CO_2} + D_{dCO_2}) \nabla \rho_{dCO_2} \quad K_{CO_2} = \frac{M_{CO_2}}{RT} \cdot H_{CO_2}$

$D_{gCO_2} = \frac{\phi_{cg} \cdot D_0^g}{\Omega} \frac{(1 - S_{cg})^4}{1 + l_m / 2(r_m - t_m)} \quad D_{dCO_2} = \frac{\phi_{cg} S_{cg}}{\Omega} D_0^d$

Oxygen

$J_{O_2} = -(D_{gO_2} \cdot K_{O_2} + D_{dO_2}) \nabla \rho_{dO_2} \quad K_{O_2} = \frac{M_{O_2}}{RT} \cdot H_{O_2}$

$D_{gO_2} = \frac{\phi_{cg} \cdot D_0^g}{\Omega} \frac{(1 - S_{cg})^4}{1 + l_m / 2(r_m - t_m)} \quad D_{dO_2} = \frac{\phi_{cg} S_{cg}}{\Omega} D_0^d$

Momentum transport with mass transport is neglected

**Mass, energy, and momentum sink terms**

Moisture

$Q_p = Q_{pd} + Q_{hyd} = \rho_l \frac{\partial(\phi_{tot} S_{tot})}{\partial t} - W_{pow} \frac{\partial \beta}{\partial t}$

$\beta = \sum p_i \cdot \frac{\overline{Q}_i}{\overline{Q}_{i,\infty}} \cdot w_i$

Chloride ion

$Q_{Cl} = \frac{\partial C_{bound}}{\partial t}$

Heat energy

$Q_H = W_{pow} \sum p_i \overline{H}_i$

$\overline{H}_i = \gamma_i \cdot \beta_i \cdot \mu_i \cdot H_{i,T_0} \exp\left[-\frac{E_i(\overline{Q}_i)}{R}\left(\frac{1}{T} - \frac{1}{T_0}\right)\right]$

$\overline{Q}_i \equiv \int \overline{H}_i dt$

$\beta_i = f(\omega_{free}, \eta_i) \quad \eta_i = 1 - \left(1 - \frac{\overline{Q}_i}{\overline{Q}_{i,\infty}}\right)^{\frac{1}{3}}$

Carbon dioxide

$Q_{CO_2} = \frac{\partial(C_{CaCO_3})}{\partial t} = k_{reac} [Ca^{2+}][CO_3^{2-}]$

Oxygen

$Q_{O_2} = -\phi_{cg} S_{cg} \frac{M_{O_2} i_{corr}}{z_{O_2} F} \cdot \frac{A_{bar}}{V_{elem}}$

Momentum generation by product/loss of mass is neglected

Fig. A Conservation laws, mass/energy transport, and sink term: Mass and energy conservation laws to govern the thermo-physics of materials, as well as the balance equations ruling mechanics of structures. These conservation laws must be satisfied in all material systems and so they apply to the field of concrete materials. Figure a also shows the mass, energy, and momentum transport terms, and sink term, for the above equations.

State, compatibility, and constitutive law of mass, energy, and solid

Thermodynamic equilibrium of mass

Moisture

$$P_l = \frac{\rho_l RT}{M_l}\ln RH = -\frac{2\gamma}{r_c}$$

$$t_a = \frac{0.525\times10^{-8}\cdot RH}{(1-RH/RH_m)(1-RH/RH_m+15RH)}$$

$$RH_m = \exp\left[\frac{-\gamma M_l}{\rho_l RT(r-t_a)}\right]$$

$$S_{ads} = \int_{r_c}^{\infty}\left[1-\left(\frac{r-t_a}{r}\right)^2\right]dV$$

$$S_{cnd} = \int_0^{r_c} dV = 1-\exp(-Br_c) = S_c$$

$$S_{lr} = RH$$

(*Wetting stage*)

$$S_{cnd} = S_c + \int_{r_c}^{\infty}\left(\frac{S_c}{V}\right)dV = S_c[1-\ln S_c]$$

$$S_{lr} = RH^{0.05}$$

(*Drying stage*)

$$S_{cnd} = S_c + \int_{r_c}^{\infty}\left(\frac{S_{r_{min}}}{V}\right)dV = S_c - S_{r_{min}}\ln S_c$$

$$S_{lr} = S_{max}RH^{0.05}$$

(*Drying to wetting stage*)

$$S_{cnd} = S_c + \int_{r_c}^{r_{max}}\frac{S_c}{V}dV = S_c\left[1+\ln S_{r_{max}} - \ln S_c\right]$$

$$S_{lr} = 1+(RH-1)\left(\frac{S_{min}-1}{RH_{min}-1}\right)$$

(*Wetting to drying stage*)

$$S_{cg} = S_{ads} + S_{cnd}$$

$$S_{tot} = \frac{\phi_{lr}S_{lr} + (\phi_{gl}+\phi_{cp})S_{cg}}{\phi_{tot}}$$

$$\omega_{free} = \frac{\phi_{cp}S_{cp}(1-V_g)}{W_{pow}}$$

Chloride ion $\quad C_{bound} = C_{chem} + \beta_{Cl}\cdot C_{Cl}$

Carbon dioxide

$$\rho_{gCO_2} = \frac{M_{CO_2}}{RT}\cdot H_{CO_2}\cdot\rho_{dCO_2}$$

Oxygen

$$\rho_{gO_2} = \frac{M_{O_2}}{RT}\cdot H_{O_2}\cdot\rho_{dO_2}$$

Ion equilibria in pore solutions

$$H_2O \leftrightarrow H^+ + OH^-$$

$$H_2CO_3 \leftrightarrow H^+ + HCO_3^- \leftrightarrow 2H^+ + CO_3^{2-}$$

$$Ca(OH)_2 \leftrightarrow Ca^{2+} + 2OH^-$$

$$CaCO_3 \leftrightarrow Ca^{2+} + CO_3^{2-}$$

$$K_{sp}^1 = [Ca^{2+}][CO_3^{2-}] \quad K_{sp}^2 = [Ca^{2+}][OH^-]^2$$

$$K_w = [H^+][OH^-] \quad K_a = \frac{[H^+][HCO_3^-]}{[H_2CO_3]} \quad K_b = \frac{[H^+][CO_3^{2-}]}{[HCO_3^-]}$$

$$K_c = \frac{[Ca^{2+}][OH^-]^2}{[Ca(OH)_2]} = \frac{[Ca^{2+}]}{[Ca(OH)_2]}\cdot\frac{K_w^2}{[H^+]^2}$$

$$[H^+] + 2[Ca^{2+}] + 2[H_2CO_3]_s + [HCO_3^-]_s = [OH^-] + [HCO_3^-]_c + 2[CO_3^{2-}]_c$$

Electric potential of corrosion cell and current density equation

$$E_{Fe} = E_{Fe}^{\ominus} + (RT/z_{Fe}F)\ln h_{Fe^{2+}}$$

$$E_{O_2} = E_{O_2}^{\ominus} + (RT/z_{O_2}F)\ln(P_{O_2}/P^{\ominus})/h_{OH^-}^z = E_{O_2}^{\ominus} + (RT/z_{O_2}F)\ln(P_{O_2}/P^{\ominus}) - 0.06pH$$

$$E_0 = E_{O_2} - E_{Fe}$$

$$\eta^a = (2.303RT/\alpha z_{Fe}F)\log(i_a/i_0)$$

$$\eta^c = -(2.303RT/(1-\alpha)z_{O_2}F)\log(i_c/i_0)$$

$$i_L/z_{Fe}F = O_2^{sup} \qquad i_{corr} = i_L$$

$$R_{corr} = \phi S\frac{M_{Fe}\cdot i_{corr}}{z_{Fe}F}$$

Fig. B　State and compatibility law of mass, energy, and solid (No. 1): Figures B and C show the state, compatibility, and constitutive laws of mass, energy, and solid concrete. These equations are also modeled by considering the specific characteristics of concrete.

State, compatibility, and constitutive law of mass, energy, and solid

Pore structure formation

$$V_s=\frac{\alpha W_{pow}}{1-\phi_{ch}}\left(\frac{1}{\rho_p}+\frac{1}{\rho_w}\cdot\frac{\beta}{W_{pow}}\right)\qquad \phi_{lr}=(t_w s_l \rho_g)/2$$

$$\alpha=\sum p_i\cdot(\overline{Q}_i/\overline{Q}_{i,\infty})\qquad \phi_{gl}=V_s\phi_{ch}-\phi_{lr}$$

$$\phi_{cp}=1-V_s-(1-\alpha)\frac{W_{pow}}{\rho_p}\qquad \phi_{tot}=\phi_{lr}+\phi_{gl}+\phi_{cp}$$

$$\rho_g=\rho_p\rho_w(1+\beta/W_{pow})(1-\phi_{ch})/(\rho_w+\beta\rho_p)$$

$$A\delta_m^3+B\delta_m^2+C\delta_m+D=0$$

$$A=\{n(1-\phi_{in})+3(1-\phi_{ou})\}/\{3(n+3)\}$$

$$B=\{n(1-\phi_{in})+2(1-\phi_{ou})\}r_0/(n+2)$$

$$C=\{n(1-\phi_{in})+(1-\phi_{ou})\}r_0^2/(n+1)^2$$

$$D=-(\alpha r_0^3/3)[\phi_{in}+\beta\rho_p/\rho_w]$$

Before contact

$$\phi_{ou}=1.0$$

After contact

$$\phi_{ou}=1-(X+Y)/Z\qquad \delta_{max}=kr_o$$

$$X=-n(1-\phi_{in})\left[\frac{k^3}{3(n+3)}+\frac{k^2}{n+2}+\frac{k}{n+1}\right]$$

$$Y=\frac{\alpha}{3}\left[\phi_{in}+\beta\frac{\rho_p}{\rho_w}\right]\quad Z=\frac{k^3}{n+3}+\frac{2k^2}{n+2}+\frac{k}{n+1}$$

$$s_l=510f_{pc}+1500f_{sg}+3100f_{fa}$$

$$\zeta=19.0f_{pc}+1.5f_{sg}+1.0f_{fa}$$

$$SA_c=\frac{3\delta_m}{\zeta r_{eq}^3(1-\phi_{in})}(A\delta_m^2+B\delta_m+C)$$

$$SA_i=2\phi_i\int_{m}^{\infty}B_i\exp(-B_i r)\,d\ln r\qquad SA_g=W_s\cdot sa_g$$

$$\phi(r)=\phi_{lr}+\phi_{gl}\cdot V_{gl}+\phi_{cp}\cdot V_{cp}\qquad V_i=1-\exp(-B_i r)$$

Strength, elastic modulus, and deformability against pore pressure

$$f_c'=a\exp(-b\cdot V_{pore})\quad a,b:\text{constant}$$

$$V_{pore}=\phi_{cp}\cdot\exp(-B_{cp}\cdot r_{50})\quad r_{50}=50\times10^{-9}[\text{m}]$$

$$E=8.5\cdot10^3 f_c'^{1/3}\qquad f_t=0.27f_c'^{2/3}$$

Unrestrained strain due to temperature

$$\varepsilon_T=\alpha_c\Delta T$$

Unrestrained shrinkage strain due to pore pressure

$$\varepsilon_{sh}=\phi_{tot}S_{tot}f_{pow}\frac{P_l}{E_{sh}}$$

$$E_{sh}=E/3$$

Compatibility equation

$$\varepsilon_{xx}=\frac{\partial u}{\partial x}\qquad \varepsilon_{xy}=\frac{\partial v}{\partial x}+\frac{\partial u}{\partial y}$$

$$\varepsilon_{yy}=\frac{\partial v}{\partial y}\qquad \varepsilon_{yz}=\frac{\partial w}{\partial y}+\frac{\partial v}{\partial z}$$

$$\varepsilon_{zz}=\frac{\partial w}{\partial z}\qquad \varepsilon_{zx}=\frac{\partial u}{\partial z}+\frac{\partial w}{\partial x}$$

Constitutive law of solidifying concrete

$$\varepsilon_o=\rho_{ag}\varepsilon_{ag}+\rho_{cp}\varepsilon_{cp}\qquad \varepsilon_{cp}=f(\sigma_{cp})$$

$$\sigma_o=\rho_{ag}\sigma_{ag}+\rho_{cp}\sigma_{cp}\qquad \varepsilon_{ag}=\frac{1}{3.K_{ag}}\sigma_{ag}$$

$$\sigma_{cp}(t)=\int_{t'=0}^{t}S_{cp}(t',t)\,d\alpha(t')$$

$$\tau_o(t)=\int_{t'=0}^{t}\tau_o(t',t)\,d\alpha(t')$$

$$(1+\frac{E_v}{E_e})S(t)+\frac{C}{E_e}\frac{dS(t)}{dt}=E_v\bar{\varepsilon}'_{cp}(t)+C\frac{d\bar{\varepsilon}_{cp}(t)}{dt}$$

$$(1+\frac{G_v}{G_e})S_{ij}(t',t)+\frac{V}{G_e}\frac{dS_{ij}(t',t)}{dt}=G_v\cdot e'_{ij}(t',t)+V\frac{de'_{ij}(t',t)}{dt}$$

$$\sigma_{cp}=\sigma'_{cp}+\beta_f\cdot P_l$$

Fig. C State and compatibility law of mass, energy, and solid (No. 2)

Table I shows input values needed for the calculation. In addition to the list in Table I, the x, y, and z-coordinates of each node and the element type/shape of the analytical target are required for the analysis. These input values determine the initial value at time=0, and also determine some of the material constants shown in Table II. Other material constants are taken from experimental results. By solving the above governing equations under given material constants and initial conditions, the variables shown in Table III are obtained in 3D space and the time domain.

Appendix Table I List of input values

| Symbol | Description | Symbol | Description |
|---|---|---|---|
| $P_l$ | Initial pore pressure [Pa] | $V_g$ | Aggregate volume per unit volume [kg/m$^3$] |
| $C_{cl}$ | Initial concentration of Cl ions [mol/l] | $f_{pc}$ | Weight fraction of Portland cement |
| $T$ | Initial temperature [K] | $f_{sg}$ | Weight fraction of blast furnace slag |
| $\rho_{gCO_2}$ | Initial concentration of gaseous carbon dioxide[kg/m$^3$] | $f_{fa}$ | Weight fraction of fly ash |
| $\rho_{gO_2}$ | Initial concentration of gaseous oxygen [kg/m$^3$] | $p_{sp}$ | Amount of organic admixture |
| $p_i$ | Mass ratio of chemical component of cement ($C_2S$, $C_3S$, $C_4AF$, C3A, gypsum) | $BF$ | Blain fineness index |
| $\rho_p$ | Density of powder materials [kg/m$^3$] | $A_{bar}$ | Surface area of reinforcement in a finite element |
| W/P | Water-to-powder ratio | $V_{elem}$ | Volume of reinforcement in a finite element |
| $W_{pow}$ | Powder weight per unit volume [kg/m$^3$] | | |

Table II List of material constant

| Symbol | Description | Symbol | Description |
|---|---|---|---|
| $\rho_l$ | Density of liquid [kg/m$^3$] | $\overline{Q}_{i,\infty}$ | Maximum theoretical specific heat of component i [kcal] |
| $\rho_v$ | Density of saturated vapor [kg/m$^3$] | $W_{pow}$ | Powder weight per unit volume [kg/m$^3$] |
| $\rho_c$ | Heat capacity [kcal/K·m$^3$] | $\gamma_I$ | Reduction factor representing the retardation effect on hydration of fly ash and organic admixture |
| $D_0$ | Vapor diffusivity in free atmosphere [m$^2$/s] | $\mu_i$ | Coefficient representing the effect of mineral composition ($C_3S$, $C_2S$) on hydration rate |
| $D_{cl}$ | Chloride ion diffusivity in pore solution phase [m$^2$/s] | $T_0$ | Reference temperature (=293 K) |
| $D_0^g$ | Diffusivity of $CO_2$ in a free atmosphere | R | Gas constant [J/mol·K] |
| $D_0^d$ | Diffusivity of dissolved $CO_2$ in pore water | $k_{reac}$ | Reaction rate coefficient of carbonation |
| $H_{CO_2}$ | Henry's constant for carbon dioxide | $K_W$ | Equilibrium constant of concentration for water dissociation |
| $M_{CO_2}$ | Molecular mass of carbon dioxide | $K_a$ | Equilibrium constant of concentration for dissociation of $H_2CO_3$ and $HCO_3^-$[mol/l] |
| $\Omega$ | Parameter representing tortuosity of pore (=$(\pi/2)^2$) | $K_b$ | Equilibrium constant of concentration for dissociation of $HCO_3^-$ and $CO_3^{2-}$ [mol/l] |
| $\eta_i$ | Viscosity under ideal conditions [Pa·s] | $K_c$ | Equilibrium constant of concentration for dissociation of $Ca(OH)_2$[mol/l] |
| $l_m$ | Mean free path of a gas molecule [m] | $K_{sp}^1$ | Solubility-product constant of the calcium carbonate [mol] |
| $G_{max}$ | Maximum additional Gibbs energy for the activation of flow (=3500kcal/mol) | $K_{sp}^2$ | Solubility-product constant of the calcium hydroxide [mol] |
| $K_H$ | Heat conductivity [kcal/K·m. sec] | $F$ | Faraday constant [C/mol] |
| $p_i$ | Mass ratio of chemical component $i$ of cement | $z_{Fe}$ | Number of electric charge of Fe |
| $w_i$ | Amount of water consumed by chemical component $i$ due to hydration [mol/mol] | $z_{O_2}$ | Number of electric charge of oxygen |
| $E_{Fe}^{\ominus}$ | Referential standard cell potential of Fe at 25C (V, SHE) | $\zeta$ | Ratio of volume to the external surface area of a typical hydrate products [nm] |
| $E_{O_2}^{\ominus}$ | Referential standard cell potential of $O_2$ at 25C (V, SHE) | $f_{pc}$ | Weight fraction of Portland cement |
| $P^{\ominus}$ | Atmospheric pressure | $f_{sg}$ | Weight fraction of blast furnace slag |
| $M_{Fe}$ | Molecular mass of Fe | $f_{fa}$ | Weight fraction of fly ash |
| $i_0$ | Exchange current density | $t$ | Equivalent spherical cell radius [m] |
| $\rho_p$ | Density of powder materials [kg/m$^3$] | $sa_g$ | Specific surface area of hydrates [m$^2$/kg] |
| $\rho_w$ | Density of chemically hydrated products [kg/m$^3$] | $r_m$ | Minimum radius of pores [m] |
| $\phi_{in}$ | Specific porosity of inner products (=0.28) | $M_l$ | Molecular mass of liquid [kg/mol] |

| | | | |
|---|---|---|---|
| $t_w$ | Thickness of interlayer porosity (=2.8(10^{-10}[m]) | $\gamma$ | Surface tension of liquid [N/m] |
| $s_l$ | Specific surface area of interlayer porosity [$m^2/kg$] | $V_g$ | Volume of aggregate per unit volume [$m^3/m^3$] |
| $N$ | Parameter representing a generic pattern of deposition of products around particle | $f_{pow}$ | Coefficient of linear expansion of concrete [1/K] |
| $R_0$ | Radius of powder particle [m] | $\nu$ | Poisson's ratio |

Table III List of variables

| | | | |
|---|---|---|---|
| $T$ | Time [s] | $\eta_e$ | Effective non-ideal viscosity of the pore fluid[$N(s/m^2$] |
| $P_l$ | Pore pressure [Pa] | $G_e$ | Additional energy for activation of flow |
| $C_{cl}$ | Concentration of chloride ions [mol/l] | $H_d$ | Fictitious humidity parameter |
| $T$ | Temperature [K] | $r$ | Pore radius [m] |
| $\rho_{gCO_2}$ | Concentration of gaseous carbon dioxide [$kg/m^3$] | $r_c$ | Pore radius in which the equilibrated interface of liquid and vapor is created [m] |
| $\rho_{dCO_2}$ | Concentration of dissolved carbon dioxide [$kg/m^3$] | $V$ | Normalized pore volume [$m^3/m^3$] |
| $\rho_{gO_2}$ | Concentration of gaseous oxygen [$kg/m^3$] | $N_k$ | Knudsen number |
| $\rho_{dO_2}$ | Concentration of dissolved oxygen [$kg/m^3$] | $\boldsymbol{u}_w$ | Velocity vector of pore solution phase |
| $J_w$ | Flux of moisture [$kg/m^2 \cdot s$] | $J_l$ | Flux of liquid [$kg/m^2 \cdot s$] |
| $J_{cl}$ | Flux of chloride ions [$mol/m^2 \cdot s$] | $D_{gCO_2}$ | Diffusion coefficient of gaseous $CO_2$ in a porous medium [$m^2/s$] |
| $J_H$ | Flux of heat [$kcal/m^2 \cdot s$] | $D_{dCO_2}$ | Diffusion coefficient of dissolved $CO_2$ in a porous medium [$m^2/s$] |
| $J_{CO_2}$ | Flux of gaseous and dissolved carbon dioxide [$kg/m^2 \cdot s$] | $D_{gO_2}$ | Diffusion coefficient of gaseous $O_2$ in a porous medium [$m^2/s$] |
| $J_{O_2}$ | Flux of gaseous and dissolved oxygen [$kg/m^2 \cdot s$] | $D_{dO_2}$ | Diffusion coefficient of dissolved $O_2$ in a porous medium [$m^2/s$] |
| $D_w$ | Moisture conductivity [$kg/Pa \cdot m \cdot s$] | $Q_H$ | Heat generation term |
| $K_l$ | Liquid conductivity [$kg/Pa \cdot m \cdot s$] | $Q_p$ | Sink term for moisture balance |
| $K_v$ | Vapor conductivity [$kg/Pa \cdot m \cdot s$] | $Q_{pd}$ | Term representing bulk porosity change effects |
| $\eta$ | Viscosity of fluid under non-ideal conditions [$N(s/m^2$] | $Q_{hyd}$ | Term representing water consumption due to hydration |
| $Q_{cl}$ | Term representing the reduction of free chlorides | $S_c$ | Degree of saturation due to condensed water in virgin wetting path |
| $Q_{Cl}^{cb}$ | Term representing the extrication of bounded chlorides due to carbonation | $r_{max}$ | Pore radius of the largest pores that experienced a complete saturation in the wetting history [m] |
| $Q_{CO_2}$ | Term representing the consumption rate of carbon dioxide | $S_{rmax}$ | Highest saturation experienced by the porous media in its wetting history |
| $Q_{O_2}$ | Term representing the consumption rate of oxygen | $r_{min}$ | Pore radius of the smallest pores that experience emptying out in the drying history [m] |
| $\beta$ | Amount of chemical combined water per unit weight of hydrated powder materials [kg/kg] | $S_{rmin}$ | Lowest saturation of porous media in its wet-dry history |
| $\overline{Q}_i$ | Accumulated heat generation of chemical component $i$[kcal] | $S_{max}$ | Highest saturation experienced by interlayer porosity in its wetting history |
| $\overline{H}_i$ | Heat generation rate of clinker component $i$ [$kcal/kg \cdot s$] | $S_{min}$ | Lowest saturation experienced by interlayer porosity in its drying history |
| $H_{i,T_o}$ | Reference heat rate of i-th component at temperature $T_o$[$kcal/kg \cdot s$] | $RH_{min}$ | Minimum RH experienced in its drying history |
| $E_i(\overline{Q}_i)$ | Activation energy of component i[kcal. $K/kg \cdot s$] | $S_{cg}$ | Degree of saturation of gel and capillary pores |
| $\beta_i$ | Reduction of probability of contact between unhydrated compounds and free pore water | $S_{lr}$ | Degree of saturation of interlayer porosity |
| $\omega_{free}$ | Amount of free water | $S_{tot}$ | Degree of saturation of total pores |

| | |
|---|---|
| $\eta_i$ | Non-dimensional thickness of cluster |
| $V_s$ | Volume of hydrated products [$m^3/m^3$] |
| $W_s$ | Weight of hydrated products [$kg/m^3$] |
| $\alpha$ | Average degree of hydration |
| $\rho_g$ | Dry density of gel products [$kg/m^3$] |
| $\phi_{cp}$ | Capillary porosity |
| $\phi_{gl}$ | Gel porosity |
| $\phi_{lr}$ | Interlayer porosity |
| $\phi_{tot}$ | Total porosity |
| $\phi_{ou}$ | Porosity at outermost boundary of the expanding cluster |
| $\delta_m$ | Cluster thickness |
| $SA_c$ | Specific surface area of capillary porosity [$m^2/m^3$] |
| $SA_g$ | Specific surface area of gel porosity [$m^2/m^3$] |
| $B_i$ | Porosity distribution parameter |
| $RH$ | Relative humidiry |
| $t_a$ | Thickness of adsorbed layer [m] |
| $RH_m$ | Humidity required to fully saturate a pore |
| $S_{ads}$ | Degree of saturation due to adsorbed water |
| $S_{cnd}$ | Degree of saturation due to condensed water |
| $X_i$ | External force in i-direction |
| $u$ | Displacement in x-direction |
| $v$ | Displacement in y-direction |
| $w$ | Displacement in z-direction |
| $\varepsilon_T$ | Unrestrained strain due to temperature |
| $\sigma_{cp}$ | Mean volumetric stress of cement paste |
| $C_{bound}$ | Amount of bound chlorides [mol/l] |
| $C_{chem}$ | Amount of chemically bounded chlorides [mol/l] |
| $\beta_{Cl}$ | Equilibrium coefficient of free and bound chlorides |
| $C_{CaCO_3}$ | Concentration of $CaCO_3$ [mol/l] |
| $i_{corr}$ | Current density of corrosion cell [$A/m^2$] |
| $E_{Fe}$ | Standard cell potential of Fe (V, SHE) |
| $E_{O_2}$ | Standard cell potential of $O_2$ (V, SHE) |
| $E_0$ | Electric potential of corrosion cell (V) |
| $\eta^a$ | Charge transfer at anode [V] |
| $\eta^c$ | Charge transfer at cathode [V] |
| $i_a$ | Current density at anode [$A/m^2$] |
| $i_c$ | Current density at cathode [$A/m^2$] |
| $i_L$ | Limit current density under $O_2$ diffusion control [$A/m^2$] |
| $\varepsilon_{sh}$ | Unrestrained shrinkage strain due to pore pressure |
| $E_{sh}$ | Deformability against capillary stress [Pa] |
| $f'_c$ | Compressive strength [Pa] |
| $V_{pore}$ | Volume of capillary pores above 50nm |
| $E$ | Elastic module [Pa] |
| $f_t$ | Tensile strength [Pa] |
| $K_{ag}$ | Volumetric stiffness of aggregate |
| $\varepsilon_0$ | Mean volumetric strain on concrete |
| $\sigma_0$ | Mean volumetric stress of concrete |
| $\varepsilon_{ag}$ | Mean volumetric strain on aggregate |
| $\varepsilon_{cp}$ | Mean volumetric strain on cement paste |
| $\sigma_{ag}$ | Mean volumetric stress of aggregate |

# 第 5 章 Chapter 5

# 预制装配式混凝土结构的研究进展

**黄小坤，田春雨**

（中国建筑科学研究院，北京 100013）

**摘　要**：简要总结了我国预制装配式混凝土结构的应用现状；重点阐述了预制装配式混凝土结构体系以及框架结构、剪力墙结构、构件连接构造等关键技术及研究进展；对我国预制装配式混凝土结构的研究及应用前景进行展望。

**关键词**：预制装配式混凝土结构；结构体系；框架结构；剪力墙结构；连接构造

## 5.1　概述

我国预制混凝土结构研究和应用始于 20 世纪 50 年代，主要借鉴了前苏联等国家的技术体系；直到 20 世纪 80 年代，在工业与民用建筑中一直有着比较广泛的应用。在 20 世纪 90 年代以后，由于种种原因，预制混凝土结构的应用尤其是在民用建筑中的应用逐渐减少，经历了一个相对低潮阶段。随着国民经济的持续快速发展、节能环保要求的提高、劳动力成本的不断增长，近十年来，我国在预制装配式混凝土建筑方面的研究逐渐升温，多家单位开展了部分技术研究，并在万科企业股份有限公司、黑龙江宇辉建设集团、瑞安房地产有限公司、江苏中南建设集团股份有限公司等开发的项目中得到了一定规模的示范和应用（图 1）。

(a)

（b）

图 1　预制剪力墙结构的工程应用

(a) 外墙预制（北京万科）；(b) 全预制装配（宇辉集团）

近年来，中国每年竣工的城乡建筑总面积约 20 亿 $m^2$，其中城镇住宅超过 6 亿 $m^2$，大量采用了钢筋混凝土结构体系。长期以来，我国混凝土建筑主要采用现场施工的传统生产方式，设计建造粗放、现场作业条件差、劳动强度高、建筑材料损耗及建筑垃圾量大、建筑质量不稳定、建筑全寿命周期能耗高、工业化程度低，与国家推行的节能、减排、环保政策不协调。

采用预制装配式混凝土结构，可以有效节约资源和能源，提高材料在实现建筑节能和结构性能方面的效率，减少现场施工对场地等环境条件的要求，减少建筑垃圾和对环境的不良影响，提高建筑功能和结构性能，有效实现“四节一环保”的绿色发展要求，实现低能耗、低排放的建造过程，促进我国建筑业的整体发展，实现预定的节能、减排目标。目前，我国所发展的预制装配式混凝土结构，在建筑设计、构件生产、安装施工及结构受力模型、构件连接构造等方面均优于我国 20 世纪七、八十年代发展的传统装配式建筑结构，在建筑外观质量、节能效果、综合经济效益等多方面具有接近于、等同于现浇混凝土结构的性能。

本文主要对国内预制装配式混凝土结构的关键技术问题及研究现状进行总结，对其前景进行展望。

## 5.2　预制装配式混凝土结构体系

### 5.2.1　结构体系分类

我国 20 世纪七、八十年代采用的预制装配式混凝土结构主要是装配式大板住宅体系，以及预制圆孔板、大型屋面板、槽形板等预制构件的应用。由于在结构受力模型、构件连接方式、建筑结构设计、构件生产、安装施工等方面存在的问题，这些结构在抗震安全性、建筑物理性能、建筑功能等方面不同程度地存在着一些问题，在 20 世纪 90 年代已经逐渐地被淘汰。

目前所发展的预制装配式混凝土结构，应该是完全满足国家现行有关标准（包括抗震规范）的要求，甚至比现浇结构具有更好的安全性、适用性和耐久性的建筑结构体系。从国内外的研究和应用经验来看，可采用预制装配式框架结构、预制装配式剪力墙结构、预制装配式框架—现浇剪力墙（核心筒）结构体系。结构中承重构件可以全部为预制构件或者预制与现浇构件相结合。其中，预制装配式剪力墙结构可以分为全预制剪力墙结构、部分预制剪力墙结构和适当降低结构性能要求、适用于低多层建筑的剪力墙结构（以下简称为预制装配式大板结构）。

预制装配式框架结构及预制装配式框架—现浇剪力墙（核心筒）结构中的框架，梁、柱全部采用预制构件，预制承重构件之间的节点、拼缝连接均按照等同现浇结构要求进行设计和施工。预制装配式框架具有和现浇结构等同的性能，结构的适用高度、抗震等级与设计方法与现浇结构基本相同；可以结合预制外挂墙板应用，实现主要结构接近 100％的预制化率，尽量减少现场的湿作业。从结构分析、结构性能、构件生产及施工安装等方面考虑，预制装配式框架结构是最简单、最适合的结构体系，其瓶颈是我国现行规范中关于框架结构的最大适用高度偏低。

部分预制剪力墙结构主要指内墙现浇、外墙预制的结构，该结构目前在北京万科企业有限公司的工程中已经示范应用。由于内墙现浇，结构性能和现浇结构类似，适用高度较大、适用性好；采用预制外墙可以与保温、饰面、防水、门窗、阳台等一体化生产，充分发挥预制结构的优势。该体系的适用高度可参照现行有关标准中的现浇结构并适当降低。该结构体系是我国目前阶段较为实用的一种体系。

全预制剪力墙结构指全部剪力墙采用预制构件拼装装配。该结构目前在南通建筑工程总承包有限公司、黑龙江宇辉建设集团的工程中已经示范应用。预制墙体之间的拼缝连接可基本等同于现浇结构或者略低于现浇结构，需要通过设计计算满足拼缝的承载力、变形要求；对于连接构造低于现浇结构要求的，应在整体结构分析中考虑拼缝的不利影响。该结构体系的预制化率高，但拼缝的连接构造比较复杂、施工难度较大，目前的研究工作和工程实践还不充分，在地震区（尤其是高设防烈度区）的推广应用还需要进一步的研究工作。

参照日本和我国 20 世纪的经验，结合我国城镇化及新农村建设的需求，可研究开发一种新型的多层预制装配式剪力墙结构体系即预制装配式大板建筑体系。该结构体系可主要用于 6 层以下的低、多层建筑，预制墙板之间的拼缝构造可不按照等同现浇要求（比如只连接部分钢筋），施工简单、速度快，适用于各地区大量的多层住宅建设。这种结构体系尚需要进一步的研究、总结和完善。

### 5.2.2 结构布置及构件拆分要求

与现浇结构相比，预制装配式结构的平面布置、立面布置宜更加规则、均匀、简单，并应具有良好的整体性。平面长宽比不宜过大，局部突出或凹入部分的尺度也不宜过大；竖向抗侧力构件的截面尺寸和材料性能宜自下而上逐渐减小，避免抗侧力结构的侧向刚度和承载力竖向突变；竖向承重构件宜上下对齐，使结构传力途径直接、明确。

相关预制结构构件（柱、梁、墙、板）及建筑构件的划分，应遵循受力合理、连接简单、少规格、多组合、方便运输和施工，能组装成形式多样的结构系列原则。

### 5.2.3 结构体系的抗震性能研究及减震、隔震技术研究

新型预制装配式混凝土结构在国外已有成功的应用经验，如日本已有许多超过 100m 的预制混凝土超高层建筑，且经受了多次强震考验；在国内地震区基本没有应用经验和强震的考验；由于预制板等构件在国内地震中有不佳的安全性表现，使国内学术界及工程界对预制装配式混凝土结构的抗震安全性产生一定的担忧，也制约了预制混凝土结构的推广应用。因此，需要对结构整体的抗震安全性进行充分的试验研究及理论分析，提出适宜的抗震设计方法及抗震构造措施。

研究减震、隔震技术在预制混凝土结构中应用，可有效减小结构地震效应、减小构件截面尺寸、提高结构抗震性能，对预制混凝土结构的推广应用具有重要意义。

## 5.3 预制装配式框架结构的研究

### 5.3.1 关键技术

预制装配式框架结构，是指柱全部采用预制构件、梁采用预制叠合梁、楼板采用预制

叠合楼板的装配整体式结构体系。预制框架也可与现浇混凝土剪力墙、核心筒组成框架—剪力墙结构、框架-核心筒结构体系。框架结构中，梁、柱宜采用简单的“一”字形预制构件，便于生产、运输及安装；预制梁、柱构件在节点区拼接成等同于现浇的整体。该结构体系的关键技术主要有以下几点：

（1）框架梁柱节点的设计及构造

装配式结构中预制构件之间或者预制构件与现浇构件之间存在节点或接缝，当采取可靠的构造措施及施工方法保证这些节点或接缝的承载力、刚度和延性不低于现浇结构，使装配整体式结构的整体性能与现浇混凝土结构基本相同时，可将预制装配式结构称为等同现浇混凝土结构，并采用与现浇结构相同的方法进行设计设计。对于预制装配式框架结构，应主要通过保证梁柱节点的性能使其与现浇结构等同；否则，结构设计分析方法将变得非常复杂，难以在设计单位推广，成为制约其应用的最大问题。

梁柱节点的连接可分为干式连接和湿式连接。采用干式连接法，可能实现承载力及刚度与现浇结构类似，但是其延性及恢复力性能难以与现浇节点等同，因此不能应用于等同现浇的预制框架结构中；采用湿式连接，即节点区主筋及构造加强钢筋全部连接，节点区采用后浇混凝土、灌浆材料等将预制构件连为整体，可实现与现浇节点性能的等同。

（2）构件的新型配筋设计技术研发

高强及大直径钢筋的应用将减少构件中配筋数量，进而简化节点及拼缝处的连接构造，降低施工难度，加快施工进度，对促进预制装配式结构的推广应用具有重要意义。500MPa 级热轧带肋钢筋在混凝土结构中的应用研究已取得大量成果，在即将发布的国家标准《混凝土结构设计规范》GB 50010 中已有充分反映；新型配筋设计方法，如适当放松现行规范[1,2]中混凝土构件的纵向钢筋间距、箍筋肢距、箍筋间距等设计方法，还需要进行大量的研究工作。

### 5.3.2　研究进展

近年来，中国建筑科学研究院、同济大学、哈尔滨工业大学、东南大学等单位均进行过预制装配式框架结构的研究。

中国建筑科学研究院与万科企业股份有限公司合作，对预制框架结构的套筒浆锚连接技术、梁端摩擦剪机理、预制拼接叠合梁、预制拼装柱、预制拼装梁柱节点、预制叠合板进行了比较系统的试验研究（图 2）。试验中所采用的构件及节点设计技术主要基于日本前田建设公司所采用的装配式框架结构技术；所有试验均采用足尺试件，试件中柱的截面尺寸为 600mm×600mm～800mm×800mm，梁截面尺寸为 350mm×600mm，板厚为预制层 70mm＋现浇层 70mm。

通过试验研究，初步总结了预制装配式框架结构的构件及节点受力性能、设计技术及构造要求；试验结果表明，梁柱节点可以基本实现等同现浇结构的要求。在施工技术、构造措施上还需要进行进一步的研究，以适应国内现行有关规范及现阶段施工技术水平等的要求。

同济大学、东南大学、广州大学等单位也做出了重要的研究成果[3～7]，为预制装配式框架结构技术的发展和应用提供了技术储备。

图 2 预制装配式框架结构系列试验研究
(a) 预制拼装梁柱节点研究；(b) 结合面抗剪试验研究；(c) 预制拼装柱研究；
(d) 预制拼装叠合梁研究；(e) 预制拼装叠合板研究

## 5.4 预制装配式剪力墙结构的研究

### 5.4.1 关键技术

预制装配式剪力墙结构的关键结构技术主要有以下几点：

(1) 结构体系的研究

与国外不同的是，剪力墙结构是我国目前住宅建筑的主要结构形式。预制装配式剪力墙结构体系应紧密结合住宅产业化的要求，吸收国外装配结构的先进理念，在现有研究的基础上进行完善。主要研究方向包括预制外墙或者预制叠合外墙、内墙现浇的结构体系，全装配剪力墙体系，外部挂板和轻质填充墙体系，简化连接的多层预制装配式大板结构体系等；各种结构体系的适用高度及范围；预制工厂加工的墙体单元率与墙体连接部位现浇率的合适比例关系等。

（2）预制墙片节点及接缝连接设计技术研究

与装配式框架结构相比，装配式剪力墙结构中存在更大量的水平接缝、竖向接缝，这些接缝的受力性能直接决定结构的整体性能。不同规格钢筋的连接方法以及接缝的设计方法、构造要求、施工工法都需要进行系统的研究。

（3）装配式混凝土剪力墙结构体系的抗震性能研究

在节点及接缝研究的基础上，对装配式剪力墙结构的整体抗震性能进行研究。结合大比例剪力墙模型、装配式剪力墙空间结构模型试验与理论分析、计算机仿真模拟，研究基于新型装配式剪力墙结构的整体破坏机制、承载能力、刚度退化、变形恢复能力等抗震性能；开展新型装配式住宅结构体系结构足尺装配单元墙体和节点模型、带楼板的剪力墙平面模型的试验研究，验证新型装配式住宅结构体系的单元和局部承载能力、变形能力和抗震性能；研究适用于新型装配式剪力墙结构的填充墙体单元的耗能减震技术；基于大比例结构模型的拟动力试验，研究主体构件与填充墙体连接节点的耗能减震设置方案、实施效果和改进措施；针对新型预制装配式结构体系的特点，研究其减震设计方法。

（4）多层预制装配式大板结构体系研究

预制装配式大板结构体系主要用于 6 层以下的住宅建筑。预制墙板之间的拼缝构造可不按照等同现浇要求，可进行适当简化。这种结构体系类似于以前的装配式大板建筑，但是构造上有所区别，施工更简单，安全性、实用性和建筑物理性能更好。需要对这种结构体系的连接构造、结构性能进行系统的试验研究和理论分析，确定其设计方法和适用范围。

### 5.4.2　研究进展

中国建筑科学研究院、清华大学、东南大学、哈尔滨工业大学、合肥工业大学、安徽建筑工业学院等多家单位开展了预制剪力墙结构墙片及模型的试验研究工作（图 3）。通过研究，在结构的连接构造、抗震性能、设计方法、施工工法等方面取得了很多成果，部分省市还编制了地方标准；在万科集团、南通建筑工程总承包有限公司[8]、上海瑞安集团、黑龙江宇辉集团等的开发项目中开展了部分工程的示范应用（图 1）。

中国建筑科学研究院、万科集团等针对剪力墙结构预制外挂墙板的形式（内嵌式、外挂式）、连接构造方法、对整体结构性能的影响、弹性及弹塑性行为等进行了初步分析研究，取得了部分成果[9]，为进一步研究工作奠定了基础。

中国建筑科学研究院还进行了低矮剪力墙新型配筋方式的部分研究工作（图 4），研究表明对低轴压比、低剪应力剪力墙，适当扩大配筋间距、降低配筋率后，墙片基本呈现弯剪型破坏，具有规定的承载力和延性[10]。这部分工作还要进一步深入和完善，以促进其在预制装配式低矮剪力墙结构中的应用。

图 3　预制剪力墙结构试验研究

(a) 5 层 1/2 缩尺 L 形大板剪力墙结构；(b) 叠合板式钢筋混凝土剪力墙；
(c) 预制拼装剪力墙振动台试验研究（中国建筑科学研究院）试验研究（合肥工业大学等）试验研究（清华大学）；
(d) 墙片及 3 层足尺模型试验（哈尔滨工业大学）

图 4　低配筋率混凝土剪力墙性能试验研究

## 5.5 预制装配式结构的连接设计

### 5.5.1 一般要求

预制装配式混凝土结构依靠节点及拼缝将预制构件连接成为整体。连接节点的选型和设计应注重结构概念，通过合理的连接节点与构造，保证构件的连续性和结构的整体稳固性，使整个结构具有规定的承载能力、刚度和延性，以及良好的抗风、抗震和抗偶然荷载的能力，避免因偶然因素出现连续倒塌。

节点和连接应同时满足施工阶段和使用阶段的承载力、稳定性和变形性能要求；在保证结构整体受力性能的前提下，应力求连接构造简单、传力直接、受力明确、具有可操作性；所有构件承受的荷载和作用，应有可靠的传向基础的连续传递路径。承重结构中节点和连接的承载能力和延性不宜低于同类现浇结构，亦不宜低于预制构件本身，应满足“强剪弱弯，更强节点”的设计理念；预制构件的连接部位还应满足耐久性和防火、防水等要求。

### 5.5.2 连接的传力机理及承载力设计

节点、接缝压力可通过后浇混凝土、灌浆或座浆直接传递；拉力应由各式连接筋、预埋焊接件传递。不同的接缝具有不同的剪力传递途径：

(1) 对于剪力墙竖缝剪力，弹性阶段（裂缝前），主要靠界面黏结强度及混凝土键槽或者粗糙面的抗剪强度传递；弹塑性阶段（开裂后），主要为连接筋、销键等传递。当混凝土界面的黏结强度高于构件本身混凝土的抗拉、抗剪强度时，可视为等同于现浇混凝土，新旧混凝土可直接参与剪力传递。

(2) 剪力墙水平接缝及框架柱接头，轴压应力和弯矩产生的压应力的静摩擦力，是主要的剪力传递方式；连接筋、销键是保证节点、接缝具有较高剩余抗剪强度和延性的关键要素。

(3) 框架梁、连系梁接头，主要靠界面黏结强度及混凝土键槽或者粗糙面的抗剪强度、销键、连接筋及弯矩压应力的静摩擦力共同传递剪力。

对于装配式混凝土结构节点、接缝，应进行受剪承载力计算。当节点、接缝灌缝材料（如结构胶）的抗压强度、黏结抗拉强度、黏结抗剪强度均高于预制构件本身混凝土的抗压、抗拉及抗剪强度时，节点、接缝配筋又高于构件配筋时，可不进行节点、接缝连接的受剪承载力计算，只需按一般现浇结构验算构件本身斜截面受剪承载力即可。

装配式混凝土结构节点、接缝受压、受拉及受弯承载力，可按现行国家标准《混凝土结构设计规范》GB 50010 的相应规定计算，其中节点、接缝混凝土等效抗压强度，可取实际参与工作的构件和后浇混凝土中的较低值。当节点、接缝所配钢筋及后浇混凝土强度高于构件本身且符合构造规定时，可不必进行节点、接缝的受压、受拉及受弯承载力计算。

### 5.5.3 连接的构造要求

装配式结构中，节点及接缝处的钢筋连接宜采用机械连接、套筒浆锚连接及焊接连接，也可采用搭接浆锚连接。剪力墙竖缝处，钢筋一般锚入现浇混凝土中；剪力墙水平接缝及框架柱接头，钢筋宜采用套筒浆锚连接或者搭接浆锚连接；框架梁接头与框架梁柱节点处，水平钢筋宜采用机械连接或者焊接。

连接节点应采取可靠的防腐蚀措施，其耐久性应满足工程设计使用年限要求。所有外露金属件，包括连接件和预埋件的设计均应考虑环境类别的影响，并进行防腐防锈处理；有防火要求时应采取可靠的防火措施。

(1) 套筒浆锚连接

采用套筒浆锚连接时，套筒抗拉承载力应大于被连接钢筋的抗拉承载力；套筒长度由砂浆与连接筋的握裹能力而定，要求握裹承载力不小于连接筋抗拉承载力。套筒浆锚连接钢筋一般不需搭接。套筒净距不宜小于 25mm。连接筋与套筒位置应完全对应，误差不得大于 2mm。连接筋插入套筒后压力灌浆，待浆液充满全部套筒后，停止灌浆并做好养护。

中国建筑科学研究院对日本、中国台湾产的套筒及配套高强砂浆进行了套筒灌浆连接的试验研究（图 5）。结果表明，套筒灌浆连接技术可以满足钢筋连接的要求。进口的套筒需要进行适当的改造并国产化，以更适应国内钢筋的构造并降低成本，目前北京万科公司在这方面已进行了富有成效的工作。

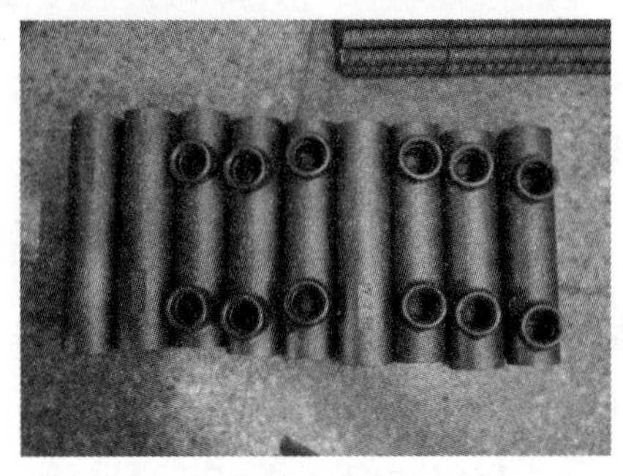

图 5 钢筋套筒灌浆连接试验研究

(2) 搭接浆锚连接

采用搭接浆锚连接时，钢筋搭接长度与钢筋外形和直径、灌浆材料性能、搭接孔形状、混凝土约束情况等有关，应根据连接构造通过试验研究确定；无试验依据时，非抗震设计时一般不小于 $25d$，抗震设计时一般不小于 $30d$，$d$ 为连接筋直径。锚浆孔的边距不宜小于 $5d$，净距不应小于 $(30+d)$ mm，孔深应比锚固长度长 50mm。在锚固区，锚孔及纵筋周围宜设置螺旋箍筋加强约束。连接筋插入锚孔后采用压力灌浆，待浆液充满全部锚孔后，停止灌浆并做好养护。

(3) 焊接连接

预制装配式混凝土结构中，钢筋的焊接连接是不可避免的，主要是对框架梁等水平结构构件。采用焊接连接时，应深入研究钢筋焊接质量的检测技术、设备和方法，还应进一步研究大直径钢筋的焊接技术。

## 5.6 展望

新型预制装配式混凝土结构的研究和应用已经取得了一定的进展，但是还存在关键技术不完备、不系统，缺乏支撑规模化应用的集成技术及技术标准，规模化推广受到很大限制。

目前预制装配式混凝土结构的发展迎来了良好的机遇：国家政策支持，符合建设资源节约型、环境友好型社会的要求；采用预制装配式结构可缩短总工期，加快资金周转速度，提高资金使用效率，提高建筑性能和工程质量，降低使用过程的管理成本、维修成本，提高建筑的科技含量，可大幅度减少现场施工中的模板、钢筋、混凝土工程量和劳动力用量，较好地应对劳动力成本上升的问题，可实现外装修、内装修与混凝土受力构件一次制作，提高使用功能和施工质量，可大幅度减少建筑垃圾，降低噪声污染，节约用水，可大幅度改善作业条件、降低劳动强度。

为了更好地进行这方面的工作，争取国家的大力支持，目前由中国建筑科学研究院牵头正在组织申报国家“十二五”科技支撑计划项目，主要针对预制装配式框架结构及剪力墙结构两种体系的关键技术进行比较系统的研究工作，并结合构配件生产技术及安装施工工艺、技术经济政策的研究，进行预制混凝土建筑结构的技术集成与示范应用，实现以改变房屋建造方式为主题，以提高工程质量、缩短工期、减少资源浪费和环境污染为目标，推动新型预制装配式混凝土建筑体系在我国的推广应用，加快建筑生产的工业化、产业化发展进程。

## 参考文献


[1] 中华人民共和国国家标准《混凝土结构设计规范》GB 50010—2002. 中国建筑工业出版社，2002，北京.

[2] 中华人民共和国国家标准《建筑抗震设计规范》GB 50011—2001(2008 版). 中国建筑工业出版社，2008，北京.

[3] 吕西林，范力，赵斌. 装配式预制混凝土框架结构缩尺模型拟动力试验研究. 建筑结构学报，2008，29(4)：58-65.

[4] 薛伟辰，杨新磊，王蕴，窦祖融. 六层两跨现浇柱预制梁框架抗震性能试验研究. 建筑结构学报，2008，29(6)：29-32.

[5] 朱宏进. 预制预应力混凝土装配整体式框架结构(世构体系)节点试验研究. 东南大学硕士论文，2006.

[6] 潘其健. 预制预应力混凝土框架结构抗震能力的试验研究. 东南大学硕士论文，2006.

[7] 李楠，张季超，楚先锋，刘波. 预制混凝土结构后浇整体式梁柱节点抗震性能试验研究. 工程力学，2009(S1)：41-44.

[8] 张军，侯海泉，董年才，龚俊杰，郭正兴. 全预制装配整体式剪力墙住宅结构设计及应用. 施工技术，2009，38(5)：22-24.

[9] 万科公司 PC 外墙性能分析报告. 建研科技股份有限公司，2010，北京.

[10] 孔慧. 配筋率对混凝土剪力墙结构性能影响的试验研究. 中国建筑科学研究院硕士学位论文，2010，北京.

# 第 6 章 Chapter 6

# 混凝土结构耐久性的使用寿命预测研究新进展*

金伟良，许　晨，金立兵
（浙江大学结构工程研究所，杭州 310058）

**摘　要：**对新建或已建的混凝土结构进行耐久性的寿命预测时，传统的寿命预测方法常常未能考虑实际暴露环境对研究对象的作用效应而导致预测结果的可信度不足；而室内人工气候模拟加速试验方法虽能模拟实际环境的作用效应，但由于研究对象现场性能劣化数据较少或甚至没有，因此仍然无法可靠地建立室内实验和实际暴露环境两者之间的劣化加速关系。如何有效地建立室内快速实验与实际暴露环境和性能劣化两者之间的相关性，这已成为国内外混凝土结构耐久性研究的难点和关键问题。为此，本文提出了一种新的混凝土结构耐久性使用寿命预测方法，通过引入与研究对象处于相同或相似暴露环境且具有较长服役年限的混凝土结构参照物，借助于参照物大量的现场性能劣化数据以及研究对象和参照物的室内模拟加速试验，预测研究对象在实际暴露环境下的性能劣化参数，从而对研究对象作出可靠的使用寿命预测。最后，结合杭州湾跨海大桥的工程实例，对此新方法进行了详细阐述与分析。

**关键词：**混凝土结构耐久性；氯离子；氯离子扩散系数；表面氯离子浓度；氯离子阈值；人工气候模拟加速试验；使用寿命预测

# STATE-OF-THE-ART ON SERVICE LIFE PREDICTION OF CONCRETE STRUCTURES

W. L. Jin，C. Xu，L. B. Jin
(Institute of Structural engineering，Zhejiang University，Hangzhou，310058，China)

**Abstract**：Traditional methods of life prediction，such as empirical formula models or numerical simulation methods，which take no consideration of the effect of physical environmental action，lead to incredible predicted results. Although effect of physical environmental action can be simulated through accelerating indoor artificial climate simulating experiment，relationship of performance deterioration between physical environmental and indoor test can still not be established，because deterioration data of newly built or even under building structures in physical environmental are little. How to effectively establish this relationship has been the difficult and key issues. So a new method is put forward with introducing an reference structure loa-

* 基金项目：国家自然科学基金重大国际合作项目（50920105806），国家科技支撑项目（2006BAJ03A04，2006BAJ03A02）

cating just near the object structure. Due to the same exposure environment, by detecting on the reference structure, lots of performance deterioration imformation can be got. Furthermore, indoor artificial climate simulating experiment is carried out simultaneously. Finally, the performance deterioration imformation of the object structures can be predicted according to the similarity relationship. To demonstrate the proposed method, example of Hangzhou Bay Bridge is provided and the results are compared with those obtained by using other methods.

**Keywords**: concrete structure durability; chloride ion; chloride diffusion coefficient; surface chloride concentration; chloride threshold; artificial climate simulating experiment; service life prediction

## 6.1 引言

所谓混凝土结构的耐久性，是指混凝土结构及其构件在可预见的工作环境及材料内部因素的作用下，在设计要求的使用年限和正常维修条件下，能够抵御长期性能劣化并保持其安全性和适用性的能力[1]。混凝土结构耐久性包含三个基本要素，即环境、功能、经济。环境是指结构处于某一特定环境（包括自然环境、使用环境、长期作用）中，并受其侵蚀作用；功能是指结构耐久性是一个结构多种功能（安全性、适用性等）与使用时间相关联的多维函数空间。因此，并不能简单地将耐久性理解为结构适用性的一部分，耐久性贯穿于结构整个设计使用寿命，并且影响着结构的安全性与适用性，是结构在使用寿命期间性能的动态表现；经济是指结构在正常使用过程中不需要大修。从工程结构全寿命期管理（Life Cycle Management）[2,3]来看，若预测的结构使用寿命远大于实际寿命，往往会使后期的维护加固工作还来不及开展时结构已经出现了耐久性失效；若预测的结构使用寿命远小于实际寿命就会过早以及频繁地进行后期维护加固工作；以上两种情况均会导致工程结构寿命期成本的大大增加。因此只有在可靠预测工程结构的使用寿命基础上，才能在经济最优的原则和符合结构目标可靠度水平的条件下，对工程结构在设计、施工、运营及老化等时期进行各项管理决策[4]。目前，耐久性问题普遍集中在海工混凝土结构钢筋锈蚀引起的保护层开裂和脱落，而诱发钢筋锈蚀的主要原因就是海水及沿海大气中的氯离子侵蚀。通常，对于建成已久的海工混凝土结构物，通过对结构物进行现场的耐久性检测可以获得充分可靠的劣化数据从而对结构的剩余寿命进行有效的预测。但是，对于新建或在建海工混凝土结构物，由于研究对象的现场劣化数据较少甚至没有，无法对结构物使用寿命进行可靠预测。传统的寿命预测方法，如经验公式法或数值模拟方法[5~10]，其模型参数往往基于一般材料通过室内试验得到，并未考虑实际暴露环境（包括荷载）对结构物的作用历程，导致模型预测结果可靠性大大降低。近些年逐渐发展起来的室内人工气候模拟加速试验方法[11~14]，虽能模拟实际环境的作用效应，但由于研究对象现场劣化数据较少甚至没有，仍然无法可靠建立两个环境之间的加速劣化时间关系。

## 6.2 需要回答的问题

大规模的混凝土基础工程建设迫使我们需要回答以下两个问题：一、如何确保能达到工程设计的使用寿命？二、工程建造过程和完工后又如何预测能否达到或大于 100 年的设计寿命以及对剩余寿命的实时监测？在此，本文主要针对第二点问题进行回答。由混凝土结构耐久性的定义可知，混凝土结构耐久性能的退化与外界环境的持续作用是不可分割的，脱离实际环境来预测结构使用寿命是不可靠的。然而，对于某一在建或新建的结构物，其在实际环境作用下的相关耐久性损伤信息很少，特别是对于当今实际工程中广泛应用的高性能混凝土来说是几乎检测不到的。因此，若要对结构进行寿命预测就必须通过相关的室内加速模拟试验。虽然，室内模拟试验能对实际环境作用下的相关劣化过程进行有效地加速模拟，但是仍然无法对结构寿命进行预测，原因有如下：(1) 室内加速模拟环境较为单一，只能对与主要劣化机理相关的主要的劣化因素进行模拟；(2) 由于实际环境条件下的劣化信息很少，仍然无法准确建立室内环境对于实际环境的劣化加速关系；(3) 室外环境瞬息万变，无法考虑整个外界环境作用历程。针对以上分析，提出了相应的解决方法——多重环境时间相似试验方法（Multi-Environmental Time Similarity，METS）[15~17]。通过选取与研究对象具有相同或相似环境且具有一定使用年限的参照物，利用参照物和研究对象环境条件的相似性，因而研究对象和参照物的性能劣化具有相似性；通过对参照物进行现场检测试验以及研究对象和参照物对应的试件模型进行室内加速试验研究，基于经典的相似理论预测研究对象在实际环境作用下的劣化机理方程控制参数，从而对研究对象进行有效地寿命预测如图 1 所示。METS 方法的创新之处在于引入了与研究对象处于相同或相似暴露环境的参照物。该参照物通常已有较长服役年数并已出现了相关的耐久性失效问题，通过对参照物的检测可以获得充分的劣化数据，并且可以认为相关的劣化参数在长时间的暴露时间下已经达到稳定。如此借助于人工气候模拟加速试验便能对研究对象进行有效的寿命预测。

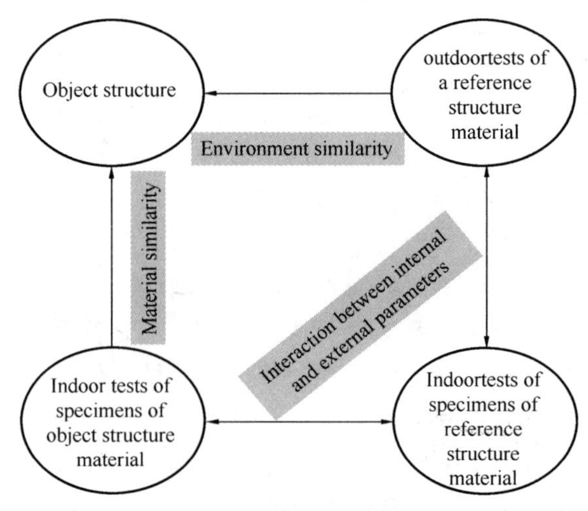


图 1 多重环境时间相似方法原理图

## 6.3 混凝土结构使用寿命预测新方法—METS

根据图 1，利用 METS 方法对结构进行寿命预测的基本步骤如下：

(1) 选取与研究对象具有相同或相似环境条件的具有一定使用年限并已出现相关耐久性失效的参照物，分析相关的劣化机理及影响该机理的自然环境因素，选择描述该劣化机理的控制方程，如式 (1) 所示：

$$y = f(\alpha_i, \alpha_j, \cdots, x, t) \tag{1}$$

式中：$y$ 为劣化量；$f$ 为劣化机理控制方程；$\alpha_i$，$\alpha_j \cdots$ 为劣化机理控制参数。以氯盐侵蚀为例，式中 $f$ 即为 Fick 第二定律的解析式；$\alpha_i$，$\alpha_j \cdots$ 为混凝土氯离子扩散系数，混凝土表面氯离子浓度等参数；$y$ 为引起钢筋锈蚀的临界氯离子浓度。

（2）收集研究对象、参照物的荷载作用与环境、气象、水文资料，并运用力学数学方法对各参数进行分析、计算，得到现场条件下研究对象和参照物荷载作用、环境作用的主要参数资料。

（3）根据主要劣化参数劣化机理的研究，对（2）中得到的主要参数资料进行人工气候加速模拟，得到室内环境下的试验主要控制参数。

（4）制作与研究对象、参照物相同材料组成与结构组成的试件模型，并在试验室进行室内加速试验。

（5）通过对参照物进行现场检测及对参照物和研究对象试件模型进行室内加速试验所得到的实测数据，基于经典的相似理论预测研究对象在实际环境作用下的劣化机理控制参数：

$$\gamma_i = \frac{\alpha_k}{\alpha_k^{oi}} = \frac{\alpha_k^{ro}}{\alpha_k^{ri}} \qquad k = 1,2,\cdots,i,j,\cdots \tag{2}$$

式中，$\alpha_k$ 为研究对象在实际环境作用下的劣化机理控制参数；$\alpha_k^{oi}$ 为研究对象试件模型在室内加速试验环境下达到稳定的劣化机理控制参数；$\alpha_k^{ro}$ 为参照物在实际环境作用下达到稳定的劣化机理控制参数；$\alpha_k^{ri}$ 为参照物试件模型在室内加速试验环境下达到稳定的劣化机理控制参数。需要指出的是，室内加速试验的劣化机理应与实际环境保持一致。

（6）将式（2）中预测得到的研究对象在实际环境作用下的劣化机理控制参数 $\alpha_k$ 代入式（1）中，求解方程便能对研究对象进行寿命预测。

以上介绍可以看到，METS 方法中，实际环境对于研究对象的作用历程其实是通过参照物来反映。参照物与研究对象处于相同暴露环境，由于试验所选择的参照物具有较长服役年限，通过现场检测获得的耐久性损伤信息其实是结构物在服役期间环境耦合作用的综合表现；在假定相同区域的环境作用信息每年变化不变的基础上，可以认为步骤（5）中得到的研究对象在实际环境中的预测值已经包含了相关的环境作用历程。此外，步骤（5）中为了简化计算，近似做了线性假定。但是，对于不同的研究问题，室内环境与室外环境下控制参数之间的关系可能是非线性的，因此式（2）根据具体问题可以是非线性的。以下将以海工混凝土受氯离子侵蚀为例详细介绍该试验方法在实际工程中的应用。

## 6.4　基于 METS 方法的氯盐侵蚀环境下混凝土结构使用寿命预测

### 6.4.1　海工混凝土结构劣化机理分析

当今，海工混凝土结构的耐久性失效主要是钢筋锈蚀引起的混凝土保护层开裂，而对于海工混凝土结构物引起钢筋锈蚀的直接原因主要为氯离子的侵蚀。对海工建筑物处于水下区的部分通常使用 Fick 第二扩散定律[18]来描述氯离子在混凝土中的传输机理。而在海洋环境的干湿交替区域，混凝土表面浅层的对流区范围内，氯离子主要以对流和扩散耦合的方式侵蚀[19,20]。目前用于计算干湿交替区域下氯离子侵蚀的简化方法较多采用欧洲混凝土结构耐久性 DuraCrete 提出的经验方法[21]，该方法认为 $0 \sim \Delta x$ 范围内的对流扩散区域中主要发生由孔隙液流动造成的氯离子对流，而其余区域内以氯离子浓度扩散作为主要渗

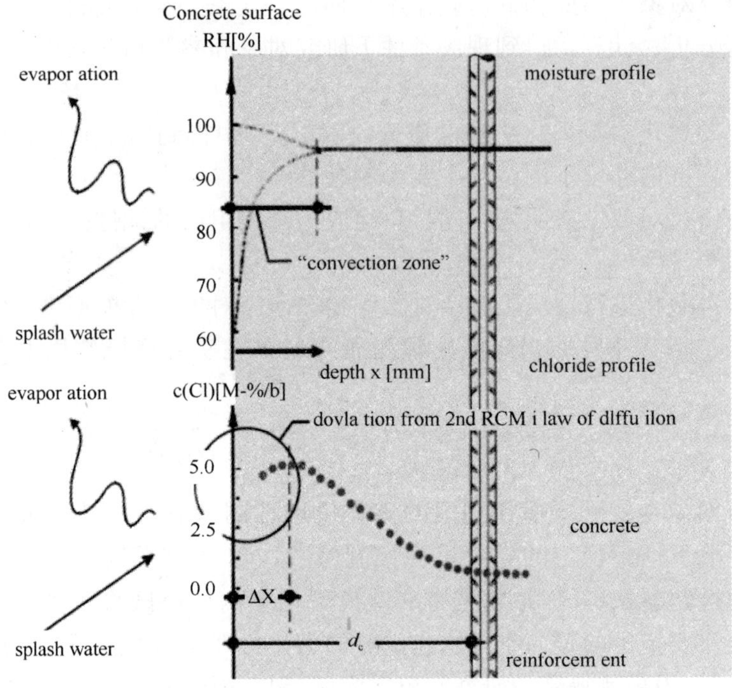


图 2 对流扩散区域示意图（引用自 DuraCrete）

透方式，如图 2 所示。

由上述分析可知，无论对于水下区还是水位变动区氯离子在混凝土中的传输机理均可简化由 Fick 第二定律来表示，如式（3）所示：

$$\frac{\partial C}{\partial t}=D\frac{\partial^2 C}{\partial x^2} \tag{3}$$

式中，$C$ 为氯离子浓度，$x$ 为扩散深度，$t$ 为扩散时间。考虑边界条件：

$$\begin{aligned} C(x,t=0)&=C_i \qquad 0<x<\infty \\ C(x=0,t)&=C_s \qquad 0<t<\infty \end{aligned} \tag{4}$$

得到式（3）的解析解为：

$$C(x,t)=C_i+(C_s-C_i)\left(1-erf\frac{x}{2\sqrt{Dt}}\right) \tag{5}$$

式中，$C_s$ 为表面氯离子浓度，对于水位变动区部分可视为 $\Delta x$ 处的氯离子浓度；$D$ 为氯离子表观扩散系数。

### 6.4.2 模型建立

如 5.4.1 节分析可知，混凝土结构使用寿命预测需明确四个劣化机理控制参数，分别为引起钢筋锈蚀的氯离子临界浓度 $C_{cr}$，对流扩散耦合作用区长度 $\Delta x$，混凝土氯离子扩散系数 $D$ 以及表面氯离子浓度 $C_s$。其中，表面氯离子浓度 $C_s$ 定为 $\Delta x$ 深度处的氯离子浓度；临界氯离子浓度 $C_{cr}$ 可参考相关规范确定。将上述各参数代入式（5）得到混凝土使用寿命的计算表达式：

$$t_{cr} = \frac{(x-\Delta x)^2}{4D} \cdot \left(erf^{-1}\left(1-\frac{C_{cr}}{C_s}\right)\right) \tag{6}$$

根据对参照物的现场检测，以及对参照物及研究对象试件模型的室内加速试验研究，基于经典的相似理论可以预测研究对象在实际环境作用下的劣化机理控制参数，如式（7）～式（9）所示：

$$D = D^{oi} \times \frac{D^{ri}}{D^{ro}} \tag{7}$$

$$C_s = C_s^{ro} \times \frac{C_s^{oi}}{C_s^{ri}} \tag{8}$$

$$\Delta x = \Delta x^{oi} \times \frac{\Delta x^{ro}}{\Delta x^{ri}} \tag{9}$$

式中：$D^{oi}$，$C^{oi}$，$\Delta x^{oi}$分别为室内加速试验环境下研究对象试件模型的氯离子扩散系数，表面氯离子浓度和对流扩散耦合作用区长度的最终稳定值；$D^{ri}$，$C^{ri}$，$\Delta x^{ri}$分别为室内加速试验环境下参照物试件模型的氯离子扩散系数，表面氯离子浓度和对流扩散耦合作用区长度的最终稳定值；$D^{ro}$，$C^{ro}$，$\Delta x^{ro}$分别为实际环境作用下参照物的氯离子扩散系数，表面氯离子浓度和对流扩散耦合作用区长度的最终稳定值。

将式（7）～式（9）代入式（6）便能对研究对象进行寿命预测，如式（10）所示。

$$t_{cr} = \frac{\left(x-\frac{\Delta x^{oi}\Delta x^{ro}}{\Delta x^{ri}}\right)^2 D^{ro}}{4D^{oi}D^{ri}} \cdot \left(erf^{-1}\left(1-\frac{C_{cr}C_{in}^{ri}}{C_{in}^{ro}C_{in}^{oi}}\right)\right) \tag{10}$$

### 6.4.3　工程应用

1. 工程背景

本文以杭州湾跨海大桥现浇墩身为研究对象对其进行寿命预测。对杭州湾在役混凝土结构腐蚀状况的调查表明，沿海湾混凝土结构腐蚀十分严重，90％的损坏是由于环境恶劣、保护层不足，氯离子渗透导致钢筋锈蚀引起的。中性化、碱骨料反应、硫酸盐侵蚀、冻融破坏、海洋生物等不是混凝土结构劣化的主要原因。宁波港某 10 万吨级矿石中转码头，建成时是全优工程，仅使用了 11 年后，桩帽、水平撑、梁板等钢筋已经胀裂，约 5cm 的混凝土保护层内水溶性氯离子含量已达 0.8％左右，钢筋部位的浓度大大超过引发锈蚀的临界浓度[22,23]。从调查结果可以认定，影响本工程混凝土结构耐久性的主导因素是氯离子侵蚀引起的钢筋锈蚀。杭州湾跨海大桥的设计使用寿命为 100 年，桥墩混凝土保护层厚度为 6cm，混凝土配合比如表 1 所示。

**现浇墩身混凝土的配合比**　　**表 1**

| 每方混凝土各种材料用量（kg） | | | | | | | |
|---|---|---|---|---|---|---|---|
| 水泥 | 矿粉 | 粉煤灰 | 砂 | 石子 | 水 | 减水剂 | 阻锈剂 |
| 126 | 168 | 126 | 735 | 1068 | 145 | 5.04 | 8.4 |

2. 参照物选取与室内加速模拟试验

由于杭州湾跨海大桥与乍浦港同属杭州湾海域，且二者相距不远，环境条件极为相似，为研究杭州湾跨海大桥的混凝土结构耐久性，本文选取乍浦港区为参照物。两者的大

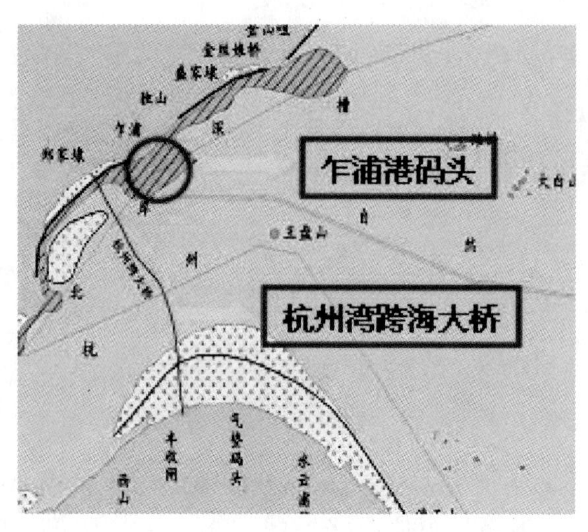


图 3 杭州湾跨海大桥与乍浦港码头地理位置关系图

致地理位置如图 3 所示。目前乍浦港主要包括陈山码头、一期工程、二期工程、三期工程等。由于一期工程建造年代最早，本文选取乍浦港一期工程进行现场检测，检测时间为 2006 年 10 月，距建成时间为 201 个月，经过如此长时间的暴露，可以认为氯离子扩散系数、表面氯离子浓度以及对流扩散耦合作用区长度均已达到稳定。混凝土配合比见表 2。

在室内浇筑与研究对象及参照物相同配合比的试块进行加速模拟试验[13]。试验主要有 3 个控制参数，分别为干湿时间比、溶液浓度、溶液温度。其中干湿时间比参数的确定是通过统计杭州湾海域 3.00m 标高处全年的海水浸润风干时间比，经过统计定为浸泡 3.5 小时，风干 48 小时；溶液氯离子浓度定为海水中氯离子浓度的 6 倍；浸泡时溶液温度定为 40℃。

**乍浦港一期二泊位混凝土配合比** **表 2**

| 配合比 | | | |
|---|---|---|---|
| 水 | 水泥 | 砂 | 石 |
| 0.45 | 1 | 1.62 | 2.43 |

3. 寿命预测

(1) 参照物检测

于 2006 年 10 月对乍浦港码头一期二泊位立墙的 3.00m 标高处进行钻孔取粉。现场钻孔取粉时采用 7mm×10=70mm 的取样深度，经过对取样粉样进行 RCT 分析，得到各深度处的自由氯离子含量（占混凝土质量百分比）。使用 matlab 对测试结果进行最小二乘拟合，拟合公式为 Fick 第二定律解析式，如式（5）所示。表观氯离子扩散系数 $D^{ro}=1.218\times10^{-6}\mathrm{mm}^2/\mathrm{s}$ 。此外，在取粉同时，于相同标高处进行取芯操作。在室内试验室对芯样以 1mm 间距进行分层碾磨，确定对流扩散耦合作用区 $\Delta x^{ro}=10\mathrm{mm}$ ，根据此非扩散距离确定表面氯离子浓度 $C_s^{ro}=0.5605\%$。

(2) 室内加速模拟试验

①氯离子扩散系数的时变性

由于扩散系数与表面氯离子将随时间变化并最终达到稳定，因此需考虑参数的时变性。因此，在室内试验中每隔一定时间需对试件模型中氯离子含量做定期检测。图 4 为参照物试件模型的氯离子含量测试结果，图 5 为桥墩试件模型（QHLA0）的测试结果。

将每次测试得到氯离子浓度分布曲线进行拟合得到不同时刻的氯离子扩散系数，然后根据式（11）对扩散系数进行拟合，便可得到研究对象与参照物的氯离子扩散系数时变公式，如式（12），式（13）所示：

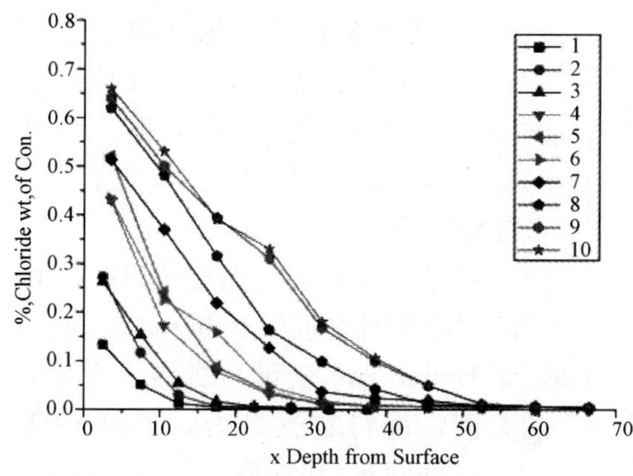


图 4　参照物试件模型的氯离子浓度分布

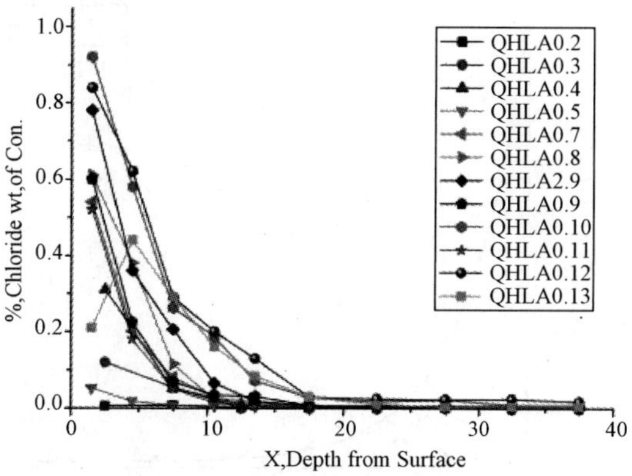

图 5　桥墩试件模型（QHLA0）的氯离子浓度分布

$$D(t) = D_{28} \cdot \left(\frac{0.0767}{t}\right)^{n}, t \leqslant 30a$$
$$D(t) = D_{28} \cdot \left(\frac{0.0767}{30}\right)^{n}, t > 30a \tag{11}$$

$$D^{ai}(t) = 4.083 \times 10^{-6} \cdot \left(\frac{0.0767}{t}\right)^{0.6006}, t \leqslant 30a$$
$$D^{ai}(t) = 4.083 \times 10^{-6} \cdot \left(\frac{0.0767}{30}\right)^{0.6006}, t > 30a \tag{12}$$

$$D^{ri}(t) = 26.78 \times 10^{-6} \cdot \left(\frac{0.0767}{t}\right)^{0.3312}, t \leqslant 30a$$
$$D^{ri}(t) = 26.78 \times 10^{-6} \cdot \left(\frac{0.0767}{30}\right)^{0.3312}, t > 30a \tag{13}$$

式中，$D_{28}$ 为第 28 天的表观氯离子扩散系数（$mm^2/s$），$n$ 为时间衰减系数。并假设氯离子扩散系数在暴露 30 年后已经达到稳定[24]。

②对流扩散耦合作用区长度确定

通过对混凝土试件模型在不同暴露时间检测得到的氯离子含量随深度的变化曲线转变为氯离子含量与 Boltzmann 变量（$\varphi = 0.5 * x/\sqrt{t}$）的关系曲线（$C \sim \varphi$ 曲线）。Tumidajski[25]、Dhir[26]采用下面的数学模型来拟合氯离子含量与 Boltzmann 变量 $\varphi$ 的关系：

$$C = C_0 \cdot \exp(-k\varphi) \tag{14}$$

式中，$C$ 为氯离子浓度（%），$C_0$ 为氯离子浓度最大值（%），$k$ 为拟合系数。

利用 Boltzmann 变量的方法可以充分利用检测数据拟合求得氯离子浓度随时间与距离表面距离的关系式，既可以减少因试验操作或检测造成的误差，又可以解决短期试验获得数据量不足的问题。

利用式（14），可计算出任意时刻、任意深度（如 $x = \Delta x$）处的氯离子浓度。为清楚表明表面氯离子浓度随时间的积累效应，将 $x = \Delta x$ 处的氯离子浓度随时间的积累规律转化为随时间的指数关系式，如式（15）的形式。

$$C_s = a + b\exp(1 - t^{-c}) \tag{15}$$

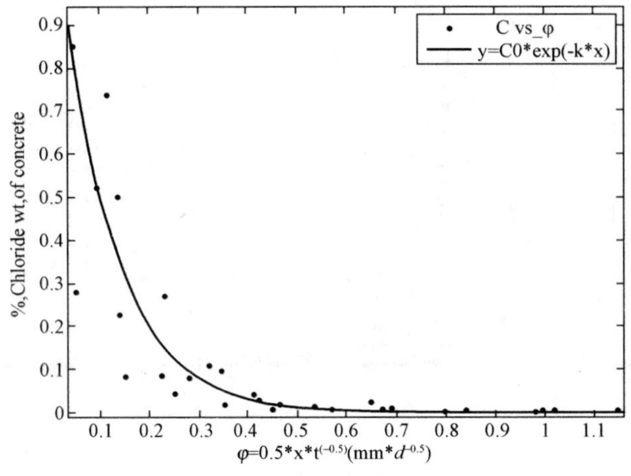


图 6 QHLA0 试件的 Boltzmann 拟合曲线

式中，$a$ 为表面氯离子浓度初始值；$b$ 为表面氯离子浓度累计值； $c$ 表示表面氯离子浓度的累积速率。由式（15）可知，$a+b$ 即为表面氯离子浓度最终稳定值。

对现浇墩身试件模型 QHLA0 取样的深度与时间进行 Boltzmann 转换，得到 Boltzmann 变量与氯离子含量 $C$ 的关系，进行数据拟合，得到 QHLA0 试件的 $C_0$ 和 $k$ 值，如图 6 所示，可得 $C_0=1.289, k=9.7$ 。将上节测得的流扩散耦合作用区长度代入 Boltzmann 变量，即 $\Delta x^{\alpha i}=2.5\text{mm}$，便求得对流区深度 $\Delta x^{\alpha i}$ 处的氯离子浓度随时间的关系式，通过对 $t=0\sim1000\text{d}$ 内的每隔 30d 取一点进行曲线拟合，可得到：$a=0.09671$，$b=0.75707$，$c=-0.00428$如图 7 所示。由此，分别可以得到桥墩试件模型与参照物试件模型表面氯离子浓度随时间变化关系，如式（16），式（17）所示。

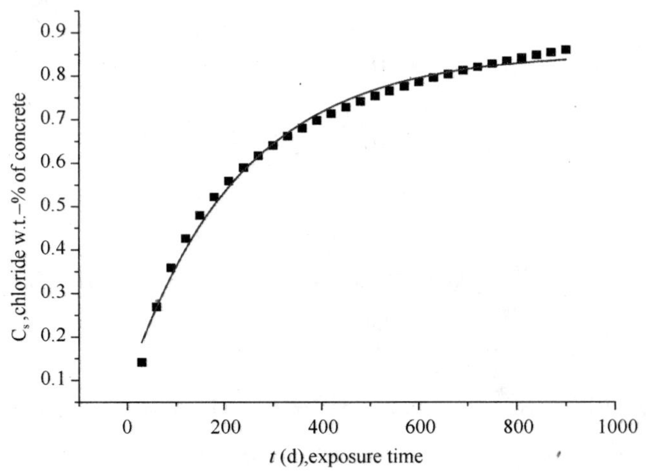


图 7 QHLA0 试件表面氯离子浓度拟合曲线

$$C_s^{\alpha i}=0.09671+0.75707\exp(1-t^{-0.00428}) \tag{16}$$

$$C_s^{ri}=0.09807+0.83979\exp(1-t^{-0.00573}) \tag{17}$$

（3）模型预测

①研究对象（桥墩）在实际环境作用下的氯离子扩散系数预测

根据式（12）和式（13），考虑到 30 年后氯离子扩散系数已稳定，将 $t=30$ 代入二式中，便能得到

$$D^{\alpha i}=4.083\times10^{-6}\cdot\left(\frac{0.0767}{30}\right)^{0.6006}=0.11325\times10^{-6}$$

$$D^{ri} = 26.78 \times 10^{-6} \cdot \left(\frac{0.0767}{30}\right)^{0.3312} = 3.70879 \times 10^{-6}$$

最后，结合 5.4.3 节 3.（1）中得到的 $D^{ro} = 1.218 \times 10^{-6} \text{mm}^2/\text{s}$，代入式（7），便能得到

$$D = 1.218 \times \frac{0.11325}{3.70879} \times 10^{-6} = 0.03719 \times 10^{-6} \text{mm}^2/\text{s}$$

②研究对象（桥墩）在实际环境作用下的表面氯离子浓度预测

根据式（16）和式（17），便能计算得到室内加速环境下桥墩试件模型与参照物试件模型的表面氯离子浓度最终稳定值，分别为 $C_s^{oi} = 0.85378\%$ 和 $C_s^{ri} = 0.93786\%$。最后，结合 5.4.3. 节 3（1）中得到的 $C^{ro} = 0.5605\%$，代入式（8）便能得到 $C_s = 0.5605\% \times \frac{0.85378\%}{0.93786\%} = 0.5102\%$。

③研究对象（桥墩）在实际环境作用下的对流扩散耦合作用区长度预测

将检测得到的 $\Delta x^{ro} = 10\text{mm}$ 与 5.4.3 节 3.（2）②中检测得到的 $\Delta x^{ri} = 5\text{mm}$ 和 $\Delta x^{oi} = 2.5\text{mm}$ 代入式（9），便能得到 $\Delta x = \Delta x^{oi} \times \frac{\Delta x^{ro}}{\Delta x^{ri}} = 2.5 \times \frac{10}{5} = 5\text{mm}$。

④研究对象（桥墩）寿命预测

根据文献［18］确定氯离子临界浓度为 $C_{cr} = 0.05\%$，将上面计算得到的劣化机理控制参数 $D = 0.03719 \times 10^{-6} \text{mm}^2/\text{s}$，$C_s = 0.5102\%$，$\Delta x = 5\text{mm}$ 代入式（6），得到

$$t_{cr} = \frac{(x - \Delta x)^2}{4D} \cdot \left(erf^{-1}\left(1 - \frac{C_{cr}}{C_{in}}\right)\right) = \frac{(60-5)^2}{4 \times 0.03719 \times 10^{-6}} \cdot \left(erf^{-1}\left(1 - \frac{0.05}{0.510}\right)\right)$$

$$\times \frac{1}{365 \times 24 \times 3600} = 110a$$

⑤与其他寿命预测方法比较

在 METS 方法进行寿命预测之前，曾经对海上混凝土结构的耐久性寿命进行了三次预测。

第一次寿命预测主要为初步设计提供混凝土耐久性设计的依据，采用 MCSLPS 和 N. P. LEE 寿命评估系统[27]，除氯离子扩散系数和混凝土保护层厚度采用推荐值外，其余计算参数均取自有关文献。

第二次寿命预测利用基于 Fick 第二定律和 Monte-Carlo（M-C）[28]方法的失效概率数值模拟对杭州湾跨海大桥混凝土结构主要构件的寿命预测，预测模型中考虑的参数有的不是现场的实测数据，而是根据国外统计资料取值[29]。

第三次寿命预测由浙江大学于 2008 年 4 月根据现场检测结果，利用基于 Fick 第二定律和 Monte-Carlo（M-C）方法的失效概率数值模拟对杭州湾跨海大桥混凝土结构主要构件进行了寿命预测，计算中模型参数的取值多数来自现场实测数据，比经验取值更加可靠、准确[30]。

对本文基于 METS 方法考虑氯离子扩散系数与表面氯离子浓度时变性的寿命预测结果与其他各种方法的预测结果进行比较如表 3 所示。可以看到 METS 与其他预测方法基本相近，特别是与 MCSLPS 方法及第三次根据现场实测参数取值采用 Monte-Carlo 方法的预测结果相一致，从而验证了该 MEST 方法的有效性。可见，虽然 METS 方法的计算非

常简单，但是预测结果是较为可靠的。

**METS 方法与其他方法比较** **表 3**

| 预测方法 | MCSLPS | N. P. LEE | Monte-Carlo (1) | Monte-Carlo (2) | METS |
|---|---|---|---|---|---|
| 预测结果（a） | 110 | 149 | 99 | 110 | 110 |

## 6.5 讨论

以上的介绍可以发现，METS 理论的精髓在于，通过引入参照物及建立相关室内加速试验将理论模型与实际环境的作用紧密联系在了一起，有效解决了以往方法无法对在建或新建结构进行可靠寿命预测的缺陷。基于 MTES 理论的寿命预测方法包括了两个重要环节：(1) 参照物的选取。对参照物选取是最为基础也是最重要的一个环节，合理地选择参照物将使后面的工作事半功倍。因此，在选择参照物时需遵循以下几点原则：①与研究对象所处的暴露环境相一致。这是为了确保该参照物与研究对象两者的劣化机理一致，而不受其他环境因素的影响致使劣化机理发生改变。②参照物必须是具有较长服役时间的结构物。这是为了确保现场检测得到的劣化参数均已达到稳定。③确保参照物与研究对象的结构类型相一致。这一点是参照物与研究对象劣化机理保持一致的基本条件。(2) 建立室内人工气候模拟加速试验。在建立室内加速试验时也需遵循几点原则：①保证在室内加速试验条件下，参照物与研究对象模型的劣化机理与实际环境作用下的劣化机理相一致，这是建立室内外相似关系的前提。②必须达到加速劣化的试验效果，从而缩短室内加速试验时间，尽快对研究对象作出可靠的寿命预测。

此外，需要指出的是，若所选择的参照物服役年数不长，无法直接通过现场检测直接获取相关劣化参数的稳定值。但是，借助于相关室内加速试验及现场检测得到的数据便能预测参照物劣化参数在实际环境下的最终稳定值，从而对研究对象做出寿命预测。

## 6.6 结语与展望

混凝土结构耐久性研究中，材料与构件层次的研究是结构使用寿命预测的理论基础。由于混凝土结构是在不同严酷条件下服役的，绝非是单一环境因素作用下引起的损伤与劣化，而是在力学因素和环境因素双重和多重因素的耦合作用，是一个复杂的损伤叠加与交互作用过程，也是引起混凝土耐久性下降和服役寿命缩短或过早退出服役的根本原因。基于以往与未来研究成果，建立暴露于荷载与环境因素耦合作用下的混凝土结构科学的设计理论和新方法，是相当必要的。此外，在对结构进行使用寿命预测时，由于现场检测得到的相关劣化信息很少甚至没有，传统的寿命预测方法在对新建或是在建的混凝土结构进行寿命预测时往往只能通过室内加速试验，如此预测得到的寿命是不可靠的。METS 试验方法的提出有效解决了相关问题，通过引入第三方参照物以及设计相关的室内模拟加速试验，从而对研究对象作出可靠的寿命预测。

在下一步的研究工作中，将对于钢筋锈蚀引起混凝土保护层开裂的性能极限状态开展研究。在外界环境反应谱作用下，通过研究混凝土内部微环境反应谱作用下钢筋腐蚀电流的时变状态，结合 METS 试验方法的基本理论，便能实时动态预测研究对象在钢筋脱钝后的腐蚀电流变化规律，从而对钢筋保护层开裂时间作出可靠的预测。

## 参考文献


[1] 金伟良，钟小平．结构全寿命的耐久性与安全性、适用性的关系．建筑结构学报．2009．30(6)：1-7.

[2] 金伟良，吕清芳，赵羽习等．混凝土结构耐久性设计方法与寿命预测研究进展．建筑结构学报．2007．28(1)：7-13.

[3] 陈继松，于群力，胡琦忠．跨海桥梁全寿命周期可靠性管理框架．世界桥梁，2008(3)：71-74.

[4] 金伟良．混凝土结构耐久性研究的主要进展及其发展趋势．国家自然科学基金委员会"十一·五"建筑学科发展战略研究报告，2005．3.

[5] 施养杭．Monte-Carlo 法氯离子侵蚀下混凝土构件寿命预测．华侨大学学报(自然科学版)，2005，26(4)：369-372.

[6] Basheer P A M，Chidiac S E，Long A. E. Predictive models for deterioration of concrete structures. Construction and Building Materials，1996，10(1)：27-37.

[7] 余红发，孙伟，鄢良慧等．混凝土使用寿命预测方法的研究-理论模型．硅酸盐学报，2002，30(6)：686-690.

[8] 吕清芳，金伟良．碳化作用下基于可靠度的混凝土结构耐久性设计方法研究．工业建筑，2008，38(12)：84-87.

[9] 朱平华，金伟良，倪国荣．在役混凝土桥梁结构耐久性评估方法．浙江大学学报，2006，40(4)：658-667.

[10] 赵羽习，金伟良．钢筋锈蚀导致混凝土构件保护层胀裂的全过程分析．水利学报，2005，36(8)：939-945.

[11] 卢振永，金伟良，王海龙金立兵，延永东．人工气候模拟加速试验的设计方法．浙江大学学报(工学版)(已录用).

[12] Weiliang JIN，Libing JIN. Environment-based on experimental design of concrete structures. 2nd International Conference on Advances in Experimental Structural Engineering：In：Structural Engineers，2007 Vol. 23(Sup)：757-764.

[13] 卢振永．氯盐腐蚀环境的人工模拟试验方法．浙江大学，2007.

[14] 姬永生，袁迎曙．恒定气候混凝土内钢筋锈蚀速率的时变特征与机理．中国矿业大学学报，2007，36(2)：153-158.

[15] Wei-Liang Jin，Li-Bing Jin：A multi-environmental time similarity theory of life prediction on coastal concrete structural durability. International Journal of Structural Engineering. 2009. 1(1)：40-58.

[16] 金立兵，金伟良，王海等．多重环境时间相似理论及其应用．浙江大学学报，2010，44(4)：789-797.

[17] 金立兵．混凝土结构耐久性的多重环境时间相似理论与试验方法．浙江大学，2008.

[18] Stephen L A，Dwayne A J，Matthew A M，et a1. Predicting the service life of concrete marine structures：an environmental methodology. ACI Structural Journal，1998，95(2)：205-214.

[19] Anna V S，Roberto V S，Renato V V. Analysis of chloride diffusion into partially saturated con-

crete. ACI Material Journal，1993，90(5)：441-451.

[20] 张奕. 氯离子在混凝土中的输运机理研究. 杭州：浙江大学建筑工程学院，2008.

[21] DuraCrete. BRPR-CT95-O132-E95-1347 General guidelines for durability design and redesign. European Union-Brite Euram III，2000.

[22] 金伟良，许晨. 一种钢筋脱钝氯离子阈值快速测定新方法. 浙江大学学报(已录用).

[23] Ki Yong Ann，Ha-Won Song. Chloride threshold level for corrosion of steel in concrete. Corrosion Science，2007，49：4113-4133.

[24] 中国土木工程学会标准. CCES01—2004 混凝土结构耐久性设计与施工指南. 北京：中国建筑工业出版社，2005.

[25] Tumidajski P J，Chan G W. Boltzmann-Matano analysis of chloride diffusion into blended cement concrete. Journal of Materials in Civil Engineering，1996，8(4)：195-200.

[26] Dhir P K，Jones M R，Ng S L D. Prediction of total chloride content profile and concentration/ time-dependent diffusion coefficients for concrete. Magazine of Concrete Research，1998，50(1)：37-48.

[27] 吕忠达. 杭州湾跨海大桥关键技术研究与实施. 长安大学博士学位论文，2007.

[28] 金伟良，吕清芳. 混凝土结构设计耐久性环境区划标准. 东南大学学报：英文版，2007，23(1)：98-104.

[29] 浙江大学结构工程研究所. 杭州湾跨海大桥混凝土结构耐久性长期性能研究中期报告. 2008.

[30] 浙江大学结构工程研究所. 杭州湾跨海大桥混凝土结构耐久性长期性能研究总结报告. 2008.

# 第 7 章 Chapter 7

# 钢筋混凝土建筑结构的防爆设计和安全评估—从构件到建筑结构系统

# BLAST RESISTANT DESIGN OF REINFORCED CONCRETE BUILDINGS-FROM COMPONENTS TO STRUCTURAL SYSTEM

B. Li(李　兵)

School of Civil and Environmental Engineering, Nanyang Technological University Singapore 639798

**Abstract:** The design and construction of building structures to provide life safety in the face of explosions is receiving attention from engineers. In order to protect people against terrorism, buildings should not collapse catastrophically; they should not burn out of control. Protection may be achieved through structural design, with the appropriate selection of materials and structural form, together with adequate elemental and joint detailing. This paper provides an overview of blast resistant design research conducted on reinforced concrete structural members and structural systems at Nanyang Technological University (NTU), Singapore. These investigations attempt to gain a better understanding of the general behaviors of these structural components and systems when subjected to blast loading.

**Keywords:** Blast Resistant Design, Impact loading, reinforced concrete, maximum midspan deflection.

## 7.1　INTRODUCTION

Over the past 20 years, accidental or deliberate explosive incidents have revealed the vulnerability of civilian structures to the extreme dynamic loading caused by detonations. These types of incidents are characterized by its great intensity and short duration. The collapse of the World Trade Center in New York City has brought the focus of many structural engineers to develop an understanding of the behavior and resistance of structures when subjected to blast loadings. This unfortunate event has brought about an urgent need to for an increased awareness in the behavior of public buildings, especially commercial high-rise buildings, when placed within a blast environment.

Buildings in Singapore are normally designed for the onerous combination of wind and gravity loads. The inherent resistance of Singapore buildings is therefore derived from these design loads and the associated structural detailing. Substantial resiliency of a structural system is required in order to sustain the blast induced load and displacement. The required resiliency can best be achieved via the continuity, redundancy and energy absorbing capacity within the structural system and its components. This therefore points to a need for a systematic review of the provisions within the current building regulations and codes with respect to lateral load requirements for both wind, earthquake and fire effects, which will collectively provide an inherent resiliency against the progressive collapse of a building structure when subjected to blast loads.

## 7.2 LOCAL AND GLOBAL STRUCTURAL RESPONSES OF A TALL BUILDING

A 30-storey high rise commercial building, located within the central business district in Singapore, was selected for study. The selected building is a reinforced concrete (RC) structure, comprising RC frames and a shear wall core. The threat scenarios selected for this study are detonations at short and long standoff distances from the subject building. The threat scenarios were postulated based on the observed traffic pattern around the building. The frame is a two bay structure with a core wall in the middle, which is depicted in Fig. 1.

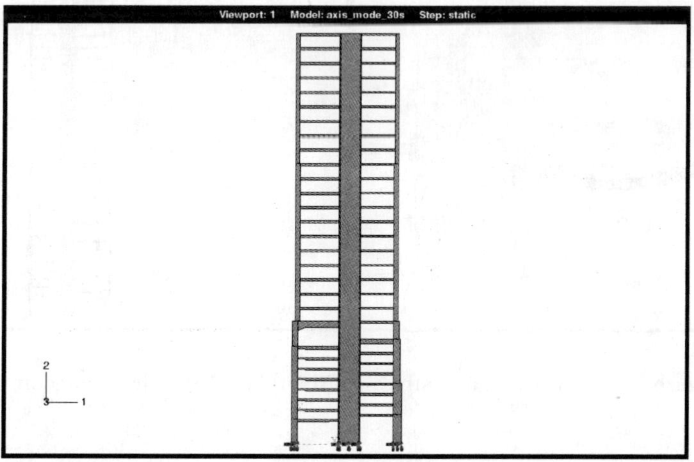


Figure 1　Finite element model of the building

### 7.2.1 Short Standoff Distance

For the short standoff distance, columns in the lower storeys were heavily damaged. For this case, progressive collapse of the damaged transverse frame may occur with the removal of the heavily damaged columns in the lower storeys. For the longitudinal frame with transfer girders on the second storey, the behavior for short standoff distance shows that if the base column beneath the transfer girder is destroyed, an extremely long span of transfer girder is formed. When loaded by the vertical service loads, this elongated transfer girder would not resist the design service loads. Hence, this disruption of the load transfer mechanism and formation of a transfer girder which spans for a long distance, can lead to progressive collapse.

Thus, a short standoff distance leads to the loss of the base column beneath the transfer girder, and possible induce a progressive collapse of the damaged transverse frame. This appears to imply a partial collapse of the portion of building around this column soon after the blast event. Hence, this column is a critical element, which requires special protection to ensure its survivability against potential threats such as vehicle bombs. The occurrence of a progressive collapse depends on a number of factors. They may include the degree of redundancy in the structural system, the ductility of members and joints, column spacing and beam span. For the building, the loss of other columns which do not support the transfer girder may not result in progressive collapse of the building. But when the transfer column is destroyed, the building would suffer disproportionate collapse. Therefore, the results of this case study indicate the vulnerable of buildings with a low structural redundancy when subjected top the threat of external blast loadings.

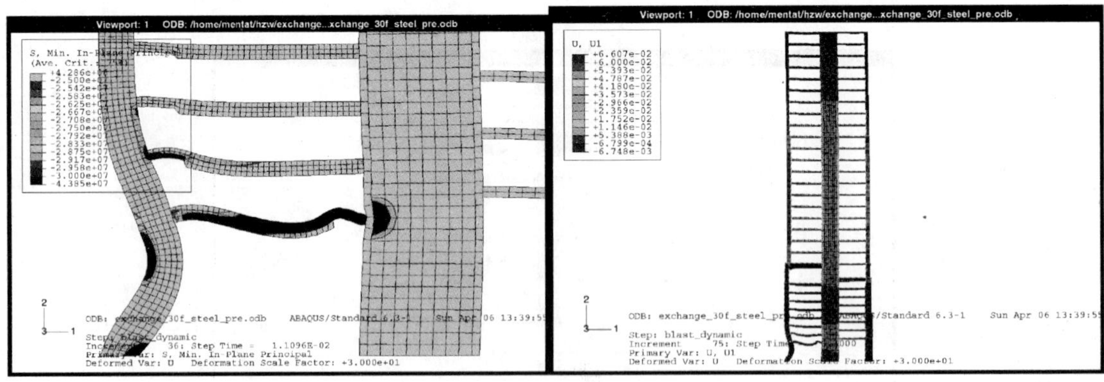


Figure 2 Deformation and stress distributions of the short standoff case

### 7.2.2 Long Standoff Distance

The building deformation at a representative time step of the long standoff distance case is shown in Figure 4. Plastic deformation was mainly concentrated at the bottom column and the beams of the second and the third stories. Most severe damage appeared in the column and beam. The beam deformation was focused at its ends where local damages occurred. Shear wall played an important role in the global response and exhibited a lateral vibration from the force transferred from the beams. The lateral vibration propagated upwards and reached the roof level. As the compressive wave propagated to the other end of the beams, the second storey beam was almost completely crushed. Crushing of concrete in the shear wall was confined to its connection with the second storey beam. The local damage index based on curvature is used to evaluate the flexural performance of structural elements under blast load. The second and the third storey beams were near complete failure. Severe damage appeared in columns on the first and the second stores, and in the beams at the fourth and the fifth stories. When compared to the case of a shorted standoff distance, the damage due to a longer standoff distance induces a global damage effect.

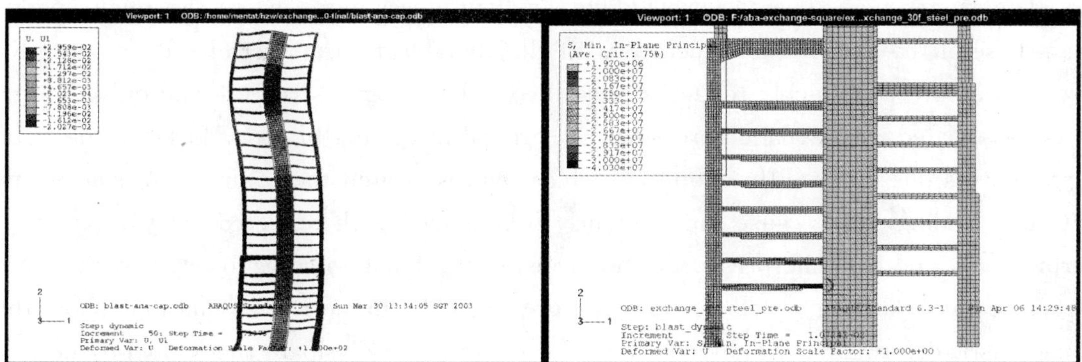

Figure 3　Deformation and stress distributions of the long standoff case

### 7.2.3 Discussion

The results of the study have indicated that buildings in Singapore would experience demands on strength and ductility which are greater than their inherent strength capacities and ductility factors for the blast event at a short standoff distance. Therefore, in such case, the potential damage to structure and consequential loss of lives may be high. More severe consequences can be expected for the blast event at a short standoff distance. The results have also shown that while the moment resistance capacity appears to be sufficient under the blast event at a long standoff distance, the shear resistance of a non-ductile section that has been detailed according to current practices has been shown to be deficient. The problem of shear resistance is most significant in the columns and beam-column joints of the building structures. Therefore, it is desirable to introduce ductile detailing as a potential improvement to provide sufficient shear resistant capacity and ductility to the structures subject to the blast event at a long standoff distance. Substantial resilience of a structural system is required in order to sustain the blast-induced load and displacement. The required resilience can best be achieved via continuity and structural redundancy. The results also suggest that a severe failure of the base column is less likely when a building is subjected to a longer standoff distance when compared to a shorted standoff distance, for an equivalent quantity of detonation. This emphasizes the important role that maintaining an adequate standoff distance can play in managing the damaging response of buildings, when subjected to the threat of external blast loadings. However, in densely populated regions where land prices can be high, maintaining a long standoff distance can be costly. Thus, the cost of standoff distance can be compared with the cost of the strengthening structural systems. From the an earlier study on the Murrah Federal Building, it was suggested that a reduction of 75 % of the floor area damaged by the 1995 bombing can be achieved by some earthquake design consideration (Hayes et al., 2005). For the 10-storey building, the additional cost would be about US$ 190 per square meter of gross floor area.

The existence of RC exterior cladding panels produces more severe dynamic responses than those of the bare frame structure with all other blast parameters held constant. Reflected pressure will quickly reduce to zero for the bare frame structure and only the drag forces associated with dynamic pressures are critical in the net blast loadings on the structure. This is due to the diffraction of the blast waves around the columns. However, the exterior cladding panels cause the structure to be loaded with reflected blast pressure, overpressure, and dynamic pressure, thus causing the blast forces subjected to the frame structure with cladding panels to be much greater. The numerical simulations show that some plastic deformation would appear to the structure to dissipate the work produced by such a large blast force, and results in the structure experiencing some levels of damage. The results obtained emphasizes that the concept of using ductile exterior cladding as a mechanism for the control of blast response appears to be somewhat unorthodox. Its potential benefits warrant further investigation. It is apparent that if proper ductile panel response is utilized, the concept should be very effective for enhancing the blast resistant performance of turbine buildings within nuclear power plants and provide them with the much needed additional level of security.

## 7.3 DEFORMATION-CONTROLLED DESIGN OF REINFORCED CONCRETE FLEXURAL MEMBERS

The design of the RC structural members against accidental or deliberate explosions is deemed necessary with an increasing emphasis on blast loading on structures. The blast load exerted on a structural member can be adequately simplified as a uniformly distributed dynamic loading, characterized by its peak pressure and duration, with the exception of very close-in explosion situations. Some level of inelastic deformation of the structural member is allowed in the blast-resistant design when subjected to severe blast loadings to dissipate energy, therefore, most of blast design guidelines with explicit consideration of inelastic deformation have been proposed in recent years. These design guidelines help to ensure that a RC member is designed such that its response is equivalent to the predefined performance level under blast loading. Thus, the proposed design procedure seeks to reconcile the differences between a possibly lower response as determined by numerical methods and a possibly higher level response as predetermined by design performance.

Both maximum displacement and displacement ductility factor should be considered in the design of a blast-resistant structure since both parameters correlate with an expected performance level of a reinforced concrete (RC) structural member during a blast event. A new deformation-controlled blast-resistant design procedure using non-dimensional energy spectra (NES) was developed by Rong and Li. The effective depth ( $d$ ) and longitudinal reinforcement ratio ( $\rho$ ) of a RC member can be determined by representing a continuous RC member as an equivalent elastic-plastic SDOF system. Subsequently, the maximum displacement and displacement ductility responses ( $y_m^{eq}$ and $\mu^{eq}$ ) can be made equivalent to the corresponding design performance targets, which are defined by its target displacement and target displacement ductility factor ( $y_t$ and $\mu_t$ )

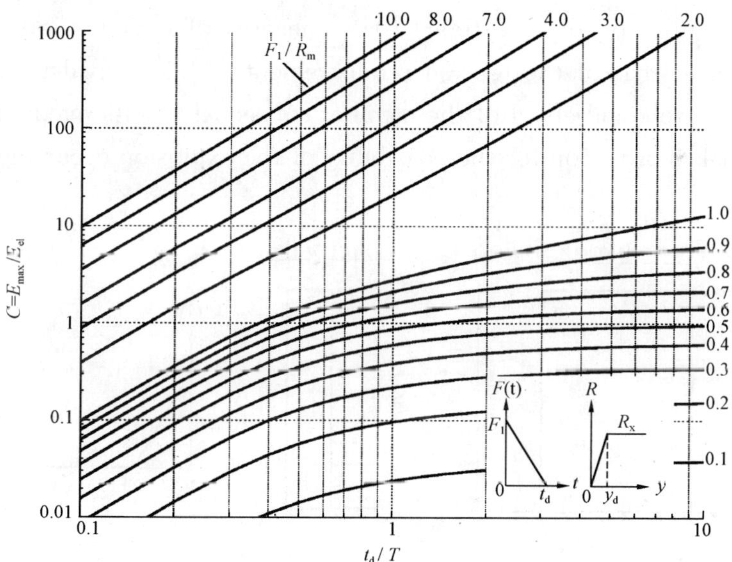


Figure 4　NES of an undamped elastic-perfectly-plastic SDOF system due to triangular load pulses with zero rise time

### 7.3.1 Deformation-Controlled Design Procedure

A non-dimensional energy spectrum (NES) is an important tool in the incorporation of the target displacements and target displacement ductility factors ( $y_t$ and $\mu_t$ ), which allows for determining design parameters of $d$ and $\rho$ for RC members. The non-dimensional energy factor ( $C$ ) is introduced into an elastic-plastic SDOF system (see Fig. 1), expressing the ratio of maximum strain energy$E_{max}$ to the ultimate elastic energy$E_{el}$ and is given in Eq. (1):

$$C = E_{max} / E_{el} \tag{1}$$

where $E_{el}=k_e y_e^2/2$ and$E_{max} = k_e y_e (y_{max} - y_e/2)$. Substituting them into Eq. (1), Cbecomes

$$C = 2\mu - 1 \tag{2}$$

A group of curves, which represent the factorCagainst the ratio $t_d/T$with respect to various$F_1/R_m$, are defined as non-dimensional energy spectra (NES) and shown in Fig. 4.

### 7.3.2 Implementation of the design procedure

To evaluate the effectiveness of applying the presented design procedure, a demonstration is given by implementing it on the design of a RC wall subjected to blast loading. The RC wall is designed to resist the blast loading perpendicular to its plane. The blast loading is simplified into a triangular pulse with the peak pressure and duration as shown in Fig. 5. The design is required to achieve the expected performance level defined by$\mu_t = 9$ and$\theta_t = 4°$. Thus, $y_t$of the wall at the free end under the given blast condition can be obtained with an approximate expression of$y_t \approx l\tan(\theta_t)$. The area of compression reinforcement is taken to be equal to that of the tension reinforcement. This equivalent of reinforcement would consider the rebound effect of the member subsequent to its maximum displacement response, and makes provision for the possibility of the explosion occurring in the opposite side of the wall.

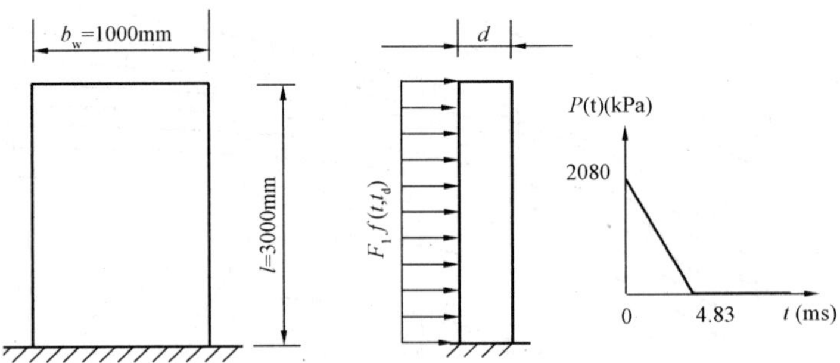


Figure 5 Sketch of a RC wall

In principle, a response analysis should be capable of controlling the actual responses of a RC member ( $y_m^{rc}$and$\mu^{rc}$ ) rather than those of the corresponding equivalent sdof system ( $y_m^{eq}$and$\mu^{eq}$ ), for a given design performance targets ( $y_t$and$\mu_t$ ). Thus, it is worthwhile determining the errors between the actual responses ( $y_m^{rc}$and$\mu^{rc}$ ) and the corresponding performance targets ( $y_t$and$\mu_t$ ) of a rc member. Application of these errors in modifying the deformation-controlled blast-resistant design procedure will actually help in controlling the responses of$y_m^{rc}$and$\mu^{rc}$.

$$\mu^{rc} = y_m^{rc}/y_e^{rc} \mu^{eq} = y_m^{eq}/y_e^{eq}$$

Rong and li (2008) discusses some numerical examples to demonstrate the analysis procedure using the modified design method. Numerical simulations of the displacement and the displacement ductility responses under blast loading are performed, and the results are compared with the performance targets. Li *et al* (2010) further verifies the method proposed by rong and li (2008) through the actual response of a rc column and beam subjected an actual denotation of explosives at a short standoff distance. The design method based on non dimensional energy spectrum provides a feasible design for the blast resistant design of members, subjected to blast loads at a short standoff distance. The method may be conservative such that the response of the member may be lower than the target level at which it was designed for.

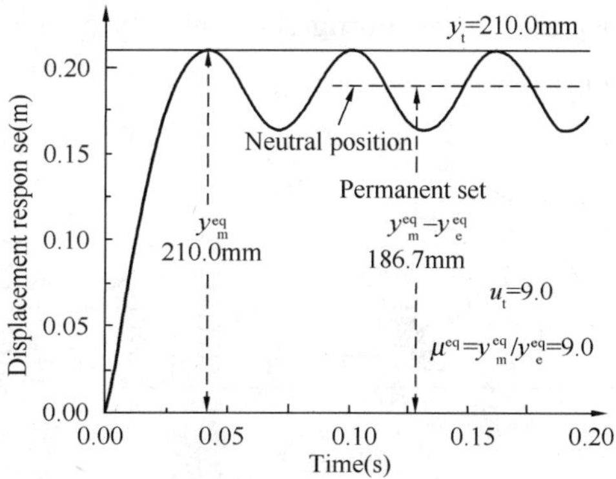


Figure 6 Deflection history of the equivalent SDOF system for the designed RC wall under the given blast loading

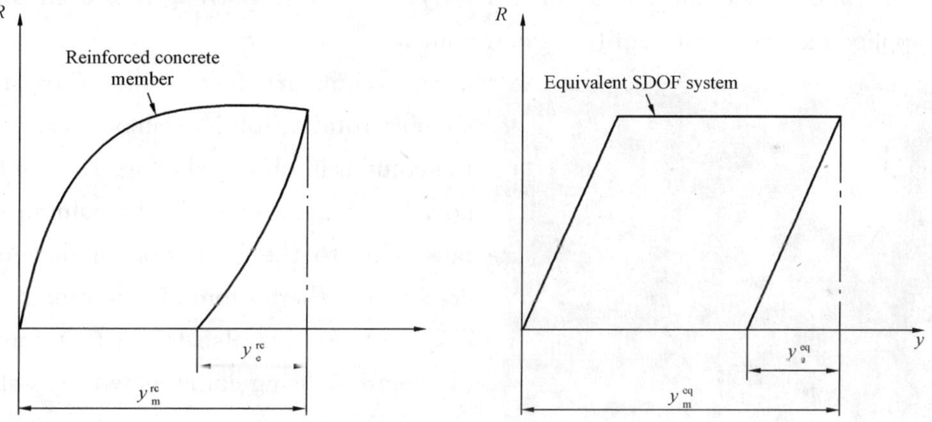


Figure 7 Response differences between the RC member and its equivalent SDOF system

## 7.4 RESIDUAL STRENGTH OF BLAST DAMAGED REINFORCED CONCRETE COLUMNS

Columns are the key load-bearing elements in frame structures. Exterior columns are probably the most vulnerable structural components in terrorist attacks. Column failure is normally the primary cause of progressive failure in frame structures. However, the current knowledge in the evaluation of residual capacity of a blast damaged reinforced concrete columns remains limited. A better understanding of residual capacity in columns would aid in the prediction of the overall performance of buildings, its resistance to progressive collapse and determining the stability of damaged buildings especially during search and rescue operations.

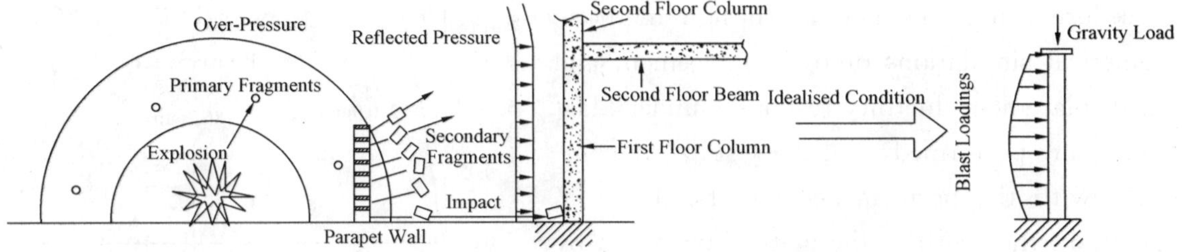


Figure 8 Blast loadings on the first floor column during a close-in explosion

RC columns representative of the base columns in 20-storey residential buildings in Singapore were studied. This means that the axial load distributed to a typical base column from the 20 storeys of load it supports, is applied to an RC column during the experiment. A set of customized experimental frames were fabricated for this (Fig. 9). Firstly, the displacement profiles of RC columns, subjected to various quantities of equivalent TNT at a short standoff distance, were determined. Following this, each displacement profile was applied experimentally on an RC column. Lastly, the residual strength of each column after the applied displacement profile, is determined.

Figure 9 Experimental set-up at NTU

Preliminary test results show that at a support rotation of 1°, minor local damage of the column is observed (Fig. 10). When support rotation exceeds 4°, the column specimen failed due to the formation of diagonal shear cracks near the bottom of the column specimen (Fig. 10). It was also found that confinement of concrete, using columns with closely spaced closed ties or spiral reinforcing, will improve confinement and shear capacity. This also improved the performance of lap splices in the event of a loss of concrete cover, which served to enhance the ductility of the column ductility.

Single-degree-of-freedom (SDOF) analysis from blast-resistant design guidelines provides engineers with simplified analytical methods to assess blast damage of RC columns. Although these simplified methods are quite useful, three-dimensional analysis, in contrast, provides a more in-depth understanding by incorporating all aspects of the response of concrete structures subjected to blast effects.

A three-dimensional nonlinear FE analysis utilizing LS-DYNA software is performed for the numerical simulations of this research. The FE model is validated through correla-

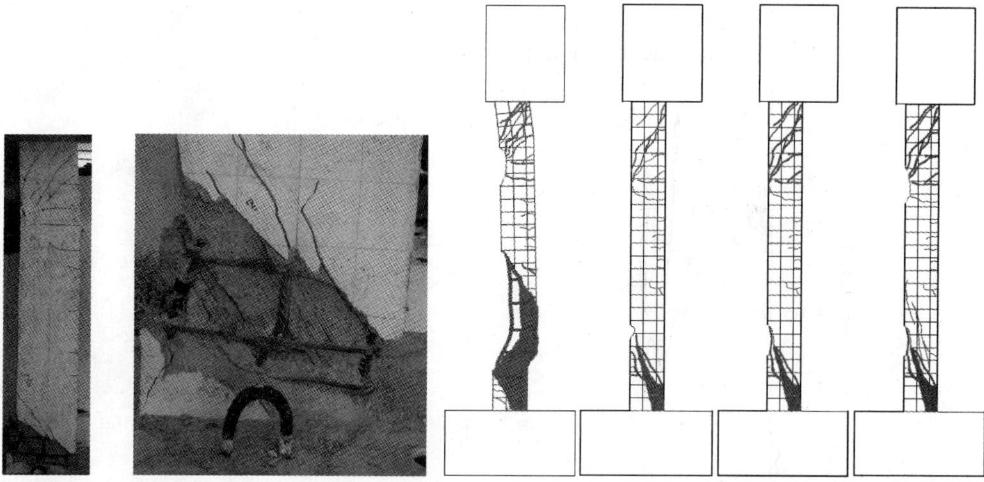

Figure 10　Local damage of the column occurs at a support rotation of 4°

ted experimental studies. The validated FE model was then analyzed under simulated blast loads and investigations were carried out on the dynamic responses and residual axial capacities of the columns (Fig. 11). An extensive parametric study was carried out on a series of 12 columns to investigate the effect of transverse reinforcement ratio, long-term axial load ratio, longitudinal reinforcement ratio, and column aspect ratio on the column responses.

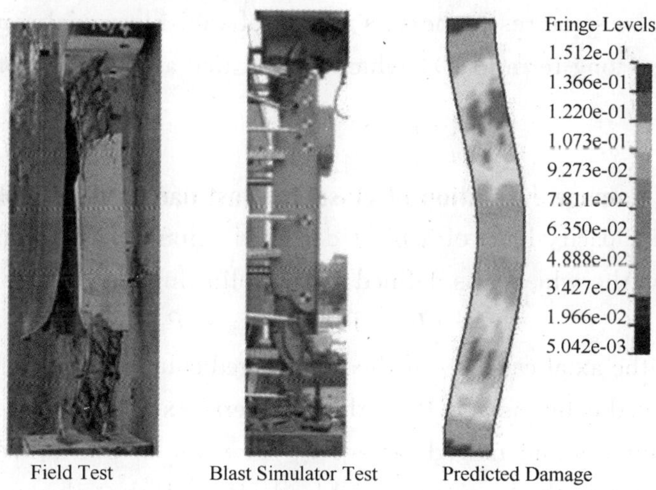


Figure 11　Comparison of numerical and experimental response of reinforced concrete columns subjected to impulsive loads

In the post-blast analysis stage, the axial load is gradually increased by applying a rigid plate attached to the top end of the column using displacement mode to capture both the residual axial capacity and the softening portion of the loading curve. One typical curve is shown in Fig. 12. It is noted that in the blast loading stage, due to the inertia effect, the axial load supported by the column is not constant but fluctuates along the deformation of the column. The sudden increase of the axial force at a displacement of 75 mm could be due

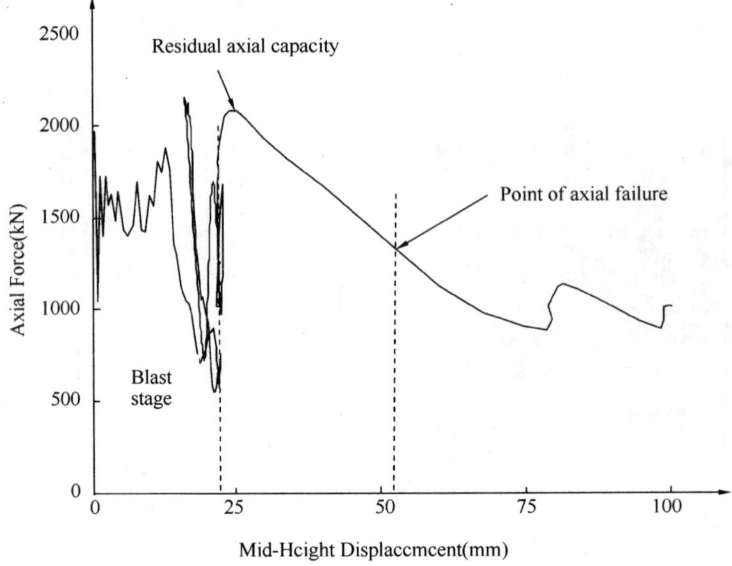


Figure 12 Axial force versus mid-height displacement

to the steel reinforcing bars reaching its strain hardening stage. The increase in the strength of the steel reinforcement at this point in time would result in an increase in the columns axial load carrying capacity. The ultimate state of reinforced concrete columns has often been defined by some researchers as a state of vanishing axial capacity to sustain the dead and live loads (long-term load), which is indicated as the point of axial failure (Fig. 12).

### 7.4.1 Post-blast damage evaluation of close -in blast damaged RC columns

Residual axial capacity level of a blast damaged column was evaluated by the ratio of residual axial strength, which was defined as the following equation:

$$\mathrm{v} = (P_r - P_L) / (P_{max} - P_L) \tag{3}$$

where $P_{max}$ is the axial capacity of the undamaged columns, $P_r$ is the residual axial capacity of the damaged columns and $P_L$ is the long-term axial load

When the column is undamaged, $P_r = P_{max}$ the value of $v$ is 1; when the column has lost the ability to sustain the long-term axial load, $P_r$=PL the value of $v$ just reaches zero, referring as the ultimate limit state of the column. As for the performance indicator, previously defined displacement to height ratio ($y_r/L$) is used.

### 7.4.2 Effect of axial load ratio

Fig. 13 shows the effect of axial load ratio on the residual axial capacity of the columns at various degrees of deformation level. The results show that at the same mid-height displacement ratio, the ratio of theresidual axial capacity is smaller in the case of larger axial loads.

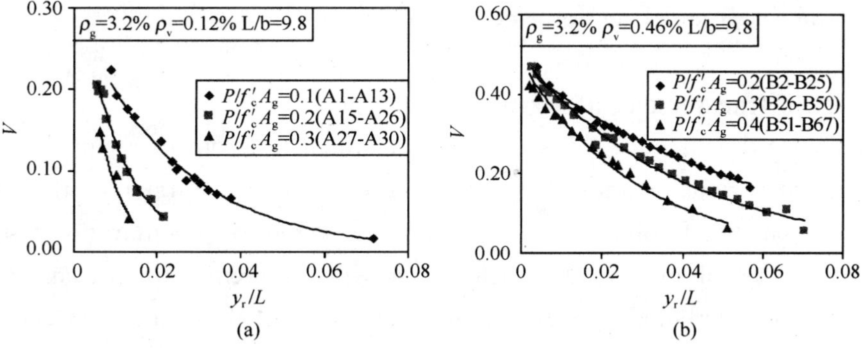


Figure 13　Effect of axial load ratio on the ratio of residual axial capacity of the blast damaged columns

### 7.4.3 Effect of longitudinal reinforcement ratio

It is usually assumed that the longitudinal reinforcement will support a portion of the axial load up to a maximum load defined by either the buckling or the plastic capacity of the reinforcing bars. In most cases the column collapse is related to the increase of axial load carried by the longitudinal reinforcing bars and their deterioration of compressive strength. Fig. 14 shows the effect of longitudinal reinforcement ratio on the residual axial capacity ratio of the columns. The results indicate that the ratio of residual axial capacity is generally

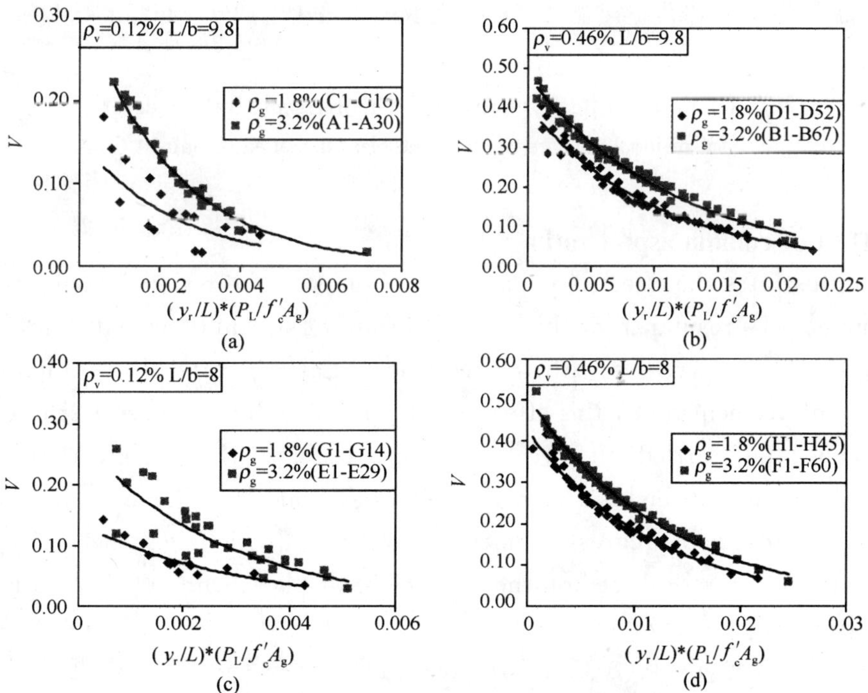


Figure 14　Effect of longitudinal reinforcement ratio on the ratio of residual axial capacity of the blast damaged columns

larger when the longitudinal reinforcement ratio increases.

### 7.4.4 Effect of transverse reinforcement ratio

Fig. 15 shows the effect of transverse reinforcement ratio on the residual axial capacity of the columns at various degrees of deformation level. The results show that the ratio of residual axial capacity of column with low transverse reinforcement ratio is significantly less than that of the column with high transverse reinforcement ratio. From these figures, it is observed that under similar axial loading conditions, axial load failure tends to occur at a relatively large displacement ratio for columns with high transverse reinforcement ratio. In contrast, columns with low transverse reinforcement ratio will tend to fail under axial loads at a smaller displacement ratio. This suggests that the displacement ratio at axial load failure is directly related to the amount of transverse reinforcement ratio.

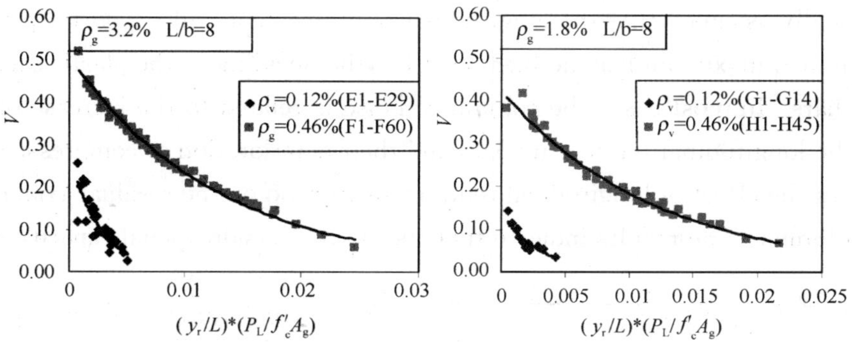


Figure 15 Effect of transverse reinforcement on the ratio of residual axial capacity of the blast damaged columns

### 7.4.5 Effect of column aspect ratio

Fig. 16 illustrates the effect of column aspect ratio on the residual axial strength ratio of the columns. The results show that at a high transverse reinforcement ratio, the residual axial capacity ratio increases with the reduction in aspect ratio. For columns with low transverse reinforcement ratio, this effect is not very clear due to the scatter in the results. The parametric study carried out revealed the significance of the parameters that affect the residual axial strength of the blast-damaged reinforced concrete column. A formula was derived through multivariable regression analysis in terms of various parameters to predict the residual axial capacity ratio based on the mid-height displacement to height ratio by fitting the parametric study results and is as follows:

$$v = [73.65\rho_v + 8.465\rho_g - 0.020879(L/b) + 0.104]e^{[89284.22\rho_v - 1308.64221\rho_g - 9.684203(L/b) - 382.12](y_r/L)(P_L/f'_cA_g)} \tag{4}$$

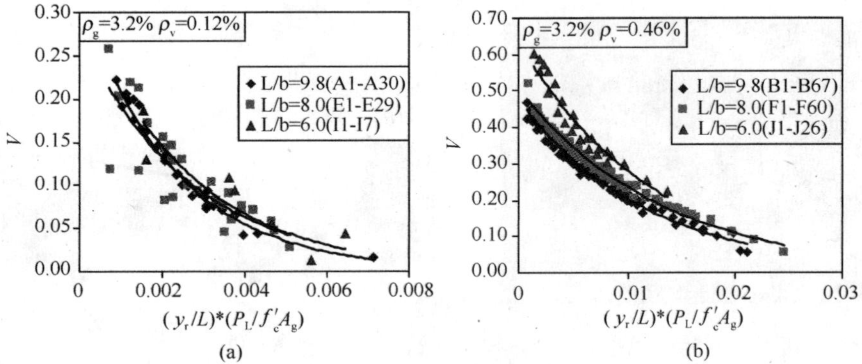


Figure 16　Effect of column aspect ratio on the ratio of residual axial capacity of the blast damaged columns

### 7.4.6　Post-Blast damage evaluation of localized blast damaged RC columns

The risks associated with suitcase bombs are of serious concern because they can be easily handled and placed within close proximity of key structural components of building structures. The most common failure mode of the structures subjected to blast loads from satchel and suitcase bombs is progressive collapse. High-fidelity physics based computer program, LS-DYNA is utilized by Wu *et al* (2010) to provide numerical simulations of the dynamic response and residual axial capacity of reinforced concrete (RC) columns subjected to blast loads. Field tests using near-field explosive charge were conducted on two RC column specimens ( see Fig 17). The test results were compared with the analytical results to validate the finite element model ( see Figs. 18 and 19). An extensive parametric study was conducted to investigate the relationship between residual axial capacity and structural and loading parameters such as material strength, column detail and blast conditions.

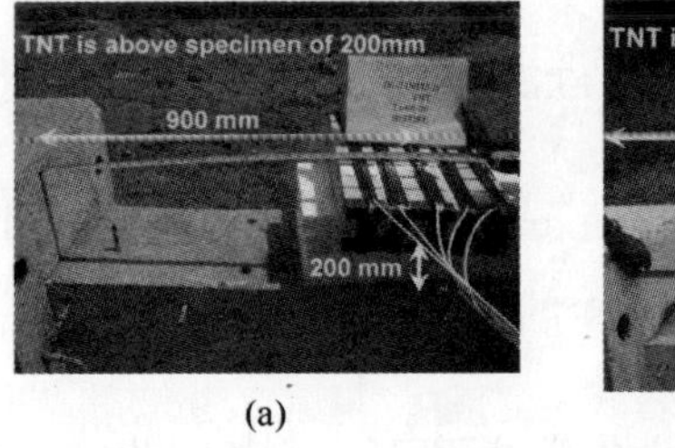

(a)

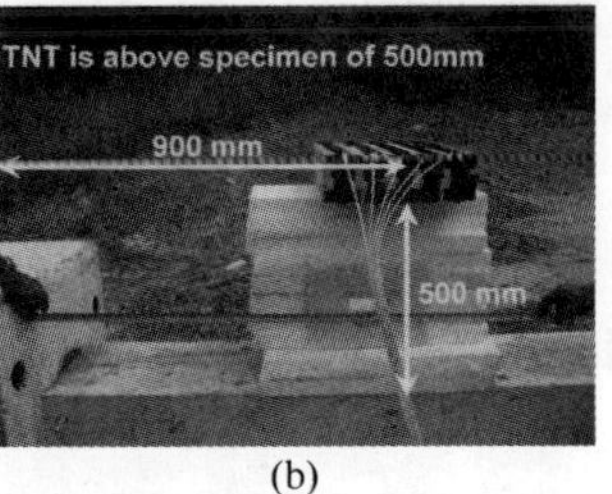

(b)

Figure 17　Test setup configuration

(a) RC column specimen 1; (b) RC column specimen 2

The parametric study revealed the significance of parameters that affect the residual axial capacity of the blast-damaged RC column. Two empirical equations based on the non-dimensional column dimension parameter were derived through multivariable regression analysis in terms of various parameters to predict the residual capacity index. The empirical

(a)

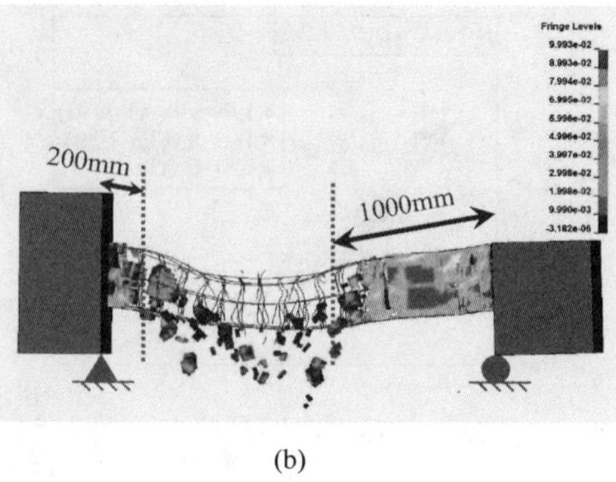


(b)

Figure 18　RC column specimen 1

(a) Explosive test; (b) Analytical results

(a)

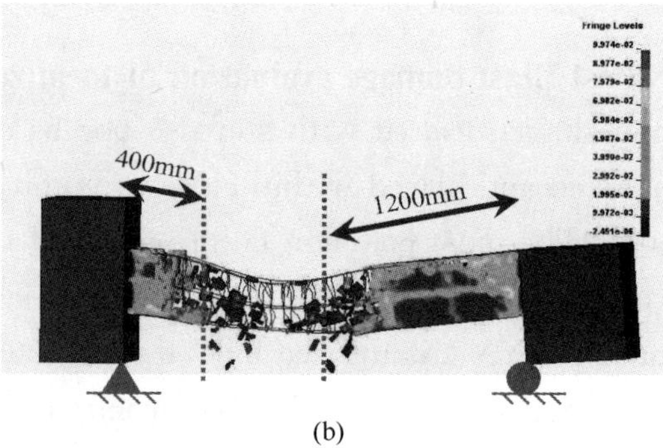


(b)

Figure 19　RC column specimen 2

(a) Explosive test; (b) Analytical results

equation for the case scenario where the TNT explosive is located at the bottom of column is expressed as follows:

$$\frac{P_{\mathrm{r}}}{P_{\mathrm{N}}}=(0.05\rho_{\mathrm{g}}+0.02\rho_{\mathrm{v}}-0.00035)\omega_{\mathrm{TNT}}^{\left(15\rho_{\mathrm{g}}-10\rho_{\mathrm{v}}-0.5\frac{P_{\mathrm{L}}}{f'_{\mathrm{c}}A_{\mathrm{g}}}-1.725\right)}\leqslant 1.0 \tag{5}$$

The other empirical equation for the case scenario where the TNT explosive is located at a height of from the footing of column is expressed as follows:

$$\frac{P_{\mathrm{r}}}{P_{\mathrm{N}}}=1.1-\left(-300\rho_{\mathrm{g}}-360\rho_{\mathrm{v}}-5\frac{P_{\mathrm{L}}}{f'_{\mathrm{c}}A_{\mathrm{g}}}+20.7\right)\omega_{\mathrm{TNT}}\leqslant 1.0 \tag{6}$$

The$\omega_{\mathrm{TNT}}$ of Eq. (5) and Eq. (6) is less than 0.04. A few examples presenting the comparison of the empirical equations with the analytical results are shown in Fig. 20. The dotted lines denote the analytical results, and the solid lines represent the empirical equations. It is observed that for most of the cases, the empirical equations provided an almost similar prediction as the analytical results.

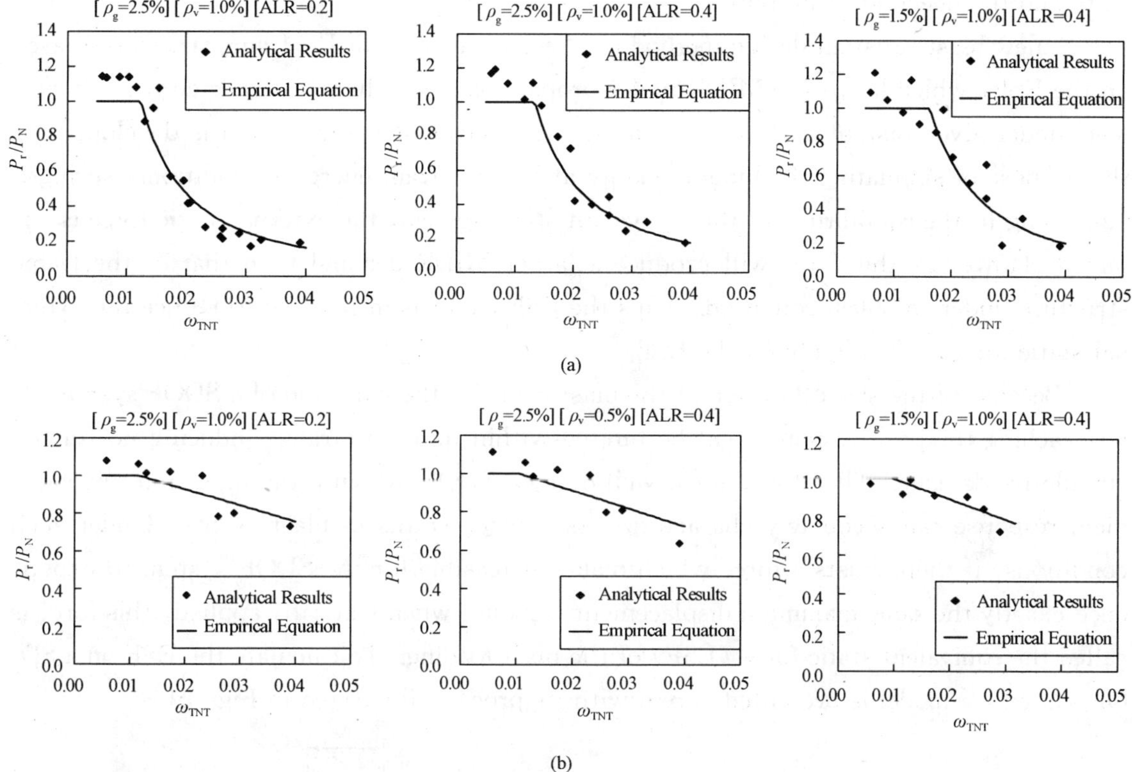


Figure 20　Comparison of analytical results with the empirical equations

(a) TNT is located at the bottom of column;

(b) TNT is located at a height of 1. 5m from the footing of column

## 7. 5 DRIFT-CONTROLLED DESIGN OF REINFORCED CONCRETE FRAME STRUCTURES

Due to an accidental severe explosion of building structures, a surrounding building may be subjected to a large-scale blast shock wave. In these kinds of relatively distant explosion conditions, the hemispherical blast wave produced may be reasonably simplified to be a planar wave and essentially parallel to the front faces of the target structure by comparing the sizes of hemispherical blast wave with the target structure. The distant blast wave loading on the structure produces a uniform lateral blast pressure on the front and rear faces and a vertical pressure on the roof. A significant side-sway response may occur inducing a certain degree of damage for a RC frame structure under such blast pressure. As a result, controlling their maximum inter-storey drift ratios (MIDR) within the predefined performance level becomes a crucial consideration in the blast resistant design of building structures. In order to ensure the side-sway response within the expected performance level, different levels of side-sway limits are specified for the design of single-storey rigid frame structures in the current design guidance according to the operational needs of the facility and the needs for reusability.

A new blast resistant design method has been developed for RC frame structures based on the ESF, which keeps the MIDR under proper control within the expected performance level under given distant surface explosions. An equivalent static system is developed for the purpose of simulating the kinetic energy at by the strain energy of additional springs. According to the equilibrium of the equivalent static system, the external static force is obtained. However, this force will produce a larger MIDR demand than that of the frame structure under the blast condition. Thus the ESF factor is introduced to reduce the external static force to finally obtain the ESF.

Because of the short duration of the blast loading, the vibration of a SDOF system after reaching the peak response will be limited within its elastic range inducing no further cumulative damage. Therefore, for a well-defined SDOF system, the maximum displacement response can adequately characterize its damage status in blast events. Under such conditions, if there exists a force which makes it feasible for the SDOF system to experience exactly the same maximum displacement response when statically applied, this force is called the equivalent static force (ESF) of the blast loading. To calculate the ESF on a SDOF system, a model is presented herein with its process illustrated in Fig. 21.

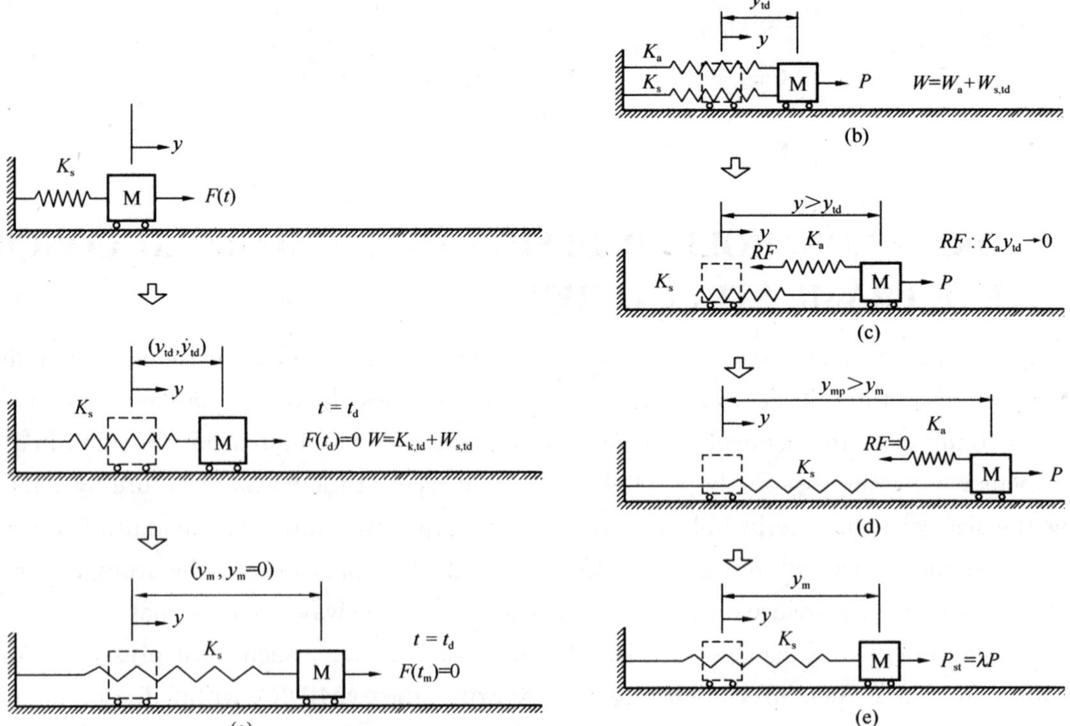


Figure 21 Process for the construction of ESF for a SDOF system

(a); (b) Equivalent static system; (c) Releasing the strain energy in spring $K_a$;

(d) Recovering to the original SDOF system plus the action of $P$;

(e) Equivalent static force

For a frame structure with n storeys, the ESF ($P_{st}$) is defined as a column vector of force, which, when applied statically to each floor level, can produce the same MIDR as that under a blast load, thus

$$P_{st} = \{P_{st,1}, P_{st,2}, \cdots, P_{st,n}\}^T \tag{7}$$

where $P_{st,i}$ represents the component of ESF at the $i$th floor level. The process of $P_{st}$ evaluation is shown in Fig. 18. At the end of the blast loading duration $t_d$, the frame structure responds with a distribution of displacement and velocity as shown in Fig. 22(a), where $y_{td} = \{y_{td,1}, y_{td,2}, \cdots, y_{td,n}\}^T$ and $y_{td,i}$ stands for the $i$th floor displacement at time $t_d$. With the increase in time, the distributed kinetic energy as a function of the velocity will be gradually transformed into further deformation reaching the MIDR. Therefore, if the kinetic energy $W_{k,i}$, which represents the kinetic energy within the part of the structure located halfway above and below the $i$th floor level at time $t_d$, is concentrated and represented by the strain energy $W_{a,i}$ of an additional uncoupled elastic spring $K_{a,i}$ at the corresponding floor level with the same deformation as $y_{td,i}$, an equivalent static system can be established in Fig. 22(b).

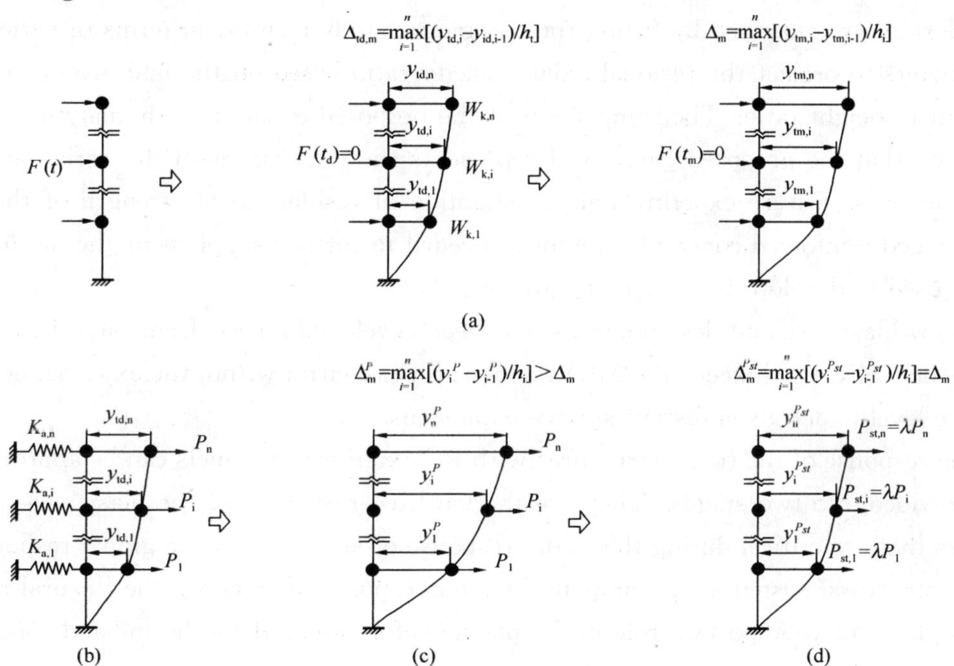


Figure 22 Process for the construction of ESF for a RC frame structure

(a) Dynamic response under the blast forec;

(b) Equivalent static system; (c) Static responses under $P$

(d) Static responses under $P_{st}$

## 7.6 CONCLUSIONS

Blast loadings on building structures can induce severe structural damage and the associated lost of lives and disruption in the service of the building. This paper seeks to comprehensively address the effects of external blast loadings on buildings and structural components. The conclusions can be summarized as follows:

(1) For blast-resistant designs, it would be more ideal that the RC member does not exhibit a brittle failure associated with shear failure during loading, and is allowed to experience flexural deformation. It has been demonstrated that the proposed procedure could incorporate the design performance criteria of maximum displacement and displacement ductility simultaneously to give a unique design of a RC member under a given blast loading on the basis of non-dimensional energy spectra.

(2) The research results show that the use of seismic detailing techniques can significantly reduce the degree of direct blast-induced damage and subsequent collapse of the reinforced concrete columns. Comparisons of the deterioration of the axial strength under different axial load ratios indicate that the ratio of residual axial strength is smaller under larger long-term axial load. The effect of axial load ratio is more critical in the case of columns with low transverse reinforcement ratio.

(3) A formula was derived by fitting the parametric study results, in terms of various parameters to predict the residual axial capacity ratio based on the mid-height displacement to height ratio. The comparison of the proposed equation with analytical results shows that the proposed curve well represents the tendency with the variation of the parameters. Future experimental investigation of residual axial strength of the blast damaged reinforced concrete columns is needed to further supplement the limited data set used to develop the proposed equation.

(4) A new blast resistant design method has been developed for RC frame structures based on the ESF, which keeps the MIDR under proper control within the expected performance level under given distant surface explosions.

(5) The response of the frame structures with RC exterior wall panels can be approximately divided into two stages. They are the localized responses of the blast-loaded members that are critical during the initial response of stage I, and the global responses of the structural system that dominate the later response of stage II. the flexural responses play a more important role in the plastic deformation of the beams and columns in the frame in comparison to their respective shear responses.

(6) It is also desirable to develop procedures for strengthening and retrofitting that can enhance the shear resistance and ductility of columns, beams and joints. These procedures can first be applied to strengthen and retrofit perimeter columns which are exposed to possibilities terrorist bomb placements and especially at locations with vehicle access.

## ACKNOWLEDGMENTS

The work reported here was carried out jointly with a number of talented and energetic graduate students including H-C Rong, Z W Huang, X. L Bao, S L Yap, A Nair and K-C Wu. Much of the basis of this paper resulted from research carried out at NTU and was funded by the Defense Science and Technology Agency (DSTA), Singapore under the overall direction of C H Lim, Director of DSTA.


## REFERENCES


[1] B Li, T-C Pan and A Nair "A Case Study of the Local and Global Structural Responses of a Tall Building in Singapore subjected to Close-in Detonations"The Structural Design of Tall and Special Buildings (To press-On line since July 2009).

[2] B Li, T-C Pan and A Nair "A Case Study of the Effect of Cladding Panels on the Response of Reinforced Concrete Frames subjected to Distant Blast Loadings" Nuclear Engineering and Design. Vol. 239, Issue 3, March 2009, Pages 455-469.

[3] B Li, H. C Rong, and T-C Pan "Drift-Controlled Design of Reinforced Concrete Frame Structures under Distant Blast Conditions-Part I: Theoretical Basis"International Journal of Impact Engineering. Vol. 34, March 2007, pp. 743-754.

[4] B Li, H. C Rong, and T-C Pan "Drift-Controlled Design of Reinforced Concrete Frame Structures under Distant Blast Conditions-Part II: Implementation and Evaluation " International Journal of Impact Engineering, Vol. 34, March 2007, pp. 743-754.

[5] H. C Rong and B Li "Probabilistic Response Evaluation for RC Flexural Members subjected to Blast Loadings" : Structural Safety. Vol. 29 March 2007, pp. 146-163.

[6] H. C Rong and B Li "Deformation-Controlled Design of RC Flexural Members Subjected to Blast Loadings" ASCE Journal of Structural Engineering, Oct 2008. Vol. 134 pp. 1598-1610.

[7] B Li, Z W Huang and C L Lim " The Verification of Non-Dimensional Energy Spectrum Based Blast Design for Reinforced Concrete Members through Actual Blast Tests " ASCE Journal of Structural Engineering, Vol. 136 No. 6, June 2010 pp. 627-636.

[8] X. L Bao and B Li "Residual Strength of Blast Damaged Reinforced Concrete Columns" International Journal of Impact Engineering Vol. 37, 2010, pp. 295-308.

[9] K Fujikake; B Li and S Soeun "Impact Response of Reinforced Concrete Beam and Its Analytical Evaluation" ASCE Journal of Structural Engineering Vol. 135 No. 8 Aug 2009 pp. 938-950.

[10] K-C Wu and B Li, K-C Tsia "Residual Axial Compression Capacity of localised Blast Damaged RC Columns" International Journal of Impact Engineering (To press).

# 第 8 章　Chapter 8

# 纳米混凝土的多功能特性

# MULTIFUNCTIONAL NANOMATERIALS FILLED CONCRETE

H. Li[1] (李惠), J. P. Ou[1,2] (欧进萍), H. G. Xiao[1] (肖会刚),
X. C. Guan[1] (关新春), B. G. Han[1] (韩宝国),

[1] School of Civil Engineering, Harbin Institute of Technology, Harbin, 150090, China;
[2] School of Civil and Hydraulic Engineering, Dalian University of Technology, Dalian, 116024, China.

**ABSTRACT**: Nanomaterials filled concrete was found to be multifunctional that has self-sensing ability and excellent mechanical properties. Self-sensing ability of nanoconcrete was based on the piezoresistivity, which was obtained by adding appropriate concentration of nano carbon black into concrete. Effect of various loading state on pieroresistivity of nanoconcrete was experimental studied, and the theoretical model was proposed to predict and modify the strain gauge factor of nanoconcrete under various loading or enviornmental conditions. Cement-based strain sensor was fabricated and used in monitoring the strain of concrete column. The other benefit on concrete was the enhancement in mechanical properties, including strength, abrasion resistance and fatigue properties. Microstructures of nanoconcrete was studied with help of SEM pictures, which showed that the hydration product of nanoconcrete was more uniform and compact than that of normal concrete.

## 8.1 INTRODUCTION

Civil infrastructures usually suffer from fatigue load, environmental corrosion or/and natural disasters. To improve the safety of infrastructures, structural health monitoring (SHM) is getting more and more important. SHM is normally performed by measuring the strain/stress of a structure's critical zone with sensors; the health of the structure can be evaluated based on the measured information, which can be used when deciding whether or not to repair the structure[1~5].

The recently developed multifunctional nanomaterials enabled concrete provides an efficient way to improve the safety of concrete structures. By incorporating some conductive nanomaterials (such as carbon nanofiber or carbon black) with concrete, the concrete can be conductive and piezoresistivity[6, 7], which is so called smart concrete. Smart concrete is a new generation of structure materials that developed from carbon fiber reinforced concrete (CFRC) in the past decade by utilizing the piezoresistivity for sensing strain[8~13] or by utilizing the relation between damage and resistance for sensing damage[14,15]. Recently, an advanced application method that using smart concrete as an embedded strain sensor was proposed. CBCC sensor is just this kind of embedment cement-based sensor[16]. The strain-sensing property, humidity insulation method and piezoresistivity model of CBCC sensor has been studied to promote the application of CBCC sensors[17,18].

Besides the self-sensing ability, another merit for concrete by incorporating nanomaterials is the improvement of microstructure and mechanical properties[19~23]. Due to an ultra-fine size, nano-particles show unique physical and chemical properties different from those of the conventional materials. This paper will introduce some progresses on utilizing nano-$SiO_2$, nano-$TiO_2$ to improve the strength, wear resistance and fatigue properties of concrete[24~27].

## 8.2 SELF-SENSING NANO-CONCRETE AND STRUCTURE

### 8.2.1 Materials and experimental methods

Carbon black (CB) of 120nm came from Liaoning Tianbao Energy Co., Ltd (Liaoning, China). The specific gravity of CB was 1.98g/cm$^3$. CB in the amount of 5%, 10%, 12%, 15%, 20% and 25% by weight of cement (i.e., 3.11%, 6.04%, 7.22%, 8.79%, 11.39%, and 13.85% by volume of composite, respectively) were used and the corresponding mixture types were called A-5, A-10, A-12, A-15, A-20 and A-25 respectively. The cement used was Portland cement (P. O42.5) from Harbin Cement Company (Harbin, China). The water-cement ratio was 0.4 for all specimens. A water-reducing agent UNF (one kind of naphthalene sulfonic acid and formaldehyde condensates) was used in the amount of 1.5% by weight of cement. The defoamer, tributyl phosphate (made in China), was used in the amount of 0.13 vol.% to decrease the number of air bubbles.

Defoamer and UNF water-reducing agent were dissolved in water, then CB was added and stirred at high speed in a mortar mixer for 3 minutes. This mixture and the cement were mixed at high speed for 2 min. After this, the mix was poured into oiled molds to form prisms of 30×40×50mm for compressive testing. After pouring, an external vibrator was used to facilitate compaction and decrease the number of air bubbles. The samples were demolded after 24h and then cured in a moist room (relative humidity 100%) for 28 days. Afterwards, the specimens were dried in an oven at 60 ℃ for 2 days to extract redundant water to eliminate the polarization effect on resistance measurement. The dried specimens were then tested at ambient temperature.

DC electrical resistance measurement was made in the longitudinal axis, using the four-probe method, in which copper nets served as electrical contacts. The copper nets were placed into the specimen when pouring the mix into molds. Four contacts were placed across the whole cross-section of 30×40mm of the specimen, these were all perpendicular to the longitudinal axis and symmetrically positioned with respect to the mid-point along the height of the specimen (i. e. , two contacts were in planes above the mid-point and two contacts were in planes below the mid-point). The outer two contacts (36mm apart) were for passing current. The inner two contacts (20mm apart) were for measuring the voltage (see Figure 1).

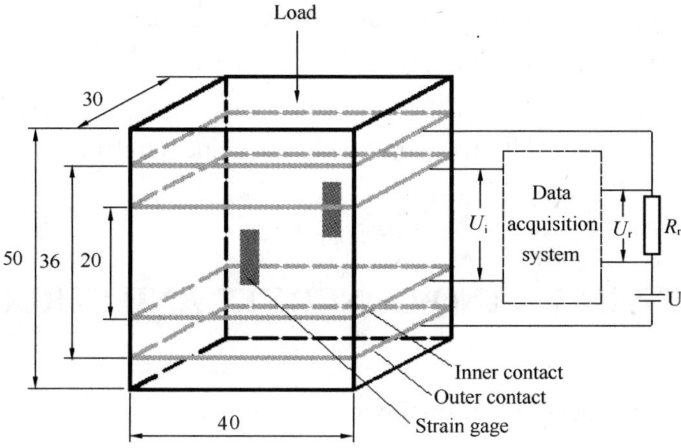


Figure 1 Schematic of the experimental set-up (mm)

### 8. 2. 2 Piezoresistivity of nano-concrete and the modeling

(1) Piezoresistivity of nano-concrete

Figure 2 shows the resistivity ($\rho$) as a function of CB volume content ($V$) of CB-filled composites. It can be observed from Figure 2 that the resistivity of the composites decreased dramatically with increasing CB content from 7. 22 vol. % to 11. 39 vol. %, i. e. from A-12 to A-20. The resistivity of the composites varied slightly outside the above range. The content range over which the resistivity varied precipitously was called percolation threshold. It was found that the tunneling effect dominated the conductivity of composites in percolation threshold. Tunneling current is an exponential function of barrier width, implying a precipitous change of resistivity upon distance between CB particles. In this paper, A-15 (so called CBCC) that following tunneling effect theory was taken for studying the piezoresistivity.

Figure 3 shows the fractional change in resistivity versus the compressive strain curves of A-15. The resistance was essentially proportional to the volume resistivity that was selected as a measurement in this study. For A-15, the resistivity decreased linearly with increasing compressive strain up to failure of the specimens except for a small perturbation over the strain range of [0.003～0.004] which indicated the occurrence of micro-cracks. The three curves for the three specimens of this mixture were almost the same, indicating that the results were repeatable. Linear fit of the experimental data showed that the relationship between the fractional change in resistivity and compressive strain was nearly linear. The fractional change in resistivity per unit strain (i. e., the strain gauge factor) was 55.28 as shown in Figure 3.

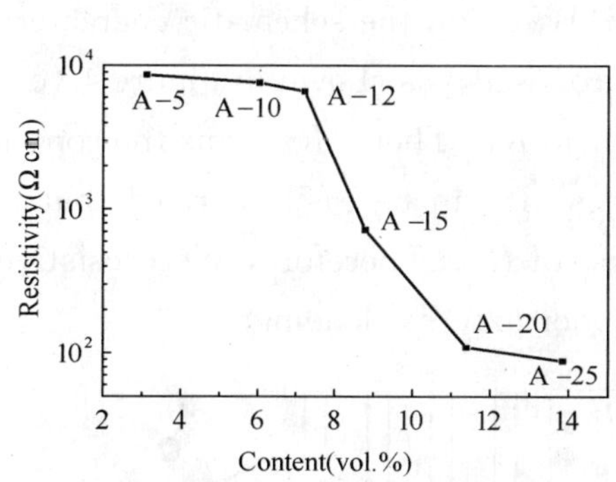


Figure 2　Logarithm of resistivity as a function of volume content of CB

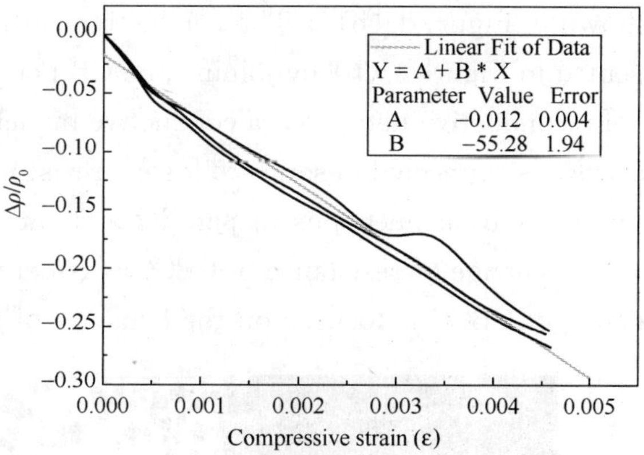

Figure 3　Fractional change in resistivity of A-15 as a function of compressive strain

(2) Modeling on piezoresistivity

According to tunneling effect theory [30], current density in a tunneling resistor formed by two adjacent CB particles at a low voltage region can be depicted by the following equation:

$$J = [3(2m\varphi)^{1/2}/2S_0](e/h)^2 U \cdot \exp[-(4\pi S_0/h)(2m\varphi)^{1/2}] \quad (1)$$

where $J$ is current density, m, e and h are the electron mass, charge on an electron and Planck's constant, respectively, $\varphi$, $S_0$ and $U$ are the height of tunneling potential barrier, tunneling width and voltage applied across tunneling resistor, respectively. $R_{t0}$ denotes the tunneling resistor and $A$ and $S_0$ denote the section area and length of the resistor, respectively. Then the resistance of the resistor can be obtained as

$$R_{t0} = k_1 S_0 \exp(k_2 S_0) \quad (2)$$

where $k_1=(2/3)(2m\varphi)^{-1/2}(e/h)^{-2}A^{-1}$ and $k_2=(4\pi/h)(2m\varphi)^{1/2}$

For a given composite, $k_1$ and $k_2$ are constants. $S_0$ can be obtained as follows:

$$S_0 = D[(\pi/6)^{1/3} V_c^{-1/3} - 1] \quad (3)$$

where $V_c$ and D are the volume concentration and diameter of CB particle, respectively. Equation (1) indicates that even a slight change of $S_0$ may cause a large change of resistance. Conductive networks in CBCC are composed of a large number of $R_{t0}$ and the change in resistance of CBCC is the integrated result of the change of each $R_{t0}$. Therefore, the piezoresistivity model of CBCC will be established based on the resistance behavior of each $R_{t0}$. The resistance of each $R_{t0}$ under strain can be quantified with equation (2); hence, the key task of the modeling is to obtain the deformation of each $R_{t0}$ under external strain.

The microstructure of CBCC was observed with SEM to study the characteristics of the conductive network of CBCC. Figure 4 (a) shows the microstructure of CBCC, in which the bright spherical objects and rounded hollowness denotes CB particles. To see the pattern of conductive network clearly, CB particles are emphasized in a white background, as shown in Figure 4 (b). The conductive path is formed by adjacent CB particles and is presented in Figure 4 (b) by joining the CB particles. Based on the schematic characteristics of a conductive network, a conductive model is proposed, as shown in Figure 4 (c). First, $R_{t0}$ is connected in series to form a resistor element $R_0$. Then, $R_0$ forms the conductive network by connections in parallel and then series. It can be easily derived that the fractional change of resistance of CBCC is equal to that of $R_0$. Therefore, piezoresistivity modeling of CBCC is focused on the behavior of $R_0$ under external loading.

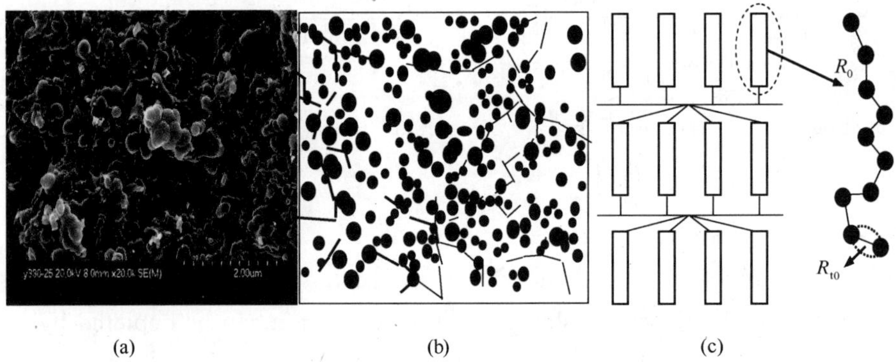


Figure 4 Schematic of conductive network in CBCC

(a) SEM picture; (b) Schematic picture; (c) conductive network model

The resistance value of each $R_{t0}$ can be calculated based on equations (1) and (2). The orientation direction of each $R_{t0}$ is assumed to be uniformly distributed by considering the infinite and randomly distributed CB particles in CBCC. Figure 5 (a) and (b) show the schematics of orientation and distribution of tunnel resistor, respectively. Assume that there are a total $4\times3\ (M+1)^2$ of $R_{t0}$ in the space and each $4\ (M+1)^2$ of $R_{t0}$ surround a coordinator. For example, $4\ (M+1)^2$ of $R_{t0}$ is first defined by $\theta_z$ and $\varphi_z$, as shown in Figure 5 (a). Both $\theta_z$ and $\varphi_z$ should be over the range of $[0, 2\pi]$; however, considering the symmetrical characteristic of $R_{t0}$ in each quadrant, $\theta_z, \varphi_z \in [0, \pi/2]$ is sufficient to represent the orientation character of $R_{t0}$ and is adopted in this paper. Therefore, $\theta_z$ and $\varphi_z$ are in

$$\theta_z, \varphi_z = [0, \pi/2M, \pi/M, 3\pi/2M, \cdots\cdots, \pi/2(M-1), \pi/2] \tag{4}$$

where $\theta_x, \varphi_x$, and $\theta_y, \varphi_y$ are defined in the same way as that of $\theta_z, \varphi_z$. Hence, the resistance behavior of CBCC can be described by the $3\ (M+1)^2$ of $R_{t0}$. As shown in Figure5 (b), the distribution of CB particles defined by above method is not uniform in space that the distribution density is higher near each coordinate, but it is uniform in calculating the effect of multi-axial strain on resistance.

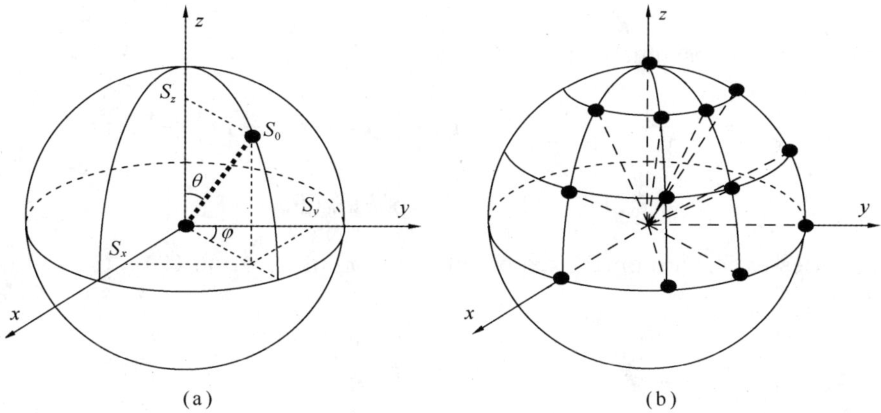


(a) (b)

Figure 5 Schematic of

(a) orientation and (b) distribution of $R_{t0}$

When CBCC is load free, the original length of $R_{t0}$ is $S_0$, and $R_0$ is the sum of 3 $(M+1)^2$ of $R_{t0}$ as

$$R_0 = 3\,(M+1)^2 k_1 S_0 \exp\,(k_2 S_0) \tag{5}$$

Then, assuming that CBCC is subjected to multi-axis strain of $\varepsilon_{mx}$, $\varepsilon_{my}$ and $\varepsilon_{mz}$ in each axis, the corresponding strain of $R_{t0}$ in each axis direction can be calculated as

$$\varepsilon_{x,y,z} = \frac{\varepsilon_{mx,my,mz}}{V_c(E_m/E_c)+V_m} \tag{6}$$

where $V_c$ and $V_m$ are the ratio of volume occupied by CB and matrix to the whole volume of composite, respectively. $E_c$ and $E_m$ are the modulus of CB and matrix, respectively. The length of $R_{t0}$ under multi-axial strain changes from $S_0$ to either $S_{\varepsilon x}$, or $S_{\varepsilon y}$, or $S_{\varepsilon z}$ as

$$\begin{aligned} S_{\varepsilon x} &= S_0\sqrt{\cos^2\theta_x\,(1+\varepsilon_x)^2+\sin^2\theta_x\,\sin^2\varphi_x\,(1+\varepsilon_y)^2+\sin^2\theta_x\,\cos^2\varphi_x\,(1+\varepsilon_z)^2} \\ S_{\varepsilon y} &= S_0\sqrt{\cos^2\theta_y\,(1+\varepsilon_y)^2+\sin^2\theta_y\,\sin^2\varphi_y\,(1+\varepsilon_x)^2+\sin^2\theta_y\,\cos^2\varphi_y\,(1+\varepsilon_z)^2} \\ S_{\varepsilon z} &= S_0\sqrt{\cos^2\theta_z\,(1+\varepsilon_z)^2+\sin^2\theta_z\,\sin^2\varphi_z\,(1+\varepsilon_x)^2+\sin^2\theta_z\,\cos^2\varphi_z\,(1+\varepsilon_y)^2} \end{aligned} \tag{7}$$

Equation (7) indicates that even under the same external strain status, the strain of each $R_{t0}$ with different orientation angle is different. Therefore, based on equations (5) and (7), the resistance of CBCC under multi-axial strain changes from $R_0$ to $R$ as

$$R = \sum_{i=1}^{M+1}\sum_{j=1}^{M+1} k_1 S_0\left[F_x\exp(k_2 S_0 F_x)+F_y\exp(k_2 S_0 F_y)+F_z\exp(k_2 S_0 F_z)\right] \tag{8}$$

where

$$F_x=\sqrt{\cos^2\left(\frac{(i-1)\pi}{2M}\right)(1+\varepsilon_x)^2+\sin^2\left(\frac{(i-1)\pi}{2M}\right)\sin^2\left(\frac{(j-1)\pi}{2M}\right)(1+\varepsilon_y)^2+\sin^2\left(\frac{(i-1)\pi}{2M}\right)\cos^2\left(\frac{(j-1)\pi}{2M}\right)(1+\varepsilon_z)^2}$$

$$F_y=\sqrt{\cos^2\left(\frac{(i-1)\pi}{2M}\right)(1+\varepsilon_y)^2+\sin^2\left(\frac{(i-1)\pi}{2M}\right)\sin^2\left(\frac{(j-1)\pi}{2M}\right)(1+\varepsilon_x)^2+\sin^2\left(\frac{(i-1)\pi}{2M}\right)\cos^2\left(\frac{(j-1)\pi}{2M}\right)(1+\varepsilon_z)^2}$$

$$F_z=\sqrt{\cos^2\left(\frac{(i-1)\pi}{2M}\right)(1+\varepsilon_z)^2+\sin^2\left(\frac{(i-1)\pi}{2M}\right)\sin^2\left(\frac{(j-1)\pi}{2M}\right)(1+\varepsilon_x)^2+\sin^2\left(\frac{(i-1)\pi}{2M}\right)\cos^2\left(\frac{(j-1)\pi}{2M}\right)(1+\varepsilon_y)^2}$$

Hence, the change in resistance of CBCC can be calculated as

$$\frac{R}{R_0}=\frac{1}{3(M+1)^2}\sum_{i=1}^{M+1}\sum_{j=1}^{M+1}\{F_x\exp[k_2S_0(F_x-1)]+F_y\exp[k_2S_0(F_y-1)]+F_z\exp[k_2S_0(F_z-1)]\} \tag{9}$$

Conducting series expansion on the exponent function in equation (9), the following equation can be obtained:

$$\frac{R}{R_0}=\frac{1}{3(M+1)^2}\sum_{i=1}^{M+1}\sum_{j=1}^{M+1}(F_x+F_y+F_z)+\frac{k_2S_0}{3(M+1)^2}\sum_{i=1}^{M+1}\sum_{j=1}^{M+1}(F_x^2-f_x+F_y^2-F_y+F_z^2-F_z)+\frac{(k_2S_0)^2}{6(M+1)^2}\sum_{i=1}^{M+1}\sum_{j=1}^{M+1}[F_x(F_x-1)^2+F_y(F_y-1)^2+F_z(F_z-1)^2]+\cdots\cdots \tag{10}$$

The first item of the right hand side in equation (10) is the inherent behavior of the integrated change in length of each $R_{t0}$ under external strain and is denoted as $\overline{S}/S_0$, which represents the resistance change trend of CBCC under multi-axial strain. $k_2S_0$ is denoted as $\beta$, which represents the sensitivity factor of CBCC and is dependent on the intrinsic property of CBCC. Therefore, $\overline{S}/S_0$ and $\beta$ represent the piezoresistivity characteristics (trend and sensitivity factor) of a composite. Considering the uncertainty in CBCC synthesis, $\beta$ is best obtained from a simple uniaxial compressive test and can then be utilized to predict the resistance behavior of CBCC under complex loading and environmental conditions. Orientation uniformity of $R_{t0}$ is dependent on the value of $M$. $M\geqslant 30$ can meet the precision requirement of calculation.

For CBCC, Based on the previously measured strain gage factor of 55.5, coefficient $\beta$ can be calculated as 337, and $k_2$ is then obtained as 3.45 $nm^{-1}$. Utilizing these parameters, the model can predict the strain gauge factors of a composite under various complex strain statuses. Besides predicting the resistance behavior of CBCC under various strain states, the model can be used to predict the resistance behavior of CBCC under various ambient conditions by transforming such conditions into strain.

### 8.2.3 Self-sensing concrete structures

#### (1) Preparation of Concrete Column with Embedded CBCC Sensor

Epoxy encapsulated CBCC sensor with shape of 30×40×50 mm was prepared with A-15, and the properties of CBCC sensor were shown in Table 1.

**Properties of CBCC sensor　Table 1**

| Strain sensing properties | | Mechanical properties | | | |
|---|---|---|---|---|---|
| Resistivity (Ωcm) | Gage factor | Strength (MPa) | Peak strain ($\mu\varepsilon$) | Modulus ($\times 10^4$ MPa) | Poisson ratio |
| 719.4±53.7 | 55.5±2.87 | 44.7±0.32 | 4400±136 | 1.44±0.06 | −0.17±0.015 |

Concrete columns were made with C40 and C80 concrete respectively for the aim of studying the strain sensing property of CBCC sensors in different strength grades of concrete matrix. Concrete columns were cast with molds of 100×100×300 mm, and a CBCC sensor was placed at the center of each column as shown in Figure 6. The longitudinal axis (resistance measurement direction) of the CBCC sensors was parallel to the 300mm side of concrete columns.

Uniaxial compressive test was performed on the 100×100mm side of a concrete column by MTS with 2, 500 kN maximum loading capacity. Two loading schedules, i. e. cyclic loading was arranged.

Figure 6　CBCC sensor and concrete column with embedded CBCC sensor

**(2) Measured Strain under Cyclic Loading**

Figure 7 shows the strain of a C40 concrete column [Figure 7(a)] and a C80 concrete column [Figure 7(b)] under cyclic loading measured by CBCC sensor and displacement transducer, respectively. Strain measured by CBCC sensor was obtained by dividing the fractional change in resistance with the gauge factor 55.5. It can be seen from Figure 7 that the strain amplitude of C40 concrete column measured by displacement transducer, and CBCC sensor is about 400 $\mu\varepsilon$ and 380 $\mu\varepsilon$, respectively, and the strain amplitude of the C80 concrete column is about 700 $\mu\varepsilon$ by displacement transducer and 630 $\mu\varepsilon$ by CBCC sensor. During the interval of load locking, the measured strain by displacement transducer and CBCC sensor both keep stable. The little discrepancy between strain measured by CBCC sensor and that by displacement transducer is caused by the accuracy of the calibrated gauge factor used here as 55.5. Considering the deformation capacity of CBCC sensor as 4400 $\mu\varepsilon$, the deformation of the embedded CBCC sensor under cyclic loading is in elastic regime, which can be verified by the zero residual strain after unloading. The experimental results indicate that the repeatability and stability of embedded CBCC sensor under cyclic elastic loading are reliable.

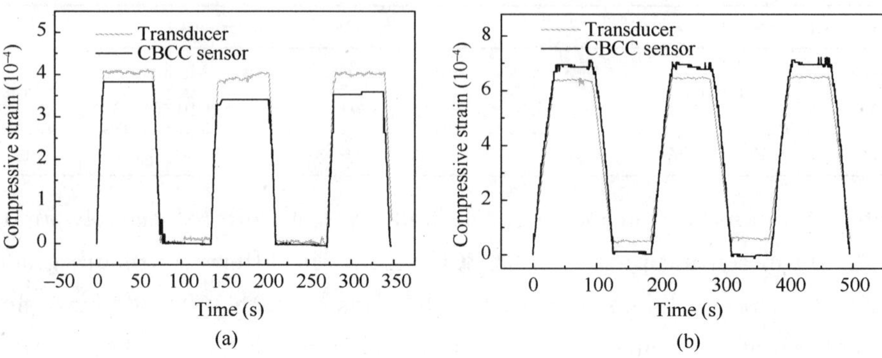


Figure 7 he strains of (a) C40 concrete column and (b) C80 concrete column under cyclic loading measured by displacement transducer and CBCC sensor. Strain measured by CBCC sensor was obtained by dividing the fractional change in resistance with the calibrated gauge factor 55. 5

# 8. 3 MECHANICAL PROPERTIES OF NANO-CONCRETE

## 8. 3. 1 Abrasion resistance of concrete containing nano-particles

### (1) Materials and experimental methods

The cement used is Portland cement (P. O42. 5) . Fine aggregate is natural river sand with a fineness modulus of 2. 4. The coarse aggregate used is crushed diabase with diameter of 5~25mm. UNF water-reducing agent (one kind of β-naphthalene sulfonic acid and formaldehyde condensates, China) is employed to aid the dispersion of nano-particles in concrete and achieve good workability of concrete. The defoamer, tributyl phosphate (made in China) is used to decrease the amount of air bubbles.

The nano-particles are purchased from Zhoushan Mingri Nano-phase Material Co. (Zhejiang, China) and their properties are shown in Table 2. The modified PP fibers are obtained from Zhangjiagang Synthetic Fiber Co. (Jiangsu, China) and their properties are shown in Table 3.

**The properties of nano-particles** **Table 2**

| Item | Diameter (nm) | Specific surface area ($m^2/g$) | Density ($g/cm^3$) | Purity (%) | Phase |
|---|---|---|---|---|---|
| $SiO_2$ | 10±5 | 640±50 | <0. 12 | 99. 9 | — |
| $TiO_2$ | 15 | 240±50 | 0. 04~0. 06 | 99. 7 | Anatase |

**The properties of PP fibers** **Table 3**

| Item | Elongation (%) | Fiber number (D) | Diameter (μm) | Length (mm) |
|---|---|---|---|---|
| Target | 40±3 | 11±0.5 | 84～92 | 15±1 |

The mixture proportions for cubic meter of concrete are given in table 4. Herein, PC denotes plain concrete. PPC1and PPC2 denote the concrete containing PP fibers in the content of 0.6 kg/m$^3$ and 0.9 kg/m$^3$, respectively. NSC1 and NSC3 denote the concrete containing nano-$SiO_2$ in the amount of 1% and 3% by weight of binder, respectively. And NTC1, NTC3 and NTC5 denote the concrete containing nano-$TiO_2$ in the amount of 1%, 3% and 5% by weight of binder, respectively.

**Mix proportions of specimens** (kg/m$^3$) **Table 4**

| Mixture no. | Water | Cement | Sand | Coarse aggregate | PP fiber | Nano-$SiO_2$ | Nano-$TiO_2$ | UNF | Defoamer | Slump (cm) |
|---|---|---|---|---|---|---|---|---|---|---|
| PC | 151 | 360 | 650 | 1260 | — | — | — | 5.4 | — | 5～6 |
| PPC1 | 151 | 360 | 650 | 1260 | 0.6 | — | — | 5.4 | — | 3～4 |
| PPC2 | 151 | 360 | 650 | 1260 | 0.9 | — | — | 5.4 | — | 2～3 |
| NSC1 | 151 | 356.4 | 650 | 1260 | — | 3.6 | — | 5.4 | 0.216 | 2～3 |
| NSC3 | 151 | 349.2 | 650 | 1260 | — | 10.8 | — | 7.2 | 0.288 | 1～2 |
| NTC1 | 151 | 356.4 | 650 | 1260 | — | — | 3.6 | 5.4 | 0.216 | 2～3 |
| NTC3 | 151 | 349.2 | 650 | 1260 | — | — | 10.8 | 7.2 | 0.288 | 2～3 |
| NTC5 | 151 | 342 | 650 | 1260 | — | — | 18 | 7.2 | 0.288 | 1～2 |

To fabricate the concrete containing nano-particles, water-reducing agent is firstly mixed into water in a mortar mixer, and then nano-particles are added and stirred at a high speed for 5 min. Defoamer is added as stirring. Cement, sand and coarse aggregate are mixed at a low speed for 2 min in a concrete centrifugal blender, and then the mixture of water, water reducing agent, nano particles and defoamer is slowly poured in and stirred at a low speed for another 2 min to achieve good workability.

Abrasion testing is conducted according to GB/T 16925—1997 (Test method for abrasion resistance of concrete and its products, China), testing equipment is ball bearing abrasion machine.

**(2) Abrasion resistance**

Table 5 shows the results of abrasion resistance of all specimens at the 28th day. It can be seen that the abrasion resistance of concretes containing nano-particles and PP fibers is remarkably improved, in particular the abrasion resistance of concrete containing nano-particles. The enhanced extent of the abrasion resistance of concrete containing nano-particles is much higher than that of concrete containing PP fibers. The side indices of abrasion resistance of all concretes are larger than their surface indices of abrasion resistance.

**The results of abrasion resistance of specimens** **Table 5**

| Mixture no. | Surface index of abrasion resistance | | Side index of abrasion resistance | |
|---|---|---|---|---|
| | Target | Enhancedextent (%) | Target | Enhancedextent (%) |
| PC | 1.19 | 0 | 1.55 | 0 |
| PPC1 | 1.42 | 19.1 | 2.42 | 55.9 |
| PPC2 | 1.60 | 34.4 | 2.62 | 69.2 |
| NSC1 | 3.06 | 157.0 | 3.71 | 139.4 |
| NSC3 | 2.39 | 100.8 | 2.93 | 89.0 |
| NTC1 | 3.34 | 180.7 | 4.24 | 173.3 |
| NTC3 | 2.95 | 147.7 | 3.72 | 140.2 |
| NTC5 | 2.27 | 90.4 | 2.88 | 86.0 |

The effectiveness of nano-$TiO_2$ in enhancing abrasion resistance increases in the order: NTC5<NTC3<NTC1 (with the decrease on nano-$TiO_2$ content). The abrasion resistance of concrete containing nano-$TiO_2$ in the amount of 1% by weight of binder increases by 180.7% for the surface index and 173.3% for the side index. Even for the concrete containing nano-$TiO_2$ in the amount of 5% by weight of binder, the abrasion resistance increases by 90.4% for the surface index and 86% for the side index. The similar results can be found for the concrete containing nano-$SiO_2$. The index of abrasion resistance of concrete containing PP fibers increases with increasing fibers content. The enhanced extent is almost the same as that presented in the related literatures. The addition of nano-particles is much more favorable to the abrasion resistance of concrete than that of PP fibers.

## 8.3.2 Flexural fatigue performance of concrete containing nano-particles

### (1) Materials and experimental methods

The materials and fabrication methods were same to that shown in section 3.1. Flexural strength testing is performed in accordance with JTJ 053-94 (Testing Methods of Concrete for Highway Engineering, China). For flexural fatigue testing, a four-point bending test method is applied with an effective span of 300 mm in a 100 kN Material Testing System (MTS). The test is carried out in load control using a continuous sinusoidal waveform with a loading frequency of 10 Hz.

The load cycle characteristic value R is defined as follows: $R = P_{min}/P_{max}$, where $P_{min}$ and $P_{max}$ refer to the minimum and maximum load of sinusoidal wave in each cycle. R is taken as 0.1 in this test.

The stress level S is defined as: $S = \sigma_p/\sigma_f$, where $\sigma_p$ and $\sigma_f$ are the flexural fatigue strength and the flexural strength, respectively. The following four stress levels are selected: 0.70, 0.75, 0.80 and 0.85 for all specimens referred in this study.

The input data for the test include the waveform, maximum and minimum load amplitude, loading frequency, maximum number of cycles. The reference for the loading stress levels is the average ultimate static flexural strength of specimens measured just before the fatigue testing. The preload of 100～200 N is put on the specimen to eliminate the error caused by poor contact. The fatigue failure numbers of specimens are recorded.

**(2) Flexural fatigue performance of concretes under the same failure probability**

The fatigue lives, as well as their enhanced extent calculated by using single-logarithm fatigue equation are almost the same as that calculated by using double-logarithm fatigue equation. The fatigue lives of various concretes increase in the order: PC<NTC3<NSC1<NTPC<PPC<NTC1.

It can be seen from Table 6 and Table 7 that the theoretic fatigue number of PPC increases by 354.07% when the stress level S is 0.85, which is approximately consistent with the test results in related literature. However, the theoretic fatigue number of NTC1 increases by 475.38% at the same stress level. When the stress level S is 0.70, the theoretic fatigue number of PPC increases by 117.46%, however, the theoretic fatigue number of NTC1 increases by 267.22%. Apparently, the concrete containing 1% nano-$TiO_2$ has much better flexural fatigue performance than the concrete containing PP fibers. The enhanced extent of fatigue life of the concrete containing 3% nano-$TiO_2$ is the lowest. Even for this concrete, the theoretic fatigue number increases by 130.77% and 41.41% at the stress levels of 0.85 and 0.70, respectively.

**Fatigue lives of concretes calculated by single-logarithm fatigue equation　Table 6**

| Mixture type | S=0.85 | | S=0.80 | | S=0.75 | | S=0.70 | |
|---|---|---|---|---|---|---|---|---|
| | Theoretic fatigue number | Enhanced extent (%) | Theoretic fatigue number | Enhanced extent (%) | Theoretic fatigue number | Enhanced extent (%) | Theoretic fatigue number | Enhanced extent (%) |
| PC | 10 | 0.00 | 45 | 0.00 | 215 | 0.00 | 1016 | 0.00 |
| PPC | 44 | 354.07 | 161 | 255.26 | 597 | 177.95 | 2209 | 117.46 |
| NSC1 | 24 | 151.90 | 96 | 111.83 | 383 | 78.13 | 1522 | 49.79 |
| NTC1 | 55 | 475.38 | 225 | 395.39 | 916 | 326.52 | 3730 | 267.22 |
| NTC3 | 22 | 130.77 | 89 | 96.01 | 358 | 66.49 | 1437 | 41.41 |
| NTPC | 33 | 245.03 | 127 | 179.59 | 487 | 126.56 | 1865 | 83.59 |

**Fatigue lives of concretes calculated by double-logarithm fatigue equation　Table 7**

| Mixture type | S=0.85 | | S=0.80 | | S=0.75 | | S=0.70 | |
|---|---|---|---|---|---|---|---|---|
| | Theoretic fatigue number | Enhanced extent (%) | Theoretic fatigue number | Enhanced extent (%) | Theoretic fatigue number | Enhanced extent (%) | Theoretic fatigue number | Enhanced extent (%) |
| PC | 10 | 0.00 | 43 | 0.00 | 203 | 0.00 | 1060 | 0.00 |
| PPC | 46 | 352.62 | 156 | 260.84 | 575 | 183.50 | 2322 | 119.05 |
| NSC1 | 25 | 151.30 | 93 | 115.08 | 369 | 82.23 | 1618 | 52.65 |
| NTC1 | 57 | 470.30 | 215 | 397.83 | 873 | 330.77 | 3912 | 269.05 |
| NTC3 | 23 | 130.10 | 85 | 98.04 | 342 | 68.80 | 1508 | 42.30 |
| NTPC | 34 | 240.68 | 121 | 180.76 | 463 | 128.51 | 1944 | 83.36 |

**(3) Mechanism of fatigue improvement**

The mechanism of nano-particles improving the flexural fatigue performance of concrete can be interpreted as follows. Supposed that nano-particles are uniformly dispersed and each particle is contained in a cube pattern, the distance between nano-particles can be specified. After hydration begins, hydrate products diffuse and envelop nano-particles as kernel. If the content of nano-particles and the distance between them are appropriate, the crystallization will be controlled to be a suitable state through restricting the growth of Ca $(OH)_2$ crystal by nano-particles. Moreover, the nano-particles located in cement paste as kernel can further promote cement hydration due to their high activity. This makes the cement matrix more homogeneous and compact. Additionally, nano-particles can act as fillers to improve the density of concrete, which leads to the porosity of concrete reduced significantly. As a consequence, the flexural fatigue performance and strength of concrete are improved evidently, as presented in previous sections of this article.

## 8.4 CONCLUSIONS

Concrete structures constitute a large portion of civil infrastructures, but their reliability is relatively low because of wide material discreteness and complex service environment. Consequently, the safety of concrete structures is an important problem being paid attention to at all times in civil engineering field. Therefore, it is necessary to take reasonable measures to monitor the state of concrete structures. In order to monitor the performance and state of concrete structures during their service periods, the information of structural state need be obtained by appropriate monitoring technologies. Now local monitoring for concrete structures is usually achieved by embedding such sensors as electric-resistance strain gauges, optic sensors, piezoelectric ceramic, shape memory alloy and fiber reinforced polymer bar in key structural positions. However, these sensors have such drawbacks as poor durability, low sensitivity, high cost, and unfavorable compatibility with concrete structures. Smart concrete has favorable piezoresistivity, great durability and good compatibility with concrete structures, etc., and it can therefore be used to develop retrofit or new installations, including traffic monitoring, weighing in motion, corrosion monitoring of rebar, strain-sensing coating.

## ACKNOWLEDGEMENT

This work is financially supported by NSFC grants No. 50525823, 50238040, 50808059, 50538020, 50278029, the Ministry of Science and Technology grant No. 2006BAJ03B05 and 2007AA04Z435, the Outstanding Young Teachers in Harbin Institute of Technology HITQNJS. 2007. 030.

# REFERENCES

[1] Scott, W. D., Charles, R. F., Michael, B. P. and Daniel, W. S. Damage Identification and Health Monitoring of Structural and Mechanical Systems from Changes in Their Vibration Characteristics: A Literature Review, (1996) Report of Los Almos Lab.

[2] Chen, J. C.. Intelligent Monitoring System for Suspension Bridge Damage Detection, Proc. Of Bridge into the 21st Century Conference, Hongkong, (1995)Oct., 25.

[3] Li, H, Ou, J. P., Zhao, X. F.. Structural health monitoring system for the Shandong Binzhou Yellow River Highway Bridge, Computer-aided Civil and Infrastructure Engineering, 21(4) (2006) 306-317.

[4] Austin, T., Singh, M., Gregson, P. J., Dakin, J. P., Powell, P. M.. Damage Asseeement in Hybrid Laminates Using an Array of Embedded Fiber Optic Sesnors, Proc. The SPIE conference on Smart Systems for Bridges, Structures and highways, Newport Beach, California, March, SPIE. 3671 (1999) 281-287.

[5] Claus, R. O., Mckeenman, J. C., May, R. G., et al.. Opitical fiber sensors and signal processing for smart materials and Structures ARO Smart Materials, Structures and Mathematical Issues workshop Proceeding, Virginia Polytechnic Institute and state University Blacksburg, VA, (1988) 15～16.

[6] Li, H., Xiao, H. G., Ou, J. P.. Effect of compressive strain on electrical resistivity of carbon black-filled cement-based composites, Cement and Concrete Composites, 28 (2006) 824-828.

[7] Xiao H. G., Li, H., Ou, J. P.. Self-monitoring properties of concrete columns with embedded cement-based strain sensors. Journal of Intelligent materials systems and structures. (2010) Accepted.

[8] Wen, S. H., Chung, D. D. L.. Carbon fiber-reinforced cement as a strain-sensing coating, Cement and Concrete Research, 31 (2001) 665～667.

[9] Reza, F., Batson, G. B., Yamamuro, J. A., Lee J. S.. Resistance changes during compression of carbon fiber cement composites, Journal of Materials in Civil, 15(2003) 476-483.

[10] Song, X. H., Zheng, L. X., Li, Z. Q.. Temperature compensation in deformation testing for smart concrete structures, Key Engineering Materials, 326-328 (2006) 1503-1506.

[11] Han, B. G., Ou, J. P.. Embedded piezoresistive cement-based stress/strain sensor, Sensors and Actuators A, 138 (2007) 294-298.

[12] Ou, J. P., Han, B. G... Piezoresistivie cement-based strain sensors and self-sensing concrete components, Journal of Intelligent material systems and structures, 20 (2009) 329-336.

[13] Yu, X., Kwon, E.. A carbon nanotube/cement composite with piezoresistive properties, Smart Materials and Structures, 18 (2009) 1-5.

[14] Chen, B., Liu, J. Y.. Damage in carbon fiber-reinforced concrete, monitored by both electrical resistance measurement and acoustic emission analysis, Construction and Building Materials, 22 (2008) 2196-2201.

[15] Wang, S., Shui, X., Fu, X., Chung, D. D. L.. Early fatigue damage in carbon-fibre composites observed by electrical resistance measurement, Journal of Materials Science, 33 (1998) 3875-3884.

[16] Xiao H. G., Li. H. A study on the application of CB-filled cement-based composites as a strain sensor for concrete structures. (SPIE) Nondestructive Evaluation and Health Monitoring of Aerospace Materials, Composites, and Civil Infrastructure. San Diego, CA, USA. Mar, 2006.

[17] Li, H., Xiao, H. G., Ou, J. P.. Electrical property of cement-based composites filled with carbon black under long-term wet and loading condition, Composites Science and Technology, 68 (2008) 2114-2119.

[18] Xiao, H. G., Li, H., Ou, J. P.. Modeling of piezoresistivity of carbon black filled cement-based composites under multi-axial strain. Sensors and Actuators A: Physical. 160 (2010) 87-93.

[19] Li, H., Xiao, H. G., Yuan J., Ou, J. P. Microstructure of cement mortar with nano-particles. Composites Part B: Engineering. 35(2) (2004) 185-189.

[20] Li, H., Xiao, H. G., Ou, J. P.. A study on mechanical and pressure-sensitive properties of cement mortar with nanophase materials. Cement and Concrete Research. 34 (3) (2004) 435-438.

[21] K. T. Lau and David Hui, The Revolutionary Creation of New Advanced Materials-Carbon Nanotube Composites, Composites: Part B 33(2002), 263-277.

[22] Zhu Yihua, Zhu Hongjie, Hu Liming Han Jinyi, Electric Properties of $Ag/Si_3N_4$ Nanostructured Composites, Journal of Inorganic Materials 11 (1996) 348-352.

[23] Ye Qing, Research on the Comparison of Pozzolanic Activity between Nano $SiO_2$ and Silica Fume, concrete 3(2001) 19-22 (in Chinese).

[24] Hui LI, Mao-hua ZHANG, Jin-ping OU. Abrasion resistance of concrete containing nano-particles for pavement. Wear 2006; 260(11-12): 1262-1266.

[25] Hui LI, Mao-hua ZHANG, Jin-ping OU. Flexural fatigue performance of concrete containing nano-particles for pavement. International journal of fatigue, 2007, (29): 1292-1301.

[26] Mao-hua ZHANG, Hui LI. The resistance to chloride penetration of concrete containing nano-particles for pavement. Proc. of SPIE (The SPIE Nondestructive Evaluation for Health Monitoring and Diagnostics Symposium——Testing, Reliability, and Application of Micro- and Nano-Material Systems IV), Vol. 6175, 61750E, March 2006, San Diego, USA.

[27] Xiao H. G., Lan C. M., Ji X. Y., Li H. Mechanical and sensing properties of structural materials with nanophase materials. Pacific Science Review. 2003, 5: 122-127.

[28] B. G. Han, X. C. Guan, J. P. Ou. Electrode design, measuring method and data acquisition system of carbon fiber cement paste piezoresistive sensors. Sensors and Actuators: A physical. (2007); 135: 360-369.

[29] Boettger H, Bryksin U V. Hopping conduction in solids. Berlin: Berlin Verlag Akademie, (1986): 108-148.

[30] Simmons J G. Generalized formula for the electric tunnel effect between similar electrodes separated by a thin insulating film. Journal applied physics (1963).; 34: 1793-1803.

[31] S. H. Wen, D. D. L. Chung. Electric polarization in carbon fiber-reinforced cement. Cement and Concrete Research (2001); 31: 141-147.

[32] H. G. Xiao. Piezoresistivity of Cement-based Composites Filled with Nanophase Materials and Self-sensing Smart Structural System. PhD Thesis, Harbin Institute of Technology, School of Civil Engineering, (2006).

[33] Wu Zhongwei, Lian Huizhen, High Performance Concrete, Beijing, China Railway Publishing Company, (1999). 49-50.

[34] Wang Xin, Tan Xunyan, Yin Yansheng, Zhou Yu, Analysis on Toughening Mechanisms of Ceramic Nano-Composites, Journal of Ceramics 2 (2000) 107-111 (in Chinese).

[35] Wu Xijun, Zhao Mingwen, Properties and Interfacial Microstructures for Nanostructured Materials,

Chinese Journal of Atomic and Molecular Physics 2 (1997) 148-152 (in Chinese).

[36] S. L. Colston, D. O'Connor, P. Barnes, Functional micro-concrete: The incorporation of zeolites and inorganic nano-particles into cement micro-structures, Journal of Materials Science Letters 19 (2000) 1085-1088.

[37] Fang Kaitai, Xu Jianlun. Statistical distribution. Beijing: Science Press, 1987. (in Chinese).

[38] Gao Zhentong. Fatigue application statistics. Beijing: National Defence Industry Press, (1986). (in Chinese).

[39] Ma Biao, Hu Chang-shun, Lu Xue-min, et al. Experimental Study of Concrete for Ultra-thin Whitetopping Pavement. Journal of Highway and Transportation Research and Development (2003); 20(6): 8-12. (in Chinese).

[40] Chen Shuan-fa. The Concrete Flexural Fatigue Property by Adding Polypropylene Fiber. Journal of Xi'an Highway University (2001); 21(2): 18-20. (in Chinese).

# 第 9 章 Chapter 9

# 混凝土动力损伤：研究进展与发展趋向

李　杰，任晓丹，黄桥平
（同济大学，上海 200092）

**提　要**：为了精细地考虑结构的响应和破坏，需要考虑结构材料在动力荷载作用下的率敏感性。本文概述了混凝土材料动力损伤的研究现状。结合作者的研究进展，着重介绍了动力损伤细观数值模拟方法和作者新近发展的随机动力损伤本构关系模型，讨论了材料率敏感性产生的物理机制，并对混凝土材料动力损伤未来研究提出了建议。

**关键词**：混凝土；动力损伤；细观模拟；随机损伤模型

# DYNAMIC DAMAGE OF CONCRETE: PROGRESSES AND TRENDS

J. Li, X. D. Ren, Q. P. Huang
（Tongji University, Shanghai 200092, China.）

**Abstract**: The elaborate structural dynamical analysis requires the appropriate consideration of material rate-dependencies. Thus the modern researches of dynamic damage models are reviewed in the present paper with the latest investigations. And then the microscopic numerical simulation methods and the stochastic dynamic damage constitutive models are introduced in the present paper. Lastly, the developing trends for the future research of dynamic damage for concrete are proposed.

**Keywords**: concrete, dynamic damage, microscopic simulation, stochastic damage model

## 9.1　导言

对重大工程结构破坏起控制作用的往往是灾害性动力荷载。地震、强风、爆炸、冲击等都可能引起结构的灾难性破坏，需要在结构设计阶段就加以合理的考虑。混凝土材料是重大工程建设中的主导性建筑建筑材料，也是一种典型的对动力加载速度较为敏感的材料，所以，在考虑动力荷载对工程结构的作用时，需要合理地考虑加载速度对混凝土本构关系的影响。

早在 1917 年，Abrams[1]对混凝土进行动载和静载压缩试验时，就发现混凝土抗压强度对加载速度存在敏感性。经过了长期的试验研究和探索，人们对混凝土材料在单轴受力状态下动力强度变化特征的认识已经比较充分。但是，由于受实验条件限制，迄今为止关于混凝土动力本构关系的实验数据仍然极为稀少[1]。尽管如此，20 世纪 80 年代以来，一些学者还是相继展开了对于混凝土动力本构关系的理论研究。例如：20 世纪 80 年代初，Bui & Ehrlacher[3]最早将损伤与断裂结合起来，建立了动力损伤断裂模型；90 年代初，Holmquist[4]对适用于金属材料的 Johnson-Cook 模型进行了修正，建立了考虑应变率和损伤的混凝土本构模型；1996 年，Dube 等人[5]参照黏塑性力学中的过应力理论，对经典的连续损伤模型进行了改造，给出了考虑加载速率影响的损伤变量表达式，并将模型应用于分析混凝土梁承受冲击荷载的问题；Cervera 等[6]对 Faria 和 Oliver 提出的率相关连续损伤模型[7]进行进一步推广，建立了相应的数值算法，并将模型应用于混凝十大坝的动力分析。在经过较长时间的探索后，研究者们逐渐清楚地认识到：要想从机理上把握混凝土的动力非线性特性，就要从细观物理入手、研究混凝土材料在动力加载条件下的损伤演化规律。

事实上，混凝土受力力学性质的基本特征，不仅表现为高度的非线性，而且表现出典型的随机性。众所周知：混凝土是由水泥、粗集料、细集料、各类掺和料组成的多相复合材料。在其形成之初，混凝土内部就具有微孔洞、微裂缝等初始缺陷。在外力作用下，这些初始损伤因应力集中而进一步发展，从而导致材料单元的应力—应变关系逐步地偏离线性关系，呈现出非线性的基本特征。同时，由于混凝土材料各组分具有随机分布的特征，因而无论是初始的损伤分布还是后续的损伤演化过程，都不可避免地具有随机性的特征。换句话说，混凝土材料构成的随机分布性质，必然导致损伤具有随机演化性质。随机的损伤演化，必然导致随机的强度表现和随机的本构关系。非线性与随机性，是混凝土本构关系的两个基本特征[8,9]。在细观层次上寻求损伤随机性的概率性反映方式，不仅有助于建立损伤演化从细观到宏观的桥梁，从而从根本上解决混凝十非线性与随机性的综合反映问题，也可以在物理机制上给予混凝土的损伤演化规律一个合理的解释。基于这一基本观点，在过去十年中，李杰等从建立细观随机断裂模型[10,11]入手、结合基于连续介质力学的双标量弹塑性损伤模型[12~15]和系列试验研究[16]，逐步建立了混凝土多维弹塑性随机损伤模型[17,18]。这类模型，不仅可以理想地反映混凝土材料所特有的强度软化、刚度退化、单边效应、拉压软化、有侧压时的强度增长等一系列特殊行为，也清晰的反映了混凝土在多维应力条件下的强度随机变化范围[9],[17]，从而，为进行结构层次的非线性分析提供了基础。

在上述工作基础上，在国家自然科学基金重大研究计划重点支持项目资助下，我们进一步研究了混凝土材料的动力损伤问题，试图将上述弹塑性随机损伤模型扩展到适用于动力加载的场合。本文，即为对我们近年来研究工作的一个阶段性小结。

## 9.2　率相关性的物理机制

混凝土材料在动力荷载作用下率敏感性的产生机理，是动力本构关系建模中的关键。

前已述及，混凝土受力行为的率敏感性，已经被众多的试验所证实。但是，混凝土率敏感性产生的物理机理却长期没有定论，现有的研究文献[19~24]大多将其归结为：黏性效应、惯性效应和微裂缝演化效应。

### 9.2.1 黏性效应

黏性效应又称作 Stefan 效应，它是在黏性流体受压缩或拉伸时由于流动产生的黏性阻力而导致的一种率相关效应。其物理模型（图 1）可表示为下述方程

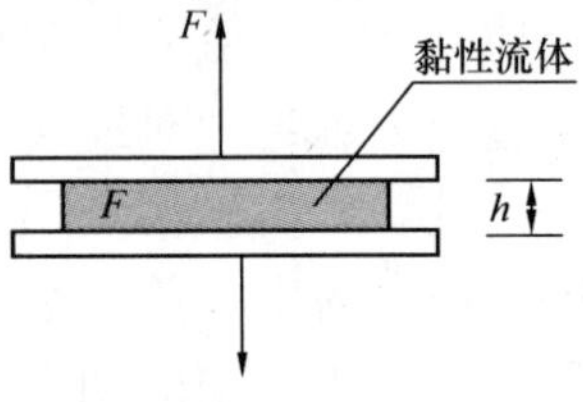

$$F = \frac{3\eta V^2}{2\pi h^5}\dot{h} \tag{1}$$

式中：$F$ 为作用力，$\eta$ 为黏性系数，$h$ 为平板间距离，$\dot{h}$ 为平板分离速度，$V$ 是黏性流体体积。

图 1 黏性效应物理模型

对混凝土而言，通常认为混凝土中的水泥基体存在与加载速率相关的能量耗散，因而在快速加载条件下表现出类似于上述模型的黏性效应。同时，一般认为黏性效应主要存在于较低的应变速率加载场合。

### 9.2.2 惯性效应

根据动力学理论，动力荷载引起的惯性力总是倾向于“抵抗”外荷载引起的变形，当惯性力随着加载速率的提高而增加到一定程度时，就可以显著提高材料的动力强度；另一方面，惯性效应引起与加载方向垂直的约束效应（图 2），使得试件处于三轴受压应力状态，从而使得材料表观抗压强度得到提高。

一般认为，惯性效应主要存在于具有较高的应变速率加载场合。

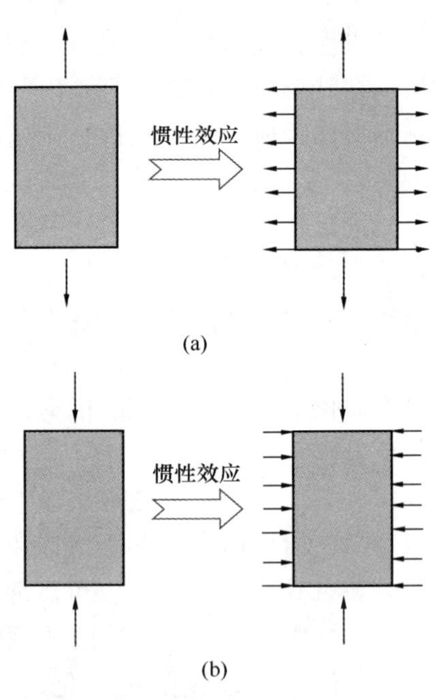


图 2 惯性效应示意图

(a) 动态受拉；(b) 动态受压

### 9.2.3 微裂缝演化效应

混凝土是存在大量初始微裂缝的材料，有研究表明：随着加载速率的提高，裂缝开展的路径变得越来越平直，穿越骨料也越来越多；同时，随着加载速率的提高，相同应力作用下混凝土试件中的裂缝减少[25]。一些学者认为，这两方面的效应都表明快速加载能够提高试件的强度。

总结既有关于混凝土率相关物理机制的研究进展可以看出，人们关于率相关物理本质的认识还处于探索阶段。事实上，混凝土动态受压和受拉强度随应变率变化的实验数据表明：在应变速率为 $10\text{s}^{-1}$左右时，率相关关系出现明显转折[2]，这可能意味着混凝土材料的率敏感性不能简单地用单一机制加以反映，不同的受力状态以及不同的加载速率都有可能“激活”不同的损伤机制，不同的损伤机制则导致不同的非线性演化机理。

## 9.3　动力损伤的细观模拟

动力作用下材料的率敏感性表现为强度的提高。这种强度的提高可以通过连续损伤模型加以描述，但由此建立的模型在本质上属于唯象学模型，不能解释动力强度提高的物理机制，也难以令人信服地给出混凝土材料的损伤演化法则。为了探求动力作用下材料损伤和破坏的物理机制，需要将研究转向细观，研究动力作用下材料细观结构的变化。此时，需要考虑材料的细观非均匀结构、细观强度的随机性、动力断裂行为、荷载施加方式等众多因素，很难以解析方法作为基本工具获得定量结果，所以，近年来的研究大多基于数值模拟。

### 9.3.1　细观数值方法

动力作用下裂缝扩展的数值模拟方法主要有两类：其一是分子动力学方法；其二是内聚单元方法。这两类方法的共同点在于将细观断裂问题离散化，采用离散模型而不是连续模型模拟细观结构的非均匀性。

在早期的研究中，Ashurst 和 Hoove[26] 最早引入分子动力学方法（MD）模拟固体的动态断裂。1994 年，Abraham 等人[27] 利用大型并行计算机系统建立了分子动力学模型，系统研究了动态断裂问题，并提出了动态断裂的不稳定性的概念。2006 年，Buehler 和 Gao[28] 基于分子动力学模型，系统研究了单一材料裂缝端部的超弹性对动态断裂的影响。

根据分子动力学模型，可将裂缝的端部模拟为原子点阵（图 3）。原子间的相互作用采用势函数来表示。Buehler 和 Gao 的工作中[28] 采用了下述势函数

$$\phi(r)=\begin{cases}\dfrac{1}{2k_1\ (r-r_1)^2} & \text{if } r<r_{\mathrm{on}}\\ a_3+\dfrac{1}{2k_1\ (r-r_2)^2} & \text{if } r\geqslant r_{\mathrm{on}}\end{cases} \tag{2}$$

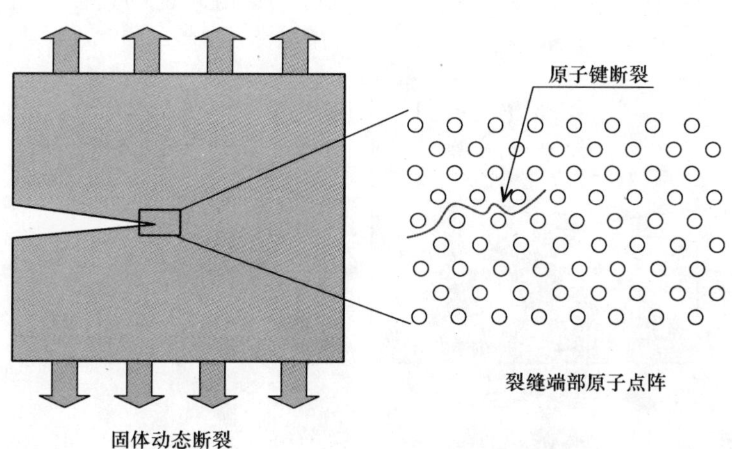


图 3　固体动态断裂的分子动力学模型

其中

$$\begin{cases} a_2 = \dfrac{1}{2k_1 \ (r_{\text{on}} - r_1)^2} - \dfrac{1}{2k_1 \ (r_{\text{on}} - r_2)^2} \\ r_2 = \dfrac{r_{\text{on}} + r_1}{2} \end{cases} \tag{3}$$

在本质上，分子动力学体系中并没有直接包含应变的概念，对于分子动力学模拟的固体体系，某个原子处对应的 Cauchy-Green 应变一般由下式估计

$$b_{ij} = \frac{1}{3r_1^2} \sum_{k=1}^{N} (x_i^l - x_i^k)(x_j^l - x_j^k) \tag{4}$$

其中 $k = 1 \sim N$ 表示遍历该原子所对应的所有紧邻原子。

基于上述过程，利用分子动力学模拟方法可以得到固体材料动态断裂行为的某些重要特征，如：在一定的动力作用下，固体中的裂缝扩展呈现出不稳定特征，开始出现分叉，与此同时，裂纹扩展的速度和能量释放过程都不再是时间的平滑函数，而是表现出显著的波动；其二，动态裂纹扩展的速率存在明显的上限值，到达上限值之后，加载速率的提升不再使裂纹的扩展速率提高；其三，裂缝端部的非线性区域对动态裂纹扩展有显著的影响。

事实上，分子动力学方法属于微观-细观层次的模拟方法。即使对于单一的匀质材料，这类方法也具有较大的局限性。由于固体中包含的原子数量非常之大，尽管当今的大型计算机系统已经具有惊人的计算能力，但目前仍不能对很小尺寸的固体进行直接的分子动力学模拟，所建立的模型不论是在时间尺度上还是在空间尺度上都做了伸缩处理，并不能与实际的固体直接对等，由此得出的结论只能用于定性研究，不能得出定量的结论。

20 世纪 90 年代中期，Xu 和 Needleman[29,30] 引入了内聚单元（cohesive element）的概念，用以在细观尺度上模拟宏观固体的动力断裂。在这类模型中，首先对固体进行有限元离散，划分得到的有限元称为体积单元（volumetric element）。与传统有限元不同的是，体积单元之间并没有共用的节点，而是相互独立的，在邻近的体积单元之间插入内聚单元（cohesive element），作为体积单元的连接，同时作为裂缝可能的路径（图 4）。

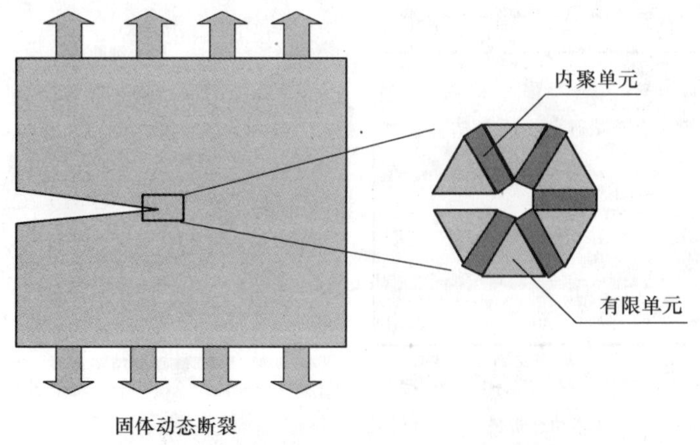


图 4　固体动态断裂的内聚单元模型

包含内聚单元的固体问题的弱形式为

$$\int_{\Omega}\sigma(u):\varepsilon(v)\mathrm{d}\Omega+\int_{S}T[w(u)]\cdot w(v)\mathrm{d}\Gamma=\int_{\partial\Omega}t\cdot v\mathrm{d}\Gamma \tag{5}$$

其中内聚应力 $T[w(u)]$ 为裂缝面相对位移 $w=u^{+}-u^{-}$ 的函数。最常用的线性函数表达式如下

$$f=f_{\mathrm{u}}-kw \tag{6}$$

其中 $f=T\cdot n$ 和 $w=w\cdot n$ 分别为法向内聚应力和裂缝面张开位移；$f_{\mathrm{u}}$ 为材料的细观抗拉强度；材料的内聚断裂能定义为 $f-w$ 曲线与坐标轴包围的图形的面积，对于线性模型有 $G_{\mathrm{c}}=f_{\mathrm{u}}^{2}/2k$。

基于内聚单元模型，Xu 和 Needleman 在数值模拟的基础上，得到了关于动力断裂的一些重要结论：第一，动力加载过程中，裂缝扩展速度会增长至一个平台而不继续增长，对于各向异性材料，这个平台大约为瑞雷波速的 90%，而对于各向同性材料，平台大约为瑞雷波速的 50%；第二，在到达了裂缝扩展速率平台之后，裂缝开始分叉，并且扩展速度出现很强的波动；第三，裂缝动态扩张条件下，其断裂能和断裂韧性不再是常数，甚至不再是连续函数。显然，这样的 3 点结论均被 12 年后 Buehler 和 Gao 所进行的分子动力学模拟研究[28]所验证。从而说明了上述细观数值模拟方法的有效性。

内聚裂缝模型虽然可以在适当的空间和时间尺度范围内给出材料动态断裂的模拟结果，但是其本身仍然存在某些问题，如：内聚力与裂缝张开位移的关系很难直接测量和确定。同时，在既有研究中，内聚裂缝模型大都基于规则网格，没有考虑不规则网格对于模拟结果的影响。

### 9.3.2　混凝土动力断裂细观模拟

通过引入不规则有限元网格考虑微观非均匀性对于宏观裂缝扩展的影响，我们建立了内聚单元模型（图 5），考虑动力加载条件下微裂缝的产生和演化过程，初步进行了混凝土材料的动力损伤细观数值模拟研究。为了尽量精细地模拟混凝土的动态断裂行为，我们建立了非常精细的有限元模型，单元数量超过 35 万，采用显示积分方法并引入并行计算策略进行数值模拟[31]。

图 5　非规则内聚单元模型

数值模拟的部分结果见图 6。从中可以明显看出静态裂缝扩展与动态裂缝扩展的根本区别：在静力作用下，裂缝的扩展具有稳定发展的趋向，而在动力作用下，裂缝端部在扩展的过程中不断分叉，形成一个裂缝群。显然，在端部应力场、扩展速度以及能量耗散等方面，裂缝群都和单条裂缝有着本质的区别，因此，不能将静力断裂的分析方法和结论简单地推广到动力断裂的分析中来。

对裂缝扩展速率的变化进行分析，结果表明（图 7）：在动力加载条件下，裂纹扩展的速度表现出显著的波动；同时，动力加载速率的提高并不能显著的提高裂纹扩展速率，裂纹的极限扩展速率一般低于瑞雷波速。

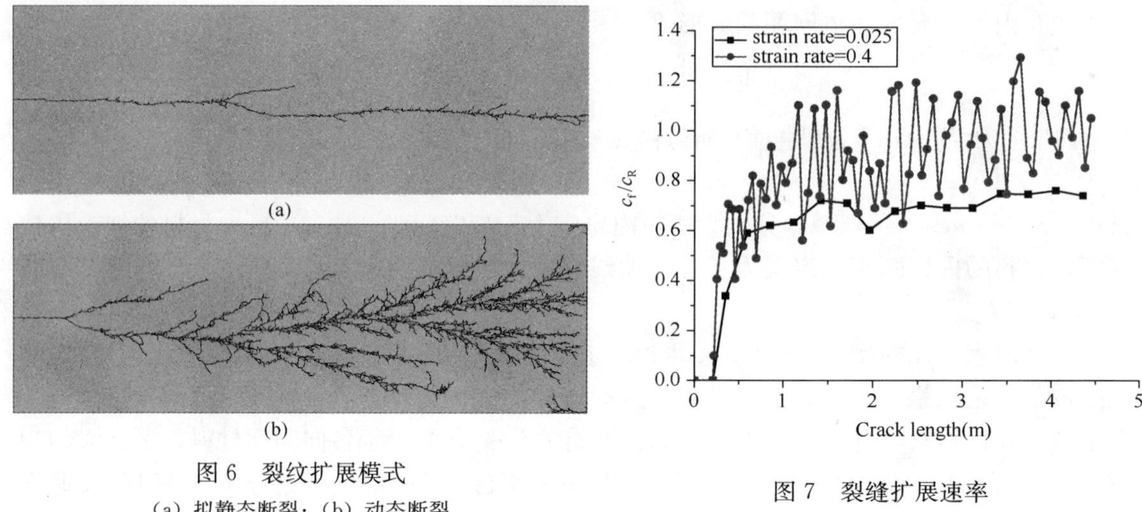


图 6 裂纹扩展模式

（a）拟静态断裂；（b）动态断裂

图 7 裂缝扩展速率

综合上述两方面结果可以得到这样的结论：在快速加载作用下，材料需要产生更多的断裂面来消耗输入能量，在可能产生断裂面的两种方式中，实际情形极可能是分叉扩展而不是加速扩展。这就从定性的角度上解释了动力强度提高的原因：动力作用下，固体内裂缝因产生局部分叉而导致更高的耗能，从而提高了材料动力强度。

## 9.4 弹塑性随机动力损伤模型

### 9.4.1 细观动力随机损伤模型

混凝土细观数值模拟的结果，从不同侧面证实，混凝土动力损伤的物理本质来源于动力断裂行为的特殊性，尽管我们尚不能十分清晰地描述这种特殊性究竟是缘于加载速度与应力波传播之间的竞争、还是源于裂纹尖端应力场在静力与动力加载条件下存在显著区别，但无可置疑的是：因损伤而导致的应力重分布过程在动力非线性过程中依然发挥关键作用。事实上，前述内聚裂缝模型的数值模拟结果恰恰证明了：细观损伤导致的动态应力重分布过程，决定着损伤演化发展过程的走向与特征，反映了混凝土裂纹扩展过程中因为裂缝相互作用所导致的复杂现象。只是由于混凝土材料的多相介质复杂性和细观数值模拟的计算复杂性，使我们暂时尚不能利用这种精细的数值多尺度模型获取损伤演化规律的定量结果。

而在另一方面，本文第一作者近年来致力发展的细观随机断裂—滑移模型[8~18]，在本质上应该属于一类串行多尺度模型。它抓住了混凝土损伤演化的要旨，用一种抽象的方式反映了应力重分布对损伤演化以及非线性过程的影响。虽然这种抽象难以说明裂缝相互作用的影响，但无论是作为一种合理的简化、还是从工程实用意义上考察，细观随机断裂—滑移模型都不失为一种有价值的选择。

大量的试验观察表明，混凝土的非线性变形来源于两种基本的物理机制：微裂缝（微缺陷）的扩展和水泥浆体的塑性滑移[8]。在变形过程中，损伤演化与塑性滑移之间相互影响、相互耦合。任何正确的混凝土本构关系，必须对这两种基本的物理变形机制作出合理的反映。一般说来，混凝土在复杂应力作用下的破坏形态可概括为三种基本形式：拉伸破坏、剪切破坏以及高静水压力下的压碎破坏[8,32]。在不考虑高静水压力导致的应变强化的前提下，混凝土材料的损伤和破坏主要源于两种不同的物理机制：受拉损伤机制和受剪损伤机制。相应的，混凝土的细观损伤也可分为受拉损伤和受剪损伤，可以分别采用受拉损伤变量 $D_t$ 和受剪损伤变量 $D_s$ 加以表述。

采用在细观尺度意义上其断裂应变服从某一概率分布的微弹簧来表征细观单元，我们先后发展了两类细观随机断裂—滑移模型[10,17]，如图 8 所示。在细观层次上，将混凝土离散为具有一定特征高度和截面积的小柱体，并用微弹簧加以表示。拉伸（或剪切）微弹簧假定为理想弹脆性材料，其极限应变为一随机变量。

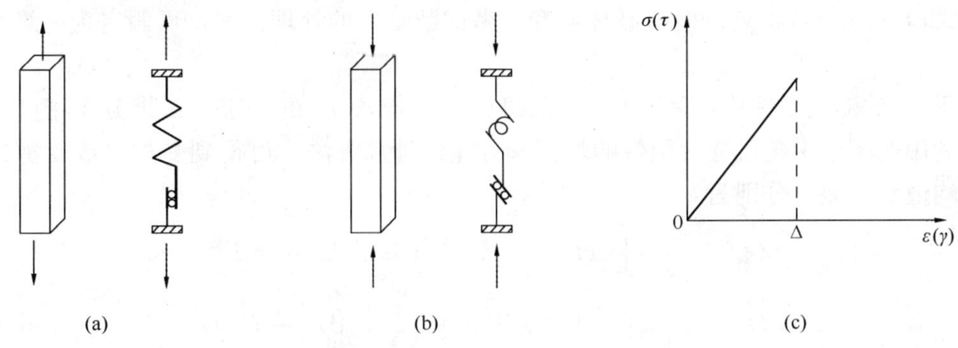


图 8　细观物理模型

(a) 拉伸单元；(b) 剪切单元；(c) 微弹簧的本构关系

在受拉伸破坏机制中，小柱体的破坏是由于骨料和水泥砂浆之间界面被拉开或集料及凝胶体中初始微缺陷扩展而产生，表现为细观受拉弹簧的随机断裂。而在以压应力为主的受剪破坏机制中，小柱体破坏起源于骨料和水泥砂浆之间界面或初始微缺陷因剪应力作用导致的界面拉开，在模型中表现为受剪弹簧的随机断裂。为了反映水泥砂浆内部以及砂浆、骨料界面间塑性滑移的影响，分别引入细观拉伸塑性变形元件和细观剪切塑性变形元件。基本的受拉损伤单元或受剪损伤单元分别由拉伸微弹簧和拉伸塑性变形元件或剪切微弹簧和剪切塑性变形元件串联组成。

由于细观受拉单元与受剪单元受力机制上的类似性，可以采用统一的方式给出其数学描述。引用 Rabotnov 的经典损伤定义损伤变量[33]，即

$$D_i = \frac{A_D}{A} \quad (i = t, s) \tag{7}$$

式中，$A_D$ 为因细观拉伸（或剪切）损伤单元破坏而导致混凝土退出工作的面积；$A$ 为无损混凝土的截面积，即试件的横截面积。

假设离散后模型中细观单元的截面积均相等，式（7）可变换为

$$D^{\pm}(\varepsilon_e) = \frac{1}{M}\sum_{i=1}^{M} H(\varepsilon_e - \Delta_i) \tag{8}$$

式中，损伤变量的角标正、负号分别表示受拉与受剪（压）损伤。$\Delta_i$ 为第 $i$ 个细观单元发

生拉伸或剪切破坏时相应的应变，可视为服从某一分布的随机变量；$H(\cdot)$ 为 Heaviside 函数，即

$$H(\varepsilon_e-\Delta_i)=\begin{cases}0, & \varepsilon_e\leqslant\Delta_i\\ 1, & \varepsilon_e>\Delta_i\end{cases} \tag{9}$$

当模型中细观单元的数目 $M$ 趋向于无穷大时，则在宏观横截面上的细观单元体可以看作一维连续体。若 $M\rightarrow+\infty$ 时式（8）的极限存在，则细观单元破坏时的应变为一连续随机场 $\Delta(x)$。不失一般性，$x$ 可认为介于 0 和 1 之间，即，$x\in[0,1]$。相应的，式（8）可以表示为如下形式[34]

$$D^{\pm}(\varepsilon_e)=\int_0^1 H\left[\varepsilon_e^{\pm}-\Delta^{\pm}(x)\right]\mathrm{d}x \tag{10}$$

式中，$\Delta(x)$ 为在位置 $x$ 处的随机破坏应变，考虑数学上的处理方便，可假定为一维均匀随机场。

由于 $\Delta_i$ 的随机场性质，$D_i(\varepsilon_e)$ 为一随机函数。若 $\Delta(x)$ 的一维、二维分布密度函数均存在，利用概率论中随机变量函数的均值和方差的计算方法，可得到受拉（或受剪）损伤变量的均值和方差，分别为

$$\mu_D(\varepsilon_e)=\int_0^{\infty}\int_0^1 H(\varepsilon_e-\delta)f_{\Delta}(\delta;x)\mathrm{d}x\mathrm{d}\delta=F(\varepsilon_e) \tag{11}$$

$$V_D^2(\varepsilon_e)=\left[2\int_0^1(1-\gamma)F_{\Delta}(\varepsilon_e,\varepsilon_e;\gamma)\mathrm{d}\gamma\right]-F^2(\varepsilon_e) \tag{12}$$

式中，$f_{\Delta}(\delta;x)$ 为破坏极限应变在位置 $x$ 处的一维概率分布密度函数；$F_{\Delta}(\varepsilon_e,\varepsilon_e;\gamma)$ 为在两个截面处的随机变量的联合概率分布函数，$\gamma=|i-j|$，为两截口处距离。

引入塑性变形元件后，拉（压）应力在单元内产生的总应变 $\varepsilon$ 由微弹簧产生的拉伸（或压缩）应变 $\varepsilon_e$ 和微裂缝面产生的塑性应变 $\varepsilon_p$ 组成，即

$$\varepsilon=\varepsilon_e+\varepsilon_p \tag{13}$$

在弹性应变 $\varepsilon_e$ 分别取受拉应变或受压应变前提下，拉伸塑性变形元件和剪切塑性变形元件的变形计算模式可以统一采用下式计算

$$\varepsilon_p=\frac{\delta}{1-D}\varepsilon_e \tag{14}$$

式中 $\delta$ 为塑性变形系数，可以利用实验结果标定。

在细观损伤层次引入加载速率影响，可以取

$$\dot{\varepsilon}_d=\mu\left(\frac{\varepsilon_e}{\varepsilon_d}-1\right)^n \tag{15}$$

式中：$\mu$ 和 $n$ 是与损伤率敏感性相关的材料参数，可以借助试验数据确定之。

### 9.4.2 多维弹塑性随机动力损伤本构关系

不失一般性，基于应变分解引入有效应力

$$\begin{cases}\sigma=C_0:\varepsilon^e\\ \varepsilon=\varepsilon^e+\varepsilon^p\end{cases} \tag{16}$$

其中 $C_0$ 为材料的初始刚度张量，$\varepsilon$、$\varepsilon^e$ 和 $\varepsilon^p$ 表示总应变、弹性应变和塑性应变。

混凝土在受拉和受压应力状态下表现出迥异的特性，为反映这一性质，引入有效应力正负分解分别表示受拉和受压的性能[35,36]，即

$$\bar{\sigma} = \bar{\sigma}^{+} + \bar{\sigma}^{-} \tag{17}$$

其中

$$\bar{\sigma}^{+} = P^{+}:\bar{\sigma}, \bar{\sigma}^{-} = \bar{\sigma} - \bar{\sigma}^{+} = P^{-}:\bar{\sigma} \tag{18}$$

$$P^{+} = \sum_{i} H(\hat{\bar{\sigma}}_i)(p_i \otimes p_i \otimes p_i \otimes p_i), P^{-} = I - P^{+} \tag{19}$$

式中 $I$ 为四阶单位张量，$\hat{\bar{\sigma}}_i$ 与 $p_i$ 为有效应力的第 $i$ 阶特征值和特征向量，$H(\ )$ 为 Heaviside 函数。

引入 Helmholtz 自由能势函数描述材料的状态，且定义 Helmholtz 自由能势的一般表达式为

$$\psi = \psi(\varepsilon^{e}, \kappa, d^{+}, d^{-}) \tag{20}$$

其中弹性应变 $\varepsilon^{e}$ 是状态变量，表征材料所处的状态；塑性变量 $\kappa$ 是表征塑性发展的内变量；$d^{+}$ 和 $d^{-}$ 分别为受拉和受压损伤标量，分别表示受拉和受压损伤的发展，亦为内变量。

在等温绝热状态下，通常可以假定材料的弹性与塑性 Helmholtz 自由能势不耦合，将 Helmholtz 自由能势分解为弹性和塑性两个部分[12～14,37]

$$\psi(\varepsilon^{e}, \kappa, d^{+}, d^{-}) = \psi^{e}(\varepsilon^{e}, d^{+}, d^{-}) + \psi^{p}(\varepsilon^{e}, \kappa, d^{+}, d^{-}) \tag{21}$$

对于弹性 Helmholtz 自由能势，不考虑受拉损伤与受压损伤的耦合，可以分解为

$$\psi^{e}(\varepsilon^{e}, d^{+}, d^{-}) = \psi^{e+}(\varepsilon^{e}, d^{+}) + \psi^{e-}(\varepsilon^{e}, d^{-}) \tag{22}$$

损伤之后的弹性 Helmholtz 自由能势可以表述为初始 Helmholtz 自由能势的折减，即

$$\psi^{e\pm}(\varepsilon^{e}, d^{\pm}) = (1 - d^{\pm})\psi_0^{e\pm} \tag{23}$$

根据有效应力的分解，初始 Helmholtz 自由能势可以定义为

$$\begin{cases} \psi_0^{e} = \dfrac{1}{2}\bar{\sigma}:\varepsilon^{e} = \psi_0^{e+} + \psi_0^{e-} \\ \psi_0^{e+} - \dfrac{1}{2}\bar{\sigma}^{+}:\varepsilon^{e} \\ \psi_0^{e-} = \dfrac{1}{2}\bar{\sigma}^{-}:\varepsilon^{e} \end{cases} \tag{24}$$

对于塑性 Helmholtz 自由能势，忽略受拉应力条件下的塑性演化，可简化为[12～14]

$$\psi^{p}(\varepsilon^{e}, \kappa, d^{+}, d^{-}) = \psi^{p}(\varepsilon^{e}, \kappa, d^{-}) \tag{25}$$

损伤后的塑性 Helmholtz 自由能势可以表述为初始塑性 Helmholtz 自由能势的折减，即

$$\psi^{p}(\varepsilon^{e}, \kappa, d^{-}) = (1 - d^{-})\psi_0^{p} \tag{26}$$

根据热力学第二定律，任何不可逆过程都需要满足下列克劳修斯—杜哈美不等式

$$-\dot{\psi} + \sigma:\dot{\varepsilon} \geqslant 0 \tag{27}$$

将前述 Helmholtz 自由能势定义代入上式，可得

$$\left(\sigma - \frac{\partial \psi^{e}}{\partial \varepsilon^{e}}\right):\dot{\varepsilon}^{e} + \left(-\frac{\partial \psi}{\partial d^{+}}\dot{d}^{+} - \frac{\partial \psi}{\partial d^{-}}\dot{d}^{-}\right) + \left(\sigma:\dot{\varepsilon}^{p} - \frac{\partial \psi^{p}}{\partial \kappa}\cdot\dot{\kappa}\right) \geqslant 0 \tag{28}$$

上述不等式成立的必要条件是

$$\sigma - \frac{\partial \psi^{e}}{\partial \varepsilon^{e}} = 0 \tag{29a}$$

$$\left(-\frac{\partial \psi}{\partial d^+}\dot{d}^+ - \frac{\partial \psi}{\partial d^-}\dot{d}^-\right) + \left(\sigma : \dot{\varepsilon}^p - \frac{\partial \psi^p}{\partial \kappa} \cdot \dot{\kappa}\right) \geqslant 0 \tag{29b}$$

由式（29a）可以给出本构关系的基本表达式

$$\begin{aligned} \sigma &= (1-d^+)\bar{\sigma}^+ + (1-d^-)\bar{\sigma}^- = (I-D):\bar{\sigma} \\ &= (I-D):C_0:(\varepsilon - \varepsilon^p) \end{aligned} \tag{30}$$

而式（29b）事实上仅仅给出了塑性内变量需要满足的基本条件，并不能直接根据这一不等式给出内变量的具体演化法则。这也恰恰是经典连续介质力学的痼疾所在!

关于塑性变形 $\varepsilon^p$，原则上可以忽略率相关效应的影响，在有效应力空间建立塑性应变演化方程[2]：

$$\dot{\varepsilon}^p = \dot{\lambda}^p\left(\frac{\bar{s}}{\|\bar{s}\|} + \alpha^p 1\right) = \dot{\lambda}^p\,(1_{\bar{s}} + \alpha^p 1) \tag{31}$$

而关于另外一组内变量—损伤变量，则可以通过引入能量等效应变的概念[2]，引用前述细观随机断裂模型所决定的损伤演化规律加以表示。事实上，根据损伤一致性条件和能量等效应变概念，可知在一般意义上存在如下损伤演化规律：

$$D^{\pm}(\varepsilon^{e^{\pm}}) = \int_0^1 H\left[\varepsilon_{eq}^{\pm} - \Delta^{\pm}(x)\right]\mathrm{d}x \tag{32}$$

其中，

$$\varepsilon_{eq}^{\pm} = \sqrt{\frac{Y^{\pm}}{Y_1^{\pm}}}\varepsilon^{e^{\pm}} \tag{33}$$

式中，$Y^{\pm}$ 为多维损伤能释放率，$Y_1^{\pm}$ 为一维损伤能释放率[9,17]。

值得特别指出：确定性损伤本构关系与随机损伤本构关系所反映的是同一个物理规律，其间唯一的区别在于损伤变量是取确定性变量还是随机变量。事实上，当以式（32）作为损伤演化准则代入式（30），得到的是一般意义上的弹塑性随机损伤本构关系。而当采用统计平均方式、以 $D$ 的均值作为损伤演化准则代入式（30）时，将得到确定性弹塑性损伤本构关系。

注意到式（11），当随机破坏应变 $\Delta_i$ 服从对数正态分布时，以损伤均值表示的确定性损伤演化准则是

$$d^{\pm} = E(D^{\pm}) = \int_0^{\varepsilon_{eq}^{\pm}} \frac{1}{\sqrt{2\pi}\zeta x}\cdot \exp\left[-\frac{(\ln x - \lambda)^2}{2\zeta^2}\right]\mathrm{d}x \tag{34}$$

式中，$\lambda, \zeta, \xi$ 是确定破坏应变随机场 $\Delta(x)$ 的概率分布参数，可以由单轴受拉或单轴受压全过程实验确定。

引用式（15）表述的动力应变与静力应变关系表达式，可以采用应变更新算法确定动力本构关系。

图 9、图 10 分别给出了利用确定性均值模型进行分析与相关试验的对比结果。其中，图 9（a）中的单轴受拉分析结果中未考虑塑性应变的率效应，只考虑了损伤演化的率效应，由此得到的结果能较好的反应动力作用下的强度提高；图 9（b）中的单轴受压分析结果同时考虑了塑性应变和损伤演化的率敏感性，所得结果在应力和应变层面上均与试验结果符合较好。图 10 中给出了单轴强度提高因子的试验结果与理论分析结果对比。可见本文模型能够很好地模拟由应变率效应所引起的混凝土的强度的提高。在图 10（b）中，同

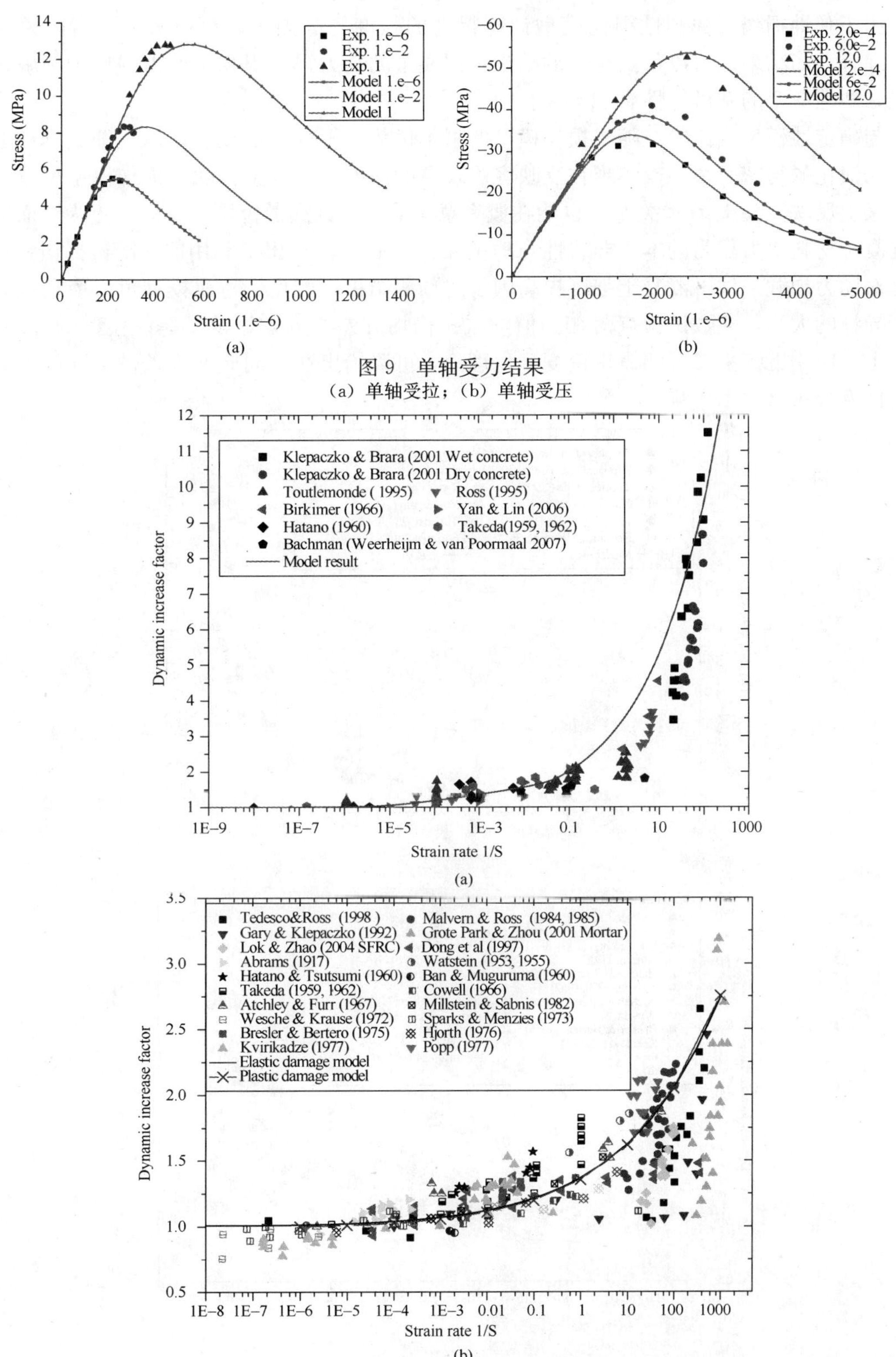


图 9　单轴受力结果

（a）单轴受拉；（b）单轴受压

图 10　强度提高因子均值比较

（a）单轴受拉；（b）单轴受压

时给出了弹性损伤与弹塑性损伤模型计算得到的强度提高因子，可以看出二者的差别不大。综合上面结果，可以得到这样的结论：混凝土的动力强度提高主要由损伤的率敏感性控制，塑性应变的率敏感性影响不大。

与确定性模型相比较，随机损伤模型的根本优势，不仅体现在它可以在细观物理上解释损伤演化的秘密所在，也体现在反映客观现象的深度与广度上：随机损伤模型可以在均值意义上反映混凝土 $\sigma-\varepsilon$ 关系，也能在概率意义上预测其离散范围，从而，更为全面地反映混凝土材料受力行为的本质非线性与随机性。图 11，即给出了利用前述随机损伤弹塑性模型对动力加载条件下混凝土强度提高因子的预测结果，可见：建议模型可以衡量应力强度变异性的大小，多数试验点落在均值加减二倍均方差之内，显示了建议模型的合理性。事实上，应用近年来发展的概率密度演化理论，可以给出在不同应变率条件下强度提高因子的概率分布（图 12）。

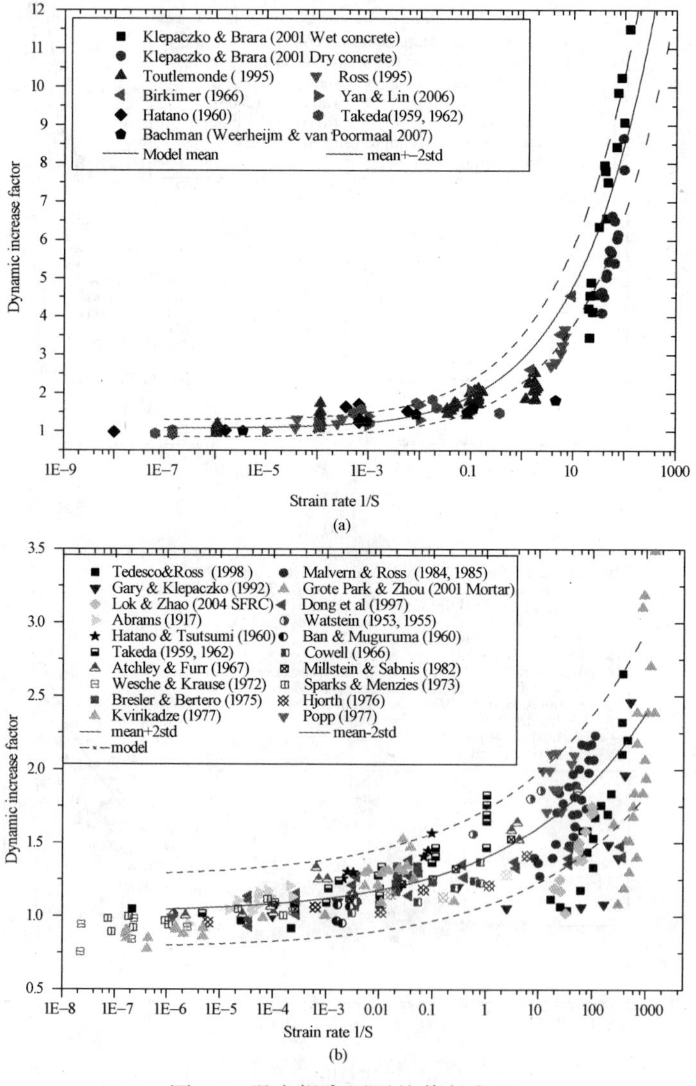


图 11 强度提高因子均值与方差

（a）单轴受拉；（b）单轴受压

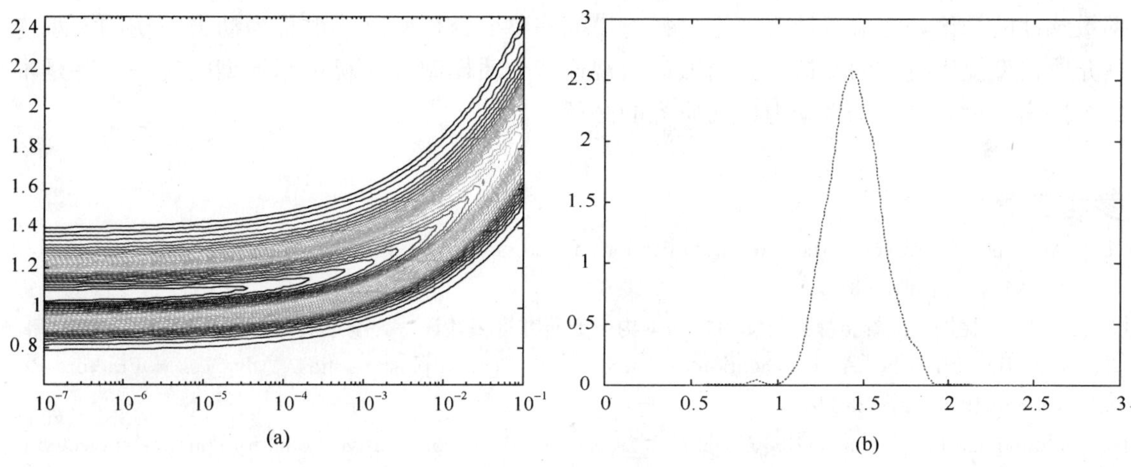


图 12　强度提高因子概率分布

（a）概率密度的演化；（b）截口概率密度

## 9.5　结论

非线性与随机性是混凝土力学行为的两个基本特征，在损伤演化过程中，受力行为的非线性与材料细观强度表现的随机性互相耦合、互为因果，而其本质规律，在于损伤导致的应力重分布。从细观层次上反映混凝土材料因损伤而导致的随机应力重分布过程，可以沿用具象的数值多尺度模型与抽象的随机断裂模型。我们近期的研究进展，说明了这两个途径的现实可行性。在这一研究进程中，深入探索混凝土材料的动力率相关效应，以期建立更为全面的混凝十动力损伤本构模型，服务于重大工程结构的精细化分析与设计，是现代高性能结构工程发展的一个不可或缺的基本环节。展望这一方向的研究，我们感到：下述趋势是值得注意的：

（1）动力损伤细观数值模拟的研究，目前还只能对非常简单的裂纹和受力状态进行模拟研究，所得的结论也大都是定性的，说明这一研究还有广阔的探索空间。其可能的发展方向包括：损伤随机场的引入——目前这一途径还没有得到必要的注意；新型数值方法的建立—目前有学者尝试分子动力学方法与内聚裂缝方法的结合；内聚裂缝模型的物理研究——从物理出发直接建立内聚裂缝模型；裂缝扩展速率上限的决定因素，即裂缝扩展速率的平台值是由那些因素决定的以及具有何种内在联系等。

（2）现行的多尺度研究大多忽略了尺度转换过程中对随机性特征的考量。因此，在逻辑上是有重大缺陷的。在实践上，也可能在理论一实验的对照环节上出现显著差距。事实上，不同尺度转换过程中的随机涨落应该构成多尺度随机物理研究的基本内容。现行多尺度研究的成功案例，大多是在尺度转换过程中随机涨落特征不明显的物理系统。混凝土材料、尤其是混凝土结构是否具有如此特征，尚待研究。

（3）连续模型与细观模型的统一是动力损伤研究发展的基本趋势。有关率敏感性的参数，目前仍然是基于试验数据的经验识别结果，缺乏必要的理论支持，这大大限制了应用的范围。事实上，率敏感性的物理机制，同样难以直接由连续介质损伤理论获得解释，而需要借助损伤的细观研究和模拟。另一方面，目前的细观模拟方法还很难得到定量的有关率敏感性的结论，也缺乏针对特定材料建模的完整研究，所以，很难为动力损伤的宏观连续介质模型提供直接的支持。如何形成宏观连续介质模型与细观物理模型的统一，既是本文的探索目标，也是未来动力损伤研究的必然趋向。

## 参考文献


[1] Abrams DA. Effect of the rate of application of load on the compressive strength of concrete. ASTM J. 1917，17：364～377.

[2] 李杰，任晓丹，混凝土静力与动力损伤本构模型研究进展述评，力学进展，2010，40(3)：284-297.

[3] Bui HD，Ehrlacher A. Propagation of damage in elastic and plastic solids. Advances in Fracture Research. 1981，2：533-551.

[4] Holmquist TJ，Johnson GR，Cook WH. A computational constitutive model for concrete subjected to large strains，high strain rates and high pressures. In：Fishler E，ed. 14th International Symposium on Ballistics，Quebec，Canada. 1993：591～600.

[5] Dube JF，Pijaudier-Cabot G，La Borderie C. Rate dependent damage for concrete in dynamics. Journal of Engineering Mechanics. 1996，10：939-947.

[6] Cervera M，Oliver J，Manzoli O. A rate-dependent isotropic damage model for the seismic analysis of concrete dams. Earthquake Engineering and Structure Dynamics. 1996，25：987-1010.

[7] Faria R，Oliver J. A rate dependent plastic-damage constitutive model for large scale computations in concrete strucrures. CIMNE Monograph no. 17，Barcelona，Spain. 1993.

[8] 李杰．混凝土随机损伤力学的初步研究．同济大学学报．2004，32(10)：1270-1277.

[9] 李杰．混凝土随机损伤力学—背景、意义与研究进展．《结构防灾、监测与控制》，中国建筑工业出版社，2008.

[10] 李杰，张其云．混凝土随机损伤本构关系研究．同济大学学报．2001，29(10)：1-8.

[11] 李杰，卢朝辉，张其云．混凝土随机损伤本构关系—单轴受压分析．同济大学学报．2003，31(5)：505-509.

[12] 李杰，吴建营．混凝土弹塑性损伤本构模型研究 I：基本公式．土木工程学报．2005，38(9)：14-20.

[13] Li J，Wu JY. Energy-based CDM model for nonlinear analysis of confined concrete structures. American Concrete Institute，2006，SP-238，209-221.

[14] Wu JY，Li J，Faria R. An energy release rate-based plastic-damage model for concrete. International Journal of Solids and Structures，2006，43(3-4)：583～612.

[15] Wu JY，Li J. Unified plastic-damage model for concrete and its applications to dynamic nonlinear analysis of structures. Structural Engineering and Mechanics，2007，23(5)：25(5)，519-540.

[16] 李杰，任晓丹，杨卫忠．混凝土二维本构关系试验研究．土木工程学报．2007，40(4)：6-14.

[17] 李杰，杨卫忠．混凝土弹塑性随机损伤本构关系研究．土木工程学报，2009，42(2)，31-38.

[18] Li J，Ren XD. Stochastic damage model for concrete based on energy equivalent strain. International Journal of Solids and Structures，2009，46(11-12)：2407-2419.

[19] Biscof PH，Perry SH. Compression behavior of concrete at high strain rates. Material and Structures. 1991，144(24)：425-450.

[20] Rossi P，van Mier JGM，Boulay C. The dynamic behavior of concrete：the influence of free water. Materials and Structures. 1992，25：509～514.

[21] 尚仁杰．混凝土动态本构行为研究．博士论文．大连理工大学．1994.

[22] Rossi P，Toutlemonde F. Effect of loading rate on the tensile behavior of concrete：description of physical mechanisms. Materials and Structures. 1996，19：116～118.

[23] Grote DL，Park SW，Zhou M. Dynamic behavior of concrete at high strain-rates and pressures：I. Experimental characterization. Int. J. Impact Engng. 2001，25，869-886.

[24] 宁建国，商霖，孙远翔．混凝土材料动态性能的经验公式、强度理论与唯象本构模型．2006，36(3)：389-405.

[25] Tang T，Malvern LE，Jenkins DA. Rate effects in uniaxial dynamic compression of concrete. ASCE Journal of Engineering Mechanics. 1992，118(1)：108-124.

[26] Ashurst WT，Hoover WG. Microscopic fracture studies in the two-dimensional triangular lattice. Phys. Rev. B. 1967. 14(4)：1465-1473.

[27] Abraham FF，Brodbeck D，Rafey RA，and Rudge WE. Instability dynamics of fracture：A computer simulation investigation. Phys. Rev. Lett. 1994. 73(2)：272-275.

[28] Buehler MJ and Gao HJ. Dynamical fracture instabilities due to local hyperelasticity at crack tips. Nature. 2006，439：307-310.

[29] Xu XP，Needleman A. Numerical simulations of fast crack growth in brittle solids. Journal of the Mechanics and Physics of Solids. 1994，42(9)：1397-1434.

[30] Xu XP，Needleman A. Numerical simulations of dynamic crack growth along an interface. International Journal of Fracture. 1995，74(40)：289-324.

[31] Ren XD，Li J. Crack Group Propagation in Dynamic Fracture，Submitted to Physical Review Letters.

[32] 宋玉普．多种混凝土材料的本构关系和破坏准则．北京：中国水利水电出版社，2002.

[33] Krajcinovic D. Damage mechanics. Second edition. Elsevier B. V. 1996.

[34] Kandarpa S，Kirkner DJ. Stochastic damage model for brittle materiel subjected to monotonic loading. Journal of Engineering Mechanics. 1996. 126(8)：788-795.

[35] LadevèzeP. Sur une théorie de l'endommagement anisotrope. Laboratoire de Mécanique et Technologie. 1983.

[36] Mazars J. Application de la mecanique de l'endommangement au comportement non lineaire et a la rupture du beton de structure. These de Doctorate d'Etat，L. M. T.，Universite Paris，France. 1984.

[37] Ju JW. On energy based coupled elastoplastic damage theories：constitutive modeling and computational aspects. International Journal of Solids and Structures，1989，25(7)：803～833.

# 第 10 章 Chapter 10

# 高层混凝土结构隔震设计的研究与实践*

**刘伟庆，王曙光，杜东升**

（南京工业大学　土木工程学院，江苏　南京　210009）

**摘　要**：针对高层混凝土结构隔震设计中的一些关键问题进行了系统的研究，并将研究成果成功应用于工程实践。首先对高层隔震结构的隔震效果进行了分析，然后对我国规范设计反映谱的地震影响系数进行了修正，为在国推广高层隔震提供了理论支持。提出了高层隔震结构上部抗侧力构件的布置原则，以减小隔震支座的拉应力，对隔震层的控制参数和设计原则也进行了深入探讨。将以上理论应用于宿迁某高层混凝土结构的隔震设计中，分析结果表明，高层混凝土隔震结构减震效果明显，上部结构与隔震层都具有很高的可靠度。最后对高层隔震结构的相关构造措施进行了简要的说明。

**关键词**：高层混凝土结构，隔震设计，地震影响系数，隔震支座布置

# RESEARCH ON THE SEISMIC ISOLATION DESIGN OF HIGH-RISE CONCRETE STRUCTURE AND ITS APPLICATION

W. Q. Liu, S. G. Wang, D. S. Du

(1. College of Civil Engineering, Nanjing University of Technology, Nanjing 210009, China.)

**Abstract**: Give a brief introduction of research and application about high-rise concrete structure which is used the base isolation technology. Some key problems in the seismic isolation design of high-rise concrete structure are presented. Firstly, discuss whether the high-rise concrete structure has a good isolation effect when use the base isolation technology. Secondly, give some recommendations for improvement of the seismic isolation design, and modify the seismic influence coefficient curves in Chinese code. Finally, some design principles on column and shear wall of superstructures are presented. And the arrangement

* 基金项目：国家自然科学基金资助项目（50908115，50608039），国家自然科学基金重大研究计划（90815017），南京工业大学青年教师学术基金.

principles of isolators are also given according to the analysis. These theories are used in the seismic isolation design of a high-rise concrete building in Suqian city. The analysis results show that this building has a good isolation effect and a high reliability both in superstructure and isolation layer. So the conclusion is that the base isolation technology is feasible in high-rise concrete structure.

**Keywords**: high-rise concrete structure, seismic isolation design, seismic influence coefficient, arrangement of isolators

## 10.1　引言

隔震技术作为一种简便、有效的减震技术已被广泛应用于各类中低层建筑中，取得了良好的减震效果。近年来，这项技术又逐步被推广到高层建筑中。1998 年，日本竹中工务店设计并建造了东京杉并花园城[1]，其主体部分共有 30 层，高度 93.1m，是日本当时最高的基础隔震建筑。2002 年底，在日本神奈川县川崎市建成了 Thousand Tower[2]，该大楼 41 层，高度 135m。2006 年 12 月，位于日本大阪的一栋超高层隔震建筑建成，该建筑地下 1 层，地上 50 层，塔屋 2 层，高度 177.4m，高宽比 5.7。在我国，基础隔震技术在高层建筑中的应用也已开始。江苏省宿迁市已建成的宿迁海关业务大楼地下 1 层，地上 17 层，高度 73.6m[3]；宿迁在建的华夏丽景地下 1 层，地上 20 层，高度 74.6m。可以预见，随着社会经济发展的需求和人们抗震意识的进一步提高，高层隔震结构将会有广阔的发展前景。

高层隔震建筑相对于多层隔震建筑具有以下主要疑难点：(1) 高层隔震是否具有隔震效果？我国规范提出的设计反应谱长周期段下降很慢，对于长周期的高层建筑，隔震后虽然周期延长，是否仍具有明显的减震效果值得探讨；(2) 当结构阻尼比大于 5%时，我国规范中地震影响系数曲线的阻尼调整系数值过于保守，这样势必会造成高层隔震结构地震响应不准确；(3) 高层建筑倾覆效应明显，隔震支座有可能出现拉应力，而通常采用的叠层橡胶支座的抗拉能力很低，隔震支座受拉问题也是隔震技术在高层建筑中推广应用的主要障碍之一；(4) 高层建筑重力荷载较大且在罕遇地震作用下竖向构件的轴力变化大，隔震支座的面压变化及其水平位移也很大，如何合理控制隔震支座的面压及稳定性也是高层隔震设计的重要内容。

本文首先对高层混凝土隔震结构的隔震效果进行了分析，解释了长周期高层建筑仍具有较好隔震效果的原因，其次对我国规范中地震影响系数曲线进行了调整，接着对高层隔震结构上部设计和隔震层设计的一些基本原则进行了研究，最后以国内目前最高的隔震建筑为例对所研究的成果进行了工程应用，并在隔震设计中引入了性能设计的理念。通过以上几个关键问题进行分析，指出我国规范中隔震设计相关规定的不完善之处，提出设计建议或改进意见，为规范的修订提供依据。

## 10.2　高层混凝土隔震结构隔震效果探讨

高层建筑本身已经具有较长的周期，高度 100m 左右高层混凝土结构的基本周期一般在 2.5s 左右，采用基础隔震技术后一般可将其周期延长到 3.5s 左右。根据我国《建筑抗震设计规范》GB 50011—2001（2008 版）（以下简称“规范”）的设计反应谱，其在长周期段下降缓慢[4]；从图 1 可以看出，在设防烈度为 8 度（0.3$g$）三类场地建造的高层隔震建筑，周期由原来的 2.35 秒延长到 4.30 秒，从 5%阻尼比的反应谱上可以算出地震影响系数从 0.0559 降低到 0.0465，认为隔震前后地震剪力比值只有 0.0465/0.0559＝0.83，按照“规范”相关条文要求，隔震后的上部结构降低半度的要求都达不到，隔震效果不明显。但事实情况却并非如此，通过对高层隔震结构进行非线性动力时程分析，探讨了长周期高层建筑进行隔震后仍具有较好减震效果的原因。

现以宿迁某高层隔震建筑为例，分析高层隔震建筑仍然具有较好隔震效果的原因。该高层建筑主楼结构地上 23 层，隔震层兼作地下室；纵向长 50.6m，横向长 26.2m，主体建筑高度约 89.8m，长边方向高宽比为 1.78，短边方向高宽比为 3.43，最高处 97.2m[5]。文献[5]对其进行了详细的非线性动力时程分析，表 1 给出了隔震与非隔震结构在计算时采用不同振型数量情况下 X 向的基底剪力。从表中可以看出，隔震和非隔震结构第 1 阶振型的基底剪力相差不大，这也就说明了为什么直接从反应谱上看长周期结构隔震效果不明显。但是，隔震结构的第 1 阶振型质量参与系数达到了 90.1%，采用多阶振型地震剪力组合后，最终的基底剪力与采用 1 阶振型基底剪力相差不大；而非隔震结构的第 1 阶振型质量参与系数只有 65.4%，通过多阶振型地震剪力组合，最终的基底剪力会远远大于第一阶振型的基底剪力[6]。

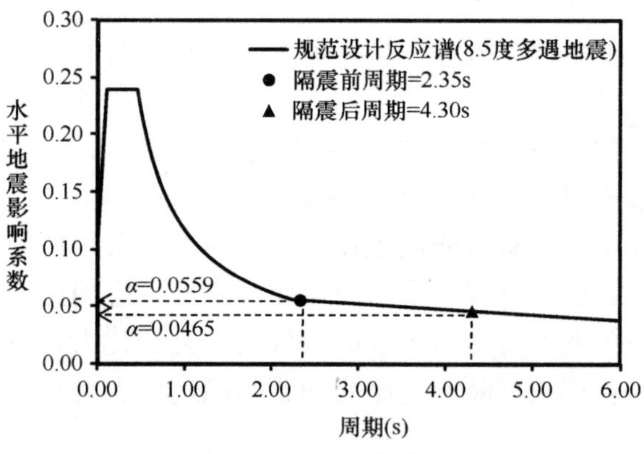

图 1　由中国规范反应谱分析高层建筑的隔震效果

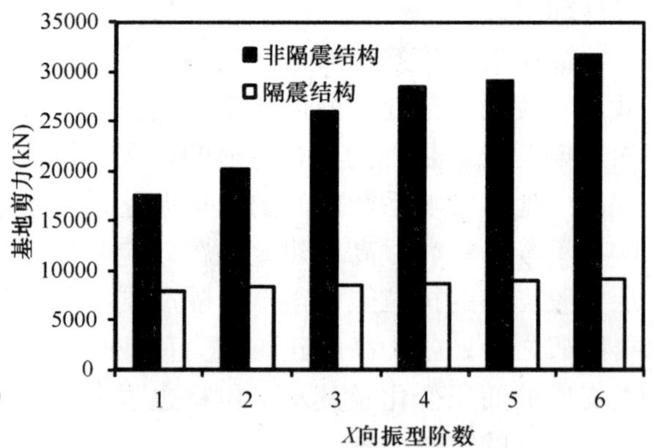

图 2　基底剪力随振型阶数增加的变化

**隔震与非隔震结构在不同振型数量下的地震剪力（X 向）　表 1**

| 采用的振型总数<br>隔震前后的周期（s） | 非隔震结构地震剪力/（kN） | | 隔震结构地震剪力/（kN） | |
|---|---|---|---|---|
| | 累计振型质量系数 | 时程结果平均值 | 累计振型质量系数 | 时程结果平均值 |
| 1(2.04/4.70) | 65.4% | 17621 | 90.1% | 7801 |
| 2(0.52/2.68) | 83.2% | 20176 | 92.7% | 8347 |
| 3(0.31/1.17) | 88.8% | 26014 | 99.5% | 8412 |
| 4(0.18/0.63) | 94.6% | 28557 | 99.7% | 8596 |
| 5(0.07/0.22) | 98.6% | 29118 | 99.8% | 8917 |
| 6(0.02/0.08) | 99.8% | 31732 | 99.9% | 9125 |

图 2 给出了隔震和非隔震结构基底剪力随振型阶数增加而变化的情况。可以看出，非隔震结构地震剪力随振型阶数增加而显著增加，从采用 1 阶振型时的 17621kN 增大到采用 6 阶振型时的 31732kN，是 1 阶振型剪力的 1.80 倍。而隔震结构的基底剪力随振型阶数的增加而变化不大，从采用 1 阶振型时的 7801kN 增大到采用 6 阶振型时的 9125kN，为一阶振型剪力的 1.17 倍。所以长周期高层建筑，在利用反应谱进行隔震效果分析时，必须要充分考虑高阶振型影响的问题，否则必然会低估减震效果。

## 10.3　高层混凝土隔震结构不同阻尼比地震影响系数修正

目前，世界上拥有专门的隔震设计规范或规程的国家和地区主要有：日本、美国、意大利、中国、中国台湾地区。这 5 种规范的隔震设计思路大致相同：(1)提出不同阻尼比下的结构加速度反应谱曲线；对隔震和消能减震结构，首先采用等效线性化方法进行迭代求解。对于隔震结构，可求出隔震层位移、基底剪力等重要数据。(2)采用非线性时程分析方法，进行隔震结构的进一步分析，以得到上部各层结构的加速度、速度、位移、层间剪力等详细动力响应。要求非线性时程分析方法和等效线性化方法的分析结果的误差较小，具有足够的可比较性。

对这 5 种隔震设计规范进行详细的比较发现，除了日本隔震设计规范以外，其他 4 种隔震设计规范，均存在等效线性化方法的计算结果比时程分析方法偏大很多的问题。对于我国规范来说，隔震结构的周期越长或阻尼比越大，两种方法的误差还会进一步增大[6]。其主要的原因可能是：(1)我国规范中长周期段的地震影响系数值偏大；(2)当结构阻尼比大于 5%时，阻尼调整系数 $\eta_2$ 的值也偏大。

我国规范可借鉴采用日本规范中的阻尼调整系数 $F_h$ 的计算方法。

$$F_h = \frac{1.5}{1 + 10(h_v + 0.8h_d)};\ F_h \geqslant 0.4 \tag{1}$$

式中：$h_v$ 为隔震层黏滞阻尼部分所提供的阻尼比，$h_d$ 为滞回阻尼部分所提供的阻尼比；当阻尼调整系数 $F_h < 0.4$ 时，取为 0.4。

相应的，我国规范中的地震影响系数 α 曲线的表达式可以调整为：

$$\alpha = \begin{cases} \left(0.45 + \dfrac{\eta_2 - 0.45}{0.1}T\right)\alpha_{max} & T \leqslant 0.1 \\ \eta_2 \alpha_{max} & 0.1 < T \leqslant T_g \\ \left(\dfrac{T_g}{T}\right)^{0.9} \eta_2 \alpha_{max} & T_g < T \leqslant 5T_g \\ [0.2^{0.9} - 0.02(T - 5T_g)]\eta_2 \alpha_{max} & 5T_g < T \leqslant 6.0 \end{cases} \tag{2}$$

式中：$\alpha_{max}$ 为水平地震影响系数最大值；$\eta_2$ 为阻尼调整系数；$T_g$ 为场地特征周期值；$T$ 为结构自振周期。

公式(2)和规范原表达式相比，取消了原来的曲线形状参数 $\eta_1$ 和 $\gamma$；只保留阻尼调整系数 $\eta_2$，令

$$\eta_2 = 1.5/(1.1 + 8.0\xi);\ \eta_2 \geqslant 0.4 \tag{3}$$

式中：$\xi$为结构阻尼比。

图 3 为采用上述表达式计算出来的不同阻尼比的地震影响系数曲线。和规范中的现有曲线相比，一个显著的区别就是对高阻尼情况下地震影响系数的折减更加合理，不同阻尼比的曲线过渡的更加自然，曲线也显得更加平滑。

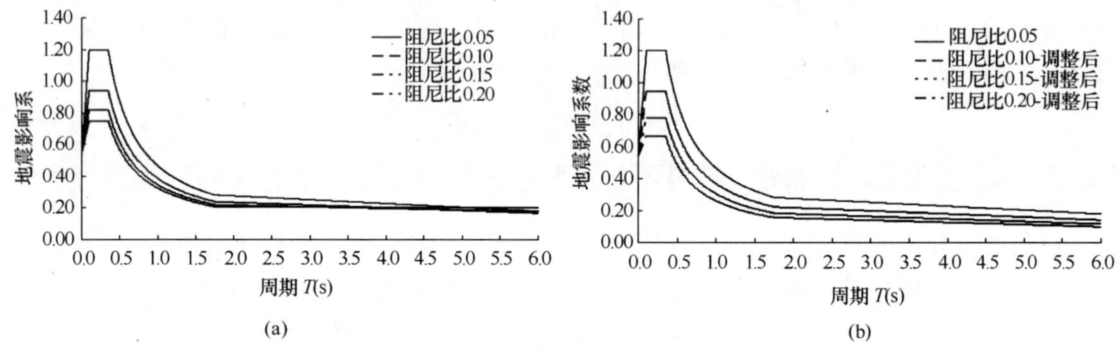


图 3 不同阻尼比结构的地震影响系数曲线

（a）调整前；（b）调整后

## 10.4 高层混凝土隔震结构上部结构的设计原则

隔震层上部结构的设计是高层隔震设计中的重要内容，合理的上部结构设计是隔震层合理设计的一个基础条件。高层建筑相对于多层建筑，倾覆效应明显，采用隔震技术后，在罕遇地震作用下，隔震支座有可能会出现拉应力，而工程中采用的叠层橡胶支座的抗拉能力较低，对于一般的滑移支座，由于倾覆效应也可能使滑移支座发生提离。隔震结构整体倾覆和隔震支座受拉问题是隔震技术在高层建筑中应用的主要障碍。

### 10.4.1 柱距布置对隔震支座竖向应力的影响

根据以往对高层隔震结构的设计和研究，提出高层隔震结构隔震层上部结构设计的基本原则：严格控制高宽比防止结构整体倾覆，合理布置竖向构件避免或者减小隔震支座的受拉。高层隔震结构高宽比的限制很多文献都进行了理论研究和试验研究[8]，但通过竖向构件的合理布置减小支座受力的问题却很少有文献涉及。隔震支座受拉的主要原因是由于水平地震荷载的倾覆力和竖向地震振动的向上作用力超过了结构本身给支座的重力荷载，所以我们可以通过增大隔震支座所承受的重力荷载范围、减小地震作用引起的倾覆力或采用具备很高抗拉能力的隔震支座三种基本方法来解决隔震支座受拉问题[9]，而前两个方法就可以通过合理布置上部结构的竖向构件来实现。竖向构件的合理布置的原则可以归纳为：柱距要大，抗震墙要整，抗震墙尽量居中。

大柱距配合大直径的隔震支座是比较合理的高层隔震设计，大柱距可以使得柱下的隔震支座承受较大的重力荷载，从而不易出现受压，这个作为隔震的基本概念设计是好理解的，但要求抗震墙尽量居中则与抗震概念设计有所违背，抗震设计要求抗震墙布置分散布置在四周以控制结构的扭转效应，但这种布置方式对抗震墙下部的隔震支座的受拉是不利的，抗震墙下的隔震支座在地震作用下轴力变化很大，而边缘四周的隔震支座承受的竖向重力荷载相对较小，所以抗震墙布置于边缘，就比布置于内部更有可能引起抗震墙端部的隔震支座受拉。

### 10.4.2　抗震墙布置对隔震支座竖向应力的影响

高层建筑中抗震墙所承受的倾覆力矩比框架大，抗震墙下的隔震支座的轴力变化更加突出。对于框架—抗震墙结构，因为边缘柱及其下隔震支座承受的竖向重力荷载相对于其他内柱较小，而倾覆力引起的轴力比内柱要大，所以边缘隔震支座比较容易产生拉应力。布置抗震墙时，抗震墙布置于边缘比布置于内部更有可能引起隔震支座的受拉。按图 4 进行两种抗震墙位置布置，图 4（*a*）中的抗震墙布置于外边缘，而图 4（*b*）中抗震墙布置在承受较大重力荷载的中间部位，这两种布置方法对结构 Y 向的抗侧刚度影响较小，共采用了 20 个 GZY800 隔震支座。对这样的 20 层结构进行时程分析，在同样的 1940 年 El Centro 地震动沿 Y 方向的地震作用下结构反应结果绘于图 5。

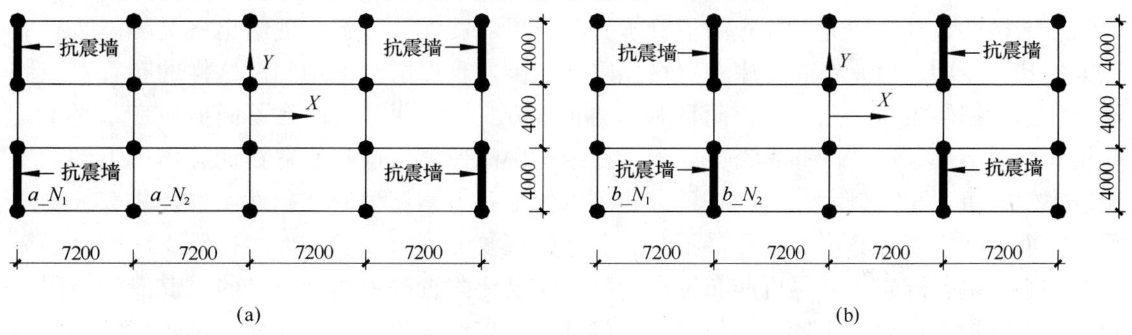


图 4　抗震墙布置图

（a）抗震墙布置于外边缘；（b）抗震墙布置于中间部位

由图 5 可知，抗震墙下隔震支座轴力变化明显比单纯框架柱下隔震支座轴力变化加剧。抗震墙布置于框架易产生拉应力的边缘和角部时（$a-N_1$ 处），使隔震支座产生了拉应力；而避开框架较易产生拉应力的位置布置（$b-N_2$ 处）推迟了拉应力的出现[9]。

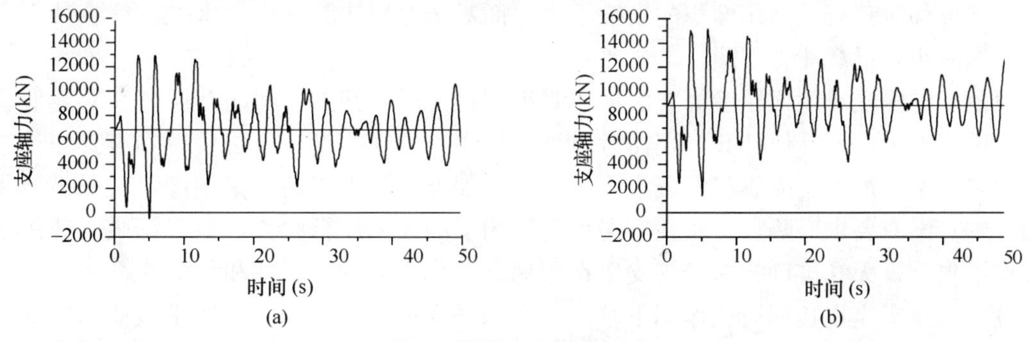


图 5　隔震支座轴力反应

（a）抗震墙布置于外边缘时隔震支座轴力响应；（b）抗震墙布置于中间部位时隔震支座轴力响应

对于一些特殊情况，可考虑开发或采用一些高抗拉能力的隔震支座。日本已建高层隔震建筑中有几个成功案例[1]：东京杉并花园城（28 层隔震建筑）采用了一种高强橡胶支座，这种隔震支座的受压和受拉能力比普通叠层橡胶支座有一定的提高；大阪 DT 办公楼（27 层隔震建筑）采用了具有较强抗拉能力的直线轨道式滑移支座。

## 10.5 高层混凝土隔震结构隔震层的设计原则

隔震层的设计是整个隔震建筑设计过程中的核心内容，但我国“规范”对隔震层设计的指导还不够完善。如何设计和检验隔震层，并使其满足安全可靠的要求还需要有一些具体的指标来衡量。本节对隔震层总体设计原则及各类隔震装置的配置原则进行了分析，并对隔震支座面压的控制给出了建议，旨在使隔震层具有更高的可靠度。

### 10.5.1 隔震装置布置原则

各种类型的隔震装置都有各自力学模型和适用条件，如何将其合理组合配置，使各自能够发挥作用从而形成安全有效的组合隔震层是进行组合隔震设计的关键问题。通过对高层隔震结构的设计和研究，并查阅了日本一些高层隔震设计方案，现给出各类隔震装置的配置的基本原则：（1）铅芯支座布置在隔震层的围，且要求对称均匀。这样的布置对控制隔震层的抗扭转效应十分有利，而且对称的布置方式可以保证支座在不同剪应变水平下具有相同的偏离率；（2）滑移隔震支座配置在结构中间位置，这些位置在地震作用下轴力变化较小但自重下轴力较大。这种布置方式可以保证其力学模型的准确性，并提供较大的水平屈服力，但需要计算隔震层在最不利工况下的回复力能力；（3）天然橡胶支座配合隔震层的偏心率进行配置，并尽可能布置在滑移隔震支座附近提供回复力；（4）黏滞流体阻尼器在大震作用下可以提供较大的阻尼力，将其布置在结构角部和边缘对控制隔震结构的扭转也是非常有效的。

### 10.5.2 隔震支座面压控制准则

隔震支座在正常使用状态下会长期承受着上部结构的重量，当发生地震时，隔震支座会发生较大的水平位移，此时需承受上部结构的重量和地震力的共同作用。对于滑移类支座，水平位移对其有效承压面积的影响不大；而对于叠层橡胶支座，水平位移会使其有效承压面积减小，承载能力下降。

规范中的隔震支座面压验算方法应用到工程设计时，其主要问题有：（1）对隔震支座的平均压应力采用设计值计算，还需考虑地震作用。这样的设计方法使得隔震支座的直径偏大很多，不单加大了隔震层的资金投入，而且加大了隔震层的水平刚度，对隔震效果也不利。（2）没有要求对隔震支座在罕遇地震作用下的面压进行验算，只对隔震支座的水平变形作了限制，无法准确把握隔震支座在罕遇地震作用下的安全性和稳定性。

叠层橡胶支座在长期荷载作用下基本没有水平变形，此时竖向轴压承载能力很高，对于第一形状系数 $S_1 \geqslant 15$、第二形状系数 $S_2 \geqslant 3$ 的条件下，一般能达到 90MPa 以上[10]。规范规定的平均压应力限值是按安全系数方法进行了折减，当采用 15MPa 作为设计限值时，安全系数达到 6 以上，所以对于长期荷载作用，只需验算结构本身的长期荷载给隔震支座产生的面压，无需再使用设计值进行面压验算，也不需考虑地震作用效应。

对于短期的地震作用，叠层橡胶支座会发生水平变形，此时承载能力降低。因此，需要对叠层橡胶支座进行短期罕遇地震作用下的面压和变形验算，这样才能准确把握其在罕遇地震作用下的安全性和稳定性。由于地震作用的短期行为，可将设计平均压应力限值适当放宽，可采用极限面压和变形关系曲线作为隔震支座极限面压和变形控制线。

日本的隔震支座面压验算方法采用长、短期两阶段进行控制，长期面压采用隔震支座所承受的重力荷载来计算，短期面压采用罕遇地震作用下的面压进行验算，包括重力荷载、水平地震及竖向地震引起附加轴力。由前面的分析可知，叠层橡胶支座的承载能力和剪应变是相关的，而对于滑移类支座，这种相关性可以忽略。图 6 给出两类隔震支座的面压控制设计范围[11]。图中最外边缘的极限面压控制线通常由隔震支座生产厂商提供。

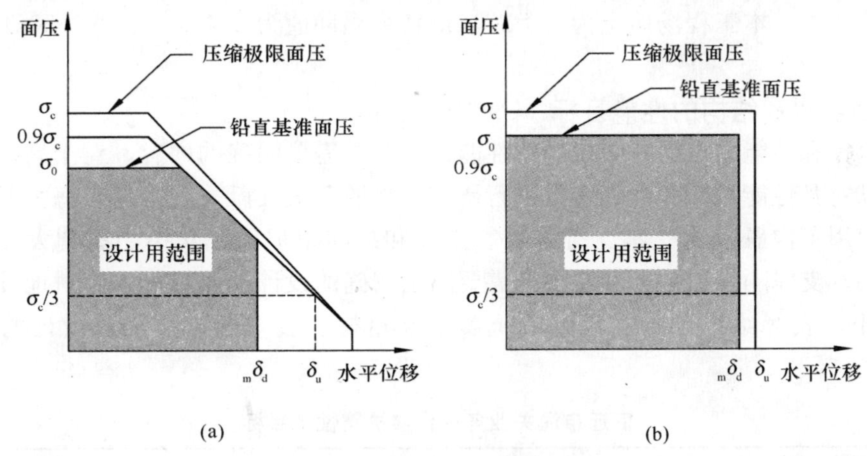


图 6　隔震支座面压控制范围
(a) 叠层橡胶支座；(b) 滑移支座

通过以上基本原则设计出的隔震层其可靠性是需要通过具体的参数来控制的，我国“规范”对相关参数没有明确的规定，现给出一些隔震层设计中需要严格控制的参数：(1) 控制隔震支座的长期面压、短期极大值、极小值面压。面压的控制在“规范”中有规定，但对应的工况不明确，也没有体现出隔震支座面压与水平剪应变之间的关系，建议采用上述方法，对隔震支座的面压进行控制。(2) 控制隔震层在不同变形水平下的偏心率。高层隔震结构在地震作用下的变形主要集中在隔震层，所以控制隔震层的扭转就可以控制整个隔震体系的扭转，我国“规范”对隔震层的偏心率没有具体要求，参照日本和中国台湾地区高层隔震设计中的相关内容，可以控制隔震层的偏心率在 3%以内。(3) 控制隔震层的屈重比，满足隔震层抗风要求。隔震层需要保证一定的屈重比水平，以确保在各类微震动和风荷载下隔震层不屈服，同时使隔震层处在最优阻尼比的范围内。(4) 严格控制隔震层的大震位移，并配合隔震支座的极大值面压，作为控制隔震支座在大震作用下的力学行为的依据，高层隔震体系隔震层大震位移的控制需要加入一定数量的黏滞流体阻尼器。(5) 控制不同输入方向下隔震支座在罕遇地震下的拉应力。(6) 对隔震上部结构抗倾覆的验算。控制以上六个参数基本上就可以保证隔震层能够具备稳定的工作状态，从而确保隔震体系的安全和可靠。

## 10.6 工程实践

本文将上述高层隔震结构设计理念应用于高度为73.6米的宿迁海关综合业务楼基础隔震设计中，为满足设定的抗震性能目标，保证隔震设计的合理性和有效性，具体的设计原则参照上文的相关内容。

宿迁海关综合业务楼位于宿迁市宿城新区，地上17层，地下1层，总高度为73.6m，现浇钢筋混凝土框架—剪力墙结构体系，建筑面积18229.5m$^2$，是目前国内最高的隔震建筑。宿迁市位于地震高烈度区，抗震设防烈度为8度，设计基本地震加速度值为0.30$g$，设计地震第一组。本工程场地土为Ⅲ类，场地特征周期值为0.45s（罕遇0.50s）。

### 10.6.1 高层隔震结构的性能目标

性能目标和性能水准是对结构进行性能设计首先需要明确的内容，对于高层隔震结构很少有文献对其性能目标和性能水准进行研究，参照相关文献[9]，宿迁市海关业务综合楼隔震系统使用了橡胶隔震支座、滑移隔震支座和黏滞阻尼器，从中小震到大震，上部结构、隔震层、支撑隔震层的下部结构都需要给出明确的设计性能目标，针对地震动发生的频度和大小，表2给出了隔震设计的抗震性能目标，表3给出了隔震设计的抗风性能目标。

**宿迁市海关业务综合楼抗震性能目标　　表2**

| 部位＼水准 | 多遇地震 | 罕遇地震 |
|---|---|---|
| 上部结构 | 最大层间位移角1/800以内，地震响应比常规结构降1度 | 弹性极限范围内最大层间位移角1/500以内 |
| 隔震装置 | 稳定变形，剪切变形在200%以内 | 稳定变形，剪切变形在200%以内 |
| 地下室和墩柱 | 短期允许应力以内 | 短期允许应力以内 |
| 地基 | 长期允许应力以内 | 长期允许应力以内 |

**宿迁市海关业务综合楼抗风性能目标　　表3**

| 部位＼水准 | 50年一遇 | 100年一遇 |
|---|---|---|
| 上部结构 | 最大加速度小于0.15m/s$^2$ | 最大加速度小于0.15m/s$^2$ |
| 隔震装置 | 铅芯橡胶支座不屈服 | 铅芯橡胶支座不屈服 |

### 10.6.2 隔震层设计

根据前文所述的隔震层配置基本原则，对该项目采用了铅芯橡胶支座、天然橡胶支座、滑移隔震支座和黏滞流体阻尼器形成了组合隔震层。各类型的隔震装置方式也基本按照前文所述的基本方法，但根据工程实际情况作了简单的调整。图 7 给出了该项目的隔震层配置情况。

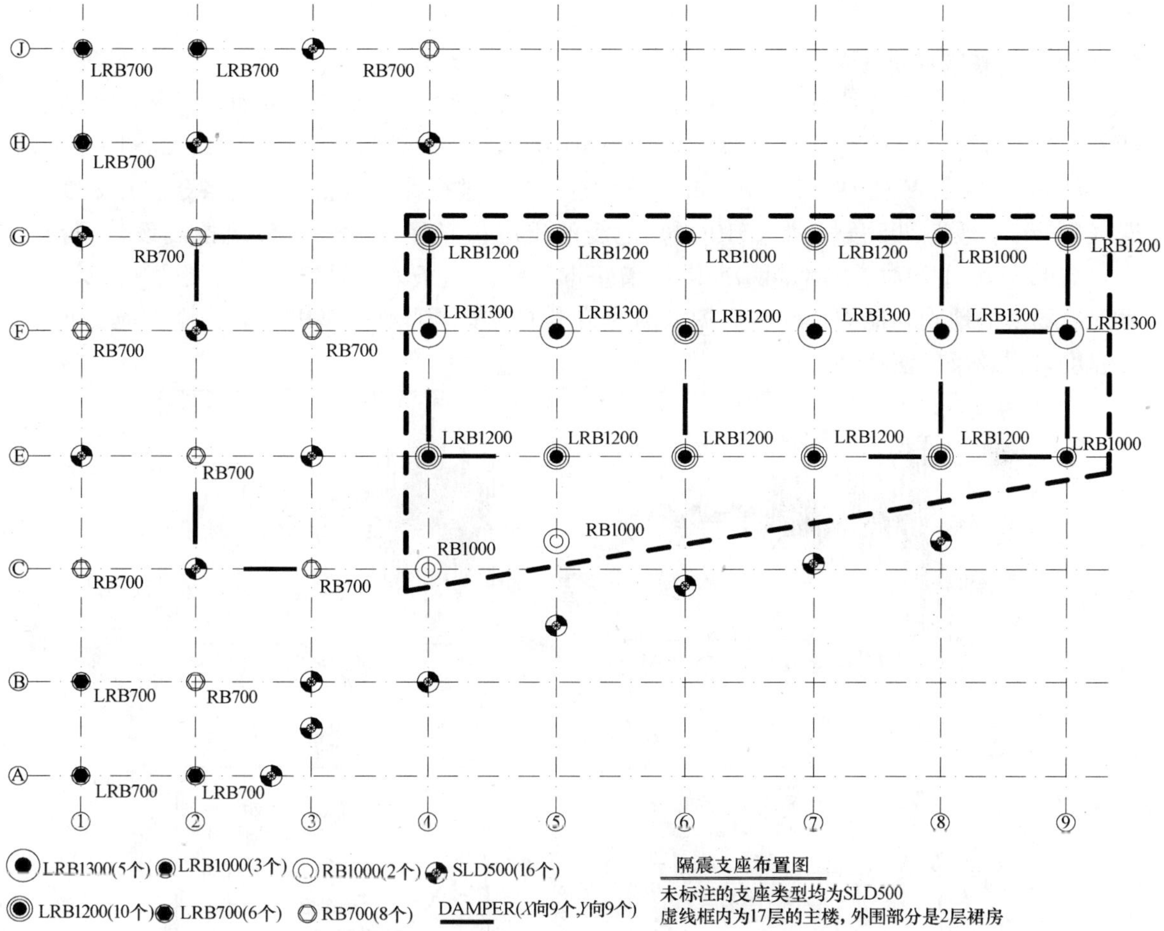

图 7　隔震支座配置图

根据隔震层的控制参数要求，隔震层必须具备屈服前刚度以满足风荷载和微振动的要求，将铅芯橡胶支座和滑移支座抗侧刚度简化为 2 线性、天然橡胶支座的水平刚度简化为线性，隔震层的水平恢复力特性由铅芯橡胶支座、天然橡胶支座和滑移支座共同组成，图 8 给出了隔震层的水平恢复力特性及抗风分析结果。

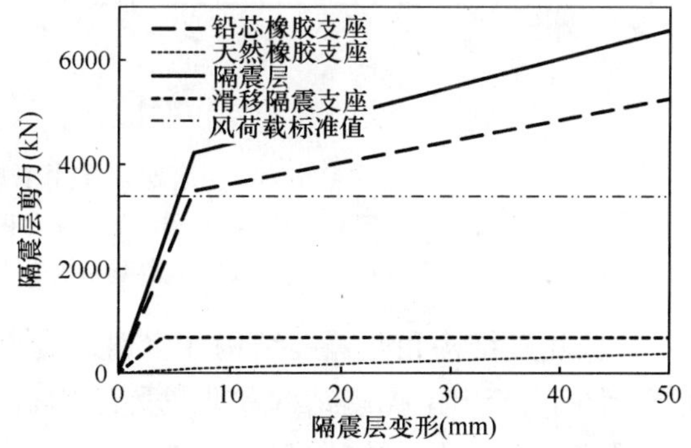


图 8　隔震层水平恢复力特性

隔震系统的偏心率也是隔震层设计中的一个重要指标，日本和我国台湾地区规范明确规定隔震系统的偏心率不得大于 3%，在进行宿迁市海关业务综合楼隔震层设计时也对隔震系统的偏心率进行了计算和控制，计算表明隔震层 $X$ 向偏心率 $R_x=1.6\%$，$Y$ 向偏心率 $R_y=0.3\%$，满足设计要求。

隔震支座的面压分析、受拉分析和倾覆分析详见文献[3]，分析结果表明该组合隔震层满足基本的控制指标，隔震层设计安全可靠。

### 10.6.3 地震响应分析

本项目的动力非线性分析共采用了六条天然地震动和两条人工地震动，六条天然地震动是 1940El Centro EW、1940El Centro NS、1952Taft EW、1952Taft NS、1968Hachinohe EW 和 1968Hachinohe NS，通过对波进行综合调整，使得各条波在 8 度多遇（110gal）地震和 8 度罕遇（510gal）地震的反应谱已经与我国《建筑抗震设计规范》GB 50011—2001 相对应的不同水准设计谱基本一致，两条人工地震动是根据宿迁市该工程附近场地的地貌和地质特性制成的。图 9 给出了 8 度罕遇地震作用下部分输入地震波的加速度时程及反应谱特性。

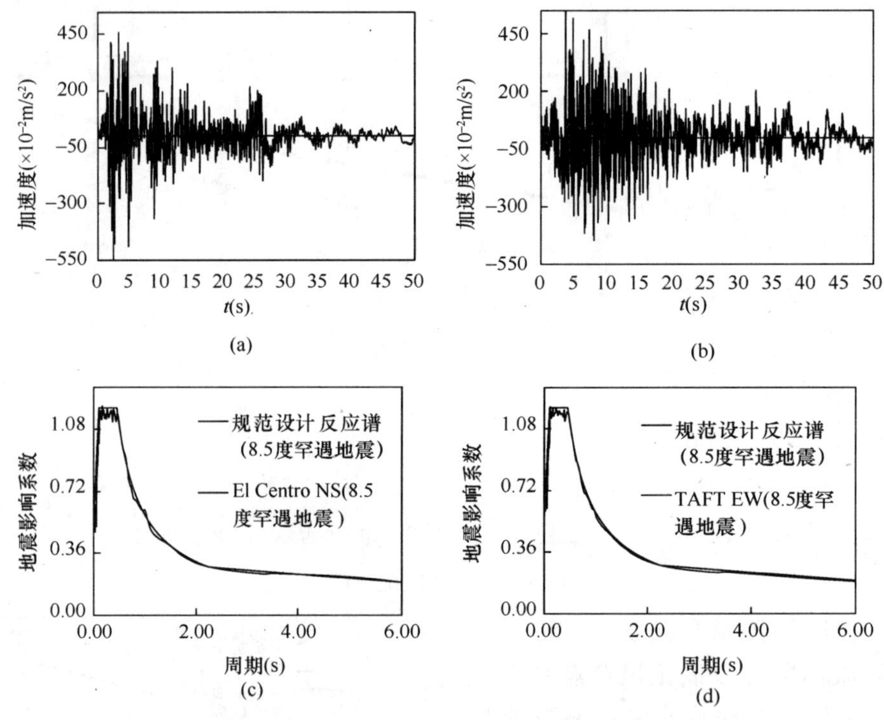


图 9 罕遇地震作用下地震波的时程及反应谱

（a）El Centro NS 地震波加速度时程；（b）Taft EW 地震波加速度时程

（c）El Centro NS 波地震影响系数曲线；（d）Taft EW 波地震影响系数曲线

比较隔震结构和非隔震结构在多遇地震作用下的响应结果，可以看出隔震系统的隔震效果。图 10 给出了隔震结构和非隔震结构 $X$ 向地震剪力和剪力系数的对比结果。

图 11 给出了隔震结构和非隔震结构 16 层部分加速度时程响应对比结果，可以看出隔震后结构顶层的加速度约为隔震前的 1/3～1/2，隔震效果非常明显。图 12 给出了隔震结构角部铅芯支座和滑移隔震支座在 8 度多遇地震作用下的滞回历程图，可以看出这些具备耗能能力的隔震装置充分发挥了其力学特性，具有较稳定的工作状态。

在地震作用下，隔震支座会吸收大量的地震能量，从而减少结构本身的粘滞耗能。结

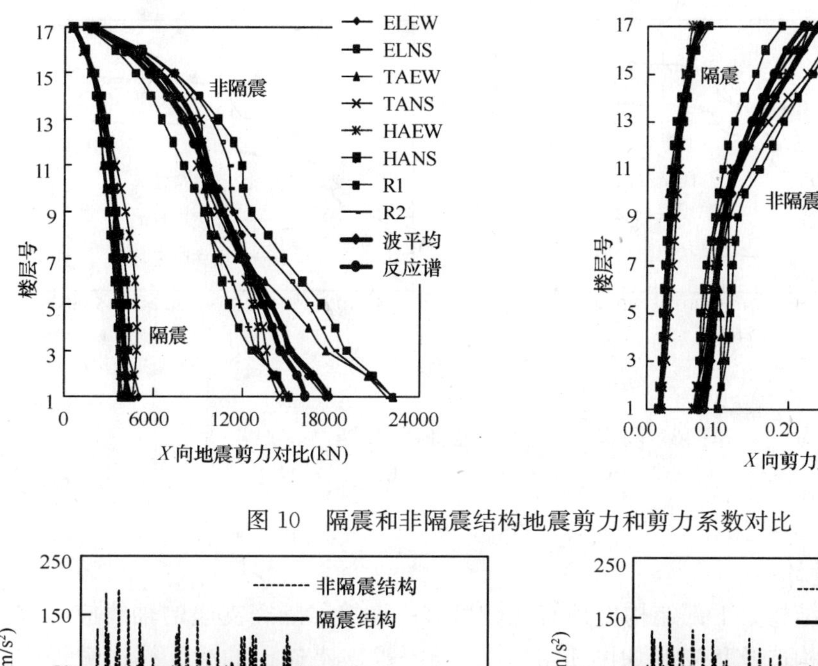


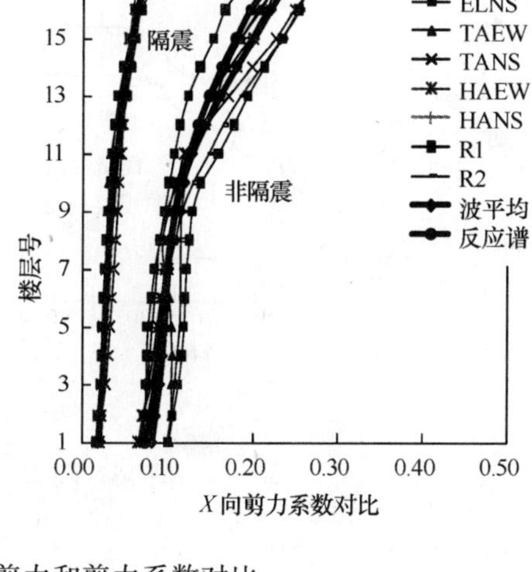


图 10　隔震和非隔震结构地震剪力和剪力系数对比

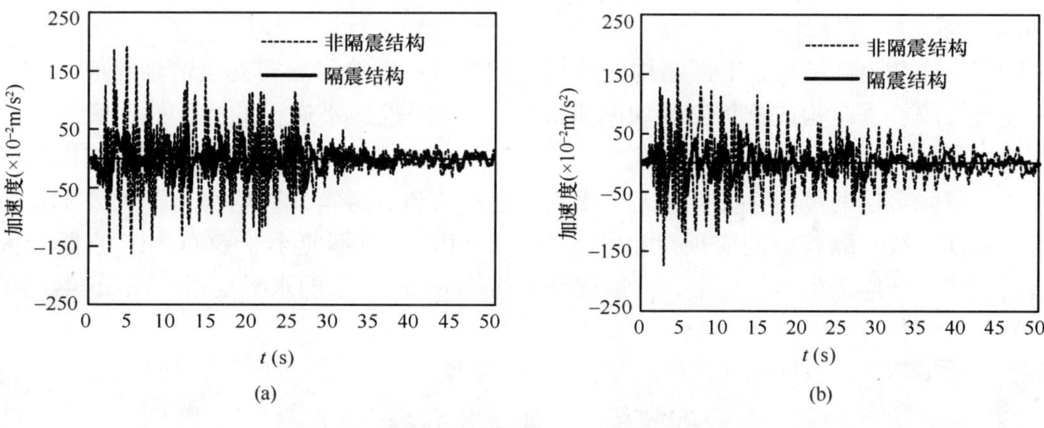


图 11　隔震和非隔震结构加速度响应对比

(a) 16 层 (EL Centro NS)；(b) 16 层 (EL Centro EW)

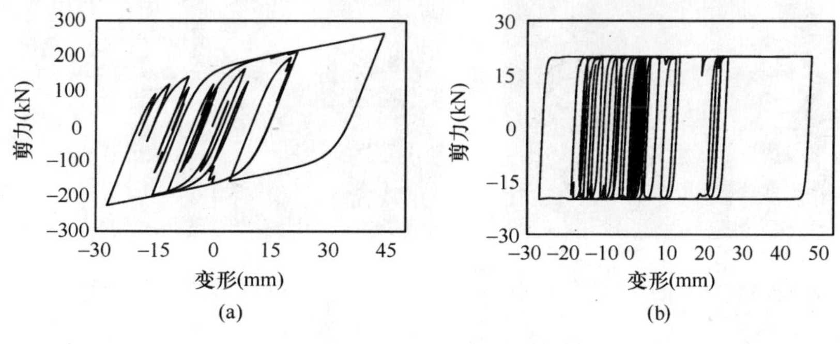


图 12　隔震支座滞回历程图（X 向）

(a) 铅芯支座（EL Centro NS)；(b) 滑移支座（EL Centro NS)

构能量时程曲线图可以直观地反映出隔震支座耗能和结构耗能随时间变化的情况，图 13 给出了隔震体系地震动作用下的能量时程曲线，可以看到，对于不同的输入地震动，隔震支座均耗散了大量的输入能量，从而减小了上部结构本身的耗能，大大提高了上部结构的抗震性能。

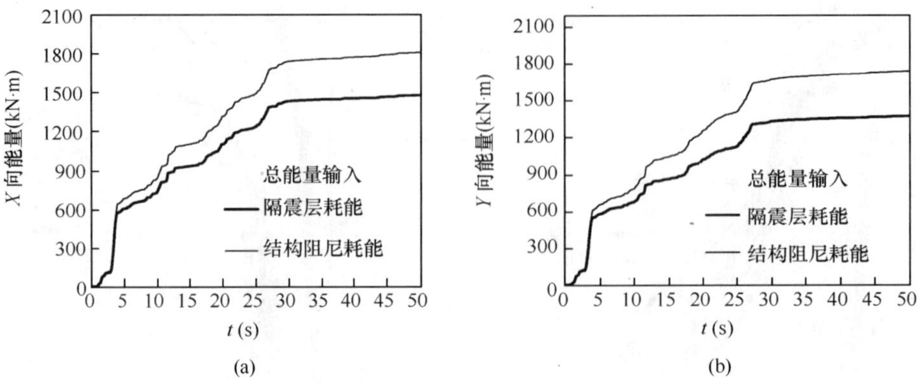


图 13 隔震结构能量时程曲线（X 向）

（a）X 向（EL-Centro NS）；（b）Y 向（EL-Centro NS）

### 10.6.4 隔震构造设计

隔震支座与上部结构、下部结构应该有比较可靠的连接，隔震支座的轴线应与柱、墙轴线重合，隔震支座构造及其安装如图 14 所示。与隔震支座连接的梁、柱、墩等应具有足够的水平抗剪和竖向局部抗压承载力，并采取可靠的构造措施，如加密箍筋或配置网状钢筋。穿过隔震层的竖向管线（含上下水管、通风管道、避雷线）应符合下列要求：直径较小的柔性管线在隔震层处应预留足够的伸展长度，其值不应小于 500mm；直径较大的管道在隔震层处应采用柔性接头，并能保证发生 500mm 以上的水平变形。图 15 给出了一种柔性导线连接方法。

图 14 橡胶隔震支座的连接

图 15 隔震层柔性导线连接图

隔震层所形成的缝隙可根据使用功能的要求，采用柔性材料封堵、填塞，以保证隔震层可以在地震下水平移动。上部结构及隔震层部件应与周围固定物脱开，与水平方向固定物的脱开距离不宜小于 500mm；与竖直方向固定物的脱开距离应取为 20mm。

隔震层顶部采用现浇钢筋混凝土梁板结构，抗震墙下托墙梁需进行专门设计及构造加强，其具体的设计方法作者将另文讨论；隔震支座附近梁、柱应考虑冲切和局部承压，加密箍筋并根据需要配置网状钢筋。隔震层顶部的纵、横梁和楼板体系应作为上部结构的一部分进行计算和设计。

## 10.7　结论

高层隔震是隔震技术发展的新趋势之一，本文针对高层隔震建筑设计中存在的主要疑点和难点进行研究，并将改进的设计理论用于工程实践，得到以下几点结论：

(1) 对于长周期高层建筑，在利用规范反应谱进行隔震前后的隔震效果分析时，必须要充分考虑高阶振型影响的修正问题，否则会明显低估非隔震结构的地震响应，从而低估结构减震效果。通常，经过合理设计的高层隔震结构，能够具有明显的减震效果。

(2) 我国规范反应谱中的长周期段结构地震影响系数值偏大，当结构阻尼比大于 5% 时，其阻尼调整系数的值也偏大，导致按照反应谱方法计算的隔震结构地震响应显著偏大。本文对规范反应谱采用的阻尼调整系数和形状参数提出了修改建议，修改后的地震影响系数曲线对长周期和高阻尼情况下的过渡更加合理。

(3) 规范规定剪力墙下隔震支座的间距尽可能小，对于高层隔震结构来说，该规定更加容易导致隔震支座受拉。本文建议上部结构采用大柱网布置方案，并将剪力墙布置在结构平面的中间部位，可大大降低隔震支座受拉的可能性。

(4) 规范中隔震支座的面压验算方法具有不完善之处，本文建议采用极限面压和变形关系曲线作为隔震支座极限面压和变形控制线，可准确把握隔震支座在地震作用下的安全性和稳定性。

(5) 将隔震技术应用于宿迁海关综合业务楼，分析结果表明：基础隔震有效地隔离了地震能量的输入，上部结构的地震反应大大降低，有效地保护了上部结构。隔震层耗能能力强，隔震效果良好。隔震层设计合理，具有很高的安全度与可靠度。所以将隔震技术应用于高层混凝土结构是可行的。

## 参考文献

[1] 何永超，邓长根，曾康康．日本高层建筑基础隔震技术的开发和应用. 工业建筑，2002，32(5)：29-31.

[2] 侯宝隆. 日本隔震技术的新发展与控震技术的实际应用. 工业建筑，2000，30(11)：74-78.

[3] 刘伟庆，王曙光，杜东升. 宿迁海关业务大楼基础隔震分析报告. 南京：南京工业大学建筑技术发展中心，2007，56-63.

[4] 中华人民共和国国家标准. GB 50011—2001 建筑抗震设计规范(2008 版). 北京：中国建筑工业出版社，2001.

[5] 刘伟庆，王曙光，程华群. 宿迁开发区商务中心基础隔震分析报告. 南京：南京工业大学建筑技术发展中心，2006，56～63.

[6] 王曙光，杜东升，刘伟庆. 高层建筑结构隔震设计关键问题. 南京工业大学学报(自然科学版)，2009，31(1)：67-72.

[7] Demin Feng, Wenguang Liu, Keiji Masuda, et al. A comparative study of seismic isolation codes worldwide. First European Conference on Earthquake Engineering and Seismology. Geneva, Switzerland, 2006.

[8] 刘文光，杨巧荣，周福霖. 大高宽比隔震结构地震反应的实用分析方法. 地震工程与工程振动，2004，24(4)：115-121.

[9] 程华群，刘伟庆，王曙光. 高层隔震建筑设计中隔震支座受拉问题分析. 地震工程与工程振动，2007，27(4)：162-164.

[10] 周福霖. 工程结构减震控制. 北京：地震出版社，1997.

[11] 和田章，久保哲夫，可児長英等[日]. 免震建築物の技術基準解説及び計算例とその解説. 東京：工学図書株式会社，2001 年 5 月.

# 第 11 章 Chapter 11

# 冲击与爆炸作用下混凝土结构的受力机理与分析研究进展

# MECHANICS AND COMPUTATION OF CONCRETE STRUCTURES SUBJECTED TO SHOCK AND BLAST LOADS - SOME RECENT PROGRESSES

Y. Lu (陆 勇)
Institute for Infrastructure and Environment, School of Engineering, The University of Edinburgh, UK.

**Abstract**: Shock and blast loads have a character of short duration and high amplitude. Consequently, the direct response of a structure is dominated by a strong stress wave process and high-frequency local dynamic effect, whereas the global structural effect is essentially an indirect process triggered by abrupt structural alterations. Although certain interaction exists, each process has a distinctive time scale and the underlying failure mechanisms differ. A clear recognition of such features is important in the assessment of the governing structural effects of a particular type of impulsive loads and the determination of an appropriate analysis approach. This paper is aimed at providing a systematic perspective of the different dynamic regimes of concrete structures responding to shock and blast loads and the varying failure mechanisms. A series of recent studies are briefly described and discussed, with a focus on the dynamic responses at the concrete material and component levels, and their intrinsic relationship with the blast load characteristics.

**Keywords**: safety shock, blast, impulsive load, concrete structure, dynamic response, failure mechanisms

## 11.1 INTRODUCTION

The response of building structures to shock and blast loading generated by a nearby explosion has been a subject of extensive studies, especially in recent years. As far as reinforced concrete structures are concerned, existing studies may be broadly divided into three categories, namely i) dynamic response of concrete material, fracture and fragmentation (e. g. , Bischoff and Perry 1991, Li and Meng 2003, Lu and Xu 2004, Xu and Lu 2006, Lu et al. 2010); ii) effect of blast load on RC structural components, especially RC columns (e. g. , Krauthammer 1984, Ghaboussi et al. 1984, Stevens and Krauthammer 1991, Krauthammer et al. 1993a, 1993b, Lellep and Torn 2005, Gong and Lu 2007, Shi et al. 2008), and iii) progressive collapse of a RC structural system (e. g. Sasani et al. 2007).

From the view point of the characteristics of the responses, categories i) and ii) may be considered as the direct response to the imposed shock and blast load, and hence are intrinsically dependent upon the dynamic characteristics of the load and the dynamic behaviour of the material and the structural components. On the other hand, category iii) is essentially an indirect effect due to abrupt alterations to the load paths as a result of the direct blast damage. Comparing to the direct effects, the indirect system response, or progressive collapse, is a low-dynamic or quasi-static process. Although certain dynamic amplification may arise from the sudden alteration to the load bearing system, e. g. , the loss of a ground storey column, this process is featured by large global deformation and high geometric nonlinearity. The subject to be discussed in this paper is concerned mainly with the high dynamic phenomena associated with shock and blast loads, and therefore only category i) and ii) responses are considered.

This paper is organized along the line of a series of studies conducted by the author and his co-workers investigating into the characteristics of blast loads, characteristics of component-wide dynamic responses, and the dynamic behaviour of concrete material under high strain rate loading. A Timoshenko beam-based nonlinear model for the analysis of RC beam-column under shock and blast loads, and a mesoscale computational model for the simulation of the underlying mechanisms governing the dynamic behaviour of the concrete material are highlighted.

## 11.2 GENERAL CHARACTERISTICS OF BLAST LOADING AND RC STRUCTURAL RESPONSE

The basic event-defining parameters for a building structure subjected to a nearby explosion are schematically illustrated in Figure 1(a), where $W$ is the charge weight, normally in kilograms of equivalent TNT, $R$ is the standoff distance in meters, $D$ is the distance of detonation above the ground surface, and $H$ and $L$ are the height and width of the building structure, respectively. Figure 1(b) depicts a propagating air blast as it encounters a solid structure. At the interface, reflection of the shock wave takes place, resulting in enhanced blast loading on the structure.

A typical blast load in open air has a time history as shown in Fig. 1(c), which may be characterized by a peak overpressure, positive phase duration and impulse. There are a number of empirical methods which may be used to determine the free-field (side-on) and reflected blast load parameters, for example Henrych (1979), Hyde (1991), and TM5-1300 (1990). The graph on the right hand side of Fig. 1(c) shows the empirical data, as recommended by TM5-1300 (1990), for the key parameters. For typical explosion scenarios with a charge weight in a range of 100-1000 kg (TNT) and at a standoff distance of around 10 meters, the peak reflected overpressure will be on the order of mega-Pascal's, while the duration is a matter of a few to tens of milliseconds.

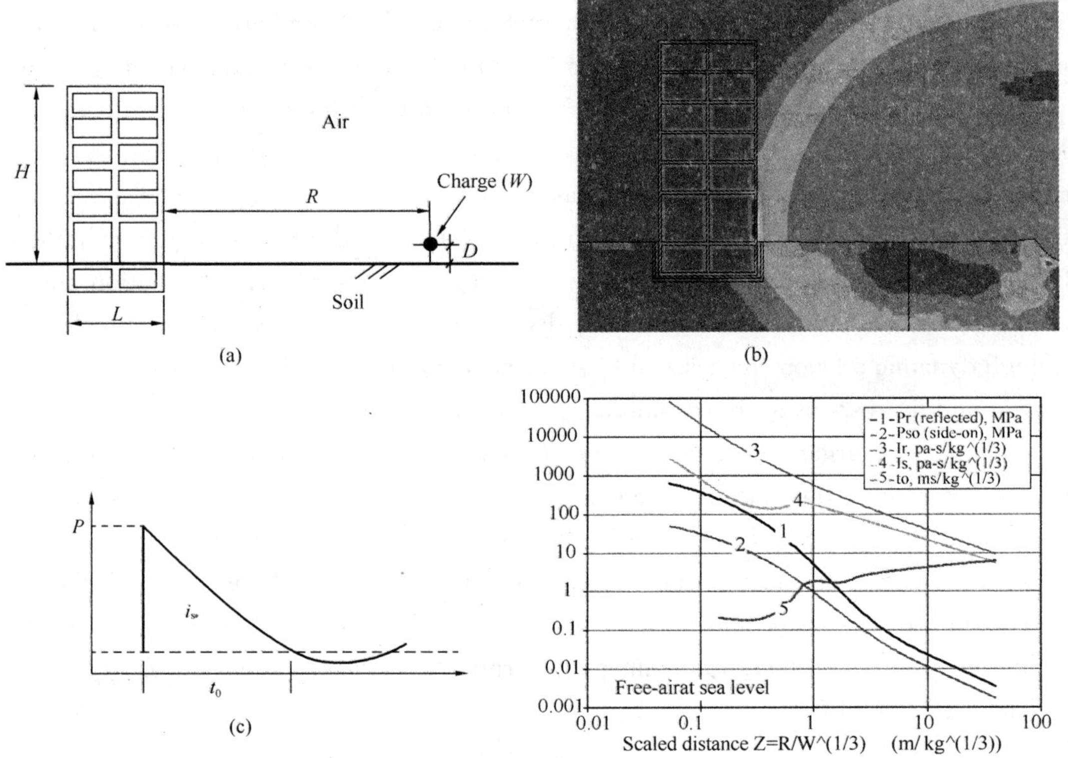


Fig. 1. Schematic of explosion effects on structure

(a) Event defining parameters; (b) Blast wave acting on structure

(c) Blast load parameters (after TM5-1300)

Because of short duration and high peak pressure, a smashing effect can be anticipated when such a blast load encounters a solid object. For building structures, this can result in severe damage on the front side of the structure, while the structural system could remain intact globally. Such a phenomenon has been evidenced in many past incidents, and can also be reproduced by a high fidelity numerical simulation, see for example Fig. 2(*a*).

To provide a quantitative evaluation of the relative significance of the local and global dynamic response in a typical building frame when subjected to a nearby explosion, a comprehensive numerical study has been carried out using a fully coupled model, in which the charge detonation, blast wave propagation, and the structural response are all explicitly included (Lu and Wang 2006). Such a model allows a direction simulation of the blast load and its interaction with the structure, as indicated in Fig. 1(*c*), and it also allows the response of the structure at both the local (material and beam-column members) and global (at floor and roof levels) to be scrutinized. Fig. 2(*b*) shows a representative damage in a RC column, along with a displacement profile along the entire height of the structure. It can be observed that severe damage occurs in front-side columns, and moreover, individual columns tend to respond in a rather independent manner with the floors acting as horizontal supports.

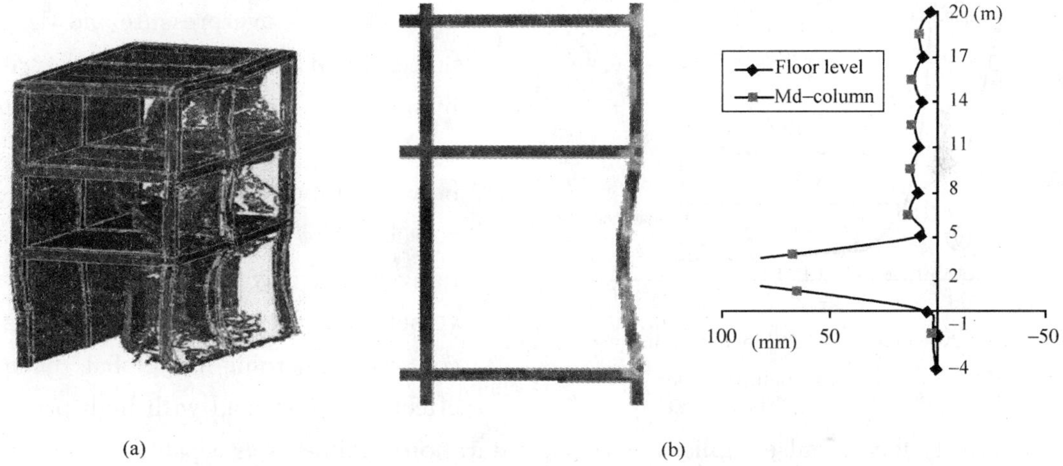


Fig. 2. Failure in a building structure subjected to blast load

(a) Simulated smashing effect of blast load on front side of structure;

(b) Damage pattern and displacement profile along frame height

When failure of one or a few load carrying components occurs, progressive collapse of the whole or partial building can follow, as depicted in an example shown in Fig. 2(*c*). As mentioned earlier, analysis of the progressive collapse can be treated separately, and this is not discussed further here.

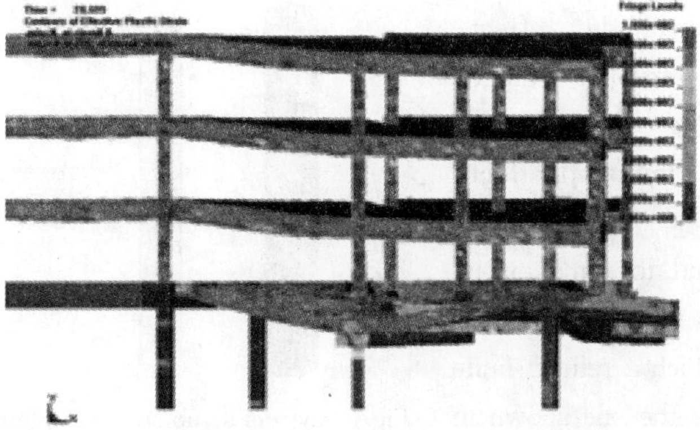

Fig. 3　Progressive collapse of partially damaged structure

# 11.3 DYNAMIC RESPONSE OF RC BEAM-COLUMN MEMBER SUBJECTED TO SHOCK-BLAST LOADING

## 11.3.1 Basic failure mechanisms

The damage in a RC beam-column member subjected to a blast load could range from a material level breach to a global level shear and flexural failure, depending upon the peak overpressure and the impulse. Such variation is best described by a pressure-impulsive ($P$-$I$) diagram, as shown in Fig. 4. Note that the ratio between impulse and the overpressure is correlated to the positive phase duration, which may also be translated inversely to represent the main frequency content concerning its global dynamic effect. A blast load with high pressure and relatively low impulse implies a very short duration, and hence is capable of generating strong stress wave effect across the thickness of a slender component like a RC column. On the other hand, a blast load with a relatively small pressure but high impulse has a longer duration, and hence will tend to induce more significant global (component-wide) dynamic response.

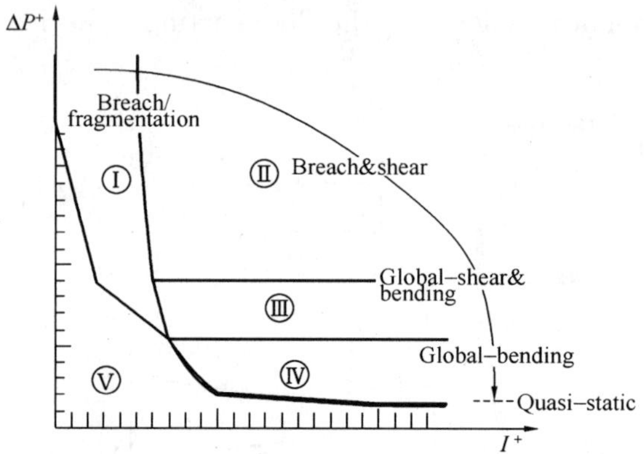


Fig. 4 A schematic of $P$-$I$ diagram and variation of response-failure modes

As indicated in Fig. 4, five distinctive $P$-$I$ combinations may be identified concerning the response of an RC component as the blast load shifts from a "shock" to a "quasi-static" character. Namely, Region Ⅰ: stress wave-dominated effect with breach of concrete material; Region Ⅱ: combined stress wave and local shear (end slide) effect; Region Ⅲ: combined global (diagonal) shear and bending effect; Region Ⅳ: global bending dominated effect; and Ⅴ: no damage.

For Region Ⅰ, and to some extend Region Ⅱ, the analysis falls into the continuum domain, for which a refined finite element model such as the one shown in Fig. 5, in conjunction with appropriate

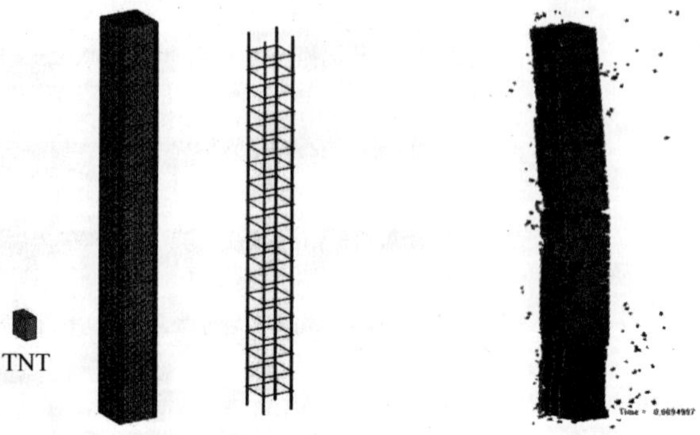


Fig. 5 Model setup and example failure pattern of RC column subjected to blast

numerical techniques to handle fracturing and spalling, would be required. Further discussion on this category of analysis will be given in Section 4. This kind of blast load may be encountered from a close-in explosion, e. g. with a scaled distance on an order of 1 m/kg$^{1/3}$.

For the majority of the remaining scenarios, a beam-column element based model may be well suited for the analysis, provided that all major failure modes are represented in the model and the numerical scheme allows them to evolve as the response develops. Use of a high-fidelity numerical simulation for this type of analysis may not be cost-effective, and moreover it does not necessarily yield a satisfactory result, especially when a cyclic response is involved and the residual structural capacity is to be reproduced.

For an RC beam-column component under shock and blast loading, it has been observed experimentally (Slawson 1984) that failure is often initiated by direct shear at the supports and occurs shortly after load application; and in some instances diagonal shear failure along the member also developed. Similar observations were made by Ross (1983). Taking into account a general flexural failure, three basic failure modes can be identified, as illustrated in Fig. 6. While flexural and diagonal shear behaviour of a RC beam-column is a classical topic and standard methods exist for the establishment of their resistance func-

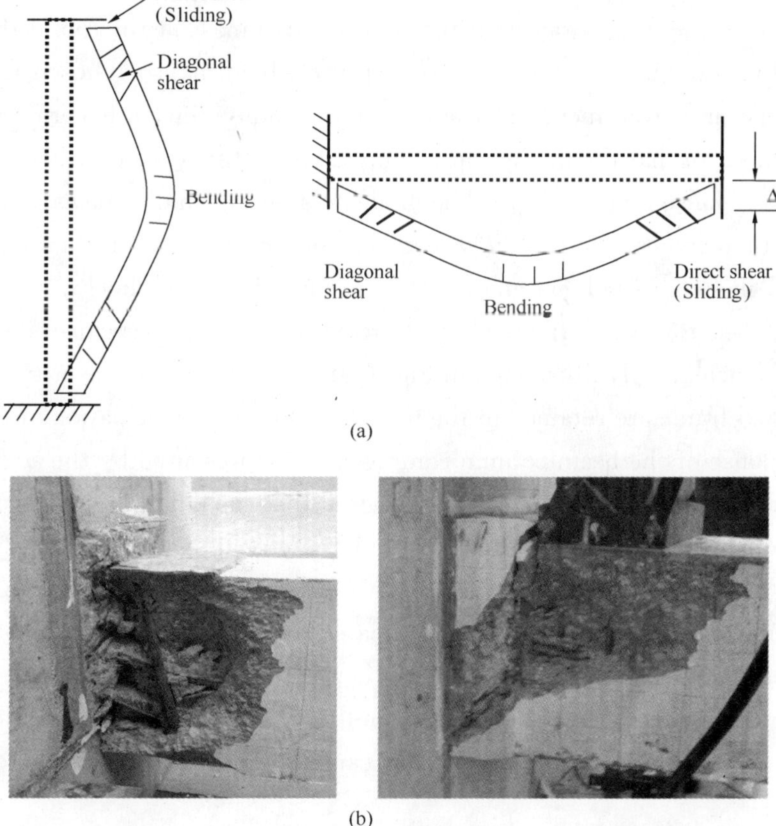


Fig. 6　Schematic of three basic failure mechanisms

(a) Three basic failure modes in RC beam-column components; (b) Typical direct shear and small shear-span failure patterns tions (moment-curvature or shear force-shear deformation relationships), relatively less amount of information is available with regard to the direct shear-sliding displacement relationship and the shear behaviour under a very small shear span (e. g., a shear span less than 0.5 times of the cross-section depth). More dedicated studies for such extreme shear behaviour are still required. Fig. 6(b) shows the direct shear failure at the sliding end and a shear failure with a small shear span from a recent experimental study. Further discussion about the experimental results will be reported separately.

### 11.3.2 A generic Timoshenko-beam based model incorporating global participation

Recognizing the fact that the dynamic response of individual beam-columns in a framed structure under impulsive loading is loosely coupled with the response of the structural system, a continuous beam-column based model with the incorporation of a concentrated mass at the member end is deemed to provide a good representation of the component response and at the same time allows for the contribution from the global system to be accounted for. A model based on these considerations, and taking into account the various possible failure mechanisms, has been developed for the analysis of RC columns subjected to shock and blast load (Gong and Lu 2007). This model is based on the Timoshenko-beam formulation. Apart from the incorporation of the concentrated mass at the top of the column and the provision for an adjustable boundary condition, which may be chosen to represent an appropriate constraint from members framing at the same joint, a generic hysteretic relationship is adopted so that the model can be applied in the analysis of other impulsive loads where a cyclic response may develop. The model has been found to be capable of capturing the characteristic response features and the governing failure modes for a range of impulsive load scenarios, including air blast and explosion-induced ground shocks.

The basic idea for the derivation of the continuous beam-column plus concentrated mass model is schematically illustrated in Fig. 7. The properties of the critical beam or column in the actual frame are retained in the model, while the participation of the global system in the response of the beam-column component is represented by the addition of an artificial mass at the member end. It should be noted that the amount of the artificial mass is an equivalent quantity which serves to represent the global system effect as if the component is within the actual frame. For the analysis of a column under a horizontal impulsive load, the amount of the mass may be approximated by the effective floor mass that is associated to the particular column.

Taking into account of the air blast load and the ground shock acceleration (where applicable), the governing equations of motion can be expressed according to Timoshenko beam theory as:

$$\frac{\partial M}{\partial x}+Q=\rho I\frac{\partial^2\psi}{\partial t^2} \tag{1}$$

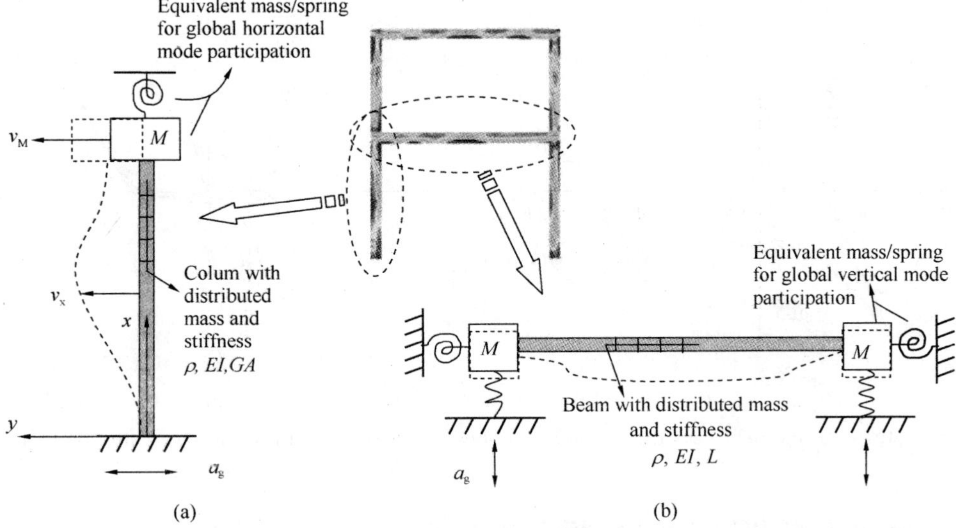


Fig. 7　Derivation of beam-column & mass-M model for impulsive response analysis (Gong and Lu 2007)

(a)Column-mass-spring model(Horizontal loading); (b)Beam-mass-spring model(Vertical loading)

$$\frac{\partial Q}{\partial x} = \rho A\left(a_g + \frac{\partial^2 w}{\partial t^2}\right) - p \tag{2}$$

where $M$ and $Q$ are bending moment and shear force, respectively, $I$ is moment of inertia, $A$ is cross-sectional area, $\rho$ is material density, $\psi$ is rotation of the cross-section due to bending, $w$ is transverse displacement of the mid-plane of the beam, $a_g$ is ground shock acceleration, and $p$ is the direct blast load.

The nonlinear solution of the dynamic equations can be sought using a finite difference method. The three mechanisms outlined in the previous section are described by their respective resistance functions (skeleton curves) along with a hysteretic rule. For the flexural nonlinearity, the resistance function can be represented by a sectional bending-moment vs. curvature relationship, which may be analyzed using a standard procedure such as the fibre (or layer) model. The diagonal shear is modeled by the generalized shear force vs. shear strain resistance function, which may be established considering the diagonal shear mechanism using a more detailed analysis, herein using a softened membrane model (SMM) proposed by Hsu and Zhu (2002). Finally, a direct shear model similar to the one used by Krauthammer (1993a) is employed, where an empirical curve comprising three branches bounded by three limit states are considered.

For the generality in representing the unloading and re-loading path during the dynamic response, a generic three-parameter hysteretic model (Park et al. 1987) is adopted. Fig. 4 shows two examples of the hysteretic loops resulted from an analysis using the present model. The achieved moment-curvature and shear force-shear strain relationships are observed as satisfying the expected hysteretic behaviour.

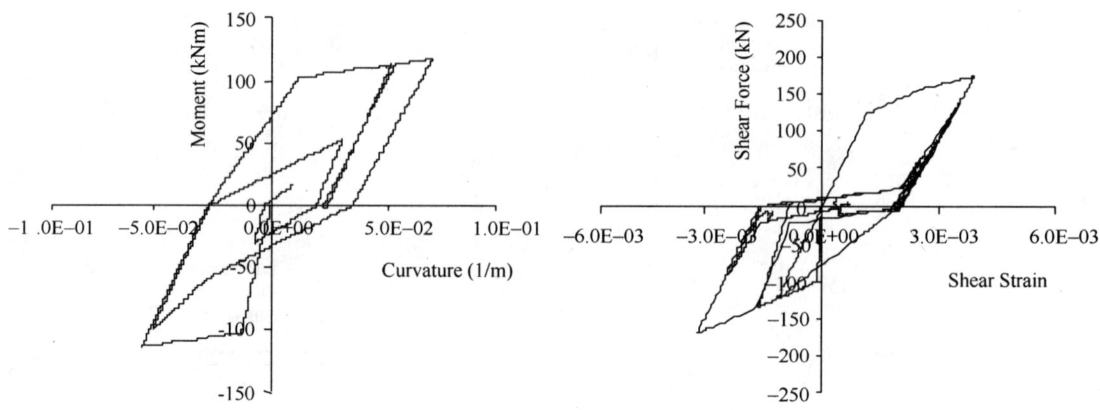


Fig. 8 Achieved moment-curvature (left) and shear force-shear strain (right) hysteretic behaviour

### 11.3.3 Analysis of RC column to air blast load

Fig. 9 illustrates example results using the above model from the analysis of a RC column in two different explosion cases, both with a charge weight of 1000 kg TNT; one has a standoff distance of 10 m, and another 30 m.

It can be observed that for the first case, which represents a close-in explosion with a scaled standoff of 1 $m/kg^{1/3}$, the response is dominated by excessively high shear deformation near both ends of the column, whereas bending remains in an elastic range. On the contrary, when the distance is increased to 30 m or in scaled term to 3 $m/kg^{1/3}$, the response appears to be dominated by a bending failure at both ends as well as at the mid-height, whereas no plastic shear occurs throughout the entire column. These results are consistent with the predictions using a refined finite element model (not shown here).

It is also worth noting from the response in the second case that a full cycle of reversed bending takes place, this indicates that the cyclic behaviour could become a factor in the blast response of a RC column.

### 11.3.4 Analysis of RC column under blast-induced ground shock

Generally speaking, the effect of an above-ground explosion on a building structure is usually dominated by the air blast load, while the effect from the ground shock is relatively insignificant (Lu and Wang 2007). However, large amplitude ground shocks may be generated in cases where the explosive charge is well coupled with the ground, for example in a buried or underground explosion scenario. Fig. 10 shows the acceleration time history and the corresponding frequency spectrum of a sample ground shock. Because of the high frequency contents in the ground shock excitation, the critical response in the structure tends to involve significant component-level (high-mode) dynamic effect, in addition to a certain degree of lower mode global dynamic participation. Such a response feature presents a special dynamic problem differing from the classical seismic response.

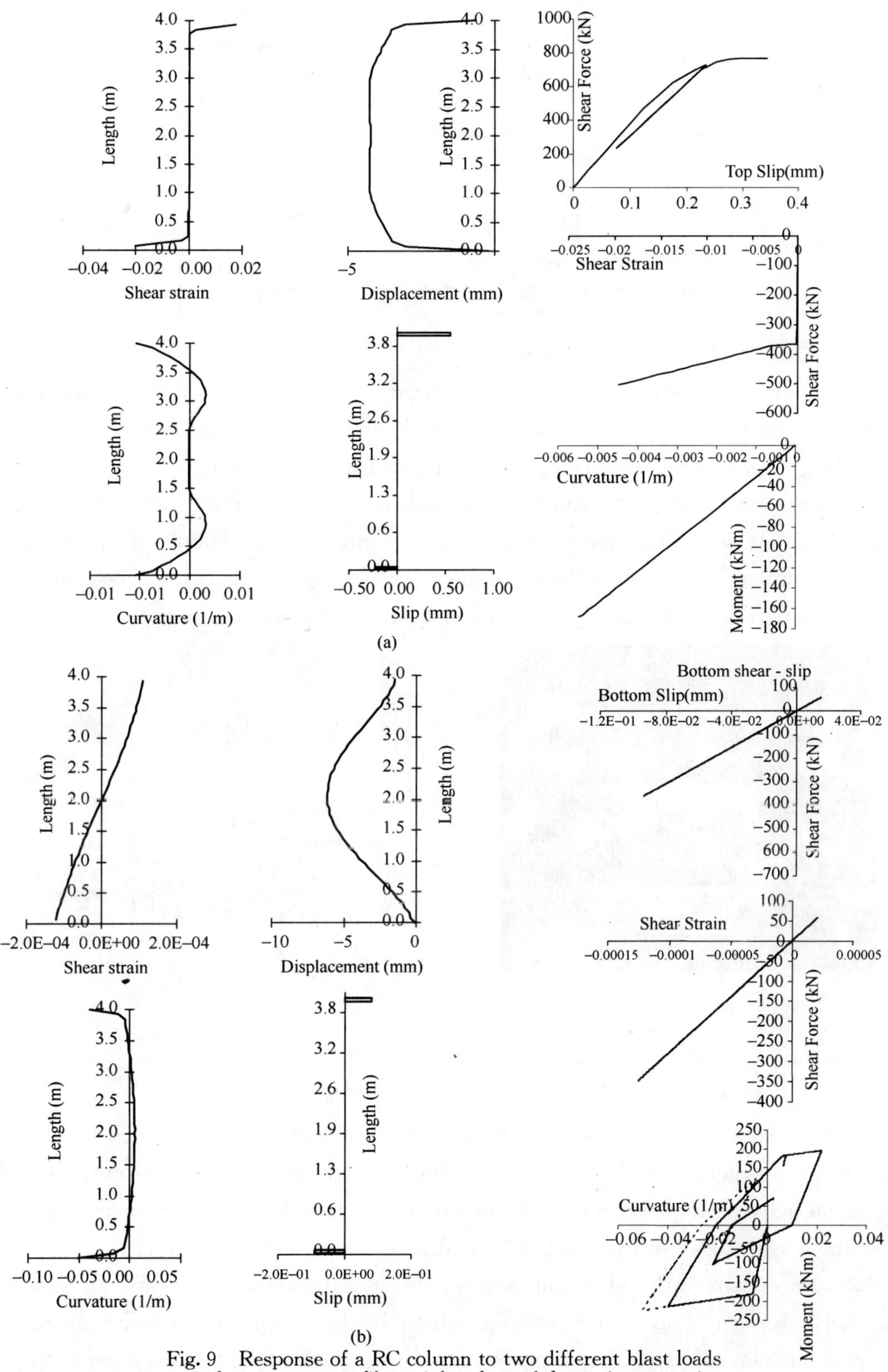


Fig. 9　Response of a RC column to two different blast loads
(Left: response profiles; right: force-deformation curves)
(a)1000kg@10m($p$~7.8MPa, $t_0$~20ms); (b)1000kg@30m($p$~0.33MPa, $t_0$~36ms)

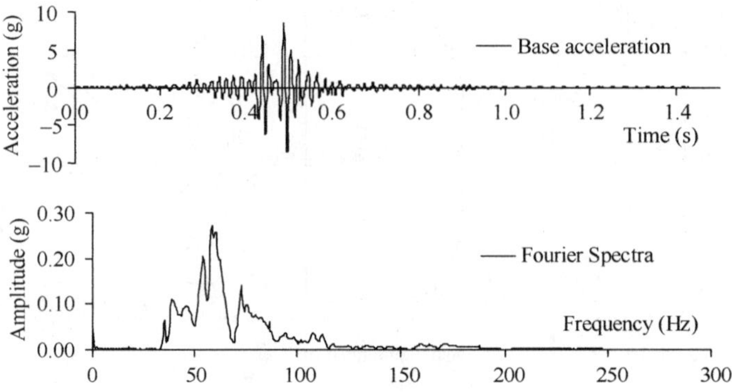


Fig. 10 Typical blast-induced ground shock time history and frequency spectrum

An experimental investigation was organized to bring insight into the dynamic response characteristics of RC frames under this class of ground shock excitation (Lu et al 2002). The experiment involved specially configured RC model structures, see Fig. 11(*a*), and the tests were conducted using a high-frequency electromagnetic shaker. Fig. 11(*b*) shows typical measured dynamic responses (accelerations) at the middle of the ground storey column and at the first floor level, respectively. A significant column resonance effect is evident, whereas the global response at the floor level is much less pronounced.

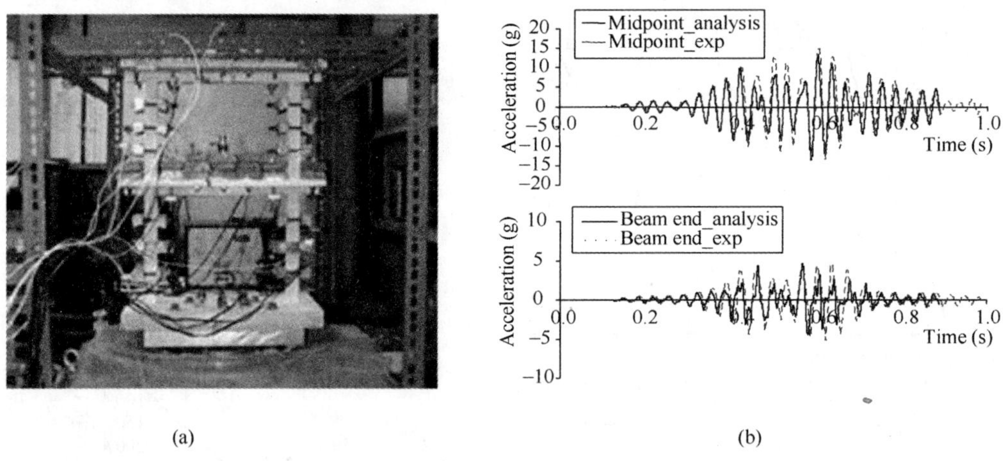


Fig. 11 Experimental setup and sample test results under simulated ground shock

(a) Test setup; (b) Middle of column (upper) and floor level accelerations

For the analysis of the critical response in the beam-columns under such ground shock excitation, the generic beam-column plus a concentrated mass model is also well suited. A comparison between the computed and measured responses for the experimental case mentioned above is included in Fig. 11(b); a good agreement can be observed.

A further example with high amplitude pulse-like ground shock, with a dominant frequency around 100Hz, is given in Fig. 12. From the figure it can be observed that under such a ground shock, apparent shear failure develops. It is also seen from the displacement profiles that there is an appreciable contribution of the global response. A parametric examination with the same ground motion

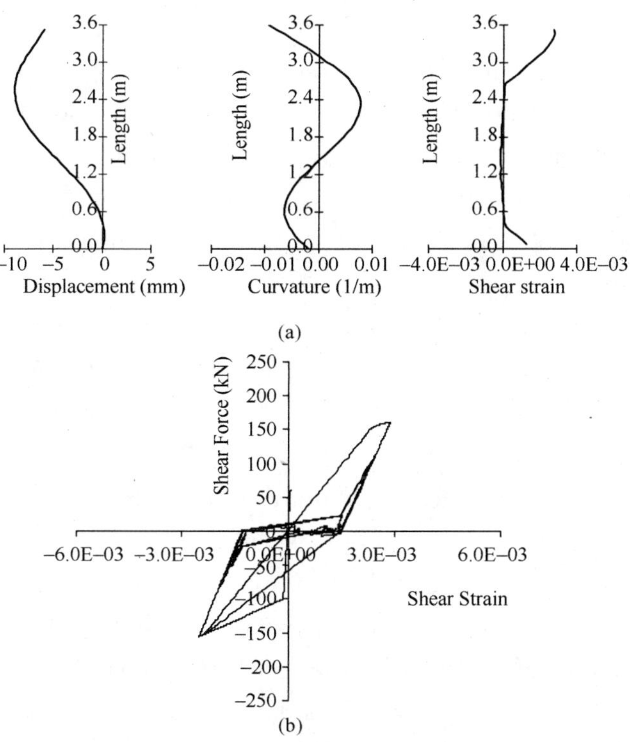


Fig. 12　Typical response to impulsive ground excitation (100 Hz & 100 g)

(a) Typical response profiles (b) Shear force-shear strain hysteresis

waveform but a varying dominant frequency from a relatively low range (10 Hz herein) to a high frequency of 200 Hz shows a clear trend of gradual shift from a global bending to a diagonal shear and then to a direct shear failure mode, as illustrated in Fig. 13.

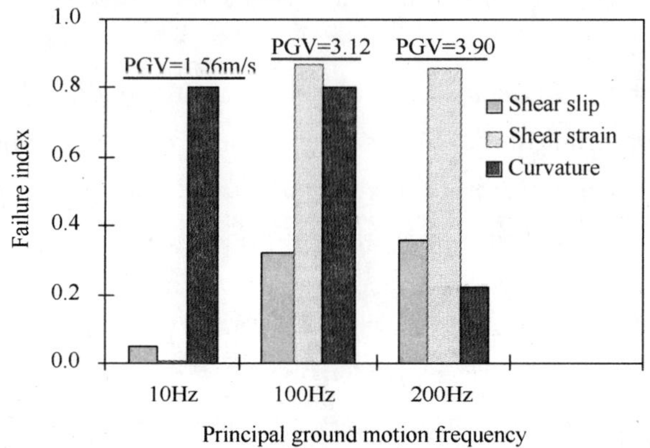


Fig. 13　Variation of failure modes with increase of ground shock frequency (Gong and Lu 2007)

## 11.4　DYNAMIC RESPONSE OF CONCRETE MATERIAL UNDER HIGH STRAIN RATE LOADING

As indicated in Fig. 4, for blast loads falling into the very high peak overpressure but relatively low impulse region (region I), local material damage could prevail due to the stress wave effect. Such an effect could degrade the (residual) capacity of the component, and in extreme situations could cause a gross disintegration of the concrete material due to intense fracturing or fragmentation. To analyze such a response and the resulting failure is beyond the capacity of a beam-column type of elements. Instead, methods in the non-linear continuum framework would be appropriate. As this category of response involves high strain rate and high pressure waves, the rate dependency of the bulk concrete material behaviour needs to be appropriately considered, whereas special schemes need be put in place to deal with large local deformation resulting from material crushing, fracturing, and fragmentation.

### 11.4.1 Pressure and rate dependence of bulk concrete behaviour and its modeling

High strain rate and high pressure are two important characteristics associated with impulsive stress waves. The bulk material behaviour of concrete is known to be strain rate and pressure dependent. For this reason, appropriate modeling of concrete for impulsive loading has attracted a lot of attention in the shock and impact engineering research community, and a number of concrete material models in this category have been proposed. A critical review of several such models has been reported by the author and his co-worker (Tu and Lu 2009). A brief summary of a few representative models using a damage-plasticity framework is given below. In particular, the behaviour of the standard RHT model and the proposed modifications for its enhancement are highlighted. It should be noted that for high dynamic problems a compaction model or equation of state (EOS) constitutes a pertinent part of the material description. The EOS can be specified independently from the material constitutive model.

*Johnson-Holmquist (JH) model* The JH concrete model (Johnson and Holmquist 1994) considers the material to be linear elastic before a prescribed failure criterion is reached. After that, damage starts to accumulate until the occurrence of a total failure, and from there onwards the material maintains a residual state. The initial and post failure surfaces are defined, respectively, as:

$$\sigma^* = [A + BP^{*N}] \times (1 + C\ln\varepsilon^*) \leqslant SMAX \tag{3}$$

$$\sigma_{pf}^* = [A(1-D) + BP^{*N}] \times (1 + C\ln\varepsilon^*) \tag{4}$$

where $\sigma^* = \sqrt{3J_2}/f_c$, $J$ = second deviatoric stress invariant, $f_c$ = uniaxial compressive strength, $P^*$ = normalized pressure ($P/f_c$), $\varepsilon^*$ = equivalent plastic strain rate. $A$, $B$, N and $C$ are constants, $SMAX$ denotes the maximum strength, and $D$ is a damage index. The model is available in some widely used hydrocodes such as LS-DYNA and AUTODYN.

As can be seen, the strain rate effect in this model is modelled by expanding the strength surface by a factor of $(1+C\ln\varepsilon^*)$. No differentiation is made in the strain rate enhancement between compression and tension in this model.

*Concrete Damage Model* (*K&C model*) This model was initially developed for DYNA3D (Malvar et al. 1997) and since then it has evolved through several enhancements. The model defines three independent strength surfaces: an initial yield surface, a maximum failure surface and a residual surface, and all the three stress invariants are considered. The general strength criterion is given by a uniform expression as:

$$\Delta\sigma = \sqrt{3J_2} = f(p, J_2, J_3) \tag{5}$$

where $\Delta\sigma$ and $p$ denote the principal stress difference and pressure, respectively, $f(p, J_2, J_3) = \Delta\sigma^c \times r'$, with $\Delta\sigma^c$ being the compressive meridian and $r'$ is a function of the Lode angle. The loading surface after yield and the post-failure surface are defined by interpolation between the respective strength surfaces. The nonlinear behaviour is controlled by an yield scale factor $\eta$, which in turn is determined by a damage function $\lambda$. The model is available in LS-DYNA.

*RHT model* Riedel et al. (1999) developed this model as an enhancement to the JH model. Similar to the K&C model, it incorporates the third invariant dependence, while an independent residual strength surface is introduced to enhance the softening representation. The failure surface is defined as a function of the normalized pressure $p^*$, strain rate $\dot{\varepsilon}$, as well as the Lode angle $\theta$,

$$Y_{\text{fail}}(p^*, \theta, \dot{\varepsilon}) = Y_c(p^*) \times r_3(\theta) \times F_{\text{rate}}(\dot{\varepsilon}) \tag{6}$$

in which $Y_c(p^*)$ represents the compressive meridian,

$$Y_c(p^*) = f_c \times [A \times (p^* - p^*_{\text{spall}} \times F_{\text{rate}}(\dot{\varepsilon})^N] \tag{7}$$

where $A$ and $N$ are two constants, $p^*_{\text{spall}} = f_t/f_c$, $F_{\text{rate}}(\dot{\varepsilon})$ represents the dynamic increase factor, $r_3(\theta)$ is a function of the Lode angle similar to $r'$ in the K&C model. The elastic strength surface is obtained by scaling the failure surface $Y_{\text{fail}}$ in the radial direction $Y_{\text{elastic}} = Y_{\text{jail}}(p^*/F_{\text{elastic}}) \times F_{\text{elastic}} \times F_{\text{cap}}(p)$, where $F_{\text{elastic}}$ is a scaling factor, $F_{\text{cap}}(p)$ is a dimensionless pressure-dependent cap function. The loading surfaces beyond the elastic strength surface are obtained by an interpolation between $Y_{\text{elastic}}$ and $Y_{\text{fail}}$, and similarly, the post-failure surfaces $Y_{\text{fracture}}$ are determined by the interpolation between the failure surface and the residual surface via a damage index D. The model is available in AUTODYN.

Experiences using RHT for a variety of explosion and impact problems revealed several issues of the model in representing the softening behaviour of concrete. Rectifying measures have been proposed both at the level of parameter settings (Tu and Lu 2009) and through modifications to the relevant elements in the model formulation (Tu and Lu 2010). Fig. 14 shows representative comparisons of the model behaviour before and after considering the proposed modifications.

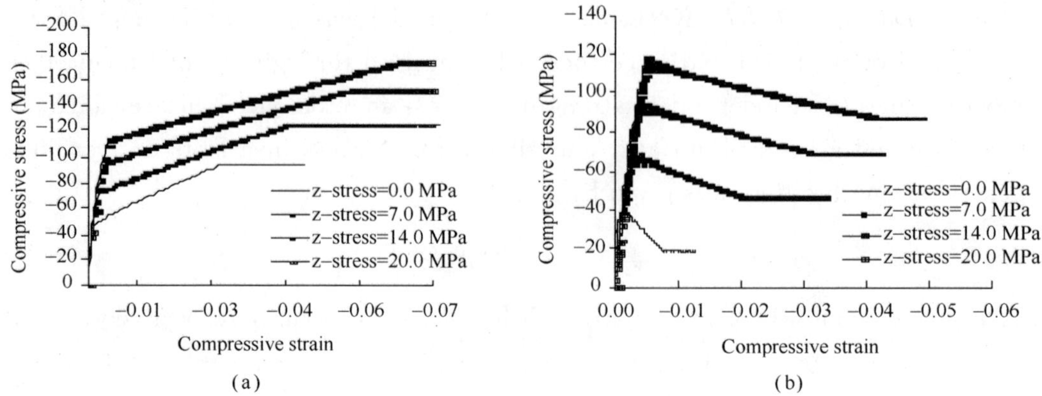


Fig. 14 Stress-strain behaviour by RHT model under tri-axial compression with different levels of confining stress (Tu and Lu 2010)
(a) Standard RHT; (b) Modified RHT

### 11.4.2 Fracture and fragmentation

The transition of a continuum solid to a discontinuous fractured or fragmented state is a characteristic phenomenon in a reinforced concrete component when subjected to the rapid stress wave effect due for example to a close-in explosion. From the modeling point of view, such a transition of state due to the evolution of fracture poses a challenge in a finite element model due to large deformations at local regions, causing mesh distortion. To tackle this problem, an appropriate use of the so-called element erosion technique may be considered, or alternatively, an adequate meshfree method may be employed. A review of the various FE-based approaches, combined FE and discrete/particle schemes, and meshfree methods for the simulation of concrete fracture and break-up under high impulsive loading can be found in Lu (2009). Herein two modeling case studies are briefly described; one concerns a RC slab under an internal explosion, using a FE model, and the other involves the simulation of a close-in detonation using a meshfree approach.

*a) RC slab under internal explosion*

This modeling study was conducted in association with an experimental investigation into RC slab response and debris launch under internal explosion. The numerical simulation was carried out using a coupled fluid (Eulerian)-solid (lagrangian FE) scheme, such that the blast loading in an initially closed and then gradually vented space can be simulated in a more realistic manner. Fig. 15 depicts the model configuration and the simulated crack patterns as compared to the test record. The crack pattern is fairly well reproduced using the modified RHT model. On the contrary, the crack pattern predicted using the RHT model with the default parameter setting fails to produce a realistic damage distribution.

*b) Simulation of fragmentation and debris throw using SPH method*

SPH (smoothed particle hydrodynamics) is a meshfree method which can be well suited for the simulation of break-up of concrete and the dispersion of fragments. Using suffi-

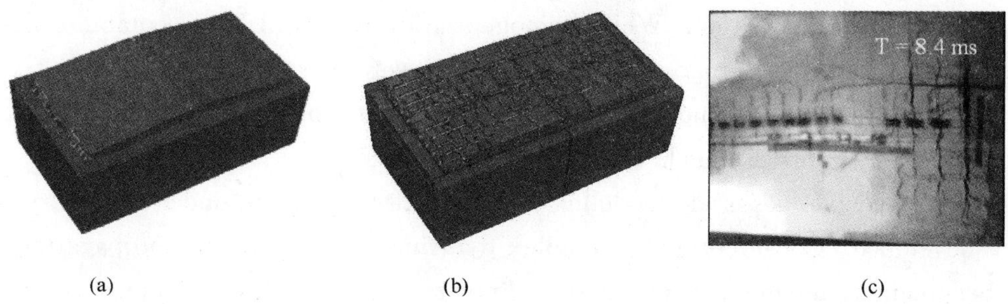


Fig. 15　Crack patterns of a RC slab under intense internal explosion (Tu and Lu 2009)
(a) Sim ulated with stan dard RHT ; (b) Simulated with modified RHT; (c) Experiment

ciently fine particles, steel bars may also be explicitly modeled for a RC member. Material models such as RHT can be employed as well for concrete in a SPH model, whereas models such as the Johnson-Cook model may be used for the reinforcing bars. Fig. 16 depicts some example results for a concrete slab subjected to a 1-kg TNT contact explosion. The slab was modelled after a field test slab, which had a dimension of 3m×3m×0.3m. Simulations using 2D and 3D models show comparable results, from which the propagation of breach, generation of the crater and formation of the fragments can be observed in detail.

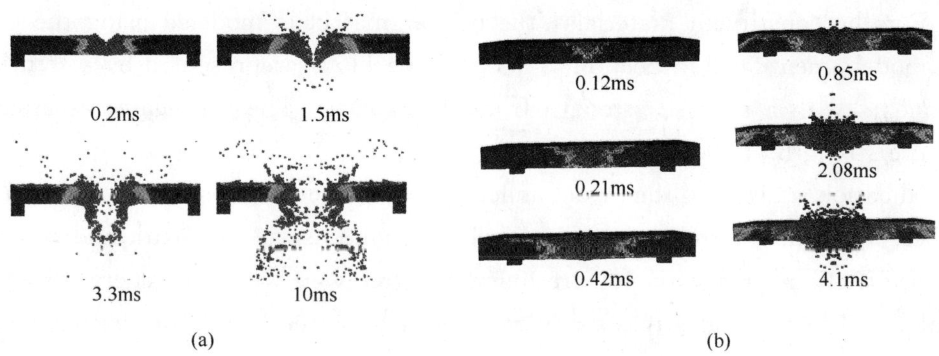


Fig. 16　Modelling of RC slab under contact explosion using SPH (Lu 2009)
(a) 2D axis-symmetrical model; (b) 3D model

### 11.4.3 Microscopic mechanisms of dynamic behaviour of concrete

Concrete is by nature a heterogeneous material with sizable aggregates embedded in the mortar matrix. The bulk behaviour of concrete is deemed to become stabilized in a sample size (or sample area within a structure) of at least a few times of the maximum aggregate particle size and under a smooth stress distribution. This argument is implicit in the homogenization treatment of concrete in a conventional analysis. However, under shock and blast loading, stress waves propagate in the material with a drastic spatial and temporal variation. The stress/strain gradient can be so large that the heterogeneities in concrete can play a much more significant role in influencing the bulk material behaviour than in a low dynamic or static situation. While homogenization may still be acceptable concerning the global behaviour at a component or structural level, it is less commendable trying to probe into the mechanisms underlying the dynamic behaviour of concrete at high strain rates using a homogenized model.

A comprehensive mesoscale modeling framework for concrete under high rate loading is being developed. Considering the complex dynamic process and the propagating stress wave phenomenon, the model is oriented for implementation with general purpose transient dynamic codes so that sophisticated numerical solvers and material models available in these codes can be readily exploited. The basic 2D mesoscale model and its pseudo 3D extension have been employed in the study of the strain rate effect on the dynamic behaviour of concrete (Lu et al 2010). The basic model setup and the main findings are summarized below.

The mesoscale model consists of three distinctive phases, namely aggregates, mortar matrix, and the interfacial transition zone (ITZ). The aggregates have random polygon shapes and follow a size distribution which can be specified. For the ITZ, two alternative schemes may be employed, using either the classical cohesive interface elements, or an equivalent layer of solid elements. Fig. 17(a) depicts a mesoscale model for a concrete cubic specimen, including the mesoscale geometrical structure and the FE mesh for the three phases. For the constituent materials, the mortar matrix is modeled using the Concrete Damage model, mentioned in Section 4. 1, and the ITZ is represented by a thin layer of solid elements with a reduced strength from the mortar. The aggregates as assumed to have failure strength of 200 MPa.

A validation analysis of the model under a quasi-static compression is shown in Fig. 17 (*b*), where the crack patterns for two loading conditions, i. e., with and without the loading face frictions, respectively, are found to agree well with the general experimental observations. The nominal stress-strain relations also agree favourably with relevant experimental results.

When the specimen is subjected to a high rate compressive loading, which is simulated in the analysis by imposing an appropriate velocity boundary condition at the loading ends, it can be observed that the damage patterns gradually changes from the static character with a few dominant cracks, to a more distributed pattern as the strain rate increases. Moreover, when the strain rate exceeds a certain limit, for example $50s^{-1}$ for a 50mm long cylinder, the failure of the mortar and ITZ tends to become a propagating process starting from the active loading face. This is visible in the example shown in Fig. 17(c). Such a propagating failure pattern allows for a direct involvement of the aggregates in resisting the propagating stress, and thus resulting in an increase in the bulk strength of concrete. It should be pointed out that such an increase is pertinent to the heterogeneous property of the concrete composite, and is additional to other "structural" contributions, particularly the inertial confinement effects as elaborated by a number of researchers (e. g. Bischoff

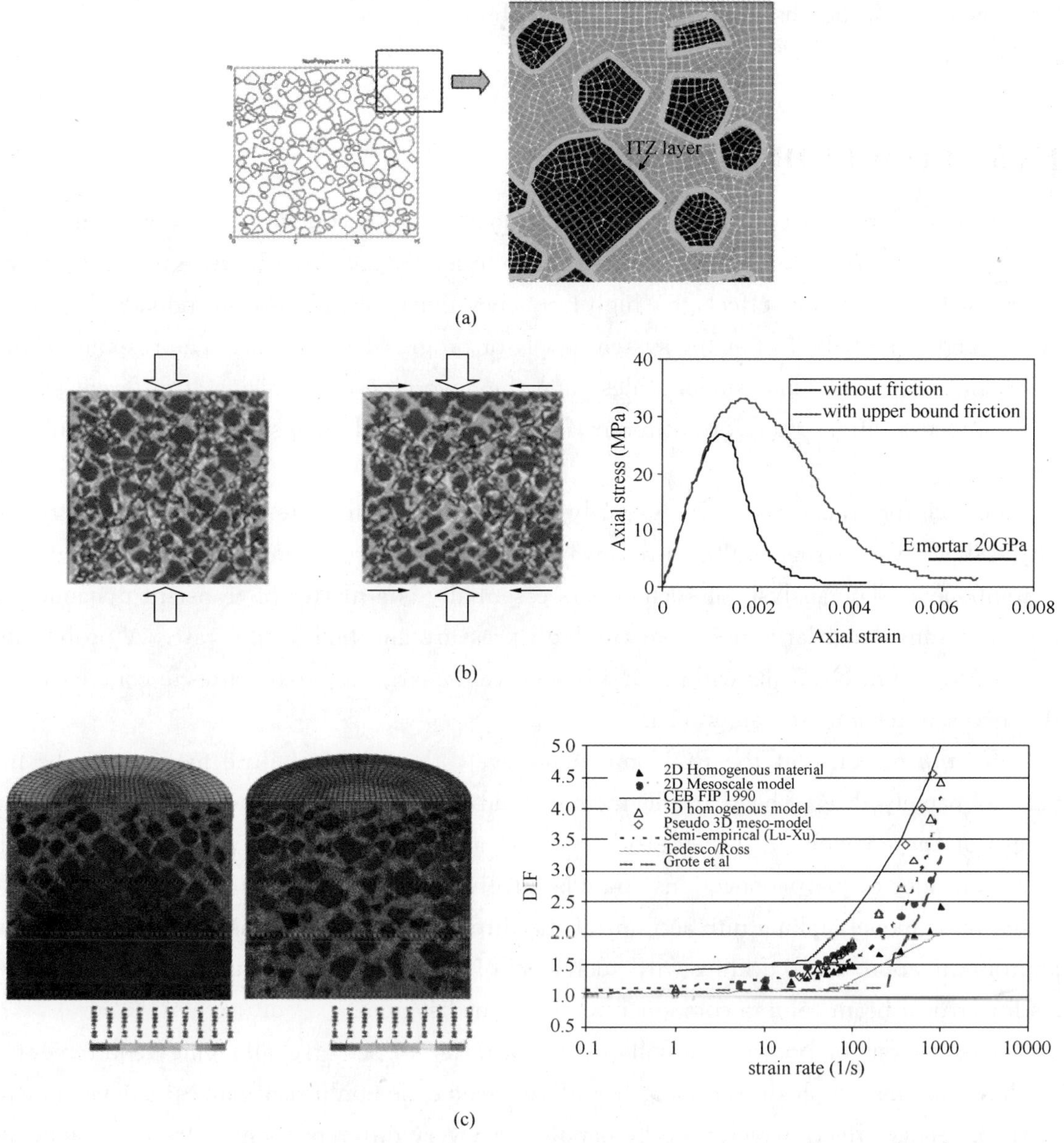


Fig. 17　Mesoscale concrete model and analyses under static and dynamic compression (Lu et al. 2010)

(a) Random mesoscale structure of concrete and FE mesh; (b) Simulation of quasi-static compression;
(c) Simulation of dynamic compression

and Perry 1991, Li and Meng 2003). Fig. 17(c) shows a collection of the numerically simulated bulk compressive strength increase in a standard concrete sample specimen using different modeling approaches in comparison with some empirical results. Note that in all the modeling analyses the constituent materials are considered to be strain rate insensitive, and therefore the observed bulk (specimen-wise) strength increase in terms of the DIF is attributable only to the dynamic structural effect (in the homogenized models) as well as the heterogeneity effect (in the mesoscale models). An appreciable amount of increase due to the heterogeneity can be observed, and this contribution tends to become larger with the increase of the strain rate.

## 11.5 CONCLUDING REMARKS

Structural response to shock and blast loading is a complex and highly dynamic process. Three distinctive effects at three characteristic stages need be recognized, namely, a) transient stress wave effect, b) high-frequency dynamic response in individual components, and c) global effect at the system level, or progressive collapse. The response characteristics, time and dimensional scales, and the governing mechanisms at different stages of the process differ; therefore different failure criteria and analysis techniques should be considered.

For reinforced concrete structures, breaching of concrete material can occur as a result of the shock wave strike and the stress wave propagation across the thickness of a structural component. The analysis of such effects essentially is a matter of transient dynamics of continuum involving large nonlinearity, high pressure and high strain rate. A probe into the underlying mechanisms will inevitably involve the large heterogeneities in concrete, and this necessitates a mesoscale model.

For the response at the RC component level, three basic failure modes may be involved, namely direct shear at the member end, diagonal shear and bending along the length of the component. On the other hand, a varying degree of system participation (through floor or joint connections) may be involved. A Timoshenko beam with a concentrated mass model, taking into account of the three failure mechanisms, is deemed to be an appropriate approach to dealing with such kind of loosely coupled dynamic response analysis for critical beam-column components in a framed system.

The analysis of progressive collapse has a different set of challenging requirements, for instance the complexities arising from large geometric nonlinearity and the development of the catenary effect, which need be handled in a very different manner from the analysis of the direct blast induced dynamic effects.

## REFERENCES

[1] Autodyn(2001). Theory Manual, v4. 2, Century Dynamics, 2001.

[2] Bischoff, P. H., Perry, S. H. (1991). Compression behavior of concrete at high strain-rates, Mater. Struct. 24, 425-450.

[3] Ghaboussi, J., Millavec, W. A. and Isenberg, J. (1984). R/C structures under impulsive loading. Journal of Structural Engineering, ASCE, 110(3), 505-522.

[4] Gong, S., and Lu, Y. (2007). Combined continua and lumped parameter modelling for nonlinear response of structural frames to impulsive ground shock. Journal of Engineering Mechanics, ASCE, 133

(11), 1229-1240.

[5] Henrych, J. (1979). The Dynamics of Explosions and Its Use. Elsevier, Amsterdam.

[6] Hsu, T. T. C., and Zhu, R. R. H. (2002). Softened membrane model for reinforced concrete elements in shear. ACI Structural Journal, 99(4), 460-469.

[7] Hyde, D. W. (1991). ConWep, US Army Waterways Experimental Station, US Army.

[8] Johnson, G. R., and Holmquist, T. J., (1994). An Improved Computational Model for Brittle Materials, High Pressure Science and Technology, S. C. Schmidt, J. W. Shaner, G. A. Samara, and M. Ross, eds., American Institute of Physics Press, New York, 981-984.

[9] Krauthammer, T. (1984). Shallow-buried RC box-type structures. Journal of Structural Engineering. ASCE, 110(3), 637-651.

[10] Krauthammer, T., Assadi-Lamouki, A. and Shanaa, H. M. (1993a). Analysis of impulsive reinforced concrete structural elements - I: Theory. Computers and Structures, 48(5), 851-860.

[11] Krauthammer, T., Assadi-Lamouki, A. and Shanaa, H. M. (1993b). Analysis of impulsive reinforced concrete structural elements - Ⅱ: Implementation. Computers and Structures, 48(5), 861-871.

[12] Lellep, J. and Torn, K. (2005). Shear and bending response of a rigid-plastic beam subjected to impulsive loading. Int. J. Impact Engineering, 31, 1081-1105.

[13] Li, Q. M. and Meng, H. (2003), About the Dynamic strength enhancement of concrete like materials in a split Hopkinson pressure bar test, Int. J. Solids and Structures 40, 343-360.

[14] LS-DYNA (2007) keyword user's manual, version 971. Livermore Software Technology Corporation. April 2007.

[15] Lu, Y. (2009). Modelling of concrete structures subjected to shock and blast loading: an overview and some recent studies. Structural Engineering and Mechanics, 32(2), 235-250.

[16] Lu, Y. and Xu, K. (2004). Modelling of dynamic behaviour of concrete materials under blast loading. Int. J. Solids and Structures, 41(1), 131-143.

[17] Lu, Y. and Wang, Z. Q. (2006). Characterization of structural effects from above-ground explosion using coupled numerical simulation. Computers and Structures. 84(28), 1729-1742.

[18] Lu, Y., Hao H. and Ma, G. W. (2002). Experimental investigation of structural response to generalized ground shock excitation. Experimental Mechanics, 42(3), 261-271.

[19] Lu, Y., Song, Z. and Tu, Z. (2010). Analysis of dynamic response of concrete using a mesoscale model incorporating 3D effects, Int. J. Protective Structures, 1(2), 197-217.

[20] Malvar, L. J., Crawford, J. E., Wesevich, J. W. (1997). A plasticity concrete material model for Dyna3D. Int. J. Impact Engineering, 19(9-10), 847-873.

[21] Park, Y. J., Reinhorn, A. M., and Kunnath, S. K. (1987). IDARC: Inelastic Damage Analysis of Reinforced Concrete Frame-Shear-Wall Structures, Technical Report NCEER-87-0008, State University of New York at Buffalo.

[22] Riedel W, Thoma K, Hiermaier S. (1999). Penetration of reinforced concrete by BETA-B-500--numerical analysis using a new macroscopic concrete model for hydrocodes. 9-th International Symposium on Interaction of the Effect of Munitions with Structures, 315-322.

[23] Ross, T. J. (1983). Direct shear failure in reinforced concrete beams under impulsive loading. A Dissertation for PH. D., Stanford University.

[24] Sasani, M., Bazan, M., Sagiroglu, S. (2007). Experimental and analytical progressive collapse e-

valuation of actual reinforced concrete structure. ACI Structural Journal, 104(6), 731-739.

[25] Shi, Y. C., Hao, H., Li, Z. X (2008). Numerical derivation of pressure-impulse diagrams for prediction of RC column damage to blast loads. Int. J. Impact Engineering, 35(11), 1213-1227.

[26] Slawson, T. (1984). Dynamic shear failure of shallow-buried flat-roofed reinforced concrete structures subjected to blast loading. Technical Report SL-84-7, Structures Laboratory, U. S. Army Engineer Waterways Experiment Station, Vicksburg, MS.

[27] Stevens, D. J., Krauthammer, T. (1991). Analysis of blast-Loaded, buried RC arch response. Part I: Numerical approach. Journal of Structural Engineering, ASCE, 117(1), 197-212.

[28] TM5-1300 (1990). Structures to Resist the Effects of Accidental Explosions, Headquarters, Department of the Army, Nov. 1990.

[29] Tu, Z. G. and Lu, Y. (2009). Evaluation of typical concrete material models used in hydrocodes for high dynamic response simulations, Int. J. Impact Engineering, 36, 132-146.

[30] Tu, Z. G. and Lu, Y. (2010). Modifications of RHT material model for improved numerical simulation of dynamic response of concrete. Int. J. Impact Engineering. 37(10), 1072-1082.

[31] Xu, K., Lu, Y. (2006). Numerical simulation study of spallation in reinforced concrete plates subjected to blast loading. Computers and Structures, 84(5-6), 431-438.

# 第 12 章 Chapter 12

# 高性能混凝土对低碳经济的影响

缪昌文
（高性能土木工程材料国家重点实验室，江苏省建筑科学研究院有限公司，江苏　南京 210008）

**提　要**：高性能混凝土材料的需求量日益增多，其使用寿命长短对能源、资源的消耗影响很大，在混凝土中掺入优质混凝土外加剂和掺合料，优化与调控混凝土的微结构，不仅可以保证混凝土建筑物的使用寿命，同时也可节省大量水泥，起到化废为宝、化害为利的作用。

**关键词**：低碳经济；高性能混凝土；耐久性；外加剂；工业废渣

# EFFECT OF HIGH PERFORMANCE CONCRETE ON LOW-CARBON ECONOMY

C. W. Miao
(State Key Laboratory of High Performance Civil Engineering Materials,
Jiangsu Academy of Building Science Co., Ltd, Nanjing 210008, China.)

**Abstract**: The demand for high performance concrete is increasing nowadays. The service life of high performance concrete has significant effect for the consumption of energy and resources. Adding good chemical and mineral admixtures to regulating and optimizing the microstructure of concrete can ensure the service life of concrete structures and save great amount of cement simultaneously. This is a good way to turn waste into assets and turn the harmful into the beneficial.

**Keywords**: low-carbon economy, high performance concrete, admixture, industrial solid waste

## 12.1　引言

随着全球人口不断增加和经济规模迅速增大，资源、能源消耗所产生的环境问题已渐渐地为人们所认识，特别是全球气候变暖对人类生存和发展带来了严峻挑战。以低能耗、低污染、低排放为基础的低碳经济成为全球各国所追求的全新经济模式，是人类社会继农业文明、工业文明之后的又一次重大进步。

我国正处于高速发展阶段，重大基础工程规模空前，城镇化高速推进。据初步测算，如要在 2020 年完成非化石能源比例达 15%的目标，核电规模至少达到 7500 万千瓦以上，水电装机规模至少达到 3 亿千瓦以上。目前，中国投入运营的高速铁路已经达到 6920km，正在建设的高速铁路有 1 万多公里，运营里程居世界首位。到 2012 年，我国高速铁路总里程将达到 1.3 万公里以上，其中时速 350km/h 的高速铁路将达到 8000km 左右。到 2020 年，中国铁路快速客运网将达到 5 万公里以上，连接所有省会城市和 50 万人口以上的城市，覆盖全国 90%以上的人口。《核电中长期发展规划（2005～2020 年）》中指出，到 2020 年核电运行装机容量 4000 万千瓦，核电年发电量 2600 亿～2800 亿千瓦时，15 年内核电项目建设资金总量约为 4500 亿元。住建部预计，在“十二五”期间，城市轨道交通建设投资将超过 7000 亿元，可能仅次于高速公路和铁路投资。这些重大工程的发展和建设无一不与混凝土材料息息相关。

然而，作为混凝土配制的主要组成材料—水泥的生产过程却是能源、资源的消耗大户，这给土木工程材料领域发展低碳经济带来了严峻的挑战。据统计，2009 年我国累计生产水泥 16.3 亿吨，已超过世界份额的 50%。生产 1t 水泥需要消耗 0.8t 石灰石，有资料表明我国用于生产水泥的石灰石矿可采储量约为 250 亿吨，只能维持水泥工业 40 年的需求[1]。2009 年我国原煤产量为 29.6 亿吨，燃烧所排放的 $CO_2$ 为 77.5 亿吨，其中水泥生产能耗约 1.8 亿吨煤，占当年全国总能耗的 6.6%，$CO_2$ 排放总量约 13.8 亿吨（包括生产水泥用碳酸钙分解排放的 $CO_2$），占当年全国 $CO_2$ 总排放量的 26.1%[2]。表 1 列出了水泥生产过程中的主要能耗[3]，其中生产过程中的高温煅烧能耗占水泥生产总能耗的 80%以上。此外水泥生产中还排放出 $NO_x$、$SO_2$ 等有害气体，以及大量粉尘。同时，制备混凝土所需的水和砂石总量为水泥量的 6.7 倍，占混凝土总质量的约 80%。这些砂石多为不可再生资源，因而混凝土制备还给自然界中岩石资源带来了不可估量的破坏作用。

**水泥生产中的能耗**　　**表 1**

| 生产步骤 | 能量(MJ/kg 水泥) | 生产步骤 | 能量(MJ/kg 水泥) |
|---|---|---|---|
| 原料提取 | 0.044 | 水泥粉磨 | 0.188 |
| 原料运输 | 0.089 | 水泥运输 | 0.133 |
| 原料破碎与粉磨 | 0.386 | 总计 | 4.882 |
| 高温煅烧 | 4.041 | | |

此外，由于现代混凝土组分复杂，胶凝材料用量大，因而早期易开裂，导致耐久性下降。混凝土因耐久性不足、性能低劣而导致过早失效的问题时有发生，有的 3～5 年就出现裂缝，有的尚未投入使用裂缝就已超过允许裂宽，许多混凝土结构在 20 年左右必须大修，而且投入的修补费用相当可观。震惊中外的重庆綦江彩虹桥等的工程事故不仅使国家遭受了巨大的经济损失，而且致多人伤亡，工程质量问题时刻敲着警钟。美国在 2005 年美国基建工程调查报告中提出，因路况不好，每年用于维修的费用高达 540 亿美元，每年将花费 94 亿美元用于维修受损的桥梁，还将耗资 101 亿美元用于维修危坝[4,5]。总之，混凝土结构的过早失效频频发生，这不仅造成了国家财富的无端大量流失，而且造成了资源与能源的极大浪费。

一方面重大工程建设对混凝土材料不断增长的需求，另一方面混凝土材料生产过程对环境带来的负荷越来越严重，如何调和两者矛盾，寻求混凝土材料低碳经济的发展模式已成当务之急。

## 12.2 高性能混凝土实现低碳经济的技术途径

高性能混凝土是在20世纪80年代末90年代初才出现的新型混凝土，美国国家标准与技术研究所与美国混凝土协会于1990年5月召开的讨论会上公开提出了高性能混凝土的概念。各国和各社会组织对高性能混凝土的界定存在较大的差异。著名混凝土专家吴中伟老先生认为[6]，高性能混凝土是一种新型的高技术混凝土，是在大幅度提高常规混凝土性能的基础上，采用现代混凝土技术，选用优质原材料，在妥善的质量控制下配制而成的。除了必须对水泥、集料和水的质量进行有效的控制外，配制高性能混凝土还必须采用低水胶比和掺加足量的矿物细掺料与高效外加剂以保证高性能混凝土的下列性能：耐久性、工作性、各种力学性能、实用性、体积稳定性和经济合理性。高性能混凝土作为现代混凝土发展的最主要方向，其具有高耐久性并掺加了化学外加剂和矿物掺合料（以工业废渣为主）等一系列特点，被认为是适应低碳经济发展战略的新材料和新技术。

### 12.2.1 提高混凝土耐久性，延长服役寿命是混凝土材料实现低碳经济的根本

耐久性支配着混凝土在整个服役过程中的“健康”和“寿命”。1997年，美国著名学者Mehta发表了题为《耐久性一影响未来的关键问题》[7]的文章，指出我们面临混凝土耐久性问题的严重性。据统计，发达国家因混凝土基础设施耐久性问题造成的年经济损失约占GDP的1.5%～2%。美国混凝土基建工程的总价为60000亿美元，今后每年由于混凝土耐久性不足产生的工程维修和重建费用估计将高达3000亿美元，对于桥梁等生命线工程，因修复、更换造成交通延误等间接费用更大，是直接修复费用的10倍。加拿大为修复混凝土耐久性劣化损失的全部基础设施工程估计需耗费5000亿美元。英国的建筑和土木工程的维修费用也高达150亿英镑，其中混凝土工程维修费用为5亿多英镑。我国现有混凝土工程因耐久性不足而出现的劣化现象也相当严重。东北的云峰水电站，大坝建成运行不到10年，溢流坝表面混凝土冻融破坏面积就高达10000$m^2$，占整个溢流坝面的50%左右，混凝土平均冻融剥蚀深度达10cm以上。天津市内的八里台立交桥，由于碱集料反应，许多潮湿部位的柱子、支撑梁发生网状开裂和顺筋开裂，部分桥台潮湿处混凝土出现胀裂。在步入21世纪的今天，我国已出现混凝土结构物维修高潮，每年所需维修费用可以高达数千亿元，耗资之巨大。

提高混凝土耐久性，延长工程使用寿命，其实是最大的节能减排，是国家实施低碳经济及可持续发展战略的关键。我国的基础设施建设规模宏大，每年投资超过20000亿元，其中混凝土工程占有极大的比重。保持并提升混凝土的耐久性，不仅每年将节省大量的混凝土结构维修重建费用，为国家带来巨大的经济效益，更为重要的是能延长结构工程的服役寿命，防止重大事故发生，社会效益显著。

混凝土耐久性等宏观性能的好坏由混凝土微结构所决定，因此必须对混凝土的微结构进行优化调控。最为简便的方法是在混凝土中掺入高性能混凝土外加剂和具有潜在活性的掺合料，实现减水、减缩、增韧、改善水化特性等目标。

（1）大减水优化混凝土的微结构

高性能减水剂通过空间位阻以及静电排斥等作用，实现了水泥颗粒在混凝土浆体中的高度分散，从而能够充分发挥拌合用水的流动作用，使混凝土能够以较小的水灰比实现所要求的流动性能。同时，通过降低水灰比，实现了对混凝土的微结构调控。图 1 和图 2 是不掺外加剂、掺萘系减水剂 FDN（减水率约 20%）和掺聚羧酸系减水剂 PCA（减水率约 30%）的水泥浆体的孔径大小与分布情况。试验研究表明，外加剂掺入水泥浆体后，改善了水泥浆体内部孔结构，掺萘系减水剂（FDN）的水泥浆体早期（3d）有害孔和多害孔明显降低，少害孔和无害孔增多；聚羧酸系减水剂（PCA）较萘系减水剂（FDN）在改善水泥浆体内部结构的作用更大。试验采用的纯水泥浆体基准试样（Ref），最可几孔径 58mm，掺萘系减水剂的最可几孔径为 40mm，较纯水泥浆体的最可几孔径减少了 18mm；掺聚羧酸系减水剂水泥浆体的最可几孔径约为 35mm，较纯水泥浆体最可几孔径减少了 23mm。由此可见，随着外加剂减水率的提高，水泥浆体的孔结构改善越来越明显，即减水率大、自身含气量小的外加剂，改善混凝土内部孔结构的效果更好。

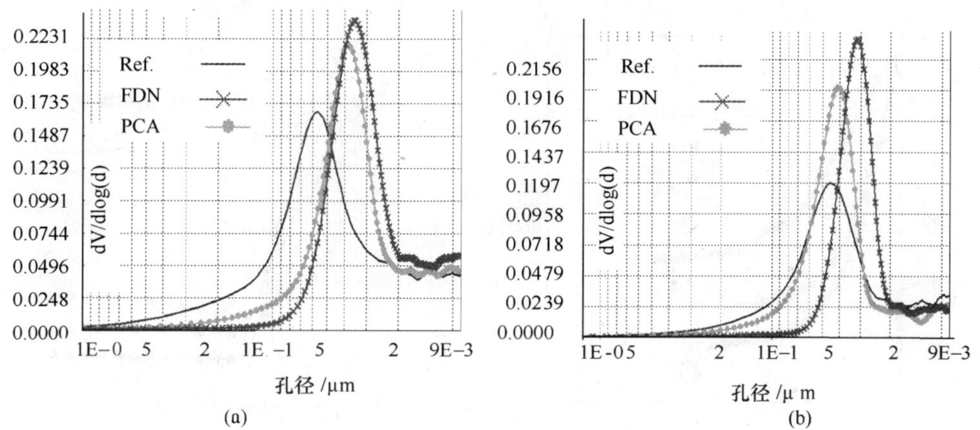


图 1　外加剂掺入水泥浆体后的孔径分布图

（a）3d 样品的孔径分布图；（b）28d 样品的孔径分布图

注：Ref—纯水泥浆体；FDN—掺萘系减水剂的水泥浆体；PCA—掺聚羧酸系减水剂的水泥浆体。

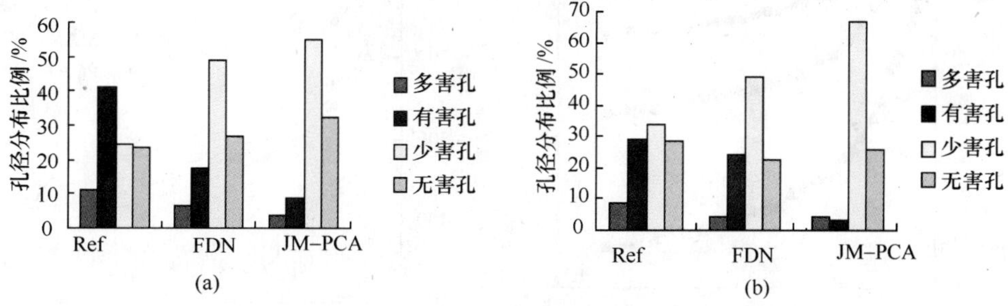


图 2　外加剂掺入水泥浆体后的孔径分布比例图

（a）3d 样品的孔径分布比例图；（b）28d 样品的孔径分布比例图

（2）降低毛细管张力，达到减缩目的

开裂的混凝土必定不耐久，良好的抗裂性是混凝土高耐久性的前提条件。收缩是引起混凝土开裂的最主要原因之一。因此，通过混凝土微结构调控，减少甚至避免混凝土开裂，是实现混凝土高服役性能的重要途径。为了改善混凝土的性能，在混凝土配制时加入一定量的减水剂，以达到减水增强的目的。但传统的萘系减水剂使用后，混凝土的收缩却会增大10%以上，加大了混凝土开裂的危险性。新一代聚羧酸类减水剂，采用接枝共聚的办法，可根据使用要求，在主链上接入相应的基团，如接入减缩基团，减小混凝土的收缩等。图3是采用分阶段、全过程混凝土变形量测系统对掺不同外加剂混凝土的各类变形进行测试的结果。由试验结果可知，不同的外加剂对混凝土的收缩影响较大。萘系减水剂（FDN）和氨基磺酸盐系减水剂（MAS）无论是凝缩还是自收缩、干燥收缩都较大。接入减缩基团的聚羧酸系减水剂不仅硬化前的塑性收缩变形（凝缩）小而且自收缩和干缩值也都较小。

同时，对几种主要高效减水剂的碱含量和表面张力进行了测试，结果见表2。

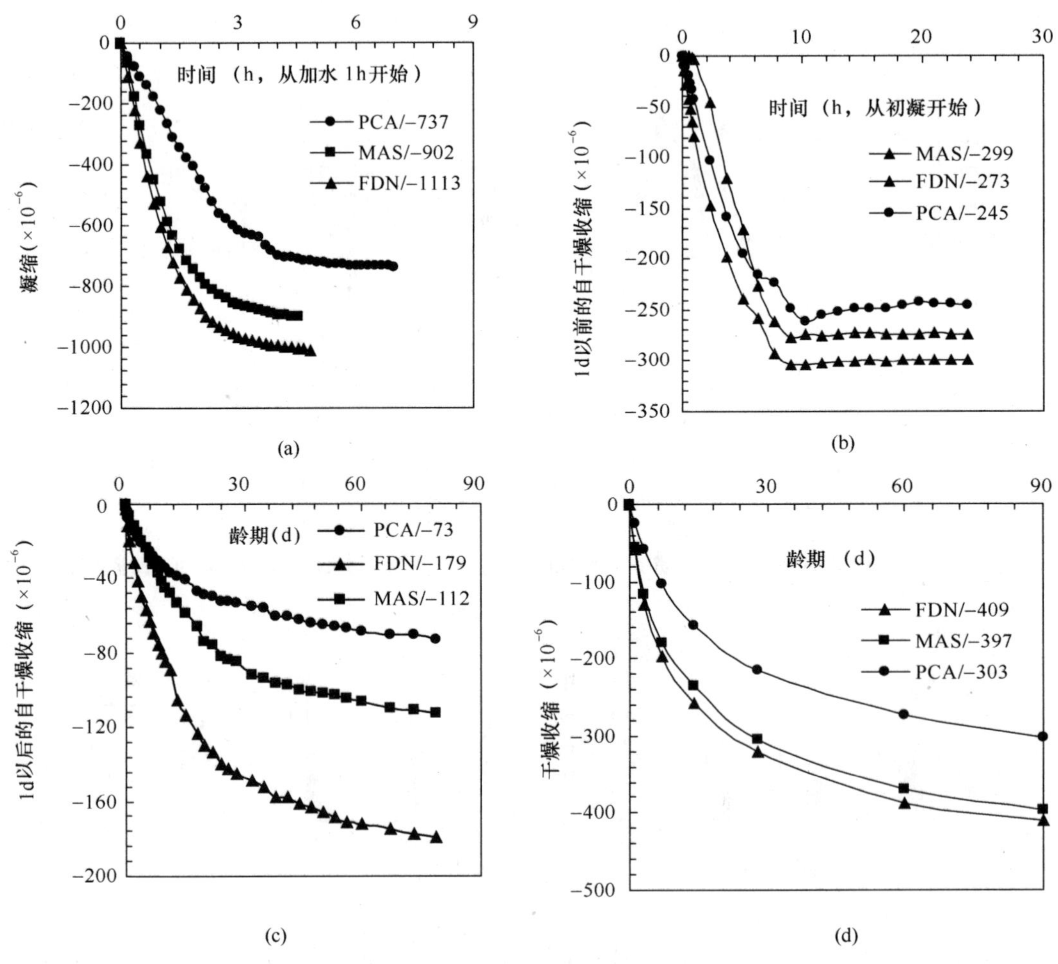


图3 外加剂对混凝土收缩的影响

（a）凝缩；（b）1d以前的自干燥收缩；（c）长龄期的自干燥收缩；（d）长龄期的干燥收缩

注：PCA—掺聚羧酸减水剂混凝土；FDN—掺萘系减水剂混凝土；MAS—掺氨基磺酸盐系减水剂混凝土。

**高效减水剂的碱含量及表面张力　表 2**

| | 掺量(%) | 减水率(%) | 碱含量(%) | 表面张力(mN/m) |
|---|---|---|---|---|
| 纯水 | / | / | / | 70.2 |
| 高浓型萘系 | 0.50 | 18 | 8.2 | 68.5 |
| 聚羧酸 | 0.20 | 23 | 0.9 | 43.0 |
| 氨基磺酸盐 | 0.50 | 22 | 8.0 | 52.0 |

注：表面张力根据掺量配成 20℃相应的水溶液测定。

由表 2 可见，在常规掺量下（即都以达到 20%左右的减水率时），作为表面活性剂范畴的几种高效减水剂均不同程度地降低了溶液的表面张力，并具有明显区别，聚羧酸系减水剂溶液表面张力最低，氨基磺酸盐减水剂次之，而掺萘系减水剂溶液的表面张力最高，只是略低于纯水的表面张力。聚羧酸系减水剂的碱含量最低，而氨基磺酸盐减水剂的碱含量与高浓型萘系接近，由于其减水率较高，因此掺入混凝土中达到相同减水率时的碱含量可较萘系低，萘系的碱含量在这三种外加剂里面相对最高。

试验结果表明具有减缩功能的 PCA 聚羧酸高效减水剂由于具有相对较高的减水率，较低的碱含量和表面张力，掺入混凝土后对于干燥收缩不仅没有增加收缩，而是降低了混凝土的干缩值。

（3）水化热的控制

温差变形是引起混凝土结构开裂的另一主要原因。控制水泥水化热和优化水泥水化热过程曲线，是大体积混凝土结构温度和裂缝控制最关键的技术措施。从图 4 水化热变化及水化放热速率曲线试验结果可知，掺缓凝型萘系高效减水剂和聚羧酸系外加剂都能减少水泥的水化热，降低并延缓了水泥的放热峰值，聚羧酸系外加剂 PCA 的效果更好。另外，在 70%水泥+30%粉煤灰的胶凝材料体系中，PCA 延缓水泥的水化放热和降低 7d 水化热值的作用更为明显；且粉煤灰对延缓水泥的水化放热和降低水化热值也有积极的贡献。因此，混凝土外加剂和矿物掺合料均能使水泥浆体的水化放热曲线分布趋于更合理，这对降低大体积混凝土的水化温升，减少温度应力具有积极的作用。

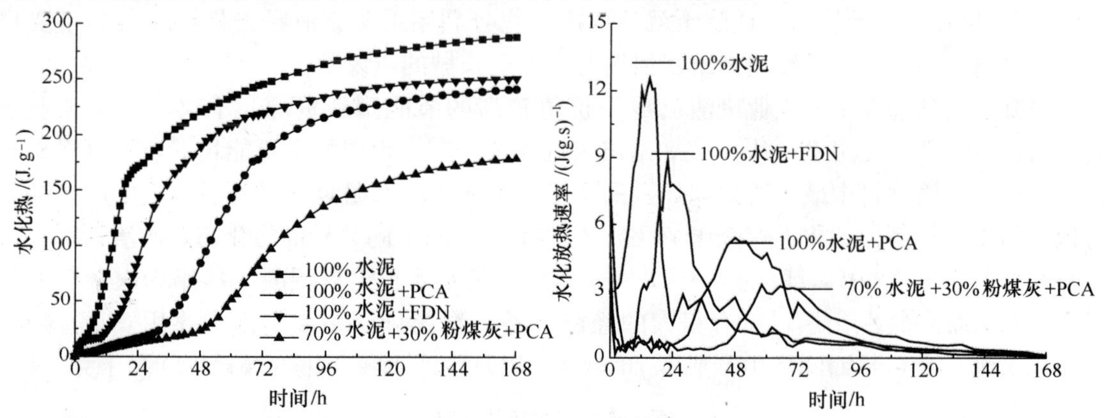


图 4　外加剂和粉煤灰掺合料对水泥水化热的影响

（4）增韧新技术

混凝土是一种脆性材料，抗折强度是抗压强度的 1/10 左右，且随着抗压强度的增加，两者比值呈现减小趋势，也即混凝土的脆性随着抗压强度的提高越来越明显。这种性能上的缺陷是混凝土耐久性难以大幅提升的重要原因之一。因此，如何提高混凝土的拉压比，增加混凝土的断裂韧性，从本质上改善混凝土材料自身的脆性，是现代混凝土研究的重点和难点。聚合物混凝土具有较高的增韧效果，但使用传统的水溶性聚合物，一是掺量高（一般为水泥用量的10%左右），二是聚合物成膜后，水泥水化便会中止，只起到了填充料的作用。基于原位增韧理论的水分散性聚合物在强碱及高离子体系中具有很好的分散稳定性。掺加0.5%～1.0%的该聚合物乳液后，混凝土中的毛细孔结构将会得到很大的改善，同时可大幅提高硬化混凝土的断裂韧性，对混凝土具有很好的增韧效果（见图5）。

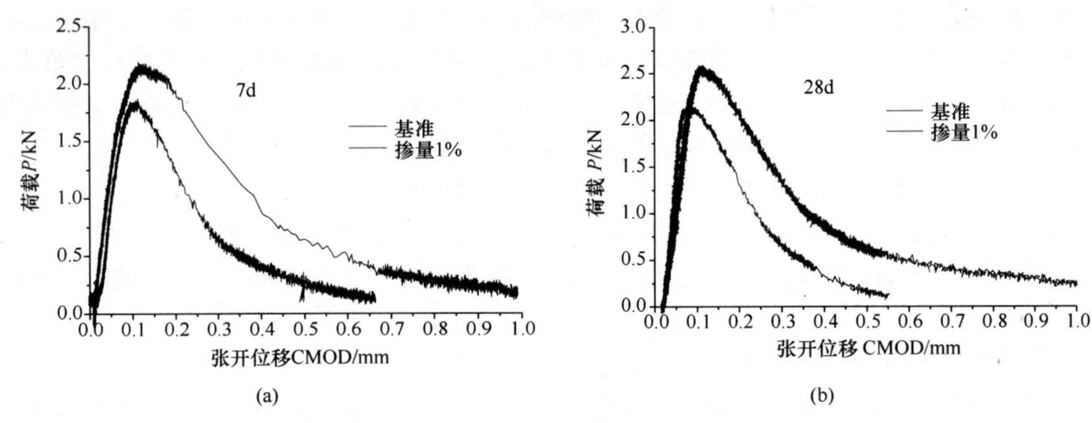


图5 混凝土荷载—挠度曲线

(a) 7d；(b) 28d

### 12.2.2 大量使用工业废渣，可以起到化废为宝、化害为利的作用

工业废渣是指工业生产过程中排放的固体废物。中国的工业部门每年排出的废渣达数亿吨，其中以冶金、能源、采矿、化工等部门的排放量最多。据2004年中国环境状况公报显示："全国工业固体废弃物产量为12亿吨，比2003年增长了20%，综合利用率为55.7%"。早在几年前，唐明述院士就曾提出：充分利用工业废渣是可持续发展的重要环节，并提出充分利用工业废渣修筑江河湖海的万里长城的建议[5,8]。

混凝土制备过程中，工业废渣已成为价值较高的节能型新资源，不仅利用了工业废渣，而且可减少水泥用量[9~11]。表3给出了混凝土生产过程中的主要能耗[3]，可以看出水泥是混凝土各原材料中最主要的能耗，采用工业废渣替代水泥对于降低混凝土生产能耗具有积极的作用。例如，按近年全国70%～80%水泥掺加不同数量的粒化高炉矿渣计算，可以节约能源20%～40%，降低成本10%～30%；又如水利工程混凝土掺用粉煤灰，取代部分水泥，既降低水化热，又有很大的经济效益，据统计，在混凝土中掺用1t粉煤灰，可取代水泥0.8t，从而降低生产成本70～80元以及生产水泥所用的煤耗100多千克。

混凝土生产过程的能耗 表3

| 构成 | 能量(MJ/kg 水泥) | 构成 | 能量(MJ/kg 水泥) |
|---|---|---|---|
| 粗集料 | 0.028 | 水 | 0.000 |
| 细集料 | 0.028 | 生产 | 0.102 |
| 水泥 | 0.735 | 总计 | 0.893 |

粉煤灰、矿渣、硅粉等作为工业废渣，已成为制备高性能混凝土不可或缺的矿物掺合料。国内许多学者经过大量的试验研究指出，粉煤灰、矿渣等工业废弃物掺入到混凝土中不仅利用了废渣、节省水泥，更重要的是可以改善混凝土的新拌性能、提高混凝土抗氯离子扩散等耐久性能[12~14]。国外 MichaelD. A. Thomas 和 PhilB. Bamforth 等[15~17]学者将掺粉煤灰、矿渣等工业废弃物的混凝土连续 8 年放置于海洋环境进行研究，结果表明掺粉煤灰、矿渣的混凝土氯离子渗透深度远远低于不掺的混凝土，并建立了氯离子扩散预测模型，模型预测结果和实际测试结果如图 6、图 7 所示。掺粉煤灰、矿渣的混凝土氯离子扩散系数比纯波特兰硅酸盐水泥混凝土氯离子扩散系数低 1～2 个数量级，在海洋环境下掺入粉煤灰、矿渣可明显延长混凝土的服役寿命。

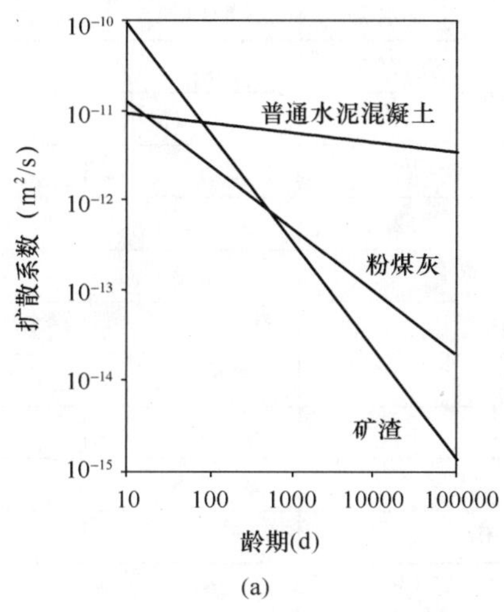

(a)

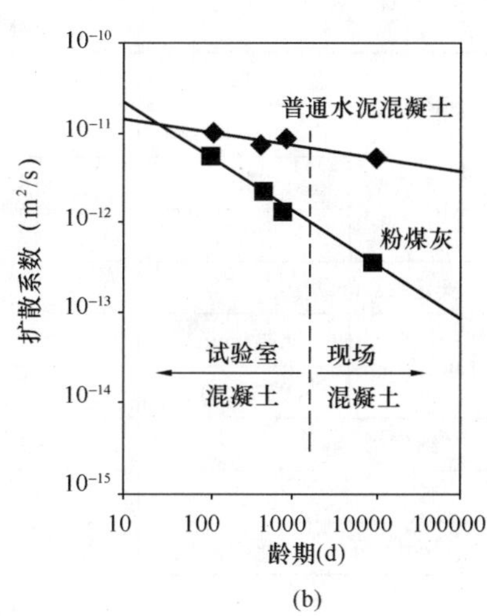

(b)

图 6　氯离子扩散系数随时间变化情况

（a）根据模型按拟合值；（b）迁移试验数据

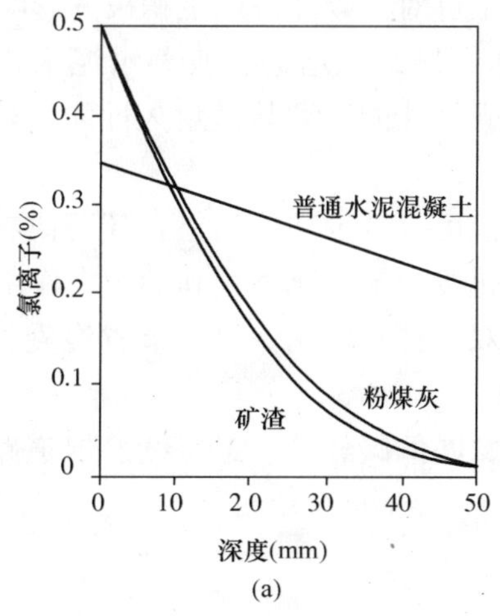

(a)

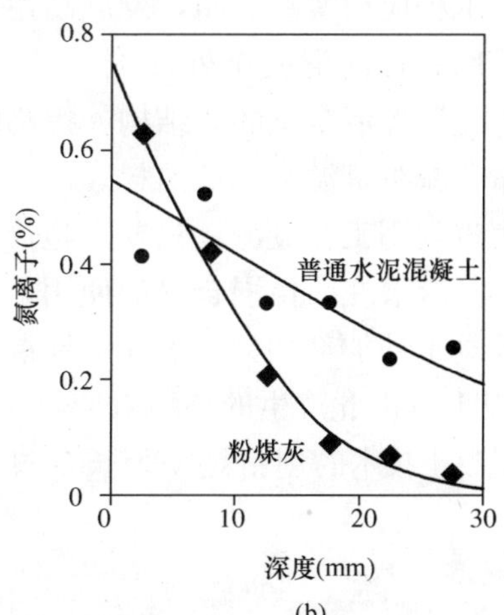

(b)

图 7　30 年养护龄期试件不同深度氯离子含量预测和实测结果

（a）预测；（b）现场数据

### 12.2.3 生态型高性能水泥基建筑材料的研究

由孙伟院士领衔研制的低熟料（15%～30%）、低能耗、低收缩、高耐久生态型高性能水泥基建筑材料，较好的克服了大掺量工业废渣水泥基材常见的早期强度低、泌水大、收缩大、易碳化、抗冻时混凝土表面易剥落等弊病，为大量利用工业废料，化废为宝、化害为利提供了新的技术途径。从表 4 可以看出，制备的生态型水泥基建筑材料 HDC-1 和 HDC-2 均能满足甚至优于现常用 P·O42.5 和 P·O52.5 水泥各项性能指标。试验研究结果表明用该材料配制的 C30～C80 混凝土具有优异的力学行为和高的耐久性能。

大掺量工业废渣高性能水泥基材料的物理力学性能　　表 4

| 项目 | | 标准值 | | 实测值 | |
|---|---|---|---|---|---|
| | | 42.5 水泥 | 52.5 水泥 | HDC-1 | HDC-2 |
| 抗折强度(MPa) | 3d | ≥3.5 | ≥4.0 | 6.3 | 6.3 |
| | 28d | ≥6.5 | ≥7.0 | 9.1 | 9.4 |
| 抗压强度(MPa) | 3d | ≥16.0 | ≥23.0 | 26.9 | 28.3 |
| | 28d | ≥42.5 | ≥52.5 | 48.8 | 53.8 |
| 安定性(mm) | | ≤5 | | 0.8 | 0 |
| 凝结时间 | 初凝(min) | ≥45 | | 287 | 224 |
| | 终凝(h) | ≤10 | | 5.37 | 4.48 |
| 细度(%) | | ≤10.0 | | 4.8 | 5.5 |
| $SO_3$(%) | | ≤3.5 | | 2.8 | 2.9 |
| MgO(%) | | ≤5.0 | | 4.2 | 4.2 |

## 12.3 结论

研究及工程应用实践表明，发展高性能混凝土可对低碳经济产生积极的影响。

（1）采用高性能混凝土外加剂，可达到减水、减缩、增韧、改善水泥水化特性等目的，从而优化调控混凝土的微结构，提高混凝土服役性能，尤其是耐久性能，延长建筑物的使用寿命，减少资源、能源消耗。

（2）大量利用工业废渣，只要措施恰当，不仅可以减少环境污染，还可改善混凝土的性能。这类具有潜在活性混合材的使用，可起到化废为宝、化害化利的积极作用。

（3）混凝土外加剂和具有潜在活性混合材的双掺技术应用，可以节省混凝土单方水泥用量 15%以上，由此产生的环境效应将不容忽视。

（4）混凝土技术的创新是一个长期过程，在发展低碳经济、走持续发展道路时代更显重要。

## 参考文献

[1] 顾真安. 院士谈建设节约型社会：加速发展节约型建材工业. 求是，2005，(12).

[2] 姚燕. 低碳经济时代水泥混凝土行业发展方向. 商品混凝土，no. 4，2010.

[3] Leslie Struble, Jonathan Godfrey, How Sustainable is Concrete? International Workshop on Sustainable Development and Concrete Technology, Beijing, May 20-21, 2004.

[4] Report card for American's Infrastructure [EB/OL]. http: //www. asce. org/reportcard/2005/Index.

[5] 唐明述. 中国水泥混凝土工业发展现状与展望. 东南大学学报(自然科学版), 2006, 36: 1-6.

[6] 吴中伟, 廉慧珍. 高性能混凝土. 北京: 中国铁道版社, 1999 年.

[7] P. K. Mehta. Durability -Critical issues for the future . Concrete International , 1997 , (7) .

[8] 唐明述. 水泥混凝土与可持续发展. 中国水泥-专家论坛, 2003, 10.

[9] P - C. Aitcin. Cements of yesterday and today : Concrete of tomorrow. Cement and Concrete Research , 2000 , (9) .

[10] 黄弘, 唐明述, 沈晓冬等. 工业废渣资源化及其可持续发展(I)-典型工业废渣的物性和利用现状. 材料导报, 2006, 5(20): 450-454.

[11] 黄弘, 唐明述, 沈晓冬等. 工业废渣资源化及其可持续发展(Ⅱ)-与水泥混凝土工业相结合走可持续发展之路. 材料导报, 2006, 5(20): 455-458.

[12] 陈正, 杨绿峰, 冯庆革等. 高性能混凝土的氯离子扩散及服役寿命研究. 建筑材料学报. 2010. 2 (13): 222-231.

[13] 金祖权, 孙伟, 张云升等. 粉煤灰混凝土的多因素寿命预测模型. 东南大学学报(自然科学版), 2005, 7(35): 150-154.

[14] 秦鸿根, 潘刚华, 李松泉. 掺粉煤灰高性能混凝土耐久性研究. 混凝土与水泥制品, 2000(5): 11-13.

[15] Michael D. A. Thomas , Phil B. Bamforth. Modelling chloride diffusion in concrete: Effect of fly ash and slag. Cement and Concrete Research. Volume 29, Issue 4, April 1999, Pages 487-495.

[16] T. Cheewaket, C. Jaturapitakkul, W. Chalee. Long term performance of chloride binding capacity in fly ash concrete in a marine environment . Construction and Building Materials, Volume 24, Issue 8, August 2010, Pages 1352-1357.

[17] W. Chalee, P. Ausapanit, C. Jaturapitakkul. Utilization of fly ash concrete in marine environment for long term design life analysis . Materials & Design, Volume 31, Issue 3, March 2010, Pages 1242-1249.

# 第 13 章 Chapter 13

# 多因素耦合作用下混凝土结构耐久性研究

**牛荻涛**

（西安建筑科技大学土木工程学院，陕西　西安 710055）

**提　要：** 在混凝土结构耐久性研究中，对于单一因素影响下的混凝土结构耐久性问题，已进行了大量理论及试验研究，取得了丰富成果，但对于多因素耦合作用下混凝土结构耐久性问题的研究还相对较少。在实际工程中，环境介质常常是多种共存，对结构造成的损伤则是多种因素相互作用的结果，因此，开展双重和多重影响因素耦合作用下的混凝土结构耐久性研究具有重要的实际意义。本文以一般大气环境与冻融环境为重点，采用试验研究与理论分析相结合的方法，研究了碳化与酸雨、中性化与荷载作用、碳化与冻融循环共同作用下混凝土的损伤演变规律，为大气环境混凝土结构耐久性评估奠定基础。

**关键词：** 混凝土结构；耐久性；耦合；大气与冻融环境

# DURABILITY RESEARCH FOR CONCRETE STRUCTURE IN THE COUPLING ACTION OF VARIOUS FACTORS

D. T. Niu

(School of Civil Engineering Xi' an University of Architecture and Technology, Xi' an 710055, China)

**Abstract:** In the field of the durability research for concrete structure, the single factor which influence the durability of concrete structures has been researched by theoretical and experimental, and the research results were rich. But the durability research for concrete structure in the coupling action of various factors was still very little. In the practice engineering, the ambient mediums usually coexist, and the structures were damaged by the action of various factors. Therefore, the durability research for concrete structures in the coupling action of various factors is very significant. In this paper, the atmospheric and freezing-thawing circumstance are the keys, studied the damage laws of the concrete structures influenced by carbonization and acid rain, neutral and load function, carbonization and freezing-thawing. This research lays the foundation for the durability assessment of concrete structures which in the atmospheric and freezing-thawing circumstance.

**Keywords:** concrete structures; durability; coupling; the atmospheric and freezing-thawing circumstance

## 13.1　引言

在混凝土结构耐久性研究中，对于单一因素影响下的混凝土结构耐久性问题，已进行了大量理论及试验研究，例如混凝土碳化，学者们对其机理和影响因素等进行了深入的研究，并且建立了较为成熟的预测模型，如阿列克谢耶夫模型[1]、Papadakis 模型[2]，牛荻涛模型[3]等。对于混凝土冻融耐久性，目前也已开展了比较深入的研究，我国学者也建立了几个混凝土抗冻性预测模型，如蔡昊模型[4]，许丽萍模型[5]等。但在实际工程当中，环境介质常常是多种共存，对结构造成的损伤则是多种因素相互作用的结果[6]。目前，国内外对于多因素耦合作用下混凝土结构耐久问题的研究还相对较少。东南大学孙伟院士针对西部严酷环境下混凝土结构耐久性问题进行了一些探讨，对复合环境因素以及力学与环境因素耦合作用下结构混凝土损伤失效规律进行了初步的研究[7~9]。

在一般大气环境中，工程结构将遭受碳化、酸雨引起的混凝土中性化以及钢筋锈蚀，而在冻融环境中，工程结构易遭受碳化和冻融循环引起的混凝土损伤。本文以一般大气环境与冻融环境为重点，对该环境影响下混凝土结构耐久性的耦合因素进行分析，采用试验研究与理论分析相结合的研究方法，研究碳化与酸雨、中性化与荷载作用、碳化与冻融循环共同作用下混凝土的损伤演变规律。

## 13.2　酸雨和碳化共同作用下混凝土中性化研究

目前的环境问题日益严重，混凝土所处的环境也越来越复杂，大大影响了混凝土结构的耐久性能。在一般大气环境中，酸雨和 $CO_2$ 是引起混凝土中性化的主要原因[10]。混凝土的碳化研究经历了长时间的积累，取得了相对完善的一系列成果，关于酸雨对混凝土的侵蚀研究只是陆续有一些文章发表，而酸雨和碳化共同作用下混凝土的耐久性能研究基本还是空白。我们开展了酸雨侵蚀与碳化共同作用下粉煤灰混凝土的中性化规律研究，深入分析了酸雨和碳化共同作用下混凝土的损伤机理。

### 13.2.1　试验方法

由于在一年内不同地区混凝土碳化和酸雨侵蚀所占比例不同，酸雨期长碳化时间短，酸雨期短碳化时间就长。所以试验时采用酸雨试验 10 次循环为一个酸雨腐蚀段，碳化时间分 3d 和 7d 二种情况，具体分类如下：

ST3：先酸雨试验 10 次，然后碳化试验 3d，一个循环为 23d，共进行 3 次大循环。

ST7：先酸雨试验 10 次，然后碳化试验 7d，一个循环为 27d，共进行 3 次大循环。

T3S：先碳化试验 3d，然后酸雨试验 10 次，一个循环为 23d，共进行 3 次大循环。

T7S：先碳化试验 7d，然后酸雨试验 10 次，一个循环为 27d，共进行 3 次大循环。

混凝土碳化试验按照《普通混凝土长期性能和耐久性能试验方法》GBJ 82—85[11]进行。模拟酸雨溶液的 $SO_4^{2-}$ 离子浓度水平为 0.15mol/L，pH 采用 4.0。试验制度为浸烘交替法，将试件浸泡于模拟的酸雨溶液中 36h，取出晾干 1h，放入 60℃烘箱中烘干 10h，冷却 1h 为一个酸雨循环，一个循环 48h。

### 13.2.2 酸雨和碳化共同作用混凝土中性化试验结果

快速碳化试验 3、6、7、9、14、21d 的碳化深度值与实验室酸雨侵蚀试验 10、20、30 次的中性化深度如表 1 所示，表 2 为酸雨和碳化共同作用试验模式测得的混凝土中性化深度。

单独碳化和单独酸雨侵蚀混凝土中性化深度/mm 表 1

| 混凝土 | 碳化时间/d | | | | | | 酸雨侵蚀循环次数 | | |
|---|---|---|---|---|---|---|---|---|---|
| | 3 | 6 | 7 | 9 | 14 | 21 | 10 | 20 | 30 |
| B1 | 0.72 | 1.22 | 1.12 | 1.27 | 1.49 | 1.62 | 0.55 | 0.80 | 0.97 |
| B2 | 1.38 | 2.22 | 2.34 | 2.65 | 3.18 | 3.60 | 0.57 | 0.89 | 1.07 |
| B3 | 2.22 | 4.13 | 4.16 | 4.92 | 6.02 | 7.42 | 0.60 | 1.07 | 1.22 |

酸雨和碳化共同作用混凝土中性化深度/mm 表 2

| 循环次数 | B1 | | | B2 | | | B3 | | |
|---|---|---|---|---|---|---|---|---|---|
| | 1 | 2 | 3 | 1 | 2 | 3 | 1 | 2 | 3 |
| T3S | 1.10 | 1.20 | 1.26 | 1.32 | 1.56 | 1.86 | 3.30 | 3.62 | 4.00 |
| T7S | 1.88 | 2.34 | 2.69 | 2.51 | 3.78 | 4.26 | 5.67 | 6.73 | 7.86 |

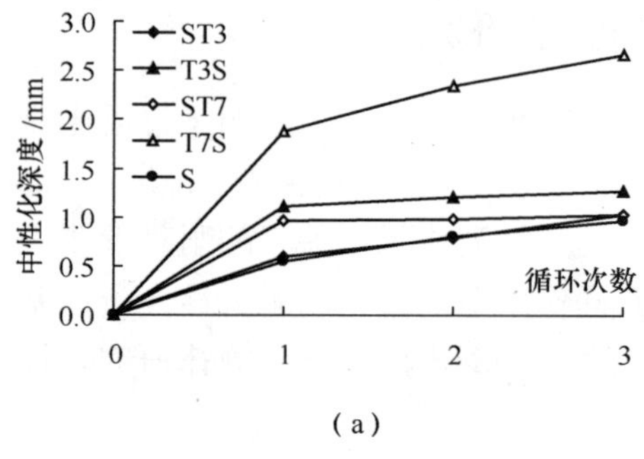

(a)

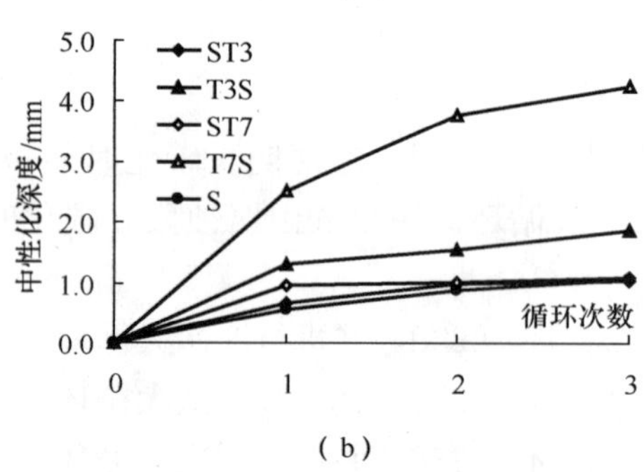

(b)

图 1 酸雨和碳化共同作用试验中混凝土的中性化规律

(a) 酸雨和碳化共同作用试验中普通混凝土中性化规律；(b) 酸雨和碳化共同作用试验中粉煤灰混凝土中性化规律

### 13.2.3 不同试验模式下混凝土中性化规律

酸雨和碳化共同作用混凝土中性化深度随时间的变化规律见图 1。随循环次数增加，各试验模式下混凝土中性化深度都在增加，不论是普通混凝土还是粉煤灰混凝土，中性化深度都有 T7S>T3S>ST7>ST3>S。在最初的共同作用循环内，混凝土中性化速度最快，之后逐渐变慢。

从试验结果来看：①先酸雨后碳化模式下，混凝土孔隙中的主要填充物是酸雨侵蚀生成的石膏，这对 $CO_2$ 的扩散起到了很大程度的抑制作用；②混凝土遭受一次酸雨侵蚀再经历一次碳化，反应的生成物 $CaCO_3$ 会继续填堵在混凝土孔隙中；③致密的石膏层吸附在混凝土孔隙内壁，阻隔了 $CO_2$ 和酸雨侵蚀介质与水泥碱性物质的反应。因此，先酸雨后碳化混凝土中性化深度发展基本停滞，先酸雨后碳化混凝土中性化过程主要是酸雨起控制作用，其对混凝土碳化过程有明显的抑制效果。

### 13.2.4　酸雨和碳化共同作用耦合程度分析

从试验结果看，先碳化后酸雨的混凝土中性化结果与实际一般大气环境的中性化过程接近，而先酸雨后碳化的混凝土试件，由于酸雨侵蚀液中 $SO_4^{2-}$ 离子浓度较大，完全抑制了碳化的发生，与真实大气环境中混凝土中性化过程不符。本文以先碳化后酸雨试验结果分析酸雨和碳化共同作用效应。

引入混凝土中性化碳化影响系数 $\lambda_T$，定义如下：

$$\lambda_T = x_{T+S}/x_S \tag{1}$$

式中：$x_{T+S}$ 为酸雨和碳化共同作用混凝土中性化深度，mm；$x_T$ 为单独碳化深度，mm。

混凝土中性化碳化影响系数 $\lambda_T$ 的计算结果列于表 3，由表 4 可见，无论是 T3S 还是 T7S，混凝土中性化碳化影响系数 $\lambda_T$ 的值都远大于 1，说明混凝土中性化耦合过程中碳化起着很大作用。

**混凝土中性化碳化影响系数 $\lambda_T$ 计算结果**　　表 3

| 循环次数 | T3S/S | | | T7S/S | | |
|---|---|---|---|---|---|---|
| | 1 | 2 | 3 | 1 | 2 | 3 |
| B1 | 2.00 | 1.50 | 1.30 | 3.42 | 2.93 | 2.75 |
| B2 | 2.32 | 1.75 | 1.74 | 4.40 | 4.25 | 3.98 |
| B3 | 4.71 | 3.38 | 3.28 | 8.10 | 6.32 | 6.44 |

为了描述酸雨和碳化共同作用时的酸雨和碳化的耦合效应，引入混凝土中性化耦合系数 $k$，定义如下：

$$k = \frac{x_{T+S}}{x_T + x_S} \tag{2}$$

耦合系数 $k$ 的计算结果见表 4。碳化和酸雨共同作用的时间与单因素试验时间的叠加值等同，可以很好的反映出酸雨和碳化共同作用时的耦合效果。

**混凝土中性化耦合系数 $k$**　　表 4

| 循环次数 | T3S/(T3+S) | | | T7S/(T7+S) | | | |
|---|---|---|---|---|---|---|---|
| | 1 | 2 | 3 | 1 | 2 | 3 | 平均值 |
| B1 | 0.87 | 0.59 | 0.56 | 1.13 | 1.02 | 1.03 | 1.060 |
| B2 | 0.68 | 0.50 | 0.50 | 0.86 | 0.93 | 0.91 | 0.955 |
| B3 | 1.13 | 0.70 | 0.65 | 1.17 | 0.95 | 0.91 | |

由表 4 可见，对于同一种共同作用试验模式，耦合系数基本上相等，T3S 模式下耦合系数平均值为 0.69，T7S 模式下普通混凝土的耦合系数平均值为 1.060，粉煤灰混凝土的耦合系数平均值为 0.955。由前面的分析知道 T3S 与实际大气环境中混凝土中性化发展过程不符，而 T7S 模式只是试验时的一种设计模式，能在一定程度上说明实际大气环境中混凝土的中性化情况。

## 13.3 承载混凝土碳化规律研究

开展了承载混凝土试件快速碳化试验，通过试验研究了混凝土碳化速度随弯曲应力水平的变化规律。

### 13.3.1 试验方法

试件尺寸采用 100mm×100mm×550mm，按照《普通混凝土长期性能和耐久性能试验方法》[11]进行混凝土快速碳化试验。试验采用拉杆加载装置施加弯曲应力模拟简支梁的受力状况。

### 13.3.2 试验结果

为了分析弯曲荷载对混凝土碳化深度的影响，试验还同时进行了一组无弯曲应力混凝土的碳化试验，混凝土碳化深度测试结果如表 5 所示。

**混凝土碳化深度试验结果/mm** **表 5**

| 混凝土 | 碳化时间 | 应力水平 | | | | | | |
|---|---|---|---|---|---|---|---|---|
| | | 0.0 | 0.2 | | 0.4 | | 0.6 | |
| | | 无应力 | 受拉区 | 受压区 | 受拉区 | 受压区 | 受拉区 | 受压区 |
| B1 | 7d | 1.12 | 1.59 | 1.12 | 2.16 | 1.06 | 2.56 | 0.90 |
| | 14d | 1.49 | 2.04 | 1.36 | 2.92 | 1.29 | 3.12 | 1.08 |
| | 28d | 1.75 | 2.62 | 1.64 | 3.55 | 1.59 | 3.69 | 1.35 |
| B2 | 7d | 2.34 | 2.83 | 2.24 | 3.22 | 2.04 | 3.64 | 1.84 |
| | 14d | 3.18 | 3.85 | 3.16 | 4.08 | 3.12 | 4.47 | 2.51 |
| | 28d | 4.01 | 4.80 | 4.00 | 5.02 | 3.98 | 5.43 | 3.23 |
| B3 | 7d | 4.17 | — | — | 5.74 | 3.90 | — | — |
| | 14d | 6.02 | — | — | 6.96 | 5.76 | — | — |
| | 28d | 8.20 | 9.32 | 6.88 | 9.97 | 6.57 | 10.61 | 6.11 |

### 13.3.3 承载混凝土碳化试验结果分析

假设有荷载情况下混凝土的碳化过程仍然服从 Fick 第一定律，并可用下式表达：

$$x_\sigma = K_\sigma \sqrt{t} = \alpha_\sigma K \sqrt{t} = \alpha_\sigma x \tag{3}$$

式中：$x_\sigma$ 为应力水平为 $\sigma$ 时混凝土的碳化深度；$K_\sigma$ 为应力水平为 $\sigma$ 时的混凝土碳化速度系数；$x$ 为无应力时混凝土的碳化深度；$K$ 为无应力时混凝土的碳化速度系数；

$t$ 为碳化时间；$\alpha_\sigma$ 为混凝土碳化应力影响系数，可以表示为

$$\alpha_\sigma = x_\sigma / x \tag{4}$$

①弯曲荷载作用下普通混凝土碳化规律

不同弯曲拉、压应力水平下，普通混凝土的碳化速度分别如图 2、图 3 所示。由图可

见，承受弯曲拉应力的混凝土碳化速度均大于无应力状态下的碳化速度，随着应力水平的增加，混凝土碳化速度逐渐增加，即弯曲拉应力加速了混凝土碳化。其原因在于：弯曲荷载增加，造成了混凝土内在缺陷（孔隙、气孔、微裂缝等）的扩展或增多，导致 $CO_2$ 扩散系数相应增大。压应力区的混凝土碳化速度小于无应力状态下混凝土碳化速度。弯曲压应力使混凝土的抗碳化能力略有增强，其原因在于：当弯曲压应力存在时，可使混凝土的微裂缝闭合或宽度减小，$CO_2$ 扩散系数减小，从而降低了混凝土的碳化速度。当弯曲压应力过大时，也可能使混凝土产生微观裂缝，加快混凝土碳化速度。

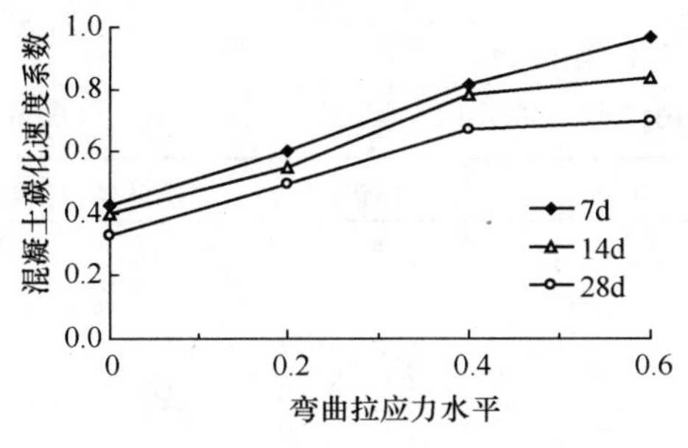


图 2　弯曲拉应力对普通混凝土碳化速度的影响

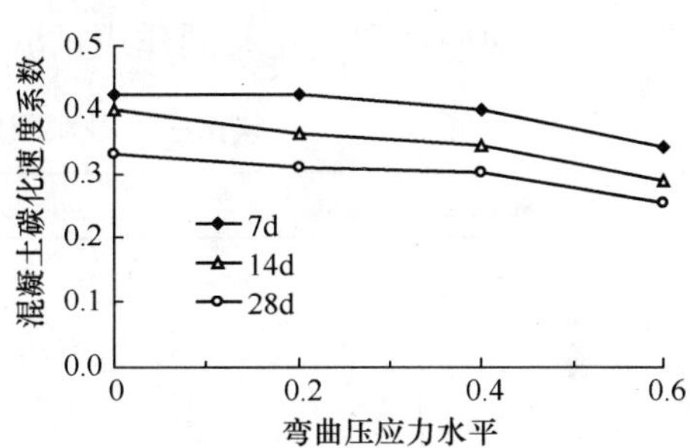


图 3　弯曲压应力对普通混凝土碳化速度的影响

②弯曲荷载作用下粉煤灰混凝土碳化规律

弯曲拉、压应力水平与粉煤灰混凝土碳化速度的关系分别如图 4、图 5 所示。从图中可以看出，与普通混凝土一样，不论弯曲拉应力水平高低，均表现出弯曲拉应力区的混凝土碳化速度大于无应力状态下的碳化速度，弯曲压应力区的粉煤灰混凝土碳化速度小于无应力状态下的碳化速度；弯曲压应力水平越大，粉煤灰混凝土碳化速度减小的也越大。

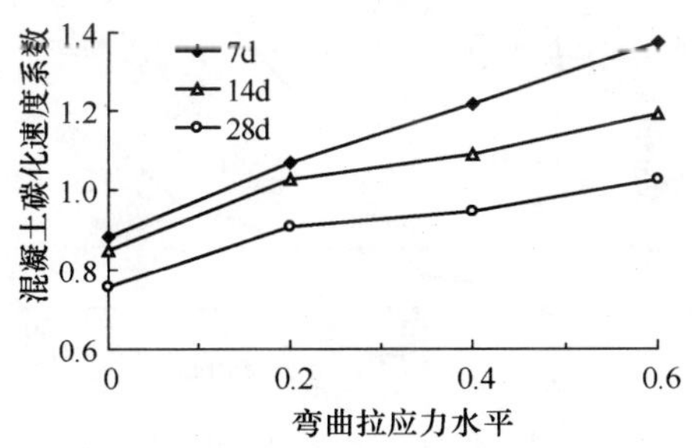


图 4　弯曲拉应力对粉煤灰混凝土碳化速度影响

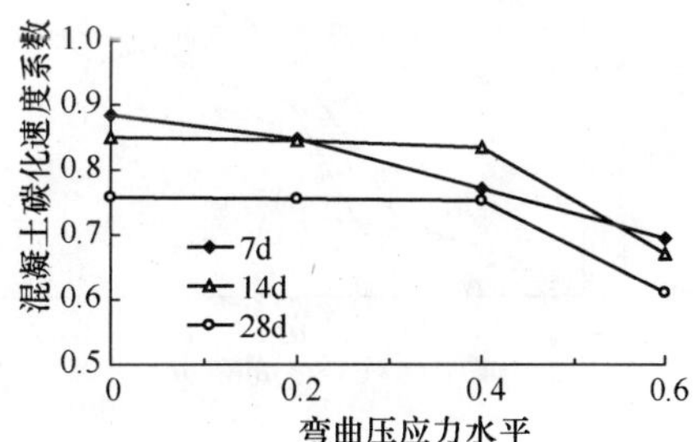


图 5　弯曲压应力对粉煤灰混凝土碳化速度影响

## 13.4　冻融—碳化共同作用下混凝土损伤试验研究

开展了冻融循环和碳化共同作用下混凝土耐久性试验，研究了混凝土遭受冻融循环和碳化共同作用下的耐久性能。

### 13.4.1 试验方法

混凝土冻融破坏和碳化的作用机理不同，发生作用后的产物也不同。一般碳化使混凝土的结构变得致密，而冻融使其变得疏松[12]。所以在试验中先进行哪个作用，都会对两个因素的复合效应带来影响。为了分析这种影响，试验模式分两种模式进行：一组为先冻融后碳化，一组为先碳化后冻融。具体分类如下：

（1）F50C：先冻融试验 50 次，然后碳化试验 7d，一个循环为 15d，共进行 4 次大循环。

（2）CF50：先碳化试验 7d，然后冻融试验 50 次，一个循环为 15d，共进行 4 次大循环。

表 6 是冻融循环和碳化共同作用试验分组情况。

**碳化—冻融双因素复合作用试验试件分组** **表 6**

| 编号 | 水胶比 | 粉煤灰掺量 | 含气量(%) | 试验模式 | 试件个数 | 强度试块个数 |
|---|---|---|---|---|---|---|
| C1 | 0.35 | 30% | 3.7 | F50C | 3+3 | 24 |
| C2 | 0.45 | | 4.4 | CF50 | 3+3 | 24 |
| C3 | | | | F50C | 3+3 | 24 |
| C4 | 0.55 | | 4.5 | F50C | 3+3 | 24 |

### 13.4.2 混凝土冻融—碳化共同作用下损伤规律分析

（1）冻融循环—碳化作用下混凝土质量损失变化规律

由图 6、图 7 可见，碳化—冻融试验模式下的质量损失要小于冻融—碳化试验模式下的质量损失，冻融—碳化试验的质量损失明显大于冻融试验的质量损失，随着冻融循环次数的增大，冻融—碳化试验下的质量损失速度加快。

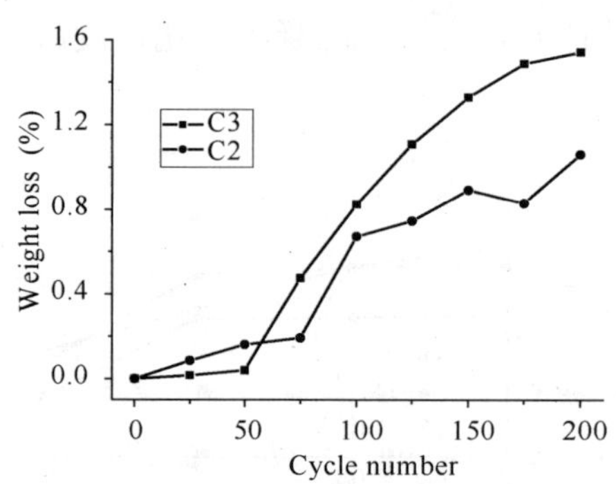


图 6 混凝土在不同试验模式下的质量损失变化规律

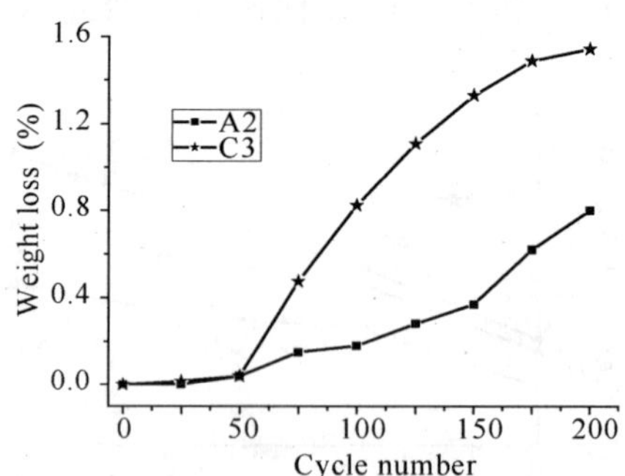


图 7 水胶比 0.45 混凝土在不同试验模式下的质量损失变化规律

（2）冻融循环—碳化作用下混凝土相对动弹性模量变化规律

由图 8、图 9 可见，冻融—碳化试验模式下，当冻融循环次数至 200 时，相对动弹性模量为 0.917，碳化—冻融试验模式下的相对动弹性模量为 0.958，表明先碳化混凝土由于内部结构变得密实要比先冻融混凝土相对动弹性模量高。冻融—碳化试验的相对动弹性模量明显小于冻融试验的相对动弹性模量。

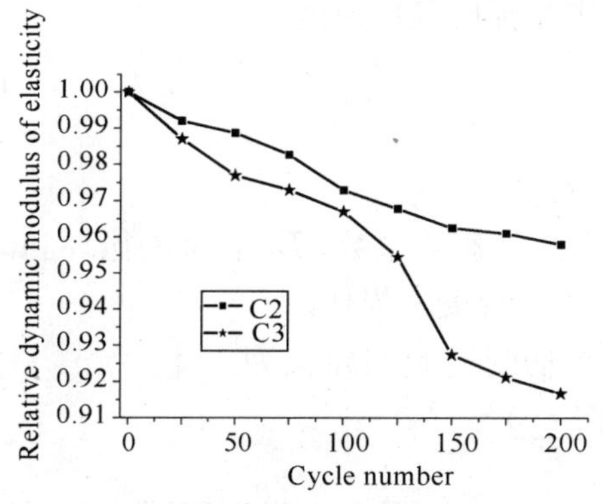


图 8　混凝土在不同试验模式下的相对动弹性模量

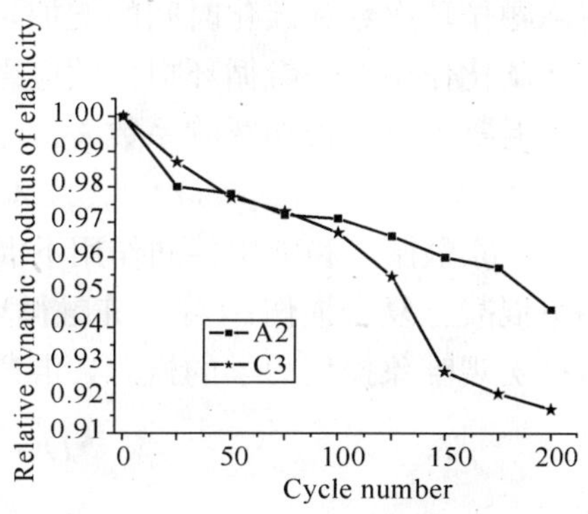


图 9　水胶比 0.45 混凝土在不同试验模式下的相对动弹性模量

(3) 冻融循环—碳化作用下混凝土抗压强度劣化规律

由图 10、图 11 可见，碳化后由于混凝土变得密实抗压强度明显增加，故碳化—冻融试验模式下在不同冻融循环次数时的抗压强度损失率也明显小于冻融—碳化试验模式的抗压强度损失率。冻融—碳化试验的抗压强度损失率明显大于冻融试验的抗压强度损失率。

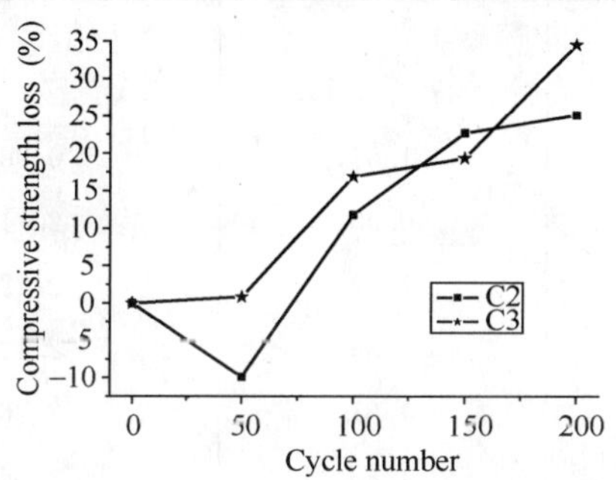


图 10　混凝土在不同试验模式下的抗压强度损失率

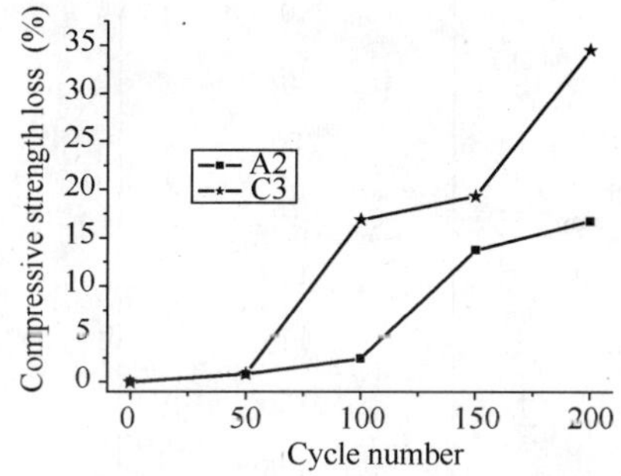


图 11　水胶比 0.45 混凝土在不同试验模式下的抗压强度损失率

(4) 冻融循环和碳化共同作用混凝土中性化规律研究

当碳化周期为 7、14、21、28d 时，碳化试验的混凝土中性化深度与冻融—碳化试验以及碳化—冻融试验下的混凝土中性化深度对比如图 12 所示。由图可见：冻融—碳化试验混凝土中性化深度最小，其次是碳化试验模式下的试件，碳化—冻融试验混凝土中性化深度最大。可以看出，无论是 F50C 试验模式还是 CF50 试验模式，混凝土碳化深度均大于碳化试验下的碳化深度；碳化试验下的碳化速率呈递减，即早期碳化深度发展较快，后期增长较慢；冻融循环和碳化共同作用下的碳化速率呈递增状态，

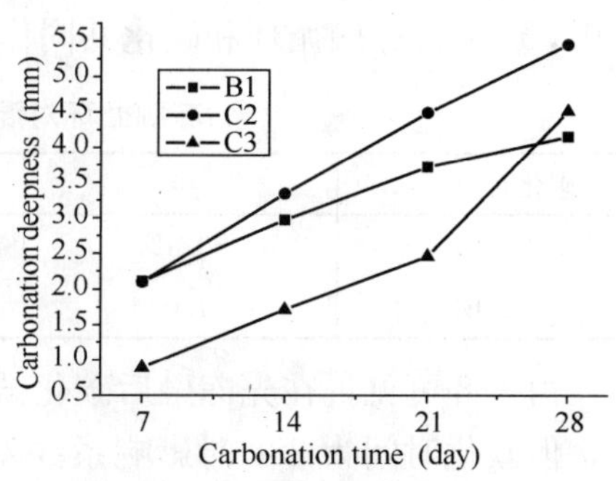


图 12　不同试验模式下随碳化时间的碳化深度变化规律

即随着冻融循环次数和碳化时间的增加，混凝土碳化深度发展速度增加。

（5）碳化作用对冻融循环损伤效应影响分析

引入混凝土复合损伤影响系数 $\lambda_C$，定义如下：

$$\lambda_C = D_F/D_D \tag{5}$$

式中，$\lambda_C$：冻融循环和碳化共同作用下混凝土复合损伤影响系数；$D_F$：冻融循环和碳化共同作用下混凝土复合损伤；$D_D$：冻融循环单独作用下混凝土损伤。

根据宏观唯象损伤力学的概念，混凝土冻融损伤变量 $D$ 可定义为：

$$D(n) = 1 - \frac{E(n)}{E_0} \tag{6}$$

式中，$D(n)$：混凝土经历不同冻融循环后的损伤；$E(n)$：混凝土经历不同冻融循环后的动弹性模量；$E_0$：混凝土未冻融的动弹性模量；$n$：混凝土经历的冻融循环次数。

混凝土冻融循环—碳化共同作用下混凝土复合损伤影响系数 $\lambda_C$ 的计算结果见表 7，可以看出混凝土碳化后损伤有所增加，且水胶比越大复合损伤影响系数越大，即损伤增长越大。进一步说明冻融循环和碳化复合作用下的损伤大于冻融循环单一作用的损伤。

**混凝土冻融循环一碳化作用下的复合损伤影响系数　　表 7**

| 编号 | | C1 | C3 | C4 |
|---|---|---|---|---|
| 冻融循环次数 | 0 | — | — | — |
| | 25 | 0.9 | 0.9 | 1 |
| | 50 | 0.97 | 1.05 | 0.96 |
| | 75 | 1.05 | 0.98 | 2.11 |
| | 100 | 1.32 | 1.12 | 2.12 |
| | 125 | 1.39 | 1.34 | 2.42 |
| | 150 | 1.11 | 1.81 | 2.26 |
| | 175 | 1.08 | 1.83 | 2.06 |
| | 200 | 1.16 | 1.51 | 2.07 |

（6）冻融循环对混凝土中性化影响分析

引入混凝土中性化冻融循环影响系数 $\lambda_F$，定义如下：

$$\lambda_F = x_{F+T}/x_T \tag{7}$$

式中，$x_{F+T}$：冻融循环和碳化共同作用下混凝土中性化深度，mm；$x_T$：碳化深度，mm。

**冻融循环对混凝土中性化影响系数 $\lambda_F$ 计算结果　　表 8**

| 碳化时间(d) | 7 | 14 | 21 | 28 |
|---|---|---|---|---|
| C3/B1 | 0.42 | 0.58 | 0.66 | 1.09 |
| C2/B1 | 1.00 | 1.12 | 1.21 | 1.31 |

由表 8 可见，在先冻融后碳化试验模式下，混凝土中性化冻融循环影响系数 $\lambda_F$ 逐渐递增，但碳化时间为 28d 时影响系数 $\lambda_F$ 才大于 1，在此之前都小于 1。这一现象说明对于先遭受冻融循环的混凝土，在试验初期冻融循环对混凝土中性化起抑制作用，但随着时间的增长，冻融循环对混凝土中性化还是起促进作用。

在先碳化后冻融试验模式下，混凝土中性化冻融循环影响系数 $\lambda_F$ 也呈逐渐递增状态，但与先冻融循环后碳化试验模式下不同的是影响系数 $\lambda_F$ 一直大于 1。表明对于先碳化混凝土，冻融循环对混凝土中性化起促进作用，并且随着时间的增长，促进作用越大。

## 参考文献


[1] 黄可信，吴兴祖等. 钢筋混凝土结构中钢筋腐蚀与保护. 北京：中国建筑工业出版社，1983.

[2] Papadakis V G，Vayens C G，*et al*. Fundamental modeling and experimental investigation of concrete carbonation. ACI Materials Journal，1991，88：363-373.

[3] 牛荻涛. 混凝土结构的碳化模式与碳化寿命分析. 西安建筑科技大学学报，1995，27(4).

[4] 蔡昊. 混凝土抗冻耐久性预测模型. 北京：清华大学博士学位论文，1998.

[5] 许丽萍，吴学礼，黄士元. 抗冻混凝土的设计. 上海建材学院学报，1993，6(2).

[6] 张誉，蒋利学等. 混凝土结构耐久性概论. 上海：上海科学技术出版社，2003.

[7] 金祖权. 西部地区严酷环境下混凝土的耐久性与寿命预测. 南京：东南大学博士学位论文，2006.

[8] 慕儒. 冻融循环与外部弯曲应力、盐溶液复合作用下混凝土的耐久性与寿命预测. 南京：东南大学博士学位论文，2000.

[9] 余红发. 盐湖地区高性能混凝土的耐久性、机理与使用寿命预测方法. 南京：东南大学博士学位论文，2000.

[10] 冯乃谦，邢锋. 混凝土与混凝土结构的耐久性. 北京：机械工业出版社，2004.

[11] 城乡建设环境保护部. GBJ 82—85 普通混凝土长期性能和耐久性能试验方法. 北京：中国标准出版社，1986.

[12] 牛荻涛. 混凝土结构耐久性与寿命预测. 北京：科学出版社，2003.

# 第 14 章 Chapter 14

# 混凝土抗震结构的高性能化及其设计理论*

**吴智深**

（东南大学城市工程科学技术研究院　南京　210096）

**提　要：** 本文在阐述目前混凝土抗震结构性能不足、高性能化必要性和发展趋势的基础上，通过对既有结构抗震加固和新建结构抗震设计的具体问题分析，提出了采用纤维增强复合材料（FRP）实现混凝土抗震结构高性能化的方法及其设计思想。首先分析对比了多种混凝土结构抗震用 FRP 材料的性能，介绍了混杂设计思想和方法对提升 FRP 材料综合性能和损伤控制可设计性的意义和具体方法。在此基础上，总结了多种不同 FRP 材料约束混凝土结构的性能，系统介绍了判断强弱约束的界限值和适用于矩形、圆形等不同截面形状以及强弱约束统一的高精度应力—应变关系模型，并在抗震加固中首次使用了玄武岩纤维，提出了地震作用下不同 FRP 约束混凝土柱承载力、延性的定量计算模型。最后，针对新建混凝土柱结构在地震作用下的安全性能，提出了损伤可控概念和采用钢-连续纤维复合筋（SFCB）作为混凝土结构主要受力部件或结构补强措施实现结构损伤可控及提高可修复性的方法，初步建立了损伤可控结构抗震设计理论。

**关键词：** 混凝土结构；抗震性能；高性能化；纤维增强复合材料（FRP）；加固；损伤可控设计；

# SEISMIC ADVANCEMENT AND DESIGN THEORY OF RC STRUCTURES

Z. S. WU

（International Institute for Urban System Engineering,
Southeast University, Nanjing, 210096）

**Abstract:** Focusing on the deficiency of current seismic design of reinforced concrete (RC) structures, the research trend and advancement of seismic design were summarized and a method of adopting fiber reinforced polymer (FRP) composites for seismic advancement of RC structures was suggested in terms of strengthening of existing structures and new seismic structures. Based on the studies by authors' research team, first, the fundamental properties of commonly used FRP composites were introduced and the method of FRP hybridization addressing enhancement of integrated behavior and seismic performance was proposed. By using the developed hybrid composite materials, the study of confinement of

---

* 资助项目：国家“973”计划（2007CB714200）和国家十一五支撑计划（2006BAJ03B07）资助。

RC columns by FRP sheets comprising the judgmental method of strong and weak confinement limit values and unified stress-strain model, which was applicable to square and circular sections were proposed, respectively. As a first attempt, the application of basalt FRP was adopted for seismic retrofit of concrete structures, and the corresponding quantitative evaluation method of bearing capacity, ductility for concrete columns was developed. Finally, the concept of enhancing recoverability and controllability of reinforced concrete structures is proposed. Moreover, the concept realized by steel-FRP composite bar (SFCB) was also investigated, and the corresponding design theory was developed.

**Keywords**: RC structure, seismic behavior, advancement, fiber reinforced polymer (FRP), strengthening, damage controllable design method

## 14.1 引言

建筑结构的抗震设计起源于 20 世纪初日本最初提出的简单抗震设计思想[1]，之后经过对地震灾害教训的不断总结，发展到目前国际上普遍认可的“大震不倒、中震可修和小震不坏”的指导思想。实践证明，现有抗震设计方法已经取得了长足的进步，建筑也获得了较好的抗震效果。然而，地震中依然有大量的建筑和桥梁结构仍然出现了倒塌和过大的永久残余变形，其中典型的由于构件延性不足或结构选型不当造成的倒塌图 1a 和 b 所示[2][3]，因震后过大残余位移而无法使用的如图 1c[4] 和 d 所示。地震对结构的破坏主要有三种形式，第一是结构延性不足，第二是因弯剪区纵筋的不连续导致的剪切破坏，第三是很多建筑发生过大永久变形，结构丧失可修复性。前两种破坏形式可以通过设计提升结构延性或改善构造措施加以保证，对于第三种结构因变形过大而不可修复的破坏形式切实反映出了目前“三水准”抗震设防思想的不足。目前的抗震设计尽管是规定了三个水准，但是由于规定比较模糊，特别是对“中震可修”没有明确的定量，从而导致在实际设计中很难控制。因此，目前抗震设计实质上还仅以保护生命安全，即“大震不倒”为单一的设计目的。基于目前的抗震设计方法，尽管可以有效地防止建筑物发生造成人生命伤害的倒塌破坏，但是由于结构变形没有得到有效地控制，建筑物不能不废弃重建，往往造成巨大的经济损失。因此，在抗震设计时，为了做到既经济又安全，不仅要防止建筑物发生倒塌坡坏，而且要有效地控制建筑物发生过大的无法修复的永久变形。基于此，从上世纪 90 年代开始，国外学者开展了基于性能（Performance-Based Design）的结构设计研究[5]，这种设计思想除了保证结构具有基本的安全性能外，在使用寿命的各个阶段还需具有良好的使用性能。这种基于性能的设计思想在结构抗震设计中得到了具体应用研究和发展，如规定既有建筑在正常使用阶段受到罕遇地震作用时，应当能够抵抗其最大水平力，并且应当在有限破坏的情况下能保持损伤发生在特定的区域，使结构在中震和大震作用后能够快速修复。日本规范[6]（JSCE 2000）更加明确的规定结构震后的可修复性应该在设计阶段就加以考虑，规定了桥梁墩柱具有可修复的评价指标为残余变形不超过柱高的 1%。上述研究表明，以高安全性能和损失可控可恢复性能等为特征的抗震高性能设计已成为混凝土结构抗震设计的重要趋势。

图1 典型震害

(a) 阪神地震中延性不足导致桥墩倒塌破坏[2]；(b) 汶川地震中桥墩倒塌破坏[3]；(c) 阪神地震中结构残余变形过大形成无法修复的损伤[4]；(d) 汶川地震中结构变形过大形成不可恢复损伤

面对大量抗震性能明显不足的既有结构，如何提高抗剪能力、极限变形能力及其综合抗震性能成为混凝土柱加固中亟待解决的问题。传统的柱抗震加固方法有加大截面法和钢套管法，加大截面法会增加结构截面尺寸及自重，减小结构使用空间；钢套管法耐久性较差，后期的维护费用高，两种方法都会引起结构刚度的增加，并且施工工艺复杂。纤维增强复合材料（FRP）以其轻质、高强、抗腐蚀及施工方便等特点，得到广泛认同与关注。利用FRP横向约束能够有效提高柱的抗剪承载力与抗震性能，已经得到实践证实：洛杉矶一幢7层旅馆，在1992年发生的7.5级Lands地震中出现了严重的倾斜裂缝，后采用GFRP进行了加固，在加固后几星期又遭遇Northridge地震却丝毫无损[7]。相关研究也证实了FRP加固可以提高柱在地震中吸收能量的能力，加固柱的滞回曲线饱满，强度退化减缓。

对于新建的重要结构在大震作用下的弹塑性阶段应能够抵抗最大水平力，同时具有较小的损伤以便能够快速修复。因此，最近的基于性能（性态）的设计规范开始注意考虑预测结构在地震作用下的残余变形，新型的结构系统和新材料也开始被引入到结构抗震设计中[8]-[15]。

国内外学者对实现结构震后较小残余位移的方法作了一些探索。在Iemura等人的研究中[9]，提出了一种无黏结筋增强混凝土（UBRC）结构作为高性能抗震结构，这种结构在荷载位移曲线上表现出稳定的屈服后刚度。UBRC结构是由传统的混凝土桥墩和无黏结高强筋构成。通过在筋材底部设置间隙的方式来使得筋材的弹性活动范围转变到大变形的区域范围。为了在大变形阶段都有稳定的屈服后刚度，就应当能够保证筋材处在弹性反应阶段。因此，不仅采用了高强材料的筋材，而且采用了无黏结的方法。

Sakai 等人[10]提出在钢筋混凝土柱的截面中央配置无黏结预应力筋材可以得到和传统柱子在荷载作用下类似的恢复力。但是卸载残余位移却小很多。为了保证预制混凝土桥墩的震后可修复性，Billington 和 Yoon[11]提出一种新的分块预制混凝土桥墩体系，采用竖向无黏结后张拉筋材和在塑性铰区使用预制工程型水泥基复合材料（ECC）块。ECC 相较于传统混凝土和许多的纤维增强混凝土展示出高的受拉延性，受拉硬化性能和耗能能力，因此在预制混凝土桥墩体系中的塑性铰区域采用 ECC 区段可以提高耗能能力，并且由于在 ECC 区段均匀分布的微裂缝，可以使得柱子可以具有更大的变形能力。

Kawashima 等人[12]的研究表明结构残余位移比显著依赖于结构的二次刚度比 $r$（屈服后刚度和初始弹性刚度之比）。同时，当桥墩的二次刚度比（屈服后刚度与初始刚度之比）超过 5%将会显著提高残余位移响应的稳定性[13]，当结构残余位移过大时，Christopous 等人[14]建议可采取的方法有使用具有硬化特征的材料或者设计具有稳定二次刚度的截面。

为实现对提高已建和新建混凝土结构抗震性能，作者研究团队利用 FRP 材料开展了一系列的研究，包括一方面对既有结构利用纤维复合材料进行抗震约束加固，柱脚嵌入式加固；另一方面，对新建结构利用钢和 FRP 的复合实现截面层次上的稳定二次刚度，实现较小的震后残余位移，进而提高结构震后的使用性能。本文将基于上述研究成果，阐述利用 FRP 材料实现混凝土结构抗震高性能化的设计思想、方法和试验验证。

## 14.2　结构增强 FRP 材料及高性能开发

纤维复合材料的商业化应用始于 20 世纪 70 年代，起初应用于体育用品领域。20 世纪 80 年代后 FRP 材料逐渐开始在土建交通领域得到应用，到 20 世纪 90 年代中期因日本阪神地震对土建交通基础设施造成了极大的灾害，纤维复合材料迅速发展成为结构抗震加固的重要手段。FRP 材料从初期的结构加固改造，到作为建筑结构材料应用于新建结构；从开始的 FRP 纤维布/板，到多种形式的纤维复合索、网格、型材，应用范围和应用形式越来越广泛。FRP 材料的发展为研究开发高性能土木工程结构设施和既有工程设施的加固补强及功能提升提供良好的选择途径。

### 14.2.1　单一纤维复合材料

常用的 FRP 材料由碳纤维（Carbon fibers）、PBO 纤维（Poly-p-phenylenebenzobis-thiazole）、玻璃纤维（Glass fibers）、玄武岩纤维（Basalt fibers）或芳纶纤维（Aramid fibers）分别与基体材料（如环氧树脂）含浸硬化后复合形成。相对于传统建筑材料（钢材，混凝土，木材）FRP 材料具有优越的力学（高比强度）和物理化学性能（轻质、耐腐蚀等），及其他优秀特性，如绝缘（除碳纤维弱导电）、耐高低温等，其密度只有钢材的三分之一到四分之一，但拉伸强度为普通低碳钢的 5 倍以上，与预应力钢丝/索相当甚至更高。碳纤维 FRP（CFRP）在各种纤维复合材料中，拥有最为突出的力学性能和化学稳定性，典型的 CFRP 抗拉强度为 3400MPa，弹性模量达到 230GPa，密度只有 1.8kg/cm$^2$，并且能够抵抗酸碱盐紫外线等各种环境腐蚀，耐高温达 600℃。PBO 纤维不仅具有和碳纤维相似甚至更高的力学性能，而且具有良好的冲击能力吸收性能，这也意味着 PBO 纤维作为干丝具有更高的可张拉性能。玻璃纤维 FRP（GFRP）力学和化学性能在常用各种纤维复合材料中相对较低，但因其相对低廉的价格得到广泛的应用，常用的 E 玻璃纤维 FRP 的抗拉强度为 1500MPa，弹性模量为 80GPa，密度为 2.6kg/cm$^2$，GFRP 虽然能够抵抗酸，盐和紫外线等腐蚀作用，但对碱作用敏感，并且适用温度范围相对较小－60～250℃，在长期荷载下会出现应力破断现象，应用面有一定的限制。玄武岩纤维 FRP（BFRP）是一种以纯天然火山岩（玄武岩为主）为原料，在 1450～1500℃高温熔融后拉丝而成的连续纤维。连续玄武岩纤维在我国已经能产业化工业化生产。该纤维具有高性价比、耐高低温（－269～650℃）、抗水损害性能好、耐紫外线、电绝缘、纤维表面呈极性、纯天然环保、防火阻燃等特点。从力学性能角度，玄武岩纤维比 E 玻璃纤维具有更高的抗拉强度（2100MPa）和弹性模量（91GPa）以及更广的适用温度范围。芳纶纤维 FRP（AFRP）的力学性能介于 CFRP 和 BFRP 之间，但长期荷载作用下会出现应力破断现象，而且化学性能中对紫外线很敏感，加上价格和 CFRP 接近，因此实际应用不如 CFRP 和 GFRP 广泛。上述几种纤维材料和其他几种土木领域应用的纤维材料典型的应力—应变关系见图 2 所示，可见相对于传统钢材，纤维复合材料具有明显的强度优势，但部分纤维材料的刚度不足或延伸率不足也是需要避免和改善的重要问题。

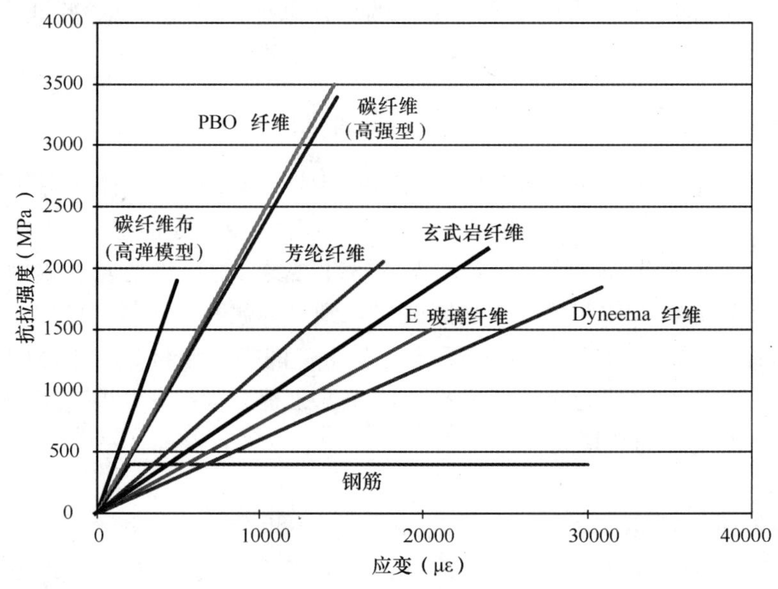


图 2　几种常用纤维复合材料力学性能

### 14.2.2　混杂纤维复合材料

为改善部分纤维材料的刚度或延伸率不足的问题，更充分的在混凝土结构抗震设计中发挥作用，作者研究团队首先提出了基于纤维复合材料混杂设计的思想，以改善各种纤维不足，达到综合高强度、高刚度和高延性的目标，如图 3 所示。通过高弹性模量、高强度和高延性纤维按一定的比例混杂，从而实现混杂纤维复合材料整体高初始刚度，高强度和充足的延性。利用这种混杂纤维复合材料增强混凝土结构，可以实现高开裂后刚度、高屈服荷载、高承载力和充足的延性，如图 4 所示[16]。

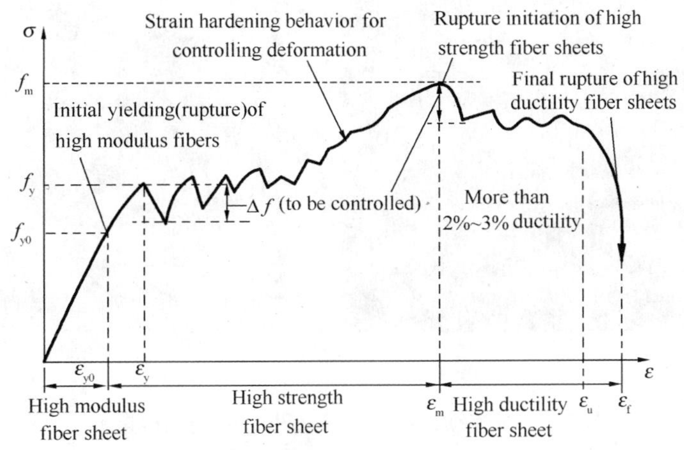


图 3　纤维混杂设计应力—应变关系图

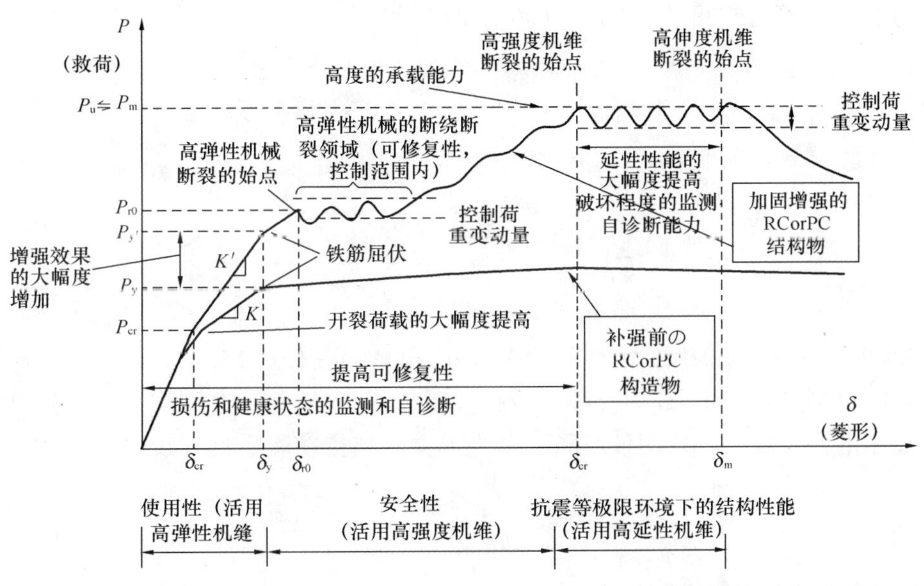


图 4　混杂纤维复合材料增强混凝土结构荷载—变形图

### 14.2.3　混杂钢-连续纤维复合材料

除多种纤维间的相互混杂外，考虑到 FRP 强度高、弹模低、延性差、耐久性好、重量轻和钢材强度低、弹模高、延性好、耐久性差、重量重等特点，两者互补性极强，复合后可以扬长避短，得到综合性能更高的钢-连续纤维复合材料，如图 5 所示。另外，线弹性的 FRP 与弹塑性的钢材复合还可以带来力学性能上的变化，如得到的钢-连续纤维复合筋（SFCB）具有稳定的二次刚度。其特点是能动地控制结构或构件的屈服后刚度（二次刚度）、震后残余变形、极限状态的破坏模式以及结构系统耗能机理，为实现“大震”不倒乃至“大震”可修的定量化设计提供了有效的途径。

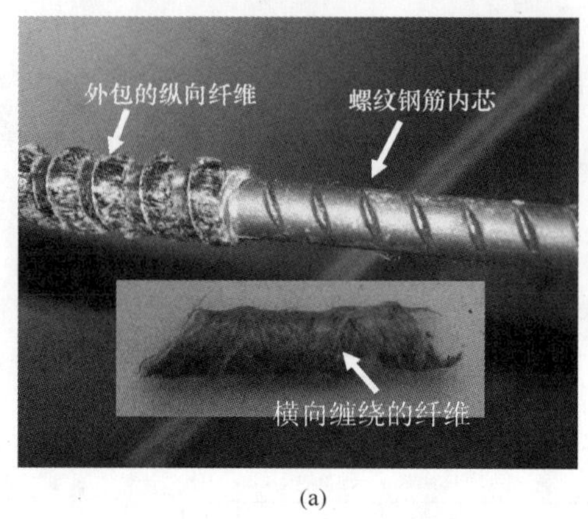

(a)

(b)

图 5　钢—连续纤维复合筋（SFCB）截面构造[17]

（a）侧视图；（b）断面图

钢筋和 FRP 复合实现稳定二次刚度应力—应变曲线如图 6a 所示。基于复合法则，SFCB 力学性能计算方法可从公式（1）和（2）[17]得到：

$$\sigma_{sf}=\begin{cases}E_{\mathrm{I}}\varepsilon_{sf}, & 0\leqslant\varepsilon_{sf}<\varepsilon_{sfy}\\ f_{sfy}+E_{\mathrm{II}}(\varepsilon_{sf}-\varepsilon_{sfy}), & \varepsilon_{sfy}\leqslant\varepsilon_{sf}\leqslant\varepsilon_{sfu}\\ f_{sfr}, & \varepsilon_{sfu}<\varepsilon_{sf}\end{cases}\tag{1}$$

$$E=\begin{cases}E_{\mathrm{I}}=(E_sA_s+E_fA_f)/A, & 0<\varepsilon\leqslant\varepsilon_y\\ E_{\mathrm{II}}=(E_fA_f)/A, & \varepsilon_y<\varepsilon\leqslant\varepsilon_{fu}\\ E_{\mathrm{III}}=0, & \varepsilon_{fu}<\varepsilon\leqslant\varepsilon_{s,max}\end{cases}\tag{2}$$

其中 $A=A_s+A_f+A_r$ 为 SFCB 截面积，$A_s$，$A_f$ 和 $A_r$ 分别为 SFCB 中钢筋、纤维和树脂的截面积。往复拉伸典型 SFCB 的荷载—应变关系如图 6b 所示，在同样的卸载应变下，SFCB 具有较小的残余应变，从而使 SFCB 增强混凝土结构在震后具有较小的残余变形。

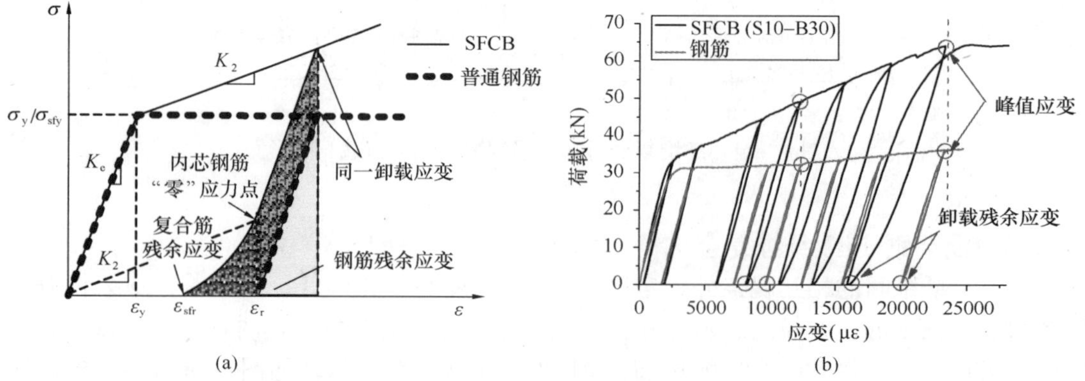


图 6　SFCB 残余应变示意

（a）SFCB 残余应变示意；（b）SFCB 和普通钢筋往复拉伸力学性能比较

## 14.3 通过 FRP 加固混凝土结构的抗震性能

如何提高既有柱的抗剪能力、加强其侧向约束、改善其抗震性能成为柱加固中亟待解决的问题。各国研究者对 FRP 约束混凝土柱抗震性能，做了大量的工作，其中以拟静力试验的开展最为广泛：Priestley 等[18]（1995）研究了 GFRP 包裹圆柱、矩形柱抗震加固的有效性，分析了钢筋搭接、原始抗剪抗弯能力对加固性能的影响，并给出了提高抗剪和抗弯承载力的设计程序；Saadatmanesh[19]（1997）等对 FRP 加固震损柱进行了试验研究，试件原型为延性较差的桥柱，试验中先加载到一定的损伤程度，然后用 FRP 加固后再进行模拟地震试验，结果表明经过修复加固的震损柱相对于未加固柱的承载力和位移延性都有所提高；Seible 等[20]（1997）提出了针对缺陷柱不同破坏模式的加固设计程序及原则，给出了详细的设计算例，并通过了试验验证，该程序得到广泛地应用；Xiao 等[21]（1997）根据约束混凝土粘结应力公式提出了防止搭接破坏的 FRP 加固要求，并进行了 GFRP 预制复合壳材的抗震加固效果的研究[22]，证实 FRP 可以阻止剪切破坏、提高滞回性能，FRP 的加固对柱的刚度影响很小；Sheikh 课题组[23]-[25]对 FRP 加固长柱抗震性能进行了全面广泛的试验，系统地研究了圆柱与方柱两种截面形式，考虑了轴压比、FRP 类型、FRP 厚度及加固方式等参数影响，认为轴力水平对加固柱的总体性能有着重要的影响；Harajli 等[26]（2004）研究了配筋率、FRP 数量等因素对 FRP 加固矩形柱抗震性能的影响，结果显示 FRP 约束可以明显增强混凝土的粘结强度，减少柱端塑性铰区混凝土的开裂与粘结破坏，作者把构件强度的提高、抗震性能的改善归因为约束对粘结强度的提高，并定义了约束指标，通过约束指标来确定 FRP 的加固量；Li[27]（2003）等先对 1 个比例为 0.4 抗剪能力不足的圆柱往复加载至脆性剪切破坏后，进行修复和 FRP 加固，对比试验表明，合适的修复与加固可以转变破坏模式，显著提高柱的抗震性能；在国内，FRP 抗震加固也得到了普遍的关注[28]-[30]，作者研究团队[31]对采用 CFRP 加固的 11 根钢筋混凝土工字形矩形截面柱（其中 6 根用碳纤维布加固）进行了试验，研究了试件在低周反复荷载作用下加固前后混凝土柱的破坏形态、极限承载力、滞回曲线和延性系数，并对影响 CFRP 加固效果的因素（柱轴压比、配箍率、CFRP 黏贴方式和层数）进行了研究，表明原有柱的轴压比和配箍率对加固效果影响很大，CFRP 的黏贴方式取决于补强的目的是抗剪补强还是延性补强，CFRP 的粘贴层数对构件的受剪承载力影响不大但对构件的延性却影响很大。

### 14.3.1 FRP 约束混凝土本构关系模型

对既有结构利用纤维复合材料进行抗震约束，FRP 约束混凝土的轴向受压性能研究是研究 FRP 约束混凝土柱抗震性能的基础。将 FRP 包裹在柱体表面可以为混凝土提供侧向约束，使混凝土处于三向受力状态，从而提高混凝土柱的轴压强度和延性性能。根据加固目标的不同，可以选择不同种类和不同用量的 FRP。在本研究中，FRP 包裹主要采用两种工法：片材湿黏法，素纱连续纤维缠绕法。

目前关于 FRP 约束混凝土的轴压性能研究已经较多。其中，大多数本构模型都是在

Richart[32]对流体静水压力下混凝土的应力应变研究的基础上发展起来的。Karbhari 和 Gao 模型[33]假设应力—应变曲线为双线性，通过简化的综合分析，建立了两种本构关系模型。Samaan et al. 模型[34]研究了外包 FRP 和 FRP 管对混凝土的约束作用，并建立了双线性模型。第二段直线段斜率为约束纤维弹性模量和未约束混凝土强度的函数。Hosotani 和 Kawashima 模型[35]通过试验回归，认为圆柱应力—应变曲线第一段为曲线，第二段为直线段。Lam 和 Teng 模型[36]为设计模型，第一段抛物线，第二段为直线。由于 FRP 层的折角和应变分布不均匀，FRP 的实际断裂应变小于单向拉伸试验的结果，所以提出了 FRP 极限应变的折减系数，并由实际断裂应变值推断本构关系。Wu et al. 模型[37]为三折线模型，主要针对有应变强化段的应力—应变关系。

圆柱和矩形柱（包括方柱）由于混凝土有效约束面积的差异，造成 FRP 对混凝土核心约束效果的不同。迄今为止，试图将圆柱和矩形柱的约束效果归纳在一起的应力—应变模型的精度都很不理想。所以，将两种柱的性能进行分别研究是非常有必要的。

14.3.1.1　FRP 约束混凝土圆柱性能

课题组吴智深、吴刚、魏洋、Fahmy 等对 FRP 约束混凝土圆柱进行了试验研究，并提出了相应的应力—应变模型。通常 FRP 对混凝土的约束可以由约束强度 $f_l$ 和约束刚度 $E_l$ 两个参数来表示和衡量。其计算式为 $f_l = 2f_f t_f / D$，$E_l = 2E_f t_f / D$。其中 $E_f$、$f_f$、$t_f$ 分别为 FRP 的弹性模量、极限抗拉强度和厚度；$D$ 为圆柱直径。

研究发现，基于 FRP 约束量和未约束混凝土强度的不同，混凝土柱的应力—应变曲线也呈现了不同的形态。如图 7 所示，若曲线中包括 AC 软化段则为弱约束混凝土柱，若曲线中包括 BD 上升段则为强约束混凝土柱。这种强弱约束的差异则导致了相应模型的不同。

（1）强弱约束的判断

为判断混凝土柱的实际加固效果，基于 FRP 约束混凝土圆柱强、弱的不同模式，必须提供一个判断运用不同模型和结论的依据。吴刚等以 FRP 约束强度 $f_l$ 和未约束混凝土强度 $f'_{co}$ 的比值 $f_l / f'_{co}$ 为变量，当该变量经由临界值 $\lambda$ 逐渐增大时，约束形式由弱向强转变。经过试验研究得出，对普通弹性模量 FRP（$E_f \leqslant 250\text{GPa}$）$\lambda$ 取为 0.13，对高弹模 FRP（$E_f > 250\text{GPa}$），则 $\lambda = 0.13\sqrt{250/E_f}$。其中 $E_f$ 为 FRP 弹性模量，单位为 GPa。

（2）FRP 弱约束混凝土圆柱的应力—应变关系

针对实际结构中，特别是一些桥墩中柱截面尺寸大而 FRP 量少的情况，吴刚等对应力—应变关系有软化段的 FRP 约束混凝土柱进行了研究。该研究采用了高强度 CFRP、高弹模 CFRP 和高延性的 Dyneema FRP 三种材料对 150mm×300mm 的混凝土圆柱体进行对比加固。

吴刚等认为峰值应力和应变是 FRP 侧向约束刚度和混凝土弹模之比及未约束混凝土强度的函数；极限应力为侧向约束强度与未约束混凝土强度之比的函数；而极限应变则与侧向约束强度和未约束混凝土强度比，为约束混凝土初始应变以及 FRP 的种类有关。根据对试验数据的回归分析，峰值应力应变和极限应力应变的计算方法被分别确定。

在计算峰值和极限两个关键点的基础上，应力—应变模型Ⅰ和模型Ⅱ被分别提出，如图 8。模型Ⅰ为全曲线模型，模型Ⅱ则由上升抛物线和下降直线构成。

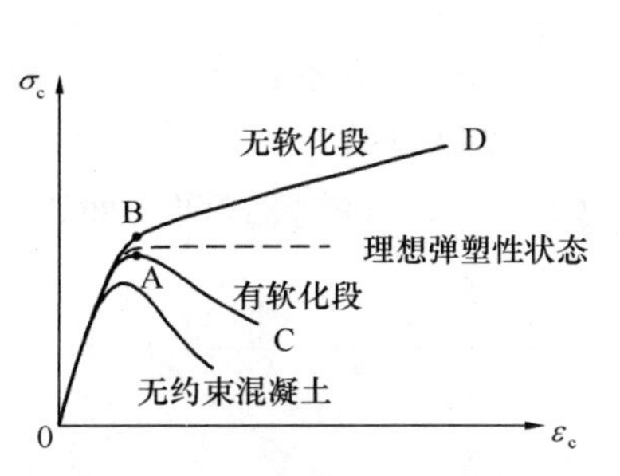


图 7　FRP 约束混凝土圆柱典型应力—应变关系曲线

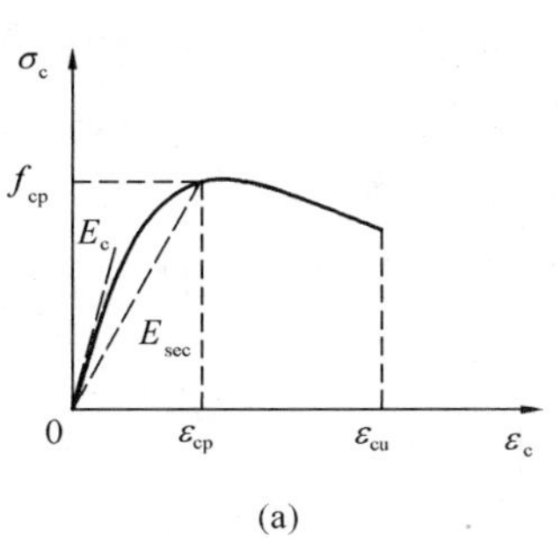
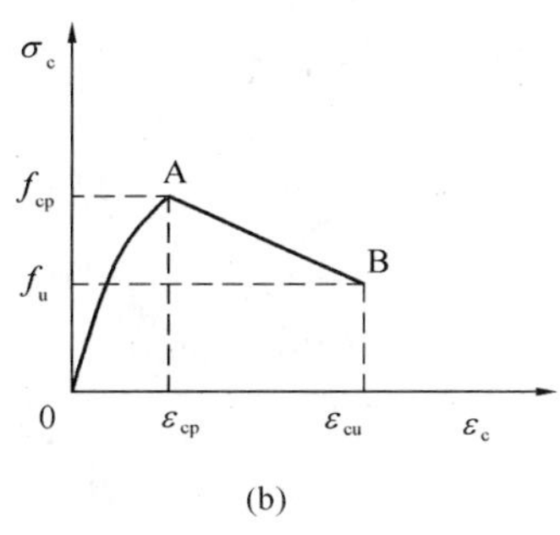


图 8　模型Ⅰ和模型Ⅱ应力—应变关系
(a) 模型Ⅰ；(b) 模型Ⅱ

该研究详情可分别参考吴刚等发表的论文，FRP 约束混凝土圆柱有软化段时的应力—应变关系研究[38]，以及 FRP 约束混凝土圆柱无软化段时的应力—应变关系研究[39]。

(3) FRP 强约束混凝土圆柱的应力—应变关系

Ⅰ. 吴刚模型

吴刚等通过对强约束效应的分析，发现外贴 FRP 和 FRP 管两种约束形式的不同，导致了其对混凝土约束效应的不同。因此，他们建议将这两种约束方式进行区分讨论。

数据分析发现 FRP 约束混凝土圆柱的极限泊松比趋向一定值。经过分析推断，这主要与 FRP 形式、侧向约束强度和未约束混凝土强度有关。于是建立了针对不同约束形式的泊松比计算方法，并在此基础上根据应变相容原理确定了混凝土极限应变计算公式。

此外，外贴 FRP 加固中发现，随着纤维布层数的增多，强度提高系数随之降低，则用多项式回归比其他学者所使用的线性回归所得到极限强度应当更为精确。

在上述分析的基础上，吴刚等建立了 FRP 约束混凝土柱无软化段的应力—应变模型，见图 9。该曲线分为三个阶段：阶段 1 类似于无约束混凝土初始阶段应力—应变关系曲线；阶段 2 为在无约束混凝土强度附近的软化和过渡区域；阶段 3 为 FRP 充分发挥作用阶段，其应力—应变关系曲线近似于直线。具体分析过程请参考吴刚等发表的论文，FRP 约束混凝土圆柱无软化段时的应力—应变关系研究[39].

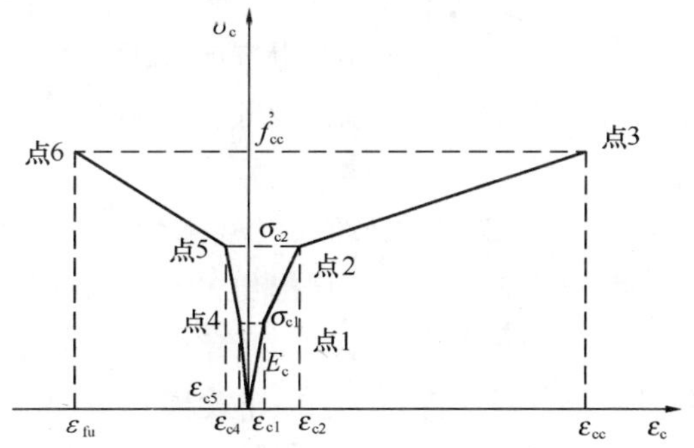


图 9　三折线应力—应变模型

Ⅱ. Fahmy 和 Wu 模型[40]

Fahmy 和 Wu 分析了现有的本构关系模型，发现他们普遍存在以下问题：(1) 均认为在相同约束刚度下，第二阶段斜率 $E_2$ 与 FRP 种类有关，(2) 对第二阶段斜率 $E_2$ 的评估精度不高。

然而，Fahmy 和 Wu 认为对于不同的 FRP 种类，在混凝土产生一定的轴向应变时，相同的约束刚度导致了相同的约束强度增量，从而导致相同的轴向应力增量，所以第二部

分的斜率 $E_2$ 理应相同。经过验证证明，现有模型基本不符合这一力学理念。于是针对强约束混凝土圆柱，提出了一种具有更高精度的应力—应变模型如图 10 所示，该模型为两段式模型。第一段为抛物线，采用 Lam 和 Teng 模型[36]的第一段抛物线形式。第二段为直线，斜率 $E_2$ 为约束刚度与混凝土强度的函数，计算公式由 Samaan[34]的公式演变而来，并通过回归分析重新定义了常数项。

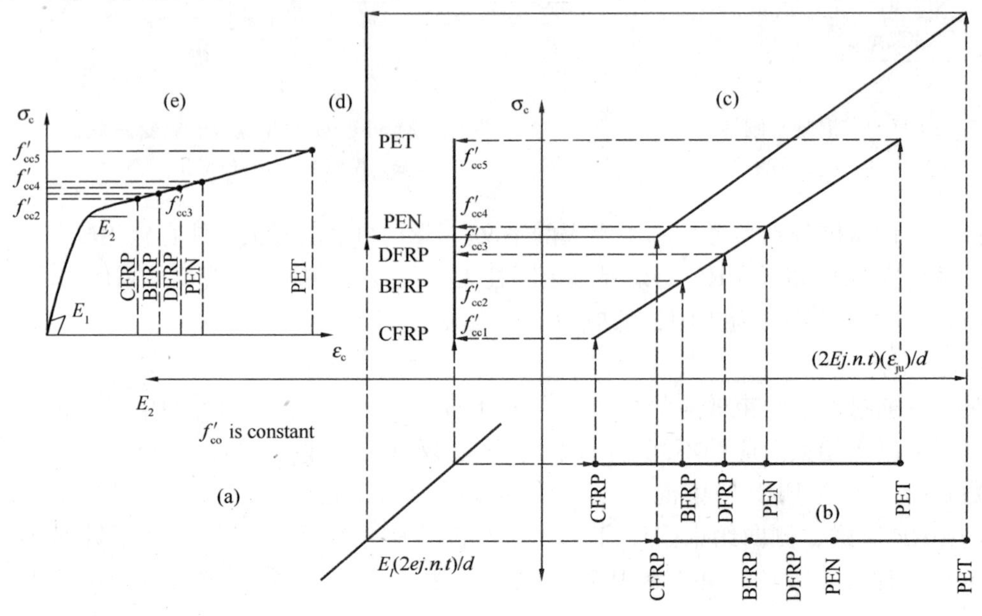


图 10 Fahmy 和 Wu 模型[40]

在与表现较好的 Lam 和 Teng 模型和 Wu et al. 模型比较后发现，对于不同混凝土强度，FRP 种类和 FRP 层数，该模型更吻合实际试验曲线，并对应变强化段刚度 $E_2$ 和极限点的确定具有显著的精度优势，如图 11 所示。在 FRP 约束混凝土圆柱有应变强化效应的范围内，该本构关系模型预测精度最高，详细分析过程请参考作者研究团队发表的论文[40]。

14.3.1.2 FRP 约束混凝土矩形柱性能

由于矩形柱中 FRP 提供的约束是不均匀的，因而 FRP 对矩形的约束作用也大大降低，混凝土的抗压强度也随着截面变化。对于非圆形截面约束混凝土的通常处理办法是得到基于平均轴向力的等效模型。现有模型中对等效混凝土圆柱约束强度或刚度的确定主要遵循两大思路。一是设定直径为 D 的等效圆柱，计算同样厚度的 FRP 套箍对该圆柱提供的侧向约束力。二是为避免采用确定等效圆柱直径不同方法造成的差异，通过相同 FRP 体积配置率直接来确定约束强度或者约束刚度。然后对该约束强度/刚度乘以形状系数 $k_s$ 进行折减，从而等同于圆柱体形式得到峰值点与极限点的应力应变情况。

(1) 强/弱约束的判断

类似于混凝土圆柱，FRP 对混凝土矩形柱也存在强/弱约束之分。由于截面的不同，应当对强/弱约束的判断方法进行修正或重新定义。

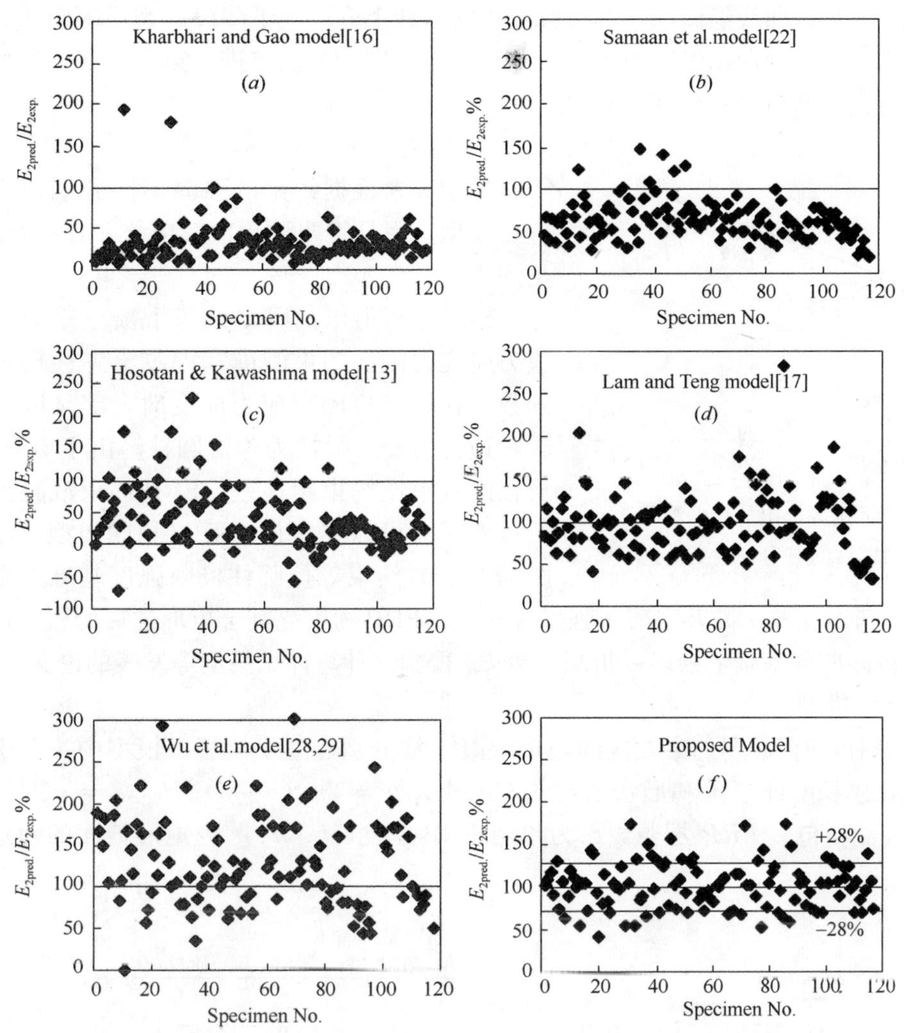


图 11　不同模型的本构关系比较[40]

Ⅰ. 吴刚方法

2.1.1 中吴刚等提出的方法仍然适用于矩形混凝土柱，只需对将计算约束强度 $f_l$ 过程中的 FRP 体积配置率 $\rho_f$ 设为矩形柱中实际配置率。

Ⅱ. 魏洋方法

魏洋等提出了新的形状系数计算形式 $k_s = r(b+h)/bh$ 。其中，$r$ 为拐角半径；$b$ 和 $h$ 分别为矩形截面宽度和高度。在此基础上进一步定义了函数 $m = k_s\rho_f f_{fu}/f'_c$ 用于判断 FRP 对混凝土矩形柱强/弱约束效应。其中，$f_{fu}$ 为 FRP 抗拉强度，$f'_c$ 为未约束混凝土强度。通过试验分析，可以取 $m=0.2$ 为临界状态。当 $m \geqslant 0.2$ 时，FRP 对矩形截面混凝土的约束为强约束；当 $m < 0.2$ 时，FRP 对矩形截面混凝土的约束为弱约束。

(2) FRP 约束混凝土矩形柱的应力—应变关系

目前对于 FRP 约束矩形柱的应力—应变模型研究存在两种思路。吴刚等认为可以将强约束与弱约束两种效应统一在一个应力—应变关系内。这两种模型的前半段由相同的抛物线组成，并且该曲线经由同一个过渡点进入直线上升段或下降段。魏洋等则认为两种约束效应模型的第一段抛物线应当不同，并且经由不同的过渡点进入第二阶段，所以强约束与弱约束应当分开讨论并分别建立模型。

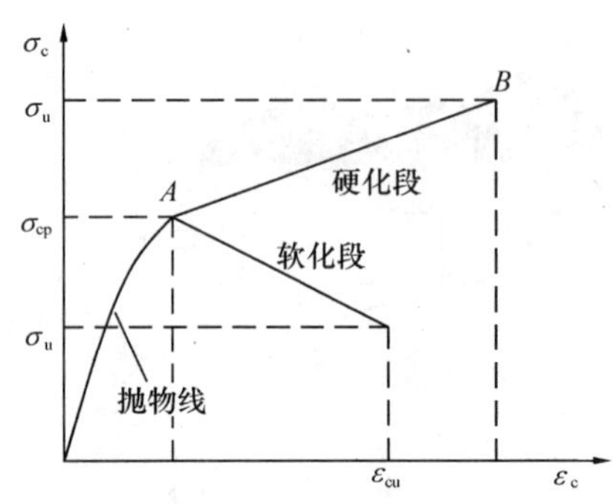


图 12 吴刚等提出的矩形混凝土柱约束后的应力—应变模型

Ⅰ. 吴刚模型

吴刚等未根据约束的强弱不同分别建立应力—应变模型，而是将两种约束效应统一在一个模型中，见图 12。

经过分析得出，FRP 约束混凝土矩形柱转折点应力应变为侧向约束刚度和混凝土弹性模量之比的函数。在计算极限点应力应变时，定义以矩形截面较长边为直径的圆柱为等价圆柱。由于 FRP 约束矩形柱的效率比约束圆柱差，FRP 约束混凝土矩形柱的极限应力可以在等价 FRP（相同类型、厚度及间距的 FRP）约束等价圆柱极限强度基础上乘以折减系数 $k_{u1}$ 得到，$k_{u1}$ 与 $r/h$ 及混凝土强度等有关。FRP 约束混凝土矩形柱后的极限应变可在等价圆柱极限应变基础上乘以一折减系数 $k_{u2}$ 得到。计算详见吴刚等发表的论文

Ⅱ. 魏洋模型

魏洋等对具有强/弱约束效应的矩形混凝土柱分别建立了模型。模型中主要引入两个关键参数，体积配纤率 $\rho_f$ 和形状系数 $k_s$，公式为 $\rho_1 = 2nt_f(b+h)/bh$，$k_s = r(b+h)/bh$。式中，$n$ 为侧向包裹 *FRP* 层数，$t_f$ 为单层 FRP 厚度，$b$、$h$、$r$ 为矩形截面的短边、长边及倒角半径。

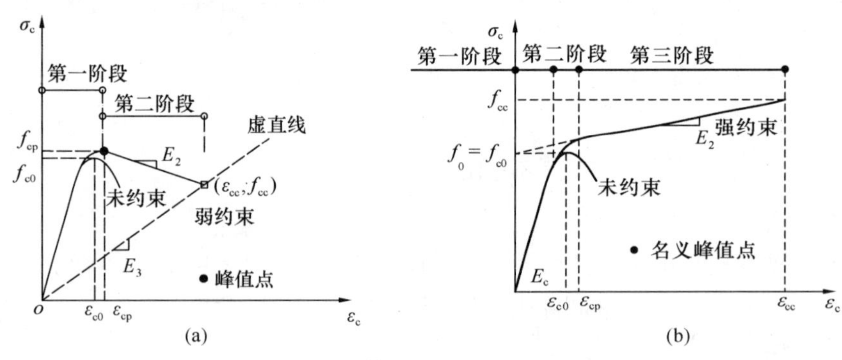


图 13 FRP 约束矩形混凝土柱模型模型

（a）模型Ⅰ-弱约束；（b）模型Ⅱ-强约束

①模型Ⅰ-弱约束模型

如图 13（a）应力—应变曲线关系由两阶段组成，第一阶段为二次抛物线，第二阶段为下降斜直线，二者以约束后的峰值点（$\varepsilon_{cp}$，$f_{cp}$）为界；2）相同混凝土强度不同 FRP 约束条件下的极限点位于同一条通过原点的虚直线上，该直线刚度（$E_3$）与 FRP 约束条件无关；3）软化段直线至其与虚直线的交点（$\varepsilon_{cc}$，$f_{cc}$）结束；4）FRP 对内部混凝土的约束效果通过峰值点、软化段坡度（$E_2$）及极限点位置来体现。详情请参照魏洋等发表的论

文，FRP 约束混凝土矩形柱有软化段时的应力—应变关系研究[41]。

②模型Ⅱ-强约束模型

本模型认为应力—应变关系曲线可由抛物线和直线两部分组成，抛物线起始点的刚度取无约束混凝土刚度 $E_c$，抛物线的终点与直线相切于 $\varepsilon_{cp}$，如图 13（b）。关键点应力、应变的计算请参照魏洋等发表的论文，FRP 强约束混凝土矩形柱应力—应变关系的研究[42]。

### 14.3.2　FRP 抗震加固混凝土桥墩

作者研究团队通过对 FRP 约束加固混凝土柱来保证结构在中震乃至大震后具有快速修复能力进行了研究。使用多种 FRP 材料（CFRP、DFRP 和 BFRP）对 RC 圆柱与方柱进行试验研究，研究参数包括轴压比、剪跨比、加载方式和 FRP 用量（图 14）。

图 14　试验相关情况（左为圆柱，右为方柱）[43][44]

图 15（a）～（k）给出了各试件的 FRP 加固前后的试验 $P$-$\Delta$ 滞回曲线。由于试验装置原因，试件 S-1.5C 部分循环加载重复，试验过早破坏，未能达到应有的延性，试件 S-2.5D 由于一侧锚杆破坏，试验中止。从图 15 中可以发现：(1) 试件在屈服以前，每次循环的残余变形很小，屈服以后，随着侧向位移、循环次数的增加，残余变形越来越大；(2) 在达到峰值荷载以后，承载能力逐渐退化，随着 FRP 加固量的增加，退化的趋势趋于缓慢，耗能能力及延性得到提高；(3) 未加固柱过早发生剪切粘结破坏，滞回环瘦小，随着加固量的增加，破坏模式逐渐从剪切粘结破坏转变为弯曲破坏，滞回环变得越来越饱满，越来越稳定，FRP 加固可以有效改善方柱的抗震性能；(4) DFRP 加固能够达到与 CFRP 加固相同效果，2.5 层 DFRP 加固试件 S-2.5D 的滞回性能表现与 1.0 层 CFRP 加固试件 S-1.0C 相似，体现在滞回环包围面积及刚度退化两个方面，但 S-1.0C 抗剪强度低于 S-2.5D，两者最终发生不同的破坏模式，前者为延性剪切破坏，后者为弯曲破坏，后者在提供相近的侧向约束刚度的同时具有更大的侧向约束强度，因此，约束试件表现出更好的延性；(5) BF 丝束缠绕对柱的抗震加固与 CFRP 同样有效：S-BF 表现出比 S-4.5C 更好的延性及耗能能力，其滞回曲线形状在中部更宽广，“捏拢”现象较轻，表明 BF 丝束缠绕加固柱的滞回曲线要优于 CFRP 布材包裹加固情况。

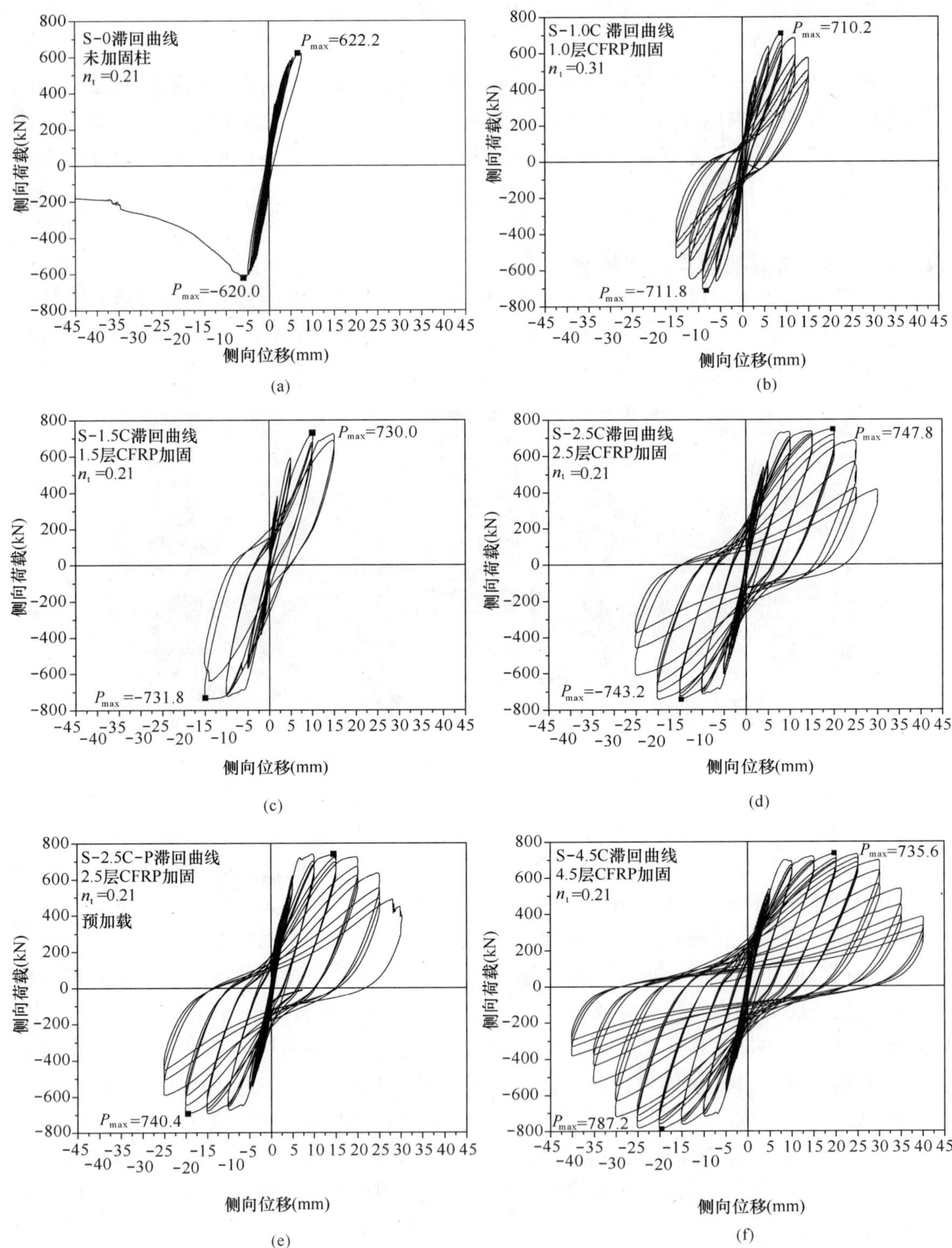


图 15 柱 FRP 加固前后的试验曲线[44]（一）

(a)S-0；(b)S-1.0C；(c)S-1.5C；(d)S-2.5C；(e)S-2.5C-P；(f)S-4.5C；

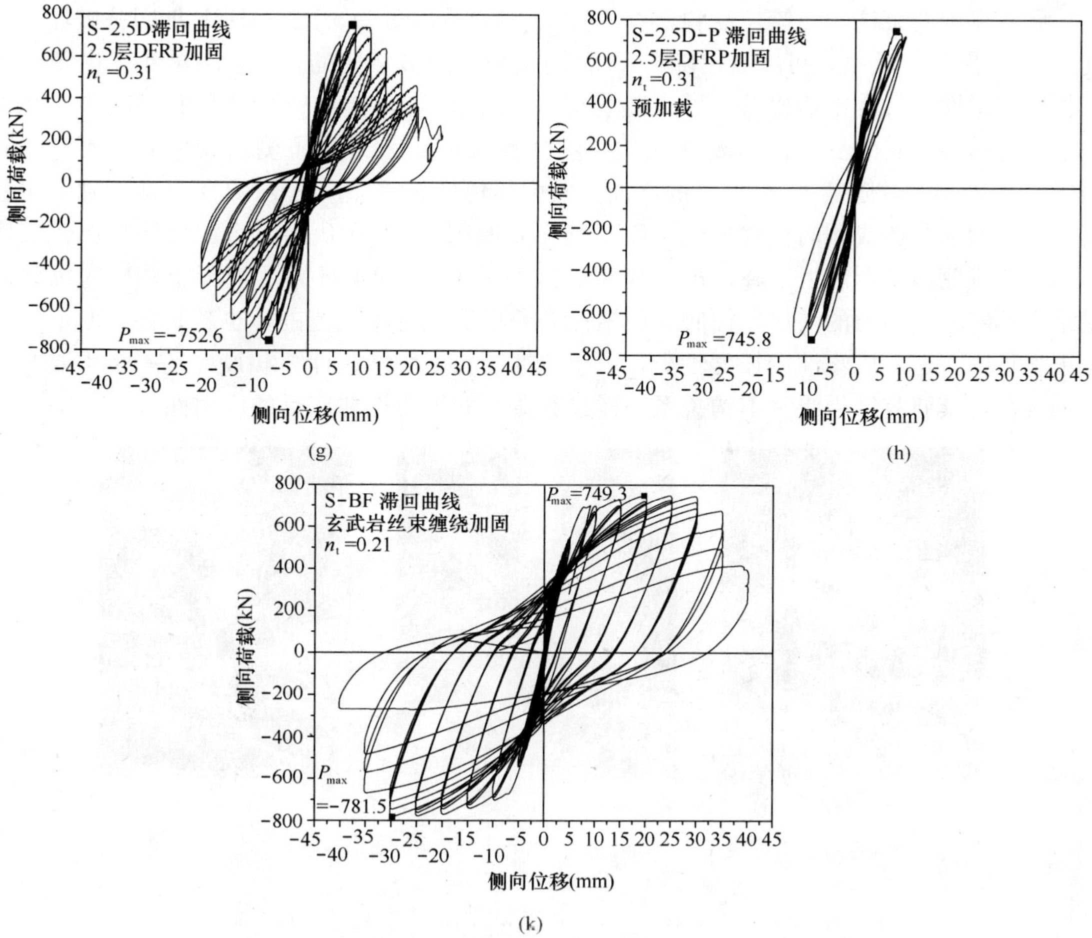

(g)　　(h)

(k)

图 15　柱 FRP 加固前后的试验曲线[44]（二）

(g)S-2.5D；(h)S-2.5D-P；(k)S-BF

试验结果[43]-[44]还表明由于横向约束使混凝土应力提高以及纵筋屈服后的强化效应导致 FRP 加固 RC 柱破坏时截面受弯承载力远大于规范规定的截面屈服强度。根据大量试验研究和理论分析，作者研究团队提出了 FRP 加固柱破坏时正截面受弯承载力计算方法；充分研究了 FRP 加固柱耗能性能的影响因素；提出了地震荷载作用下 FRP 加固柱变形能力在截面层次上的曲率延性和构件层次上顶点侧向位移定量计算方法；提出了 FRP 加固柱塑性铰长度计算方法。其中对截面破坏时受弯强度和截面屈服强度比值的参数分析结果显示，随着轴压增加，这一比值显著提高（图 16），在轴压较大时，这一比值达到 1.5 左右。这是由于混凝土强度的提高、钢筋应力强化以及中和轴位置的偏移引起截面受弯承载力增加，为了合理确定柱子承受的剪力，这一弯矩增强现象值得重视。

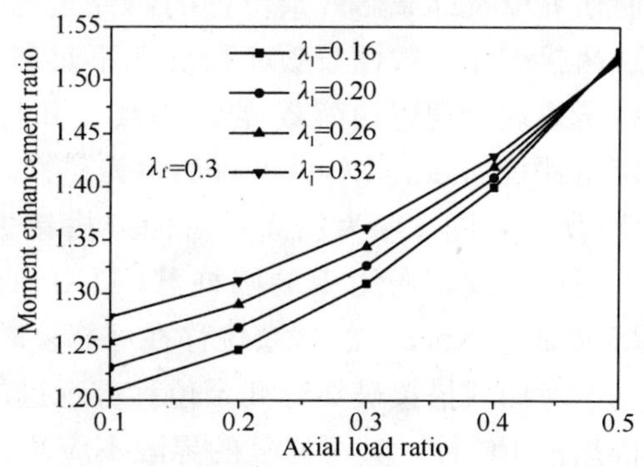


图 16　加固柱截面受弯承载力强化系数

研究发现约束柱塑性铰长度和FRP用量密切关系，图17示出了FRP加固柱破坏区长度与FRP用量的关系，可以看出随着FRP用量的增加破坏区的长度显著减小。由于破坏区长度和塑性铰长度密切相关，可见随着FRP用量增加，塑性铰长度会有所减小。塑性铰长度和加固柱侧向变形能力密切相关，塑性铰长度减小对柱子侧向变形能力不利。在FRP用量较小时，随着FRP用量的增加侧向位移角显著增加；当FRP用量增加到一定程度进一步增加会造成侧向位移角的下降。其主要原因就是FRP用量增加会使塑性铰长度下降，造成侧向变形能力下降。这一现象的机理是FRP约束提高了混凝土和柱底纵筋的黏结力，随着黏结力的增加，柱底纵筋塑性应变集中于更短的范围，也就是塑性曲率的分布更加集中，这造成塑性铰长度较小。由于塑性铰长度对结构计算和设计是一个重要和基础的参数，其研究结果将和本构关系一样具有普遍的重要性和广泛的应用性。

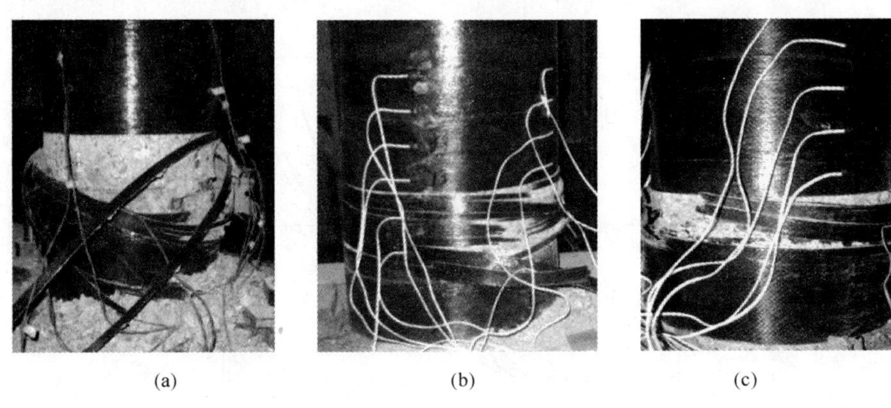

(a) (b) (c)

图17 加固柱破坏区高度

(a) 1.5层CFRP加固；(b) 2.5层CFRP加固；(c) 3.5层CFRP加固

图18示出的加固柱实测侧向位移角和FRP用量的关系，在FRP用量较小时，随着FRP用量的增加侧向位移角显著增加；当FRP用量增加到一定程度进一步增加会造成侧向位移角的下降[45]。其主要原因就是FRP用量增加会使塑性铰长度下降，造成侧向变形能力下降。这一现象表明对FRP加固RC柱，FRP用量并不是越多越好，FRP用量过大对柱子变形能力不利。

另外，通过对既有其他研究者文献中的109根各种参数的FRP加固后的混凝土柱的试验研究以及卸载后残余位移的研究表明[46][47]，FRP补强能够有效的防治三种形态的破坏：搭接破坏，塑性铰破坏和抗剪不足破坏。从加固后滞回曲线的包络线可以看出，61根柱子表现出理想的荷载-位移曲线，并且保持稳定的屈服后刚度，这就是二次刚度。为了定量评价加固后RC柱的震后可修复性，柱滞回曲线屈服后刚度终点对应的残余位移如图19所示，图19a为加固后弯曲或搭接失效的RC柱，而图19b为相应剪切失效的RC柱。可以发现，弱约束的圆柱（CH-1.5D，CL1-1C）和矩形柱（RS-R3，RS-R4，FR1，FRS和试件No.3）的残余位移在可修复限值（1%柱高）附近。其他柱子无论是加固前委剪切、弯曲或搭接破坏，残余位移都远远超出了可修复限值。因此，对于延性较高的FRP加固后的RC柱，其可修复的界限不应从屈服后刚度的终点进行定义，有必要对RC柱进行非线性推覆分析来研究加固后RC柱的残余位移及最大位移的变化过程。

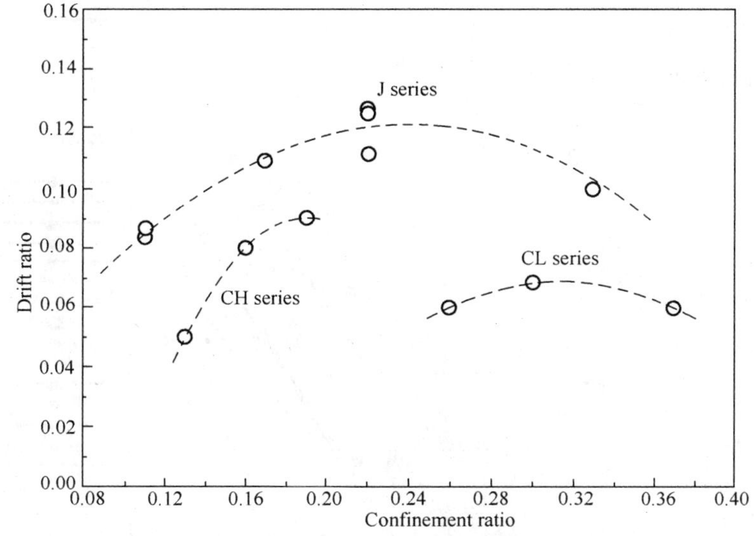


图 18　加固柱侧向位移角和 FRP 用量的关系[45]

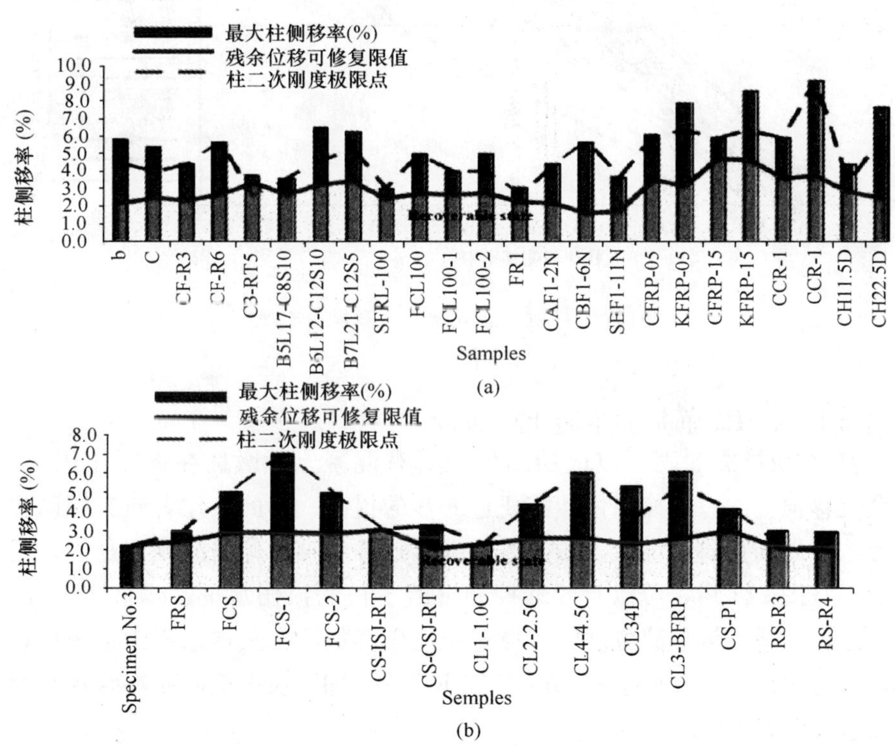


图 19　FRP 加固混凝土柱的可修复性[46]

(a) 弯曲或搭接失效；(b) 剪切失效

混凝土柱在理论强度、可修复限值和最大强度时的柱端侧移率和相应的残余位移率如图 20 所示。可以发现，虽然 1%柱高的残余位移受柱子各参数变化的影响较大，和柱端侧移率没有特定的关系，但通过统计可以发现，实现震后较高可修复性的 RC 柱的最大侧移在2%～3.5%侧移率之间，混凝土柱可修复性可以通过在这一区间的卸载残余位移来评估。

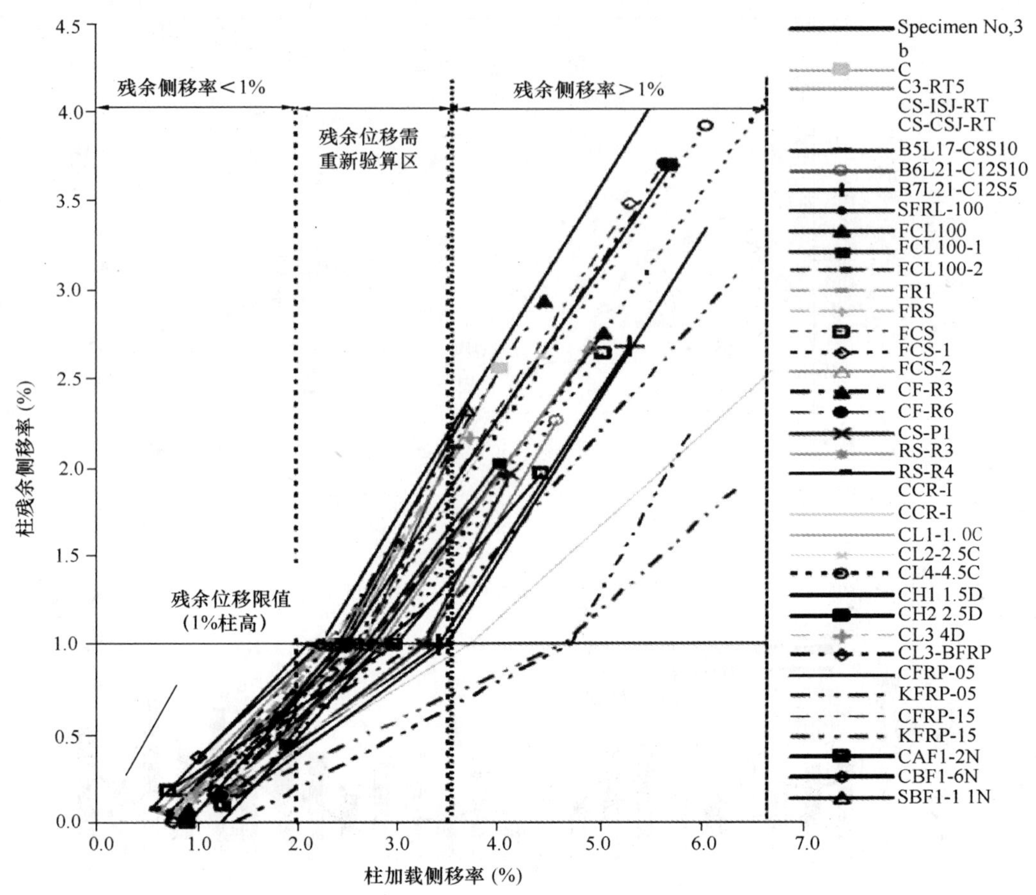


图 20 柱侧移率和残余位移率关系[46]

### 14.3.3 柱脚嵌入 FRP 筋加固混凝土桥墩

通过 FRP 约束抗震加固可以成功的保证既有混凝土桥墩柱在强震作用下的安全性，但震后残余位移损伤的有效控制依然需要进一步的提高。因此，作者研究团队提出了新型抗震加固方法，即在柱脚塑性铰区域嵌入 BFRP 筋并根据具体情况决定是否采用外包 FRP 进行组合加固（图 21）[48]，可以有效的控制混凝土柱的损伤水平、残余变形和二次刚度。值得注意的是，在适用该加固方法时，应避免发生类似于在阪神地震中部分桥墩由于纵筋切断而引起的剪切破坏，这要求柱脚嵌入 BFRP 筋的加固长度应限制在合理长度内以避免改变柱子的破坏模式。

在进行 BFRP 筋加固混凝土柱之前，进行了模拟加固方式的 BFRP 筋与混凝土的黏结性能试验，结果表明随着黏结长度的增加，试验破坏时的 BFRP 筋应力呈增加趋势，最终破坏模式从拔出破坏（20 倍直径黏结长度）转变为 BFRP 筋拉断破坏（40 倍直径），因此文献［48］推荐 60 倍直径的锚固长度可以确保嵌入的 BFRP 筋的强度得到充分发挥。

柱脚嵌入 BFRP 加固前后的 RC 柱荷载-位移曲线如图 22 所示[48]，其中包括了普通钢筋混凝土的滞回曲线。可以看出，由于柱脚嵌入 BFRP 筋，并没有改变混凝土柱的弹性刚

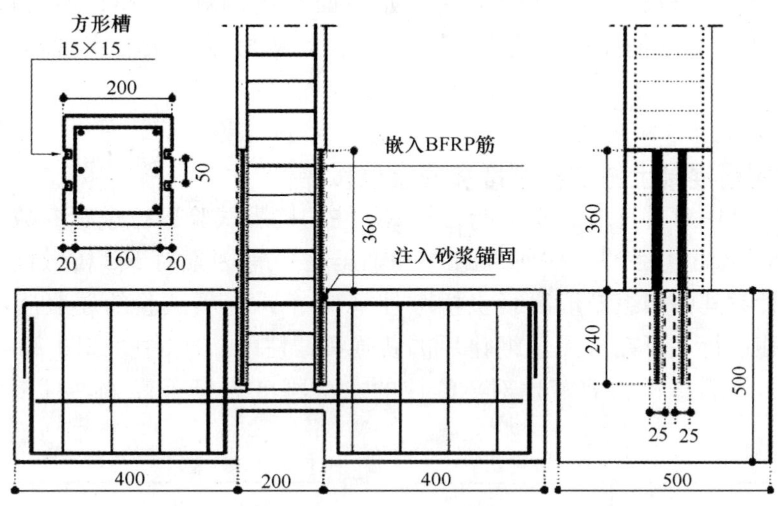


图 21　柱脚嵌入 FRP 筋加固设计[48]（mm）

度，而可以实现和 SFCB 增强混凝土柱类似的稳定二次刚度，加固柱由于 BFRP 筋的拔出而发生破坏，相应的柱顶侧移为 30mm。加固柱相对于普通 RC 柱，可以显著减小卸载残余位移，具有较高的可修复性的效果（图 22）。

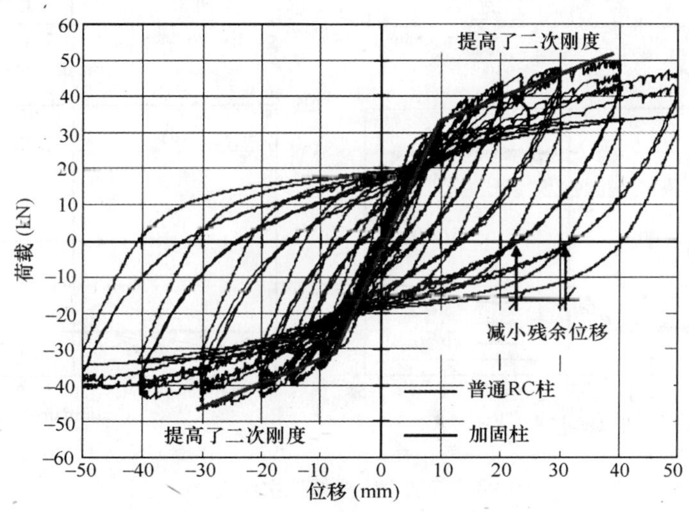


图 22　嵌入 BFRP 筋加固混凝土柱滞回曲线[48]

## 14.4　提高混凝土结构抗震性能的设计理论

为了研究提高新建结构抗震性能，从二次刚度和残余位移两个指标入手，在日本抗震规范的荷载-位移曲线基础上，提出理想抗震结构的设计理念。具体的研究内容主要包括以下三个方面，一是研究普通 RC 桥墩柱的可修复性限值，考虑的因素包括剪跨比、约束水平、纵筋直径和轴压比等；二是对于已有的 RC 柱如何进行抗震加固实现理想的二次刚度和较小的残余位移，如前面所述的利用 FRP 约束抗震加固以及柱脚嵌入式加固等；三是探讨如何利用新型复合筋从材料层次上实现新建结构的损伤控制。

### 14.4.1 普通钢筋混凝土柱二次刚度和残余位移评价

通过既有文献中对具有理想变形能力的 RC 柱的抗震试验进行进行参数分析，包括钢筋约束水平[49]、纵筋直径[50]、柱剪跨比[51]和轴压比[52]等因素对 RC 桥墩柱的抗震性能的影响，并主要针对其中四组矩形混凝土桥墩柱（13 个）抗震试验的屈服后刚度（二次刚度）和残余位移进行了研究。无量纲化后的普通 RC 柱的二次刚度如图 23 所示。可以发现，约束水平、纵筋直径和剪跨比对 RC 柱的二次刚度无显著影响，且未能达到 5%的比例。

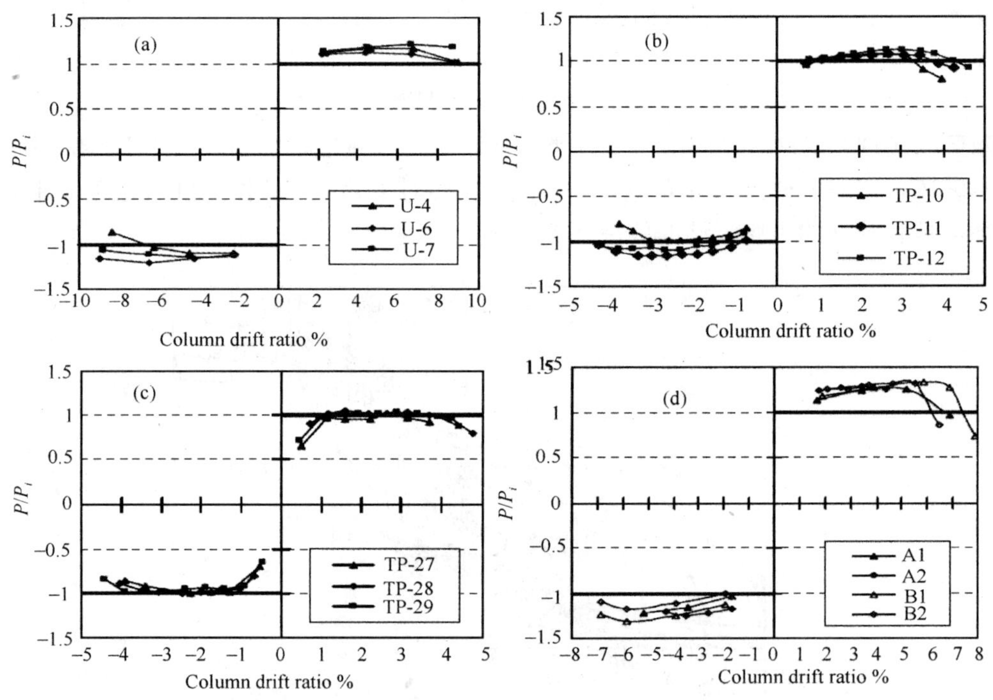


图 23 普通 RC 柱的二次刚度[47]

具有较好变形能力的不同 RC 柱的残余位移率如图 24 所示[47]，可以发现，所有的 RC 柱在达到最大水平承载力时的卸载残余位移都超出了可修复的限值（1%柱高）。因此，如何控制延性的 RC 柱在大震作用下的残余位移在可修复范围依然是一个很具有挑战性的课题。从各个参数的 RC 柱的柱端侧移率和残余位移率关系可以发现（图 24b），其卸载点位移和卸载残余位移都基本呈线性关系，随着加载位移增加，残余位移显著增加。在比较各因素对 RC 柱残余位移的侧移水平后可以发现，轴压比是一个显著影响桥墩变形能力的因素，当轴压比为 0.24 的混凝土柱残余位移在可修复范围内时的柱顶侧移可达到 3%柱高的水平，而相应轴压比为 0.1 的 RC 柱只达到了 2.64%的侧移水平。

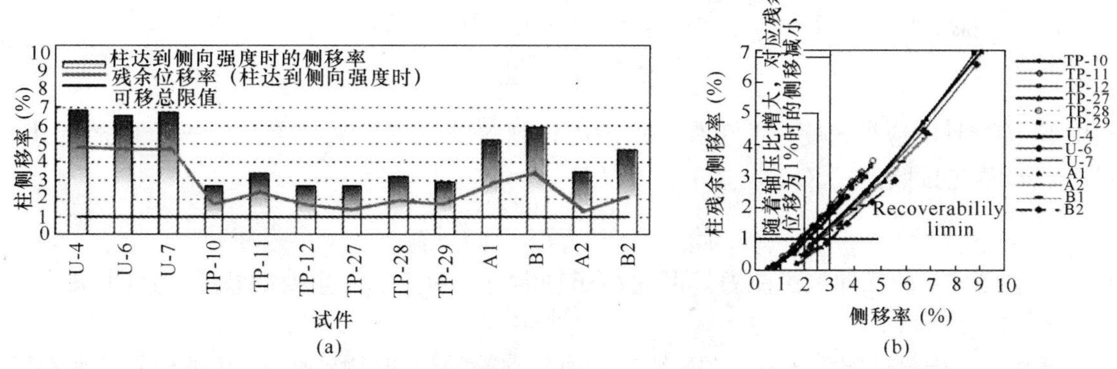


图 24　普通 RC 柱的最大侧移和相应的卸载残余位移[47]

(a) 最大变形能力及相应的残余位移；(b) 柱端侧移率和残余位移率

### 14.4.2　理想的损伤可控混凝土结构的荷载-变形关系

日本抗震规范[6]对混凝土结构的荷载-位移曲线分为三个水准（图 25），Level 1：作用于结构上的荷载应当小于结构的抗弯承载能力，并且不能出现有过大的位移，应力不能超过屈服强度；Level 2：虽然结构的构件出现塑性变形，但是仍然能够保持稳定，最大位移不能影响结构的使用功能，也不允许有震后残余变形；Level 3：结构仍然具有良好的承载能力以确保结构整体不倒塌。

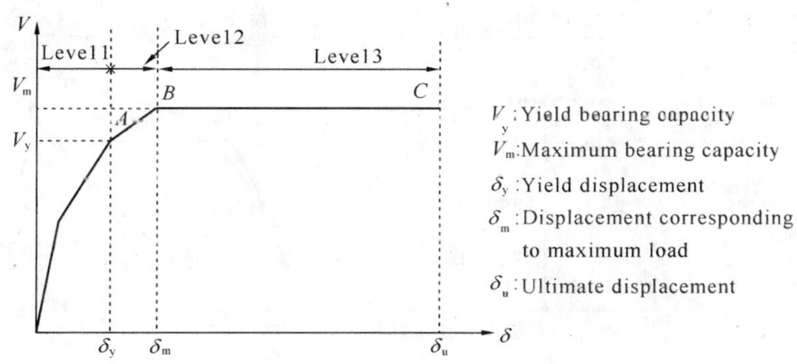


图 25　日本抗震规范中不同破坏水平的荷载-位移示意[6]

对于具体的残余位移限值计算方法如公式（3）所示：

$$\delta_R = C_R\ (\mu_R - 1)\ (1 - r)\delta_y \tag{3}$$

其中，$\mu_R = \delta/\delta_y$ 是柱子的延性系数，$C_R$ 是依赖于变量 $r$ 的一个系数。该规范并要求柱震后残余位移不应大于柱高的 1%。根据日本桥梁抗震规范，桥墩的设计延性系数 $\mu_a$ 是根据地震作用的类型来确定的，规范中考虑了两个不同水准即罕遇地震和多遇地震作用。

据此，柱子的允许位移延性系数 $\mu_a$ 按照公式（4）计算

$$\mu_a = 1 + \frac{\delta_u - \delta_y}{\alpha\delta_y} \tag{4}$$

其中 $\alpha$ 为根据不同水准的地震作用以及桥梁的重要程度不同确定的安全系数，$\delta_u$ 为柱子的

极限位移。

按照等能量原则，等效响应加速度 $S_{es}$ 根据公式（5）计算：

$$S_{es} = \frac{S_s}{\sqrt{2\mu_a - 1}} \tag{5}$$

其中 $S_s$ 为弹性响应加速度。关于如何取值的详细解释请参照文献[6]。基于以上抗震设计考虑，$\mu_R$ 值应按照公式（6）计算：

$$\mu_R = \frac{1}{2}\left\{\left(\frac{S_s W}{H_{max}}\right)^2 + 1\right\} \tag{6}$$

其中 $H_{max}$ 是柱子的横向抗侧能力，$W$ 是柱子轴向力。对于一些重要的桥梁，JSCE 规范[6]推荐 $\mu_a$ 最大值为 8。

然而，对于普通 RC 结构，在屈服后，由于钢筋的弹塑性特征，结构响应在屈服荷载之后，位移变化无法进行控制（图 25 中的 Level 3），从而造成在普通 RC 结构在中大震作用下的较大损伤和不同地震波（假设最大加速度幅值相同）激励下响应的较高离散性。

不同二次刚度的结构的荷载位移曲线及相应残余位移示意如图 26a 所示，二次刚度越大，在相同的卸载位移下相应的残余位移越小，而负的二次刚度结构卸载残余位移最大。因此，作者研究团队在日本抗震规范的荷载-位移曲线（图 25）的基础上，提出的损伤可控型的结构设计理念如图 26b 所示，其最为重要的特征是具有稳定二次刚度（屈服后刚度，图 26b 中 BC 段），在这一阶段，随着位移的增加，荷载可以持续增加，从而保持结构体系的位移相应的稳定，实现结构损伤的定量控制。

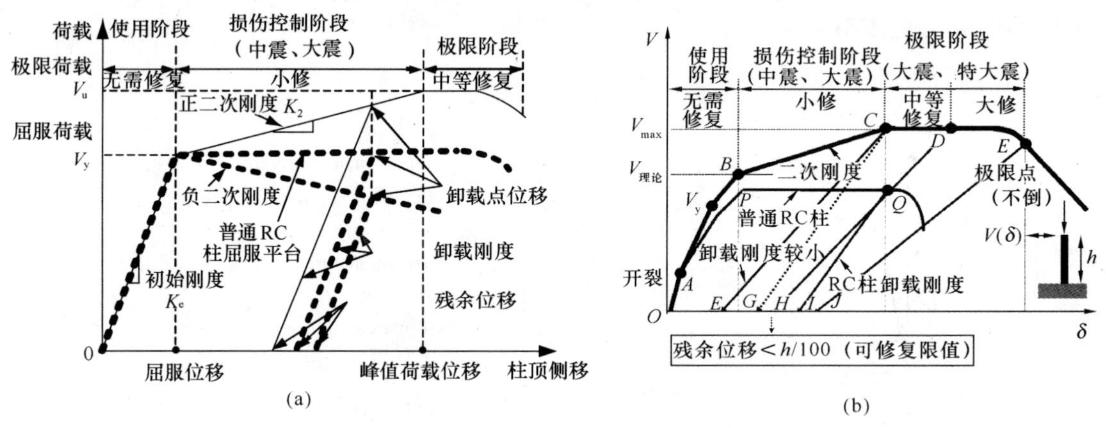


图 26 新型抗震结构的荷载—位移关系

(a) 不同二次刚度的混凝土结构残余位移示意；(b) 理想抗震结构的荷载-位移关系

新型抗震结构的荷载-位移曲线可以分为四个主要阶段。阶段 1，为 OB 段（相）；阶段 2，为 BC 段；阶段 3，为 CD 段；阶段 4，为 DE 段。阶段 1 代表了结构整体屈服之前，对于应结构的弹性阶段和开裂后的状态。小震作用下的结构响应在阶段 1 以内，保持体系的弹性阶段，结构或构件在相应的地震烈度作用后无需修复，这当于图 25 中的 Level 1 和 Level 2。图 26b 的阶段 2 代表了结构在中等地震作用下，增强纵筋进入屈服阶段，但具有明显的硬化特征，即稳定的二次刚度，结构变形可以得到有效控制，并且震后损伤（中震、大震）可以得到快速修复。阶段 3 相应于结构体系在二次刚度段之后的变形能力，从而使结构在大震作用下具有足够的延性，避免结构由于过高的二次刚度而产生太大的地震响应，结构震后可以通过替换部分单元进行修复。而在罕遇的特大震作用下，结构可能进入第 4 阶段，这一阶段的结构应能够避免倒塌，当荷载下降至极限荷载的 20%时，定义为极限阶段，并能够保持结构不倒，这和传统结构的定义相同。

值得注意的是，新型抗震体系由于可以具有更小的卸载刚度（如图 26b 中卸载刚度 CG 和 CF 的区别），进一步减小残余位移，实现较高的可修复性。

### 14.4.3　利用钢-连续纤维复合筋损伤可控型新建混凝土结构

在提出理想抗震结构设计理念的基础上，如何实现之是重要课题。作者研究团队的已有的研究表明，通过 FRP 约束抗震加固的 RC 柱，可以实现一定程度的二次刚度[46][47]，然而由于通过 FRP 约束实现的 RC 柱的二次刚度提高并不明显，而且由于尺寸效应、方柱和圆柱的不同处理方式等因素使得该途径实现的结构二次刚度具有不可设计性。

因此，吴智深、吴刚等人[53]提出以钢-连续纤维复合筋（SFCB）增强混凝土抗震结构（图 27），利用钢筋和 FRP 材料的优势互补，在实现复合筋材性稳定二次刚度（图 6）的基础上，实现 SFCB 增强结构的稳定二次刚度，并给出了 SFCB 关键构造工艺，实现了工业化批量生产。SFCB 增强结构其特点包括：①在正常使用荷载或中小地震作用下，具有与普通钢筋混凝土结构相同的强度抵抗能力，可以充分发挥 SFCB 内芯钢筋带来的高弹性模量作用；②利用外包 FRP 的高强度特性使 SFCB 增强的结构具有截面层次上稳定的二次刚度，这一特征可以预防塑性铰在柱脚小范围内集中转动形成的过大的塑性变形，实现在一个更长的区域内实现曲率的较均匀分布，减小截面的需求曲率，因而相应的减小 SFCB 中内芯钢筋塑性的应变；③用 SFCB 代替普通钢筋，不改变截面的初始屈服强度，有利于结构抗震设计时控制损伤截面的位置，强地震荷载作用下，可以控制损伤截面的位置对结构的变形能力具有重要的影响。

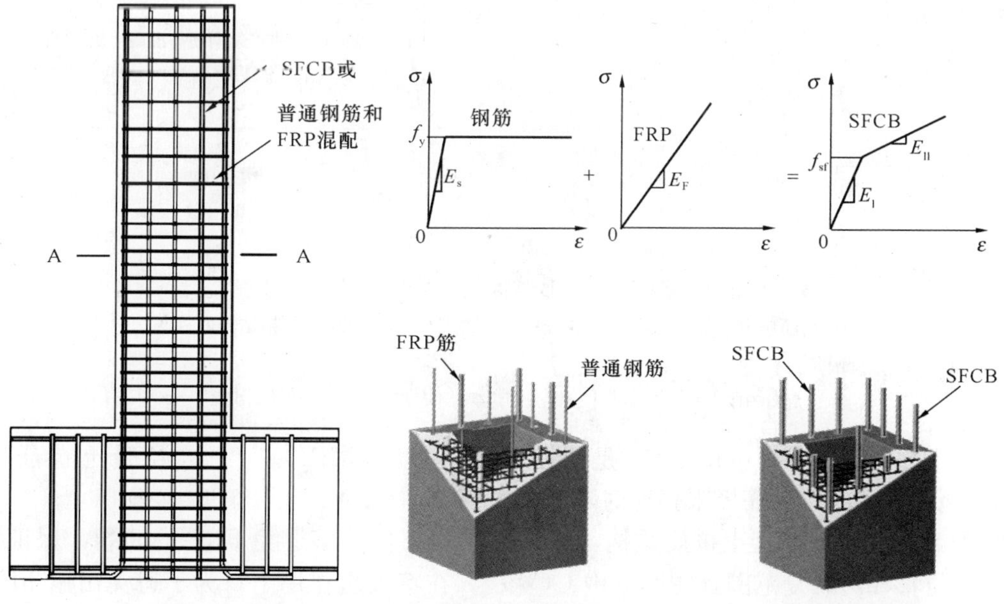


图 27　利用 SFCB 或普通钢筋和 FRP 筋混杂配筋实现混凝土结构的稳定二次刚度

利用SFCB所增强的混凝土结构具有稳定的二次刚度，这同样可以通过钢筋和FRP筋的混杂配筋来实现，然而SFCB由于外侧全部为FRP，具有高度的耐久性特征，可以适用于各种恶劣环境下。作者研究团队进行了SFCB单向和往复拉伸试验[54]、SFCB与混凝土黏结性能试验[55]和SFCB增强混凝土柱抗震性能的试验研究。黏结试验表明[55]，SFCB与混凝土的黏结强度约为相应带肋钢筋的94%；特别的，拉拔试件的破坏形态可分为钢筋内芯屈服后SFCB拔出（图28a）、钢筋内芯屈服前SFCB拔出和SFCB拉断3种。在SFCB黏结试件自由端发生轻微滑移后，复合筋的钢/纤维种类和比例、试件有效黏结长度和混凝土强度等级等因素决定了在发生SFCB拔出或拉断破坏，以及SFCB拔出之前钢筋内芯是否屈服。

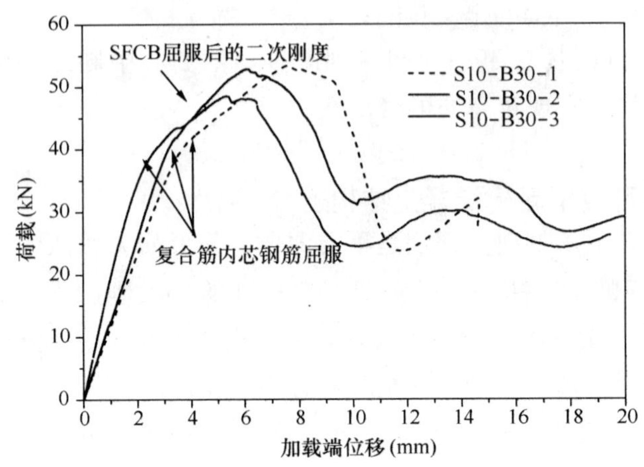


图28 SFCB与混凝土拉拔试验结果[55]

当钢筋一端完全锚固时，其荷载-滑移曲线如图29a所示[56]，其主要参数是增强纵筋的屈服强度、极限强度和对应的滑移值。钢筋在屈服时的界面滑移伸长量计算公式如下[56]：

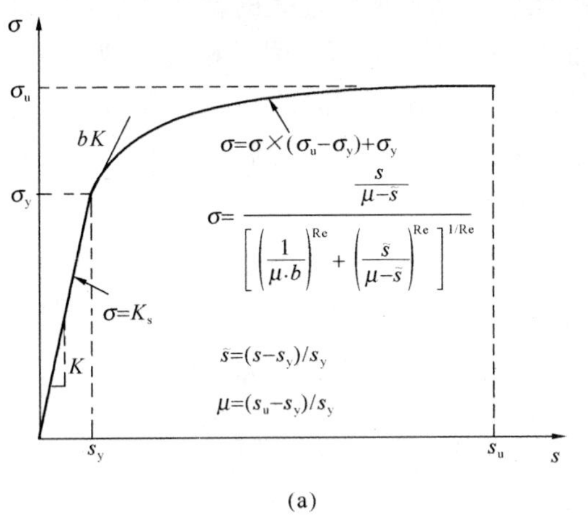


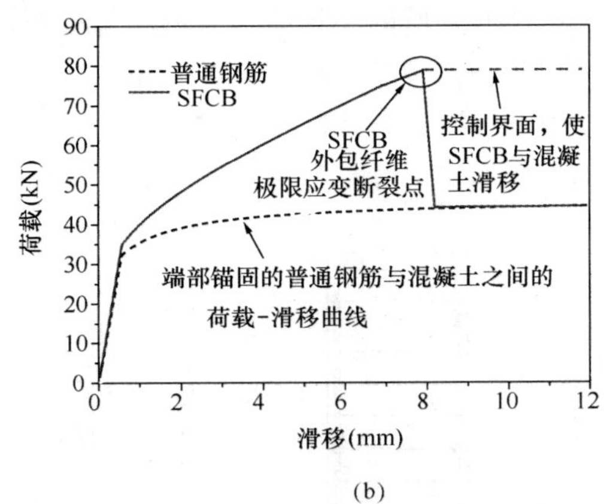


图29 钢筋、SFCB与混凝土黏结—滑移控制

(a) 钢筋应力—伸长（滑移）曲线；(b) 端部锚固的SFCB的粘结界面控制

$$s_y(\mathrm{mm}) = 2.54\left(\frac{d_b}{8437}\frac{f_y}{\sqrt{f'_c}}(2\alpha+1)\right)^{\frac{1}{\alpha}} + 0.34 \tag{7}$$

其中，$d_b$ 是纵筋直径（mm）；$f_y$ 是纵筋屈服强度（MPa）；$f'_c$ 是混凝土抗压强度（MPa）；$\alpha$ 取值0.4；纵筋极限强度时对应的滑移 $s_u$ 一般为：$s_u=(30\sim40)\ s_y$。

对于SFCB增强混凝土抗震结构，在大震作用下应能够避免FRP的断裂，保证结构震后残余位移在可修复范围（图26b中OC段）。在特大震作用下，为了避免由于FRP断裂引起的倒塌，SFCB与混凝土之间应该能够进行滑移并保持承载力水平，如图29b中

SFCB 荷载-滑移曲线峰值点后的虚线所示。具体如何实现 SFCB 与混凝土之间的这种具有高度韧性的黏结-滑移关系，有待于进一步的研究。

为了研究钢-连续纤维复合筋（SFCB）增强混凝土柱的抗震性能，共制作了截面为 300mm×300mm 的 4 个 SFCB 增强混凝土柱和 1 个 RC 对比柱（图 30），试件设计参数可见文献[57]，试验结果表明，所有柱试件均发生弯曲破坏，表现为先是柱脚附近混凝土开裂，纵筋或 SFCB 内芯钢筋屈服，SFCB 外包纤维部分（或全部）断裂，继续加载，柱脚混凝土压碎并剥落，纵筋压弯屈曲（图 31）。

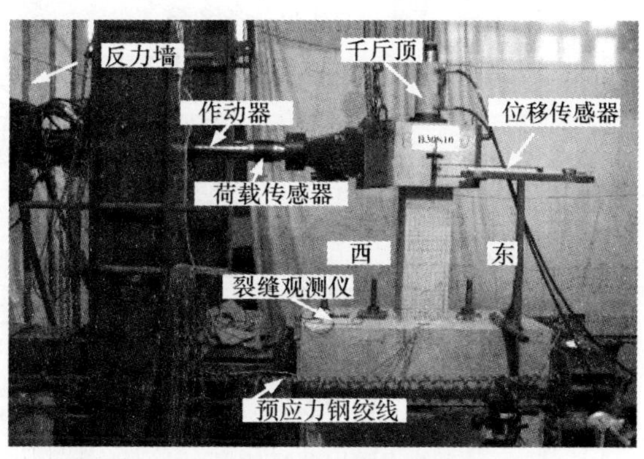


图 30　SFCB 柱构件[57]

(a)

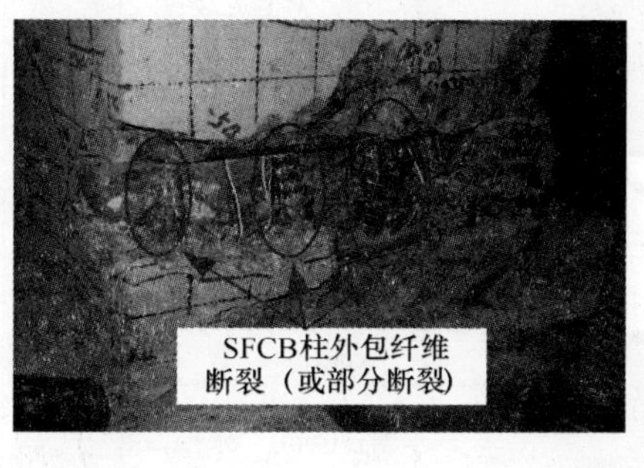


(b)

图 31　典型试验现象[57]

(a) 柱 C-S14 柱脚混凝土压溃纵筋屈曲；(b) SFCB 增强混凝土柱外包纤维断裂

SFCB 柱和 RC 对比柱荷载-位移滞回曲线（$V$-$\delta$ 曲线）如图 32[57] 所示。从按力加载阶段 SFCB 增强混凝土柱和 RC 柱荷载-位移曲线（图 32a）比较可以看出：SFCB 增强混凝土柱由于所用复合筋轴向刚度小于柱 C-S14 所用钢筋，在混凝土开裂后，SFCB 增强混凝土柱刚度比柱 C-S14 柱小。普通 RC 柱滞回曲线呈梭形，较丰满，表现出良好的延性和耗能能力（图 32b）；由于钢筋的屈服后刚度基本为“0”，柱屈服后荷载-位移曲线基本呈平台状。SFCB 增强柱的开裂荷载和 RC 柱相差不多，都为 51kN 左右，复合筋内芯钢筋屈服后，由于外包 FRP 的高强度特征，滞回曲线在正向卸载及反向加载阶段表现出稳定的二次刚度（屈服后刚度），而卸载阶段有一定的捏拢效应。柱 C-S10B20 和柱 C-S10C24 的破坏过程也与柱 C-S10B30 和柱 C-S10C40 类似，滞回曲线表现出稳定的二次刚度，卸载阶段有一定的捏拢效应。

卸载刚度决定了结构在震后的残余位移，是提高结构可修复性的重要参数。在同样卸载点，卸载刚度越小，相应残余位移越小。对五个混凝土柱按屈服荷载、屈服位移无量纲化骨架曲线如图 33a 所示。可以看出各柱在屈服位移以前的荷载-位移曲线基本重合。SFCB 增强混凝土柱由于 FRP 高强度特征，表现出稳定的二次刚度，不同卸载刚度将直接带来不同的残余位移。

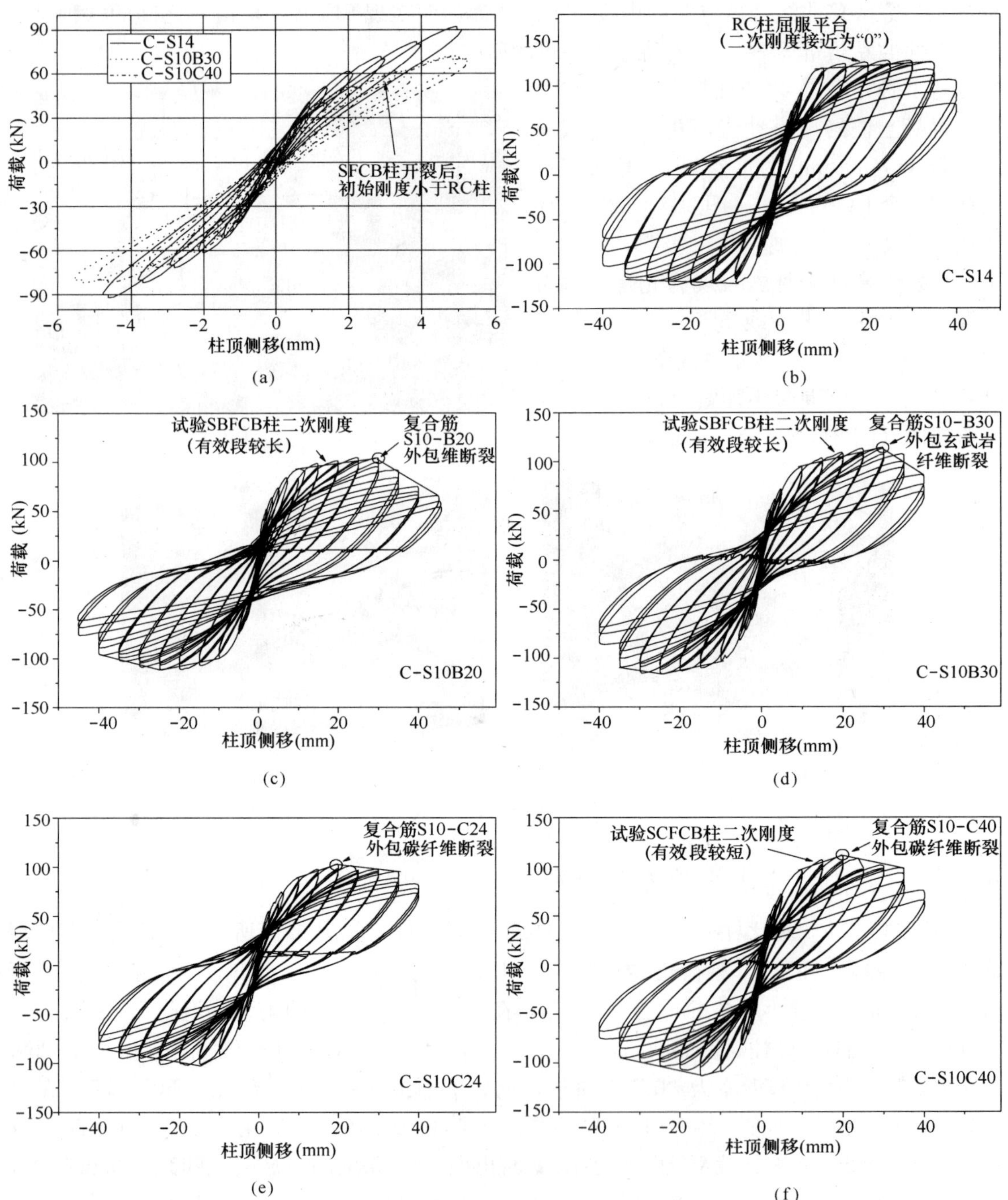


图 32　各柱滞回曲线[57]

(a) 按力加载阶段；(b) C-S14；(c) C-S10B20；(d) C-S10B30；(e) C--S10C24；(f) C--S10C40

TK 模型是钢筋混凝土结构弹塑性地震反应分析中应用最为广泛的模型，其最大的特点是考虑了卸载刚度的退化，但其未能考虑柱卸载刚度随钢筋塑性发展而退化的特征。根据 Takeda 恢复力模型（TK 模型）[58]，RC 柱卸载刚度可按公式（8）计算：

$$K_{u-TK}=\left(\frac{\delta_{max}}{\delta_y}\right)^{-0.4}\times K_1 \tag{8}$$

其中 $\delta_{max}$、$\delta_y$ 分别为柱极限位移和屈服位移；$K_1$、$K_u$ 分别为柱初始刚度和卸载刚度。

五个混凝土柱卸载刚度（含柱 C-S14 的 TK 模型计算值）如图 33b 所示。可以看出，随着柱顶侧移增大，柱卸载刚度随柱顶侧移增大而显著退化，而按 TK 模型计算的卸载刚度约为 4.5 倍屈服位移时相应值（图 33b）。因此当重要结构需要评价非弹性变形较小时的可修复性时，TK 模型将过小估计结构的残余变形，从而不够安全。

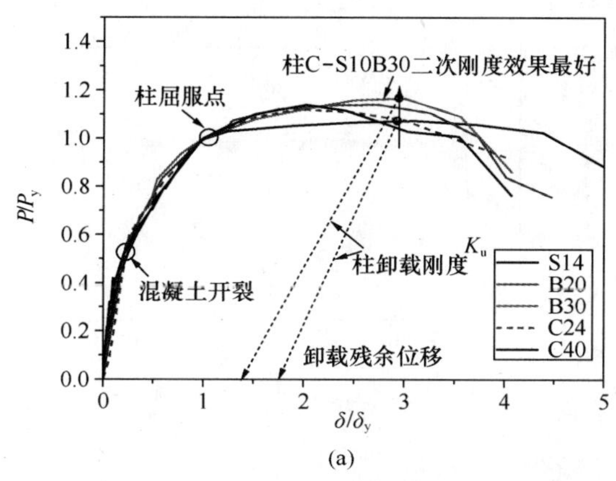

(a)

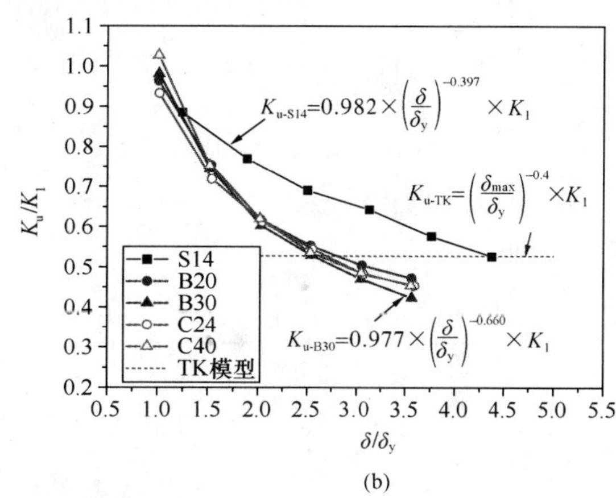

(b)

图 33　SFCB 柱无量纲化试验柱骨架曲线和卸载刚度[57]

(a) 无量纲化试验柱骨架曲线；(b) 推拉平均卸载刚度

SFCB 增强混凝土柱试件的残余侧移率见图 34，普通 RC 柱由于钢筋的塑性发展，残余位移随着加载位移的增加而增加（图 34），且最先达到日本规范[6] 的可修复性限值（柱顶加载位移 24.37mm）。

SFCB 增强混凝土柱 C-S10C40 和 C-S10B30 可修复性限值的柱顶侧移分别达到 29.76mm 和 32.80mm，是 RC 柱的 122.11% 和 134.59%；随着加载位移增加，SFCB 外包 FRP 逐渐断裂，SFCB 增强混凝土柱残余侧移率加速变大（曲线增加），相应普通 RC 柱残余侧移率与加载位移基本呈线性关系（图 34），这个前面统计的普通 RC 柱的趋势相同。

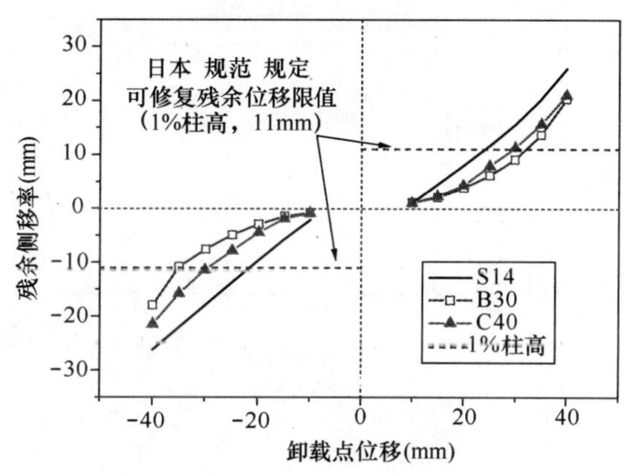

图 34　残余侧移率比较[57]

## 14.4.4　不同二次刚度单自由度体系地震反应分析

水平往复荷载作用下具有稳定二次刚度的 SFCB 柱与普通钢筋混凝土柱相比残余位移明显较小，但是在地震动荷载作用下构件的二次刚度及相应残余位移如何表现，还有待于进一步的数值模拟和振动台试验研究。作者研究团队首先基于 OpenSees 抗震分析软件，利用美国太平洋地震工程研究中心（Pacific Earthquake Engineering Research Center，简称 PEER）开发的强震数据库的收集的 173 条不同地震的将近 10000 条强震记录[59]，对动力荷载作用下具有稳定二次刚度的单自由度体系进行了理论分析。对比我国规范的场地分类与 NEHRP 的场地分类标准，确定以 NEHRP 的 C 类场地对应我国的Ⅱ类场地（主要是场地剪切波速相近）。从数据库中选出了 32 条强震记录，对这 32 条波形成的反应谱，进行了与规范反应谱的对比分析，并且从中选择与规范反应谱符合较好的 10 条地震波作为输入波进行时程分析。计算结果如图 35[60] 所示，对计算得到的最大弹塑性位移和残余位移进行了分析，得出了相关主要结论。

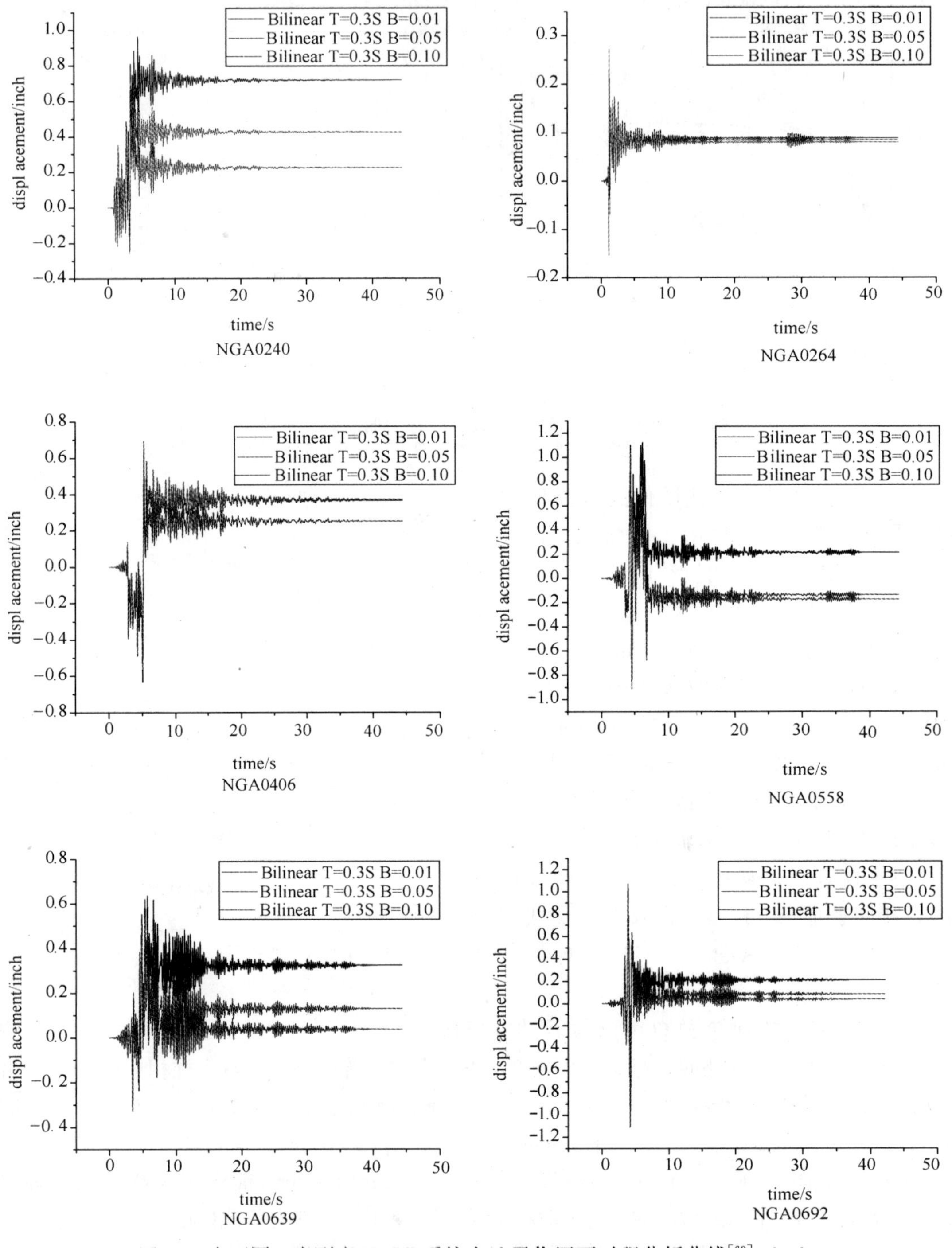


图 35 为不同二次刚度 SDOF 系统在地震作用下时程分析曲线[60]（一）

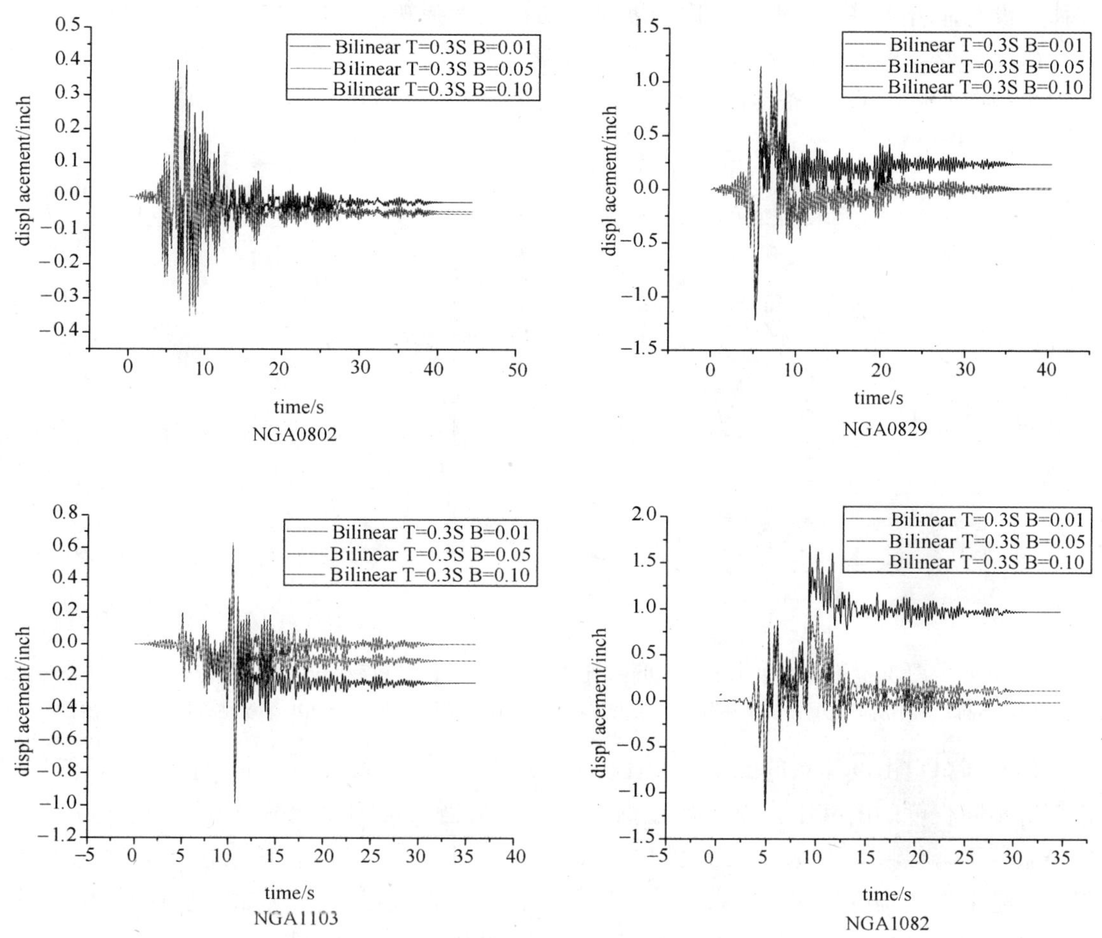


图 35　为不同二次刚度 SDOF 系统在地震作用下时程分析曲线[60]（二）

#### 14.4.4.1　最大弹塑性位移分析

对于结构在地震作用下的最大弹塑性位移研究已经进行了很多，也比较深入。现有的研究表明，结构的地震弹塑性最大位移受到了场地土、滞回特性中捏缩效应的影响较为明显，但是地震震级、震中距以及结构阻尼对弹塑性位移比的影响不是很明显。

单自由度体系在所选出的 10 条地震波作用下最大弹塑性位移如图 36a 所示，从图中可以看出：

（1）当双线性单自由度体系的二次刚度一定时，对于不同的地震波形，其地震最大弹塑性位移相差较大，并没有特别明显的规律可以遵循，这说明在采用时程分析确定结构在地震作用下的最大弹塑性位移的时候，其结果对于所选用的地震波依赖性很大，不同的波形所产生的结果相差很大。对于某一条波所得的最大弹塑性位移可能只适用于该波形。因此时程分析确定结构最大弹塑性位移的结果的正确与否是很值得探讨的问题。

（2）当单自由度体系的二次刚度比由 0.01 变化到 0.10 的时，地震作用下最大弹塑性位移总体上有稍微的减小，但是作用并不是很明显，而且在某些地震波作用下甚至有所上升，因此可以得出结论即提高结构的二次刚度，对改变结构其在地震作用下的最大弹塑性位移并没有显著效果，或者说二次刚度对于结构最大弹塑性位移并没有显著影响。

14.4.4.2 残余位移分析

结构在所选出的10条地震作用下残余位移变化如图36b所示，可以看出：

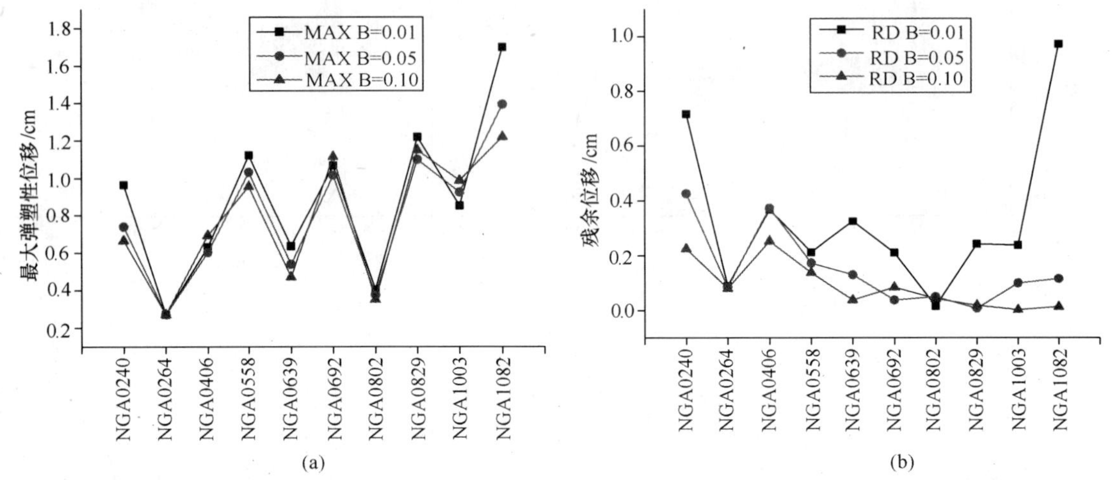


图36 不同二次刚度SDOF系统地震作用下最大位移和残余位移响应[60]

(a) 不同二次刚度SDOF系统地震作用下最大位移图；(b) 不同二次刚度SDOF系统地震作用下残余位移

(1) 当单自由度体系的二次刚度比由0.01变化到0.10的区段时，地震作用下残余位移都有不同程度的的减小，而且从数据上可以看出当二次刚度比从0.01变化至0.05时，残余位移下降很快，但是当二次刚度比从0.05变化至0.10时，残余位移虽然还是在下降，但是下降的幅度有明显的减小。所以可以得出，在进行结构的二次刚度设计时，并不是将二次刚度设计的越大越好，因为当过大增加结构二次刚度以期望减小结构的震后残余位移时，势必会带来建造成本的大比例增长。

(2) 对于不同的地震波形，当单自由度体系的二次刚度比是0.01时，震后残余位移相差很大，并没有特别明显的规律可以遵循。但是，随着二次刚度的增大，二次刚度值为0.05时，我们可以看出对于不同的波形，残余位移开始变得稳定，当二次刚度值变为0.10时残余位移已经变得相当稳定了。这就是说，结构的二次刚度越大，震后残余位移会对模拟采用的波形依赖性越小，也就是说即使采用不同的波形，计算出来的残余位移可能相差并不是很大，这就大大简化了时程分析时选波的程序，并且使得采用残余位移作为可修复性指标的可信程度大大增加。

## 14.5 结论

本文结合作者研究团队近年来开展的利用FRP提高混凝土结构抗震性能的研究成果，阐述了混凝土结构抗震高性能化的方法和设计思想，主要结论如下：

(1) 利用混杂设计理论，可以实现FRP强度刚度延性等基本力学性能的综合提升，可以解决部分纤维材料的刚度或延伸率不足的问题，实现高初始刚度，高强度和充足的延性特征。同时，利用钢和 FRP 进行复合，可以从材料层次上实现稳定的二次刚度，在和普通钢筋相同的卸载应变下，新型复合筋具有较小的残余应变。

（2）利用 FRP 抗震加固约束混凝土柱可以提高既有结构的抗剪能力和变形能力，改善其抗震性能。课题研究提出了判断 FRP 约束混凝土强/弱约束的界限值、建立了适用矩形、圆形、椭圆等不同截面形状以及强/弱约束统一的应力—应变关系模型、地震作用下 FRP 约束混凝土柱承载力和延性的定量计算模型。

（3）在 FRP 约束混凝土本构关系模型的基础上，研究了往复荷载下 FRP 约束加固混凝土圆柱和方柱的抗震性能，分析考虑了轴压比、剪跨比、加载方式和 FRP 用量等参数，并进一步统计研究了既有文献中 FRP 抗震加固混凝土柱的残余位移和二次刚度，提出了建议的设计方法和加固准则。

（4）针对现行规范对既有结构震后可修复性考虑不足，创造性地发展了高度提升混凝土结构震后可恢复性能的损伤可控结构设计理念，建立了利用 FRP 控制二次刚度和残余变形的方法及设计理论模型，并开发了钢-连续纤维复合筋增强混凝土柱，实现了结构损伤可控，提升了结构震后可修复性。

（5）对于不同的地震波形，随着单自由度体系二次刚度的增大，残余位移响应逐渐变得稳定，当二次刚度比变为 0.10 时，震后残余位移会对模拟采用的波形依赖性很小，计算出来的残余位移相差很小。

## 参考文献

[1] 龚思礼. 建筑抗震设计[M]. 北京：中国建筑工业出版社，1994.

[2] Chen W F，Duan L. Bridge engineering：seismic design[M]. Florida，USA：CRC Press，2003

[3] http：//news. sxxw. net/html/20085/29/187129. shtml

[4] http：//web. ics. purdue. edu/～braile/edumod/eqphotos/eqphotos2. htm

[5] Priestley M J N. Performance based seismic design[J]. Bulletin of the New Zealand Society for Earthquake Engineering. 2000，33(3)：325-346.

[6] Japan Society of Civil Engineering (JSCE). Earthquake resistant design codes in Japan[S]. JSCE Earthquake Engineering Committee，2000，Tokyo，Japan.

[7] 滕锦光，陈建飞，Smith S. T. FRP 加固混凝土结构[M]. 北京：中国建筑工业出版社，2005.

[8] 谢旭，布占宇，应用碳纤维筋控制桥墩地震损伤方法初探[J]，浙江大学学报(工学版)，2005，39(10)：1589-1594.

[9] Iemura H，Takahashi Y，Sogabe N. Two-level seismic design method using post-yield stiffness and its application to unbonded bar reinforced concrete piers[J]. Structural Engineering/Earthquake Engineering，JSCE. 2006，23(1)：109s-116s.

[10] Sakai J，Mahin S A. Analytical investigations of new methods for reducing residual displacements of reinforced concrete bridge columns[R]. Rep. No. PEER 2004/02. Pacific Earthquake Engineering Research Center，Berkeley，CA2004.

[11] Billington S L，Yoon J K. Cyclic response of unbonded posttensioned precast columns with ductile fiber-reinforced concrete[J]. Journal of Bridge Engineering，ASCE. 2004，9(4)：353-363.

[12] Kawashima K，MacRae G A，Hoshikuma J，Nagaya K. Residual displacement response spectrum

[J]. Journal of Structural Engineering, ASCE. 1998, 124(5): 523-530.

[13] Pettinga D, Stefano P, Christopoulos C, Priestley N. Accounting for residual deformations and simple approaches to their mitigation. [C]//First European Conference on Earthquake Engineering and Seismology. Geneva, Switzerland: 2006.

[14] Christopoulos C, Pampanin S, Priestley M J N. Performance-based seismic response of frame structures including residual deformations, Part I: Single-degree-of-freedom systems [J], Journal of Earthquake Engineering, 2003, 7(1): 97-118.

[15] Wu Z S, Fahmy M F M, Wu G. Safety enhancement of urban structures with structural recoverability and controllability[J]. Journal of Earthquake and Tsunami. 2009, 3(3): 143-174.

[16] Wu Z S., Structural strengthening and integrity with hybrid FRP composites (keynote paper), Proceedings of the 2th International Conference of CICE, 2004: 905-912

[17] 吴刚，罗云标，吴智深，胡显奇，等. 钢-连续纤维复合筋(SFCB)力学性能试验研究与理论分析[J]. 土木工程学报. 2010, 43(3): 53-61.

[18] Priestley M. J. N. and Seible F. Design of seismic retrofit measures for concrete and masonry structures [J]. Construction and Building Materials, 1995, 9(6): 365-377.

[19] Saadatmanesh H., Ehsani M. R. and Jin L. M. Repair of earthquake-damaged RC columns with FRP wraps [J]. ACI Structural Journal, 1997, 94(2): 206-215.

[20] Seible F. and Priestley M. J. N. et al. Seismic retrofit of RC columns with continuous carbon fiber jackets [J]. Journal of Composites for Construction, ASCE, 1997, 1(2): 52-62.

[21] Xiao Y. and Ma R. Seismic retrofit of RC circular columns using prefabricated composite jacketing [J]. Journal of Structural Engineering, 1997, 123(10): 1357-1364.

[22] Xiao Y., Wu H. and Matin G. R. Prefabricated composite jacketing of RC columns for enhanced shear strength [J]. Journal of Structural Engineering, ASCE, 1999, 125(3): 255-264

[23] Sheikh S. A., Yau G. Seismic behaviour of concrete columns confined with steel and fiber-reinforced polymers [J]. ACI Structural Journal, 2002, 99(1): 72-80.

[24] Iacobucci R. D., Sheikh S. A. and Bayrak O. Retrofit of square concrete columns with carbon fiber- reinforced polymer for seismic resistance [J]. ACI Structural Journal, 2003, 100(6): 785-794.

[25] Memon M. S. and Sheikh S. A. Seismic resistance of square concrete columns retrofitted with glass fiber-reinforced polymer [J]. ACI Structural Journal, 2005, 102(5): 774-783.

[26] Harajli M. H. and Rteil A. A. Effect of confinement using fiber-reinforced polymer or fiber-reinforced concrete on seismic performance of gravity load-designed columns [J]. ACI Structural Journal, 2004, 101(1): 47-56.

[27] Li Y. F. and Sung Y. Y. Seismic repair and rehabilitation of a shear-failure damaged circular bridge column using carbon fiber reinforced plastic jacketing [J]. Can. J. Civ. Eng. 2003, 30, 819-929.

[28] 张柯，岳清瑞，叶列平. 碳纤维布加固混凝土柱滞回耗能分析及目标延性系数的确定[J]. 工业建筑，2001, 31(6): 5-8.

[29] 李忠献，徐成祥，景萌等. 碳纤维布加固钢筋混凝土短柱抗震性能的试验研究[J]. 建筑结构学报，2002, 23(6): 41-48.

[30] 陶忠，高献，于清. FRP约束圆钢筋混凝土柱滞回性能的理论分析[J]. 工业建筑，2005, 35(9): 15-19.

[31] 吴刚，吕志涛，蒋剑彪. 碳纤维布加固钢筋混凝土柱抗震性能的试验研究[J]. 建筑结构，2002，32(10)：42-45.

[32] Richart FE，Brandtzaeg A，Brown RL. A study of the failure of concrete under combined compressive stresses. Bulletin，vol. 185. University of Illinois Engineering Experimental Station；1928. p. I11.

[33] Karbhari VM，Gao Y. Composite jacketed concrete under uniaxial compression-Verification of simple design equations. J Mater Civ Eng，ASCE 1997；9(4)：185-93.

[34] Samaan M，Mirmiran A，Shahawy M. Model of concrete confined by fiber composites. J Struct Eng，ASCE 1998；124(9)：1025-31.

[35] Hosotani M，Kawashima K. A stress-strain model for concrete cylinders confined by both carbon fiber sheets and tie reinforcement. J Concr Eng，JSCE 1999；620/43：25-42 [in Japanese].

[36] Lam L，Teng JG. Design-oriented stress-strain model for FRP-confined concrete. Construct Build Mater 2003；17：471-89.

[37] Wu G，Gu DS，Wu ZS，Jiang JB，Hu XQ. Comparative study on seismic performance of circular concrete columns strengthened with BFRP and CFRP composites. In：Proceedings of the Asia-Pacific conference on FRP in structures（APFIS 2007），vol. 1. Hong Kong，China；2007. p. 199-204.

[38] 吴刚，吴智深，吕志涛. FRP 约束混凝土圆柱有软化段时的应力-应变关系研究[J]. 土木工程学报. 2006，39(11)：7-14.

[39] 吴刚，吕志涛. FRP 约束混凝土圆柱无软化段时的应力-应变关系研究[J]. 建筑结构学报，2003，24(5)：1-9.

[40] Fahmy M F M，Wu Z S. Evaluating and proposing models of circular concrete columns confined with different FRP composites[J]. Composites Part B：Engineering. 2010，41(3)：199-213.

[41] 魏洋，吴刚，吴智深等. FRP 约束混凝土矩形柱有软化段时的应力-应变关系研究[J]. 土木工程学报，2008，41(3)：21-28.

[42] 魏洋，吴刚，吴智深. FRP 强约束混凝土矩形柱应力-应变关系的研究[J]. 建筑结构，2007，37(12)：75-78.

[43] 顾冬生. FRP 加固钢筋混凝土圆柱抗震性能研究[D]. 博士论文，南京：东南大学，2007.

[44] 魏洋. FRP 约束混凝土矩形柱力学特性及其抗震性能研究[D]. 博士论文，南京：东南大学，2007.

[45] Gu D S，Wu G，Wu Z S，Wu Y F. The confinement effectiveness of FRP in retrofitting circular concrete columns under simulated seismic load[J]. Journal of Composites for Construction，ASCE. 2010，14(5)：531-540.

[46] Fahmy M F M，Wu Z S，Wu G. Seismic performance assessment of damage-controlled FRP-retrofitted RC bridge columns using residual deformations[J]. Journal of Composites for Construction，ASCE. 2009，13(6)：498-513.

[47] Fahmy M F M，Wu Z，Wu G. Post-earthquake recoverability of existing RC bridge piers retrofitted with FRP composites[J]. Construction and Building Materials. 2010，24(6)：980-998.

[48] Fahmy M F M. Enhancing recoverability and controllability of reinforced concrete bridge frame columns using FRP composites[D]. Doctor of Philosophy，Ibaraki University，Hitachi，Japan，2010.

[49] Ozcebe G，Saatcioglu M. Confinement of concrete columns for seismic loading[J]. ACI Structural Journal. 1987，84(4)：308-315.

[50] Kawashima K，Shoji M，Sakakibara Y. A cyclic loading test for clarifying the plastic hinge length

of reinforced concrete piers[J]. Journal of Structural Engineering, Japan. 2000, 46(2): 767-776.

[51] Nagaya K, Kawashima K. Effect of aspect ratio and longitudinal reinforcement diameter on seismic performance of reinforced concrete bridge columns[C]. Tokyo Institute of Technology , Tokyo, Japan, 2002.

[52] Wehbe N I, Saiidi M S, Sanders D H. Seismic performance of rectangular bridge columns with moderate confinement[J]. ACI Structural Journal. 1999, 96(2): 248-259.

[53] 吴智深，吴刚，吕志涛. 钢-连续纤维复合筋增强混凝土抗震结构[P]. 中国: CN1936206A, 2006.

[54] Wu G, Wu Z S, Luo Y B, Sun Z Y, Hu X Q. Mechanical properties of steel-FRP composite bar under uniaxial and cyclic tensile loads[J]. Journal of Materials in Civil Engineering, ASCE. 2010, 22(10): 1056-1066.

[55] 孙泽阳，吴刚，吴智深，等. 钢-连续纤维复合筋(SFCB)与混凝土粘结性能试验研究[J]. 工程抗震与加固改造. 2009, 31(1): 21-27.

[56] Zhao J, Sritharan S. Modeling of Strain Penetration Effects in Fiber-Based Analysis of Reinforced Concrete Structures[J]. ACI Structural Journal. 2007, 104(2): 133-141.

[57] 孙泽阳，吴刚，吴智深，张敏. 钢-连续纤维复合筋增强混凝土柱抗震性能试验研究[J]. 土木工程学报. 正式录用，稿号(0020179).

[58] Takeda T, Sozen M A, Nielsen N N. Reinforced concrete response to simulated earthquakes[J]. Journal of the Structural Division, ASCE. 1970, 96(12): 2557-2573.

[59] http: //peer. berkeley. edu/nga/.

[60] 郝建兵. 新型混杂配筋增强混凝土柱抗震性能研究[D]. 硕士论文，南京: 东南大学，2010.

# 第 15 章 Chapter 15

# 混凝土再生混合构件的基本力学性能研究及应用*

吴　波，赵新宇

（华南理工大学　亚热带建筑科学国家重点实验室，广州 510640）

**提　要：** 介绍了混凝土再生混合构件的基本概念，以及外置型钢再生混合构件和再生混合钢筋混凝土构件基本力学性能的初步研究成果，最后给出了钢管再生混合柱的一个工程应用实例。研究结果表明：①废弃混凝土大尺度（块体/节段层次）回收利用的经济性明显优于再生骨料层次的回收利用。②新、旧混凝土的混合强度可在二者各自强度的基础上进行初步估算。③混凝土再生混合构件的力学性能总体上接近或略低于全现浇构件，现行规范有关后者的计算公式可直接或略作修正后应用于前者。混凝土再生混合构件是废弃混凝土循环利用的一种新的有效途径，具有广阔应用前景。

**关键词：** 废弃混凝土；循环利用；混合物；构件；力学性能；工程应用

# STUDY ON MECHANICAL PROPERTIES OF STRUCTURAL MEMBERS FILLED WITH DEMOLISHED CONCRETE SEGMENTS/LUMPS AND ITS APPLICATION

B. Wu[1], X. Y. Zhao[1]

(1. State Key Laboratory of Subtropical Building Science,
South China Univ. of Technology, Guangzhou 510640.)

**Abstract:** The conception of structural members filled with demolished concrete segments/lumps (DCSs/DCLs) is introduced in this paper. The mechanical properties of steel hollow section members filled with DCSs/DCLS and reinforced concrete members filled with DC-Ss/DCLs are reported herein. Finally, a real building with steel tubular columns filled with DCLs is presented. Research results show that: (a) the reuse of demolished concrete in a scale of segments/lumps is much more economical than that in a scale of recycled aggregates; (b) the strength of the mixture of new concrete and old concrete can be preliminarily assessed using the strengths of the two kinds of concretes; and (c) the mechanical

* 基金项目：广东省科技计划项目（2009B060700094）、亚热带建筑科学国家重点实验室人才培养基金项目（2008ZB15）、华南理工大学中央高校基本科研业务费重点资助项目（2009zz0046）.

properties of the structural members filled with DCSs/ DCLs are similar to or little less than those of cast-in-situ members, and the formulas in codes for the cast-in-situ members can be used directly or slightly revised for the members filled with DCSs/DCLs. It is indicated that the structural members filled with DCSs/DCLs are new effective approaches of recycling demolished concrete, and own profound development potentials.

**Keywords**: demolished concrete; recycling use; mixture; structural members; mechanical property; engineering application

## 15.1 引言

### 15.1.1 废弃混凝土循环利用的意义

随着我国经济社会的发展，建筑业也迎来了迅猛发展的黄金时期。在大量新建建筑物拔地而起的同时，大批旧有建筑物也在因拆除而不断消失，其中不乏使用年限仅 10～20 年的次新建筑物。据统计[1]，2005 年我国房屋竣工面积 150116.47 万平方米，比 2004 年增长 1.9%。以此速度估算，2009 年全国房屋竣工面积可达 161854.61 万平方米。一般来说，建筑施工过程中产生的建筑垃圾约为 500～600 吨/万平方米，这样 2009 年全国建筑施工至少产生建筑垃圾约 8092.73 万吨。在世界多数国家，建筑施工产生的建筑垃圾还不及旧有建筑物拆除产生建筑垃圾的一半[2]。照此估算，2009 年全国旧有建筑物拆除产生的建筑垃圾至少达 16185.46 万吨。通常废弃混凝土约占建筑垃圾的 48.35%[3]，即 2009 年全国旧有建筑物拆除可产生约 7825.67 万吨的废弃混凝土。

目前，我国废弃混凝土绝大部分都是直接运往郊外堆放填埋，不仅占用大量土地资源，而且耗费高额的清运和管理费用。另一方面，新建建筑物所需混凝土又完全采用新鲜水泥、石子、河砂等材料配制而成，由此造成的危害包括：①过度开山采石破坏地表植被和原有地貌，造成水土流失和生态环境恶化且易诱发地质灾害；②大量淘挖河砂毁坏河床和堤坝，危及河流安全；③水泥生产消耗大量能源并造成严重环境污染。因此，废弃混凝土循环利用已势在必行。

如何科学利用旧有建筑物拆除产生的废弃混凝土，避免或减少旧混凝土填埋和新混凝土生产所引发的各种负面影响，已成为整个建筑行业践行“环保、节能、减排”国家战略所亟待解决的问题。

### 15.1.2 再生骨料混凝土简介

目前，再生骨料混凝土（简称“再生混凝土”）在国内外已得到一定程度的研究[4-16]，但在我国建筑行业的应用还不多见。所谓再生骨料混凝土，就是将旧有建筑物拆除所得的废弃构件，经过破碎、筛分、净化等过程，获取大量骨料尺度的颗粒，然后利用其部分或全部代替天然骨料配制而成的混凝土。再生骨料按粒径大小可分为两类：再生粗骨料（粒径 5～31.5mm）和再生细骨料（粒径 0～5mm）。我国再生骨料混凝土主要应用于道路路基等场合。

由于再生骨料的生产过程繁琐且消耗大量能源，加之在再生骨料混凝土的配制过程中仍然需要耗费大量水、水泥和能源，因此有必要寻求一种更为经济简便的废弃构件回收利用策略。

### 15.1.3 混凝土再生混合构件的提出

所谓混凝土再生混合构件，就是将旧有建筑物拆除所得的废弃混凝土构件（如梁、板、柱、墙等）去除保护层、纵筋、箍筋之后的剩余部分分割成较大尺度的块体或节段，施工时放入外置型钢或预先绑扎有钢筋的模板内部，然后浇入新混凝土并混合振捣而成的结构构件。废弃混凝土构件保护层的去除，可有效剥离构件表层的碳化混凝土，并便于纵筋和箍筋的回收。废弃混凝土块体一般呈不规则形状，以凸多面体为宜，特征尺寸 0.1m 以上；废弃混凝土节段一般长约 0.5m 至数米，以施工需要为准。

混凝土再生混合构件可分为如下两个系列：

(1) 外置型钢再生混合构件系列：将废弃混凝土块体或节段与新混凝土交替放入外置型钢内部，然后混合振捣而成的结构构件。例如：圆钢管再生混合柱、矩形钢管再生混合柱、U 型外包钢再生混合梁、外置钢板再生混合墙、底钢再生混合板等。

(2) 再生混合钢筋混凝土构件系列：将废弃混凝土块体或节段与新混凝土交替放入预先绑扎有钢筋的模板内部，然后混合振捣而成的结构构件。例如：再生混合钢筋混凝土柱、再生混合钢筋混凝土梁、再生混合钢筋混凝土墙、再生混合钢筋混凝土板等。

与国内外已有的再生骨料混凝土相比，混凝土再生混合构件主要是从更为宏观的角度（即块体或节段层次而非骨料层次）对废弃混凝土构件进行回收利用，从而省略再生骨料的繁琐生产过程，节省废弃构件破碎、筛分、净化等工序所消耗的大量能源，以及与块体或节段体积相当的再生骨料混凝土配制所耗费的大量水、水泥和能源。

### 15.1.4 预期效益分析

根据前面的分析可知，2009 年全国旧有建筑物拆除产生的建筑垃圾至少达 16185.46 万吨。假设其中的 1/10 采用混凝土再生混合构件的方式予以回收利用，并统一取混凝土容重为 2280kg/m$^3$，则每年可回收利用废弃混凝土约 709.9 万立方米。据统计[17]，每立方米混凝土消耗的资源和能源如表 1 所示。由于 1kWh＝0.361kg 标准煤、1MJ＝0.034kg 标准煤[18]，据此可估算出每年回收利用 709.9 万立方米废弃混凝土所节约的资源及能源量，具体见表 2。

从表 2 中可以看出，混凝土再生混合构件的应用，可以节约大量资源和能源，同时显著减少温室气体的排放。其中每年节约标准煤约 41.5 万吨，相当于一个中型煤矿全年的产量。

按照目前每立方米新混凝土约 300 元的市场价估算，每年 709.9 万立方米废弃构件的回收利用，可为业主节省约 21.3 亿元的直接投资，经济效益巨大。

此外，由于大量采用废弃混凝土，混凝土再生混合构件对新混凝土以及相应的水泥、石子、河砂等材料的需求明显减少，环境效益明显。

**每立方米混凝土消耗的资源和能源　表 1**

| 材料 | 石灰石 (kg) | 黏土 (kg) | 铁粉 (kg) | 集料 (kg) | 水 (kg) | 电耗 (kWh) | 煤耗 (MJ) | 排放 $CO_2$ (kg) |
|---|---|---|---|---|---|---|---|---|
| 水泥 | 372.0 | 62.0 | 18.5 | — | — | 42.0 | 1004 | 254.0 |
| 集料 | — | — | — | 1800 | — | 25.2 | — | 6.2 |
| 水 | — | — | — | — | 170 | — | — | — |
| 混凝土（合计） | 372.0 | 62.0 | 18.5 | 1800 | 170 | 67.2 | 1004 | 260.2 |

**年均节约资源和能源　表 2**

| 资源节约量(万吨) | | | | | 能源节约量(万吨) | | 排放减少量(万吨) |
|---|---|---|---|---|---|---|---|
| 石灰石 | 黏土 | 铁粉 | 集料 | 水 | 电耗（标准煤） | 煤耗（标准煤） | $CO_2$ 排放 |
| 264.1 | 44.0 | 13.0 | 1277.9 | 120.6 | 17.2 | 24.3 | 184.7 |

## 15.2　外置型钢再生混合构件的基本力学性能

为深入揭示外置型钢再生混合构件的力学性能并建议其设计方法，促进该类构件的工程应用，本课题组正积极开展该类构件基本力学性能的系列研究工作。限于篇幅，下面主要介绍不同壁厚圆钢管再生混合短柱的轴压性能，以及薄壁圆钢管再生混合柱的受剪性能。

### 15.2.1　中厚壁圆钢管再生混合短柱的轴压性能

（1）试验方案

共 17 个试件，分为 7d 轴压试验和 28d 轴压试验两组。所有试件均采用外径 219mm、壁厚 4mm（径厚比 54.8）、高度 657mm 的直缝焊接钢管。12 个试件的钢管内填充有不同型式（节段或块体）的废弃混凝土，作为对比 5 个试件的钢管内全部填充现浇混凝土。这样，所有试件可分成节段型钢管再生混合短柱、块体型钢管再生混合短柱、全现浇钢管混凝土短柱三类。各试件的基本参数见表 3。对应每一工况，试件数量一般取为 3 个，分别以 A、B、C 表示。

现浇混凝土所用材料包括：矿渣硅酸盐水泥、最大粒径 20mm 碎石、中砂。废弃混凝土取自几年前华南理工大学本科教学试验多余的钢筋混凝土梁，将去除了保护层、纵筋、箍筋之后的核心部分，制备成所需节段或块体备用，具体见图 1。

图 1　节段和块体照片

利用与多余钢筋混凝土梁同时浇筑的剩余立方体试块，测得试验时废弃混凝土的抗压强度。利用与现浇混凝土同时浇筑的立方体试块，分别测取其 7d 和 28d 抗压强度。将钢管切开，按照标准试验方法进行拉伸试验以获得其屈服强度。具体材性试验结果见表 3。

**试 件 参 数** **表 3**

| 序号 | 试件编号 | $f_{c1u}$ (MPa) | $f_{c2u}$ (MPa) | $f_s$ (MPa) | 废弃混凝土型式 | $\eta$ | 实测抗压承载力 (kN) |
|---|---|---|---|---|---|---|---|
| 1 | 7-A | 23.53 | 37.1 | 294.7 | 无 | 0 | 2000 |
| 2 | 7-B | | | | 无 | 0 | 2000 |
| 3 | J-7-A | | | | 节段型 | 33.2% | 2040 |
| 4 | J-7-B | | | | 节段型 | 34.8% | 2070 |
| 5 | J-7-C | | | | 节段型 | 34.7% | 2045 |
| 6 | K-7-A | | | | 块体型 | 31.8% | 2090 |
| 7 | K-7-B | | | | 块体型 | 33.7% | 2040 |
| 8 | K-7-C | | | | 块体型 | 34.7% | 2035 |
| 9 | 28-A | 34.67 | 37.1 | 294.7 | 无 | 0 | 2150 |
| 10 | 28-B | | | | 无 | 0 | 2120 |
| 11 | 28-C | | | | 无 | 0 | 2155 |
| 12 | J-28-A | | | | 节段型 | 34.9% | 2250 |
| 13 | J-28-B | | | | 节段型 | 34.9% | 2300 |
| 14 | J-28-C | | | | 节段型 | 35.0% | 2260 |
| 15 | K-28-A | | | | 块体型 | 33.4% | 2220 |
| 16 | K-28-B | | | | 块体型 | 35.0% | 2230 |
| 17 | K-28-C | | | | 块体型 | 35.0% | 2230 |

注：$f_{c1u}$和 $f_{c2u}$分别为试验当天现浇和废弃混凝土的立方体抗压强度；$f_s$ 为钢管屈服强度。试件编号中，7 和 28 分别表示 7d 和 28d 轴压试验；J 和 K 分别表示节段型和块体型钢管再生混合短柱；$\eta$ 为混合比，即钢管内部废弃混凝土质量与全部混凝土质量之比。

制作节段型钢管再生混合短柱时，首先从钢管上端开口处灌入约 20mm 厚的现浇混凝土，然后在钢管内部放置一根预先制备的废弃混凝土节段（注：节段低于钢管开口面约 20mm），最后在节段与钢管内壁之间灌入现浇混凝土并采用插入式振捣棒不断振捣，直至钢管填满。块体型钢管再生混合短柱的制作过程与上述过程类似，只是现浇混凝土与废弃混凝土块体交替加入钢管内并不断振捣，直至钢管填满。浇筑完成后，试件上端敞口自然养护。试验前敞口处用水泥砂浆找平，然后焊好盖板。图 2 所示为试件制作过程中的典型照片。

试验在华南理工大学结构实验室的 1500t 长柱压力机上进行。每个试件半高处对称布设四个应变花，试件侧面对称设置两个量程 100mm 的位移计。施加竖向荷载时，首先采用力控制策略（注：每级荷载 100kN），接近预估的抗压承载力时改用位移控制策略。每级荷载持续时间约 1～2min。图 3 所示为加载与测量装置的示意图。

（2）试验结果与分析

各试件的最终破坏形态见图 4。7d 和 28d 轴压试验对应的荷载—变形曲线分别见图 5 和图 6。图中各曲线的纵坐标都是该曲线所对应工况下不同试件的平均值，RCSFST、

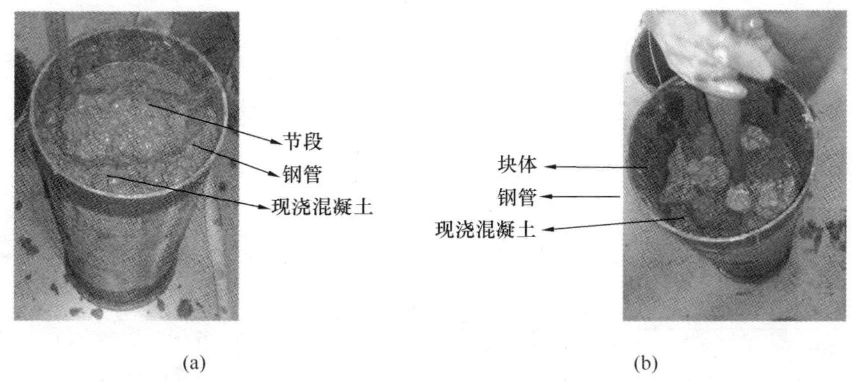


图 2　试件制作过程典型照片
(a) 节段型钢管再生混合短柱；(b) 块体型钢管再生混合短柱

RCLFST、CFST 分别表示节段型钢管再生混合短柱、块体型钢管再生混合短柱、全现浇钢管混凝土短柱。

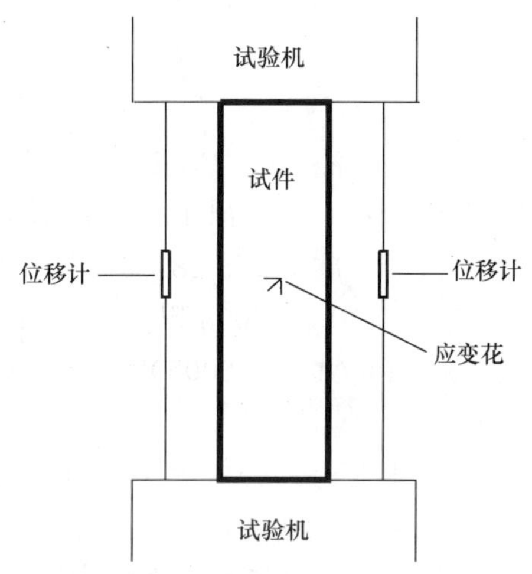


图 3　加载与测量装置

从图 5 和图 6 中可以看出：

①无论 7d 还是 28d，三类短柱的荷载—竖向变形曲线的弹性阶段几乎重合，表明三类短柱具有相同的初始刚度。

②无论 7d 还是 28d，节段型和块体型钢管再生混合短柱的抗压承载力都略大于全现浇钢管混凝土短柱。这是由于试验当天现浇混凝土的抗压强度仍低于废弃混凝土所致。可以预计，当现浇和废弃混凝土的试验当天抗压强度相等时，三类短柱的抗压承载力应十分接近，即 32%～35%废弃混凝土的采用并未导致钢管再生混合短柱的抗压承载力降低。

图 4　试件破坏形态

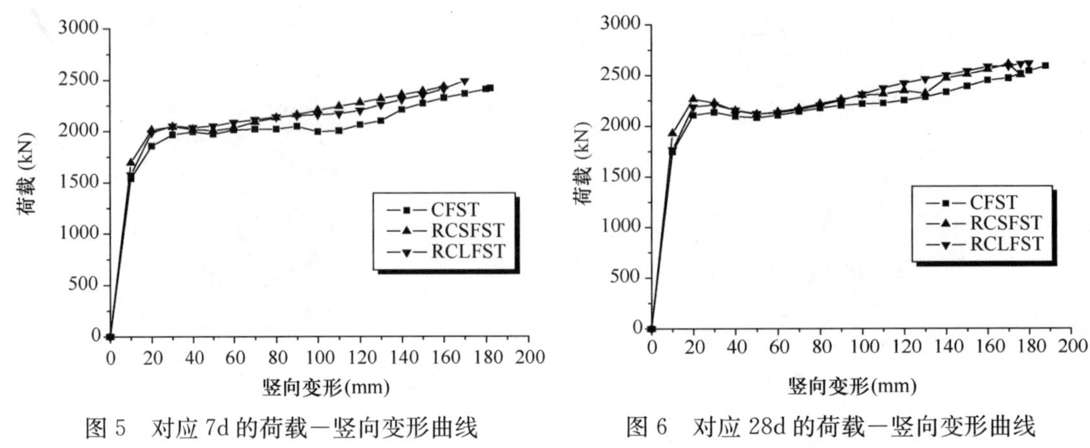


图 5 对应 7d 的荷载—竖向变形曲线　　图 6 对应 28d 的荷载—竖向变形曲线

③无论 7d 还是 28d，三类短柱的荷载—竖向变形曲线都展现出了较为相似的延性特征。

（3）计算比较

关于全现浇钢管混凝土短柱的抗压承载力计算，国内外已开展了较多研究工作，并在部分设计标准中给出了具体规定。例如，国家建材局标准 JCJ 01—89[19]、工程建设标准化协会标准 CECS 28：90[20]、电力行业标准 DL/T 5085—1999[21]、福建省工程建设标准 DBJ 13-51-2003[22]、天津市工程建设标准 DB 29-57-2003[23] 等，以及美国 ACI（2005）[24] 和 AISC（2005）[25]、日本 AIJ（1997）[26]、英国 BS（2005）[27]、欧洲 EC4（2004）[28] 等。

下面根据 JCJ 01—89、CECS 28：90、DL/T 5085—1999、DBJ 13-51-2003、ACI（2005）、AISC（2005）、EC4（2004）和 AIJ（1997）共 8 部设计标准，分别计算各试件的 28d 抗压承载力并与试验实测值进行比较，具体见表 4。计算过程中，混凝土抗压强度考虑三种取法：

①现浇混凝土的 28d 强度 $f_{c1u}$；②废弃混凝土的当前强度 $f_{c2u}$；③现浇和废弃混凝土的组合强度 $f'_{cu}=f_{c1u}\times(1-\eta)+f_{c2u}\times\eta$。根据不同设计标准计算得到的试件抗压承载力的统计精度见表 5。

从表 4 和表 5 中可以看出：

①相对于其他 6 部设计标准，依据 JCJ 01—89 和 EC4（2004）给出的三类短柱的抗压承载力计算结果与试验实测值最为接近。

②针对 JCJ 01—89 而言，在混凝土抗压强度的三种取法中，取法①的计算精度≈取法③的计算精度>取法②的计算精度。

③虽然 JCJ 01—89 的抗压承载力计算公式是针对全现浇钢管混凝土构件给出的，但同样适用于钢管再生混合构件。

根据上述分析并考虑到我国设计人员直接采用国外标准 EC4（2004）存在困难，建议中厚壁钢管再生混合短柱的抗压承载力可采用 JCJ 01—89 的公式进行计算，并根据取法③确定计算过程中混凝土的抗压强度。

计算结果与试验结果的比较　表 4

| 混凝土抗压强度 | 试件编号 | 实测抗压承载力(kN) | 实测抗压承载力/计算抗压承载力 | | | | | | | |
|---|---|---|---|---|---|---|---|---|---|---|
| | | | JCJ | CECS | DL/T | DBJ | ACI | AISC | EC4 | AIJ |
| 取法① | 28-A | 2150 | 0.968 | 0.892 | 1.071 | 1.150 | 2.082 | 1.255 | 0.935 | 1.194 |
| | 28-B | 2120 | 0.955 | 0.880 | 1.056 | 1.134 | 2.053 | 1.237 | 0.922 | 1.178 |
| | 28-C | 2155 | 0.971 | 0.895 | 1.074 | 1.153 | 2.087 | 1.258 | 0.937 | 1.197 |
| | J-28-A | 2250 | 1.013 | 0.934 | 1.121 | 1.203 | 2.179 | 1.313 | 0.978 | 1.250 |
| | J-28-B | 2300 | 1.036 | 0.955 | 1.146 | 1.230 | 2.228 | 1.342 | 1.000 | 1.278 |
| | J-28-C | 2260 | 1.018 | 0.938 | 1.126 | 1.209 | 2.189 | 1.319 | 0.983 | 1.256 |
| | K-28-A | 2220 | 1.000 | 0.921 | 1.106 | 1.187 | 2.150 | 1.295 | 0.965 | 1.233 |
| | K-28-B | 2230 | 1.004 | 0.926 | 1.111 | 1.193 | 2.160 | 1.301 | 0.970 | 1.239 |
| | K-28-C | 2230 | 1.004 | 0.926 | 1.111 | 1.193 | 2.160 | 1.301 | 0.970 | 1.239 |
| 取法② | 28-A | 同上 | 0.944 | 0.862 | 1.033 | 1.108 | 2.009 | 1.208 | 0.908 | 1.156 |
| | 28-B | | 0.931 | 0.849 | 1.019 | 1.092 | 1.981 | 1.191 | 0.896 | 1.140 |
| | 28-C | | 0.946 | 0.864 | 1.036 | 1.110 | 2.014 | 1.211 | 0.910 | 1.159 |
| | J-28-A | | 0.988 | 0.902 | 1.081 | 1.159 | 2.102 | 1.264 | 0.951 | 1.210 |
| | J-28-B | | 1.010 | 0.922 | 1.105 | 1.185 | 2.149 | 1.292 | 0.972 | 1.237 |
| | J-28-C | | 0.992 | 0.906 | 1.086 | 1.164 | 2.112 | 1.270 | 0.955 | 1.216 |
| | K-28-A | | 0.974 | 0.890 | 1.067 | 1.144 | 2.074 | 1.247 | 0.938 | 1.194 |
| | K-28-B | | 0.979 | 0.894 | 1.072 | 1.149 | 2.084 | 1.253 | 0.942 | 1.199 |
| | K-28-C | | 0.979 | 0.894 | 1.072 | 1.149 | 2.084 | 1.253 | 0.942 | 1.199 |
| 取法③ | 28-A | 同上 | 0.968 | 0.892 | 1.071 | 1.150 | 2.082 | 1.255 | 0.935 | 1.194 |
| | 28-B | | 0.955 | 0.880 | 1.056 | 1.134 | 2.053 | 1.237 | 0.922 | 1.178 |
| | 28-C | | 0.971 | 0.895 | 1.074 | 1.153 | 2.087 | 1.258 | 0.937 | 1.197 |
| | J-28-A | | 1.004 | 0.922 | 1.107 | 1.188 | 2.152 | 1.296 | 0.969 | 1.236 |
| | J-28-B | | 1.027 | 0.943 | 1.132 | 1.214 | 2.200 | 1.325 | 0.990 | 1.264 |
| | J-28-C | | 1.009 | 0.927 | 1.112 | 1.193 | 2.161 | 1.301 | 0.973 | 1.241 |
| | K-28-A | | 0.991 | 0.911 | 1.093 | 1.173 | 2.124 | 1.279 | 0.956 | 1.220 |
| | K-28-B | | 0.995 | 0.914 | 1.097 | 1.177 | 2.133 | 1.284 | 0.960 | 1.225 |
| | K-28-C | | 0.995 | 0.914 | 1.097 | 1.177 | 2.133 | 1.284 | 0.960 | 1.225 |

计算结果的统计特性　表 5

| 混凝土抗压强度 | 统计特征值 | 实测抗压承载力/计算抗压承载力 | | | | | | | |
|---|---|---|---|---|---|---|---|---|---|
| | | JCJ | CECS | DL/T | DBJ | ACI | AISC | EC4 | AIJ |
| 取法① | 均值 | 0.9966 | 0.9186 | 1.1024 | 1.1836 | 2.1431 | 1.3247 | 0.9622 | 1.2293 |
| | 变异系数 | 0.0266 | 0.0267 | 0.0267 | 0.0265 | 0.0267 | 0.0267 | 0.0266 | 0.0267 |
| 取法② | 均值 | 0.9714 | 0.8870 | 1.0634 | 1.1400 | 2.0677 | 1.2760 | 0.9349 | 1.1900 |
| | 变异系数 | 0.0266 | 0.0268 | 0.0265 | 0.0266 | 0.0266 | 0.0264 | 0.0268 | 0.0267 |
| 取法③ | 均值 | 0.9906 | 0.9109 | 1.0932 | 1.1732 | 2.1250 | 1.3128 | 0.9558 | 1.2200 |
| | 变异系数 | 0.0227 | 0.0212 | 0.0213 | 0.0209 | 0.0212 | 0.0209 | 0.0222 | 0.0218 |

### 15.2.2 薄壁圆钢管再生混合短柱的轴压性能

为在量大面广的中低层建筑中推广应用钢管再生混合构件，有必要进一步降低其造价使之与一般钢筋混凝土构件接近，此时采用薄壁钢管无疑是有益的。有鉴于此，开展了20个薄壁钢管再生混合短柱的轴压试验，比较了国内外相关公式预测试件抗压承载力的有效性，同时还在横截面积和用钢量相同的情况下对比了钢筋混凝土短柱与试件的抗压承载力。

(1) 试验方案

20个试件均采用外径200mm、壁厚1mm（径厚比200）、高度600mm的直缝焊接钢管。其中16个试件的钢管内部填充有不同型式（节段或块体）的废弃混凝土，余下4个试件的钢管内部全部填充现浇混凝土以作对比。这样，所有试件可分为节段型钢管再生混合短柱、块体型钢管再生混合短柱、全现浇钢管混凝土短柱三类。对应每一工况，试件数量取为2个，分别以A、B表示。各试件的基本参数见表6，试件编号的具体含义见图7。

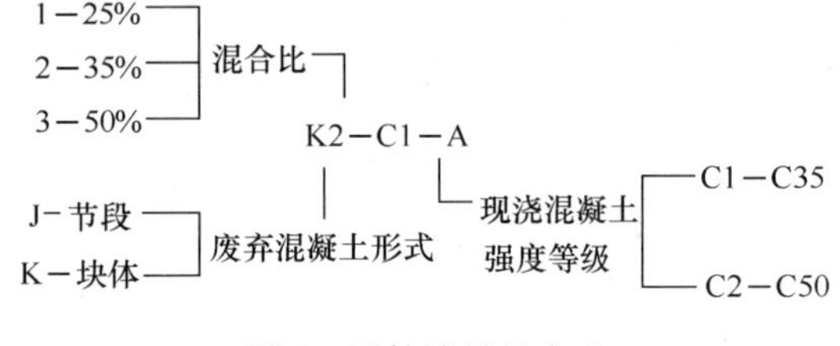


图7 试件编号的含义

现浇混凝土考虑C35和C50两个强度等级，前者采用矿渣硅酸盐水泥（P.S.A32.5），后者采用普通硅酸盐水泥（P.O42.5），二者均采用中砂和最大粒径20mm碎石。废弃混凝土取自几年前华南理工大学本科教学试验多余的钢筋混凝土梁，将去除了保护层、纵筋、箍筋之后的核心部分制备成所需节段或块体备用。试件制作过程和试验过程与前面中厚壁圆钢管再生混合短柱类似，在此不再赘述。

试 件 参 数 表6

| 序号 | 试件编号 | $f_{c1u}$ (MPa) | $f_{c2u}$ (MPa) | $f_s$ (MPa) | 试件型式 | 混合比 $\eta$ | 抗压承载力 (kN) | |
|---|---|---|---|---|---|---|---|---|
| | | | | | | | 实测值 | 平均值 |
| 1 | C1-A | 36.3 | 37.1 | 266.7 | 全现浇 | 0 | 966 | 1039 |
| 2 | C1-B | | | | 全现浇 | 0 | 1112 | |
| 3 | K1-C1-A | | | | 块体型 | 25.0% | 1048 | 1063 |
| 4 | K1-C1-B | | | | 块体型 | 24.8% | 1078 | |
| 5 | K2-C1-A | | | | 块体型 | 33.2% | 954 | 998 |
| 6 | K2-C1-B | | | | 块体型 | 35.0% | 1041 | |
| 7 | J2-C1-A | | | | 节段型 | 35.3% | 1101 | 1140 |
| 8 | J2-C1-B | | | | 节段型 | 33.6% | 1178 | |
| 9 | J3-C1-A | | | | 节段型 | 48.2% | 1131 | 1150 |
| 10 | J3-C1-B | | | | 节段型 | 49.4% | 1168 | |
| 11 | C2-A | 50.3 | 37.1 | 266.7 | 全现浇 | 0 | 1359 | 1388 |
| 12 | C2-B | | | | 全现浇 | 0 | 1417 | |
| 13 | K1-C2-A | | | | 块体型 | 24.8% | 1252 | 1271 |
| 14 | K1-C2-B | | | | 块体型 | 24.6% | 1289 | |
| 15 | K2-C2-A | | | | 块体型 | 34.6% | 1293 | 1278 |
| 16 | K2-C2-B | | | | 块体型 | 31.8% | 1262 | |
| 17 | J2-C2-A | | | | 节段型 | 34.5% | 1414 | 1386 |
| 18 | J2-C2-B | | | | 节段型 | 34.7% | 1357 | |
| 19 | J3-C2-A | | | | 节段型 | 45.3% | 1384 | 1374 |
| 20 | J3-C2-B | | | | 节段型 | 49.5% | 1363 | |

注：$f_{c1u}$和$f_{c2u}$分别为试验当天现浇和废弃混凝土的立方体抗压强度；$f_s$为钢管屈服强度；$\eta$为钢管内部废弃混凝土质量与全部混凝土质量之比。

（2）试验结果与分析

各试件的宏观破坏见图 8。图 9 所示为试件的荷载—变形曲线，图中各曲线的纵坐标都是该曲线所对应工况下 A、B 两个试件的平均值。从图中可以看出：

图 8　试件破坏形态

①试件的荷载—变形曲线由上升段和下降段组成，下降段总体呈现出荷载随位移增加开始降低明显随后渐趋平缓的趋势，从而表现出与前文所述中厚壁试件的荷载—变形曲线后期仍持续上升明显不同的特征，这主要是由于薄壁钢管对混凝土的约束作用较弱的缘故。需要指出的是，试验过程中个别试件因焊缝较早开裂而提前破坏，导致下降段相对较短。

②在其他参数相近的情况下，块体型钢管再生混合短柱、节段型钢管再生混合短柱和全现浇钢管混凝土短柱的荷载—变形曲线的弹性阶段几乎重合，表明三类短柱具有几乎相同的初始刚度。

③所用废弃混凝土相同时，现浇混凝土的抗压强度越高，块体型和节段型钢管再生混合短柱的抗压承载力都越大。

④混合比相近时，节段型钢管再生混合短柱的抗压承载力大丁块体型钢管再生混合短柱。这可能是由于节段型废弃混凝土的表面积小于相同质量的块体型废弃混凝土，前者与现浇混凝土之间出现界面缺陷的可能性小于后者的相应可能性，从而使得节段型钢管再生混合短柱的抗压承载力偏高。

⑤对于节段型和块体型钢管再生混合短柱，当混合比分别在 35%～50%和 25%～35%之间变化时，抗压承载力变化不大。

⑥当现浇混凝土立方体抗压强度大于废弃混凝土约 0～15MPa 时，混合比 25%～35%的块体型钢管再生混合短柱相比于全现浇钢管混凝土短柱的抗压承载力偏差约−8.4%～2.3%，而混合比 35%～50%的节段型钢管再生混合短柱相比于全现浇钢管混凝土短柱的抗压承载力偏差约−1.0%～10.6%。考虑到试验误差的偶然性，从方便工程应用角度出发，可近似认为在上述强度差和混合比范围内，三类短柱具有相近的抗压承载力。

（3）计算比较

不同学者对薄壁钢管混凝土的相关问题进行了探讨[29～33]。为考察现有的抗压承载力计算方法是否适用于薄壁钢管再生混合短柱，下面根据 JCJ 01—89、CECS 28 ： 90、DL/T

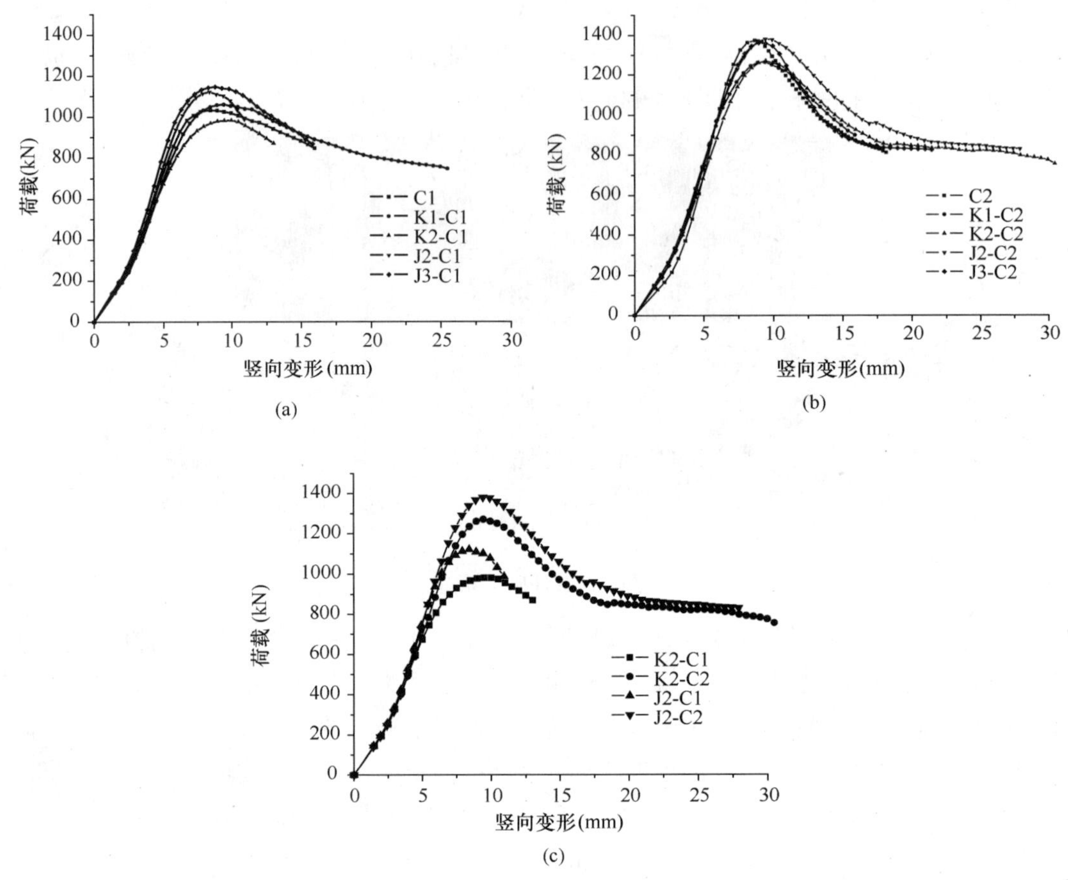


图 9 荷载—变形曲线

(a) 现浇混凝土强度等级 C35；(b) 现浇混凝土强度等级 C50；(c) 混合比约 35%

5085—1999、DBJ 13-51-2003、ACI 318-05、ANSI/AISC 360-05、EC4-2004 和 AIJ-1997 共 8 部设计标准，以及张耀春等[32]和占美森[33]的研究成果，分别计算各试件的抗压承载力并与试验实测值进行比较。为清楚起见，不同计算方法的适用范围统一列于表 7 中。

计算过程中，钢管屈服强度直接取表 6 中的实测值，混凝土立方体抗压强度取现浇混凝土和废弃混凝土的组合强度 $f'_{cu}=f_{c1u}\times(1-\eta)+f_{c2u}\times\eta$，其中 $f_{c1u}$、$f_{c2u}$和 $\eta$ 均取表 6 中的实测值。对于我国标准 JCJ 01—89、CECS 28 : 90、DL/T 5085—1999、DBJ 13—51—2003 和占美森的计算方法，取 $f_c=0.88\times0.76f'_{cu}$(注：0.88 为考虑构件混凝土强度与试块混凝土强度之间差异等因素的修正系数)[34]；对于 ACI 318—05、ANSI/AISC 360—05 和 EC4—2004，取 $f'_c=0.8\times f'_{cu}$[35]；对于 AIJ—1997，文献[36]指出 $f'_c=0.96F_c$，由此可导出 $F_c=0.83f'_{cu}$；张耀春等在文献［32］中提出计算公式时，混凝土棱柱体与立方体之间的抗压强度转换系数为 0.76，本文在采用其计算方法时也取 $f_c=0.76f'_{cu}$。其中 $f_c$ 为 150mm×150mm×300mm 的混凝土棱柱体抗压强度；$f'_{cu}$为直径 6in (152.4mm)、高度 12in (304.8mm) 的混凝土圆柱体抗压强度；$F_c$ 为直径 100mm、高度 200mm 的混凝土圆柱体抗压强度。计算结果与试验结果的比较见表 8，相应的统计特性见表 9。从表中可以看出：

**计算方法的适用范围　表 7**

| 计算方法 | 适用范围 | Q235 钢材对应的径厚比限制 |
| --- | --- | --- |
| JCJ 01—89 | 3 号钢和 16Mn 钢的含钢率宜分别采用 0.04～0.16 和 0.04～0.12；一般情况下钢管壁厚不宜小于 4mm | 25～100 |
| CECS 28：90 | 套箍指标宜限制在 0.3～3 之间；钢管外径不宜小于 100mm；壁厚不宜小于 4mm；径厚比宜限制在 20～85 $\sqrt{235/f_y}$之间 | 20～85 |
| DL/T 5085—1999 | 含钢率为 0.04～0.20；钢管外径不宜小于 100mm；壁厚不宜小于 4mm；径厚比宜在 20～100 之间 | 20～100 |
| DBJ 13—51—2003 | 钢管外径不宜小于 100mm；壁厚不宜小于 4mm；钢管外径与壁厚之比不得大于无混凝土时相应限制的 1.5 倍 | ≤150 |
| ACI 318—05 | 径厚比不大于 $\sqrt{8E_s/f_y}$ | ≤82.5 |
| ANSI/AISC 360—05 | 钢管横截面面积应不小于总组合截面面积的 1%；圆钢管径厚比应不大于 $0.15E_s/f_y$ | ≤127.7 |
| EC4—2004 | 圆钢管径厚比应不大于 90× $(235/f_y)$ | ≤90 |
| AIJ—1997 | 径厚比不大于 1.5×240/(F/98) | ≤150 |
| 张耀春 | 根据作者的试验数据[32]以及 O'Shea 和 Bridge 的试验数据[29]回归得到计算公式，两批试验的径厚比范围分别为 80～120 和 58.5～221 | 58.5～221 |
| 占美森 | 根据作者的试验数据[33]，以及张耀春等、O'Shea 和 Bridge 的试验数据[29, 32]回归得到计算公式，三批试验的径厚比范围分别为 41～80、80～120 和 58.5～221 | 41～221 |

注：钢材屈服强度 $f_y$ 和弹性模量 $E_s$ 分别取为 235MPa 和 $2.0\times10^5$MPa；$F=\min(f_y, 0.7f_u)$，其中为 $f_u$ 为钢材极限抗拉强度，表中取 $F=f_y$。

**计算结果与试验结果的比较　表 8**

| 试件编号 | 计算抗压承载力/实测抗压承载力 | | | | | | | | | |
| --- | --- | --- | --- | --- | --- | --- | --- | --- | --- | --- |
| | JCJ | CECS | DL/T | DBJ | ACI | AISC | EC4 | AIJ | 张耀春 | 占美森 |
| C1-A | 1.102 | 1.310 | 1.158 | 1.078 | 0.814 | 1.041 | 1.177 | 1.024 | 1.235 | 1.148 |
| C1-B | 0.957 | 1.138 | 1.006 | 0.936 | 0.707 | 0.905 | 1.022 | 0.890 | 1.073 | 0.997 |
| K1-C1-A | 1.020 | 1.213 | 1.073 | 0.999 | 0.754 | 0.965 | 1.090 | 0.949 | 1.143 | 1.063 |
| K1-C1-B | 0.992 | 1.179 | 1.043 | 0.971 | 0.733 | 0.938 | 1.060 | 0.922 | 1.111 | 1.033 |
| K2-C1-A | 1.122 | 1.334 | 1.181 | 1.099 | 0.830 | 1.062 | 1.199 | 1.044 | 1.257 | 1.170 |
| K2-C1-B | 1.029 | 1.223 | 1.082 | 1.008 | 0.761 | 0.973 | 1.100 | 0.957 | 1.153 | 1.072 |
| J2-C1-A | 0.973 | 1.157 | 1.023 | 0.953 | 0.719 | 0.920 | 1.040 | 0.905 | 1.090 | 1.014 |
| J2-C1-B | 0.909 | 1.081 | 0.956 | 0.890 | 0.672 | 0.860 | 0.971 | 0.846 | 1.018 | 0.947 |
| J3-C1-A | 0.949 | 1.129 | 0.999 | 0.930 | 0.702 | 0.898 | 1.015 | 0.883 | 1.063 | 0.989 |
| J3-C1-B | 0.919 | 1.093 | 0.967 | 0.901 | 0.680 | 0.870 | 0.983 | 0.855 | 1.030 | 0.958 |
| C2-A | 0.995 | 1.189 | 1.085 | 1.013 | 0.762 | 0.978 | 1.085 | 0.952 | 1.102 | 1.041 |
| C2-B | 0.955 | 1.140 | 1.041 | 0.971 | 0.731 | 0.938 | 1.040 | 0.913 | 1.057 | 0.998 |
| K1-C2-A | 1.027 | 1.226 | 1.112 | 1.037 | 0.781 | 1.002 | 1.115 | 0.977 | 1.140 | 1.073 |
| K1-C2-B | 0.998 | 1.192 | 1.080 | 1.008 | 0.759 | 0.973 | 1.083 | 0.949 | 1.108 | 1.043 |
| K2-C2-A | 0.974 | 1.163 | 1.051 | 0.981 | 0.739 | 0.947 | 1.056 | 0.924 | 1.082 | 1.017 |
| K2-C2-B | 1.004 | 1.199 | 1.085 | 1.012 | 0.762 | 0.977 | 1.089 | 0.953 | 1.115 | 1.049 |
| J2-C2-A | 0.891 | 1.064 | 0.962 | 0.897 | 0.675 | 0.866 | 0.966 | 0.845 | 0.990 | 0.931 |
| J2-C2-B | 0.928 | 1.108 | 1.001 | 0.934 | 0.704 | 0.902 | 1.006 | 0.880 | 1.031 | 0.969 |
| J3-C2-A | 0.889 | 1.061 | 0.956 | 0.892 | 0.672 | 0.861 | 0.962 | 0.841 | 0.989 | 0.928 |
| J3-C2-B | 0.895 | 1.067 | 0.961 | 0.896 | 0.675 | 0.865 | 0.967 | 0.846 | 0.996 | 0.934 |

**计算结果与试验结果之比的统计特性** **表 9**

| 统计参数 | 计算抗压承载力/实测抗压承载力 | | | | | | | | | |
|---|---|---|---|---|---|---|---|---|---|---|
| | JCJ | CECS | DL/T | DBJ | ACI | AISC | EC4 | AIJ | 张耀春 | 占美森 |
| 均值 | 0.9764 | 1.1633 | 1.0411 | 0.9702 | 0.7316 | 0.9371 | 1.0512 | 0.9178 | 1.0892 | 1.0187 |
| 变异系数 | 0.0663 | 0.0656 | 0.0638 | 0.0637 | 0.0638 | 0.0637 | 0.0645 | 0.0640 | 0.0677 | 0.0659 |

①虽然 JCJ、CECS、DL/T、DBJ、ACI、ANSI/AISC、EC4 和 AIJ 的计算方法都要求钢管径厚比明显小于本节试件取值 200，但除 CECS 和 ACI 以外，采用其余计算方法获得的试件抗压承载力计算值与实测值之比的均值仍小于 10%。

②采用占美森的计算方法获得的试件抗压承载力计算值与实测值之比的均值仅为 1.0187，优于所列其他计算方法。

③采用不同计算方法获得的试件抗压承载力计算值与实测值之比的变异系数十分接近，平均约 0.0649。

为进一步展示薄壁钢管再生混合柱的优越性，下面在横截面积和用钢量相同的情况下，就本节试件与钢筋混凝土短柱的抗压承载力进行对比。钢筋混凝土短柱的截面尺寸和高度分别取为 177.2mm×177.2mm 和 600mm，箍筋 $\phi6@100$，纵筋 $4\phi12$（注：配筋率 1.44%），相应的用钢量为 2.92kg，这与本节试件的用钢量 2.94kg 几乎相等。钢筋混凝土短柱的抗压承载力按照《混凝土结构设计规范》（GB 50010—2002）计算，计算过程中混凝土立方体抗压强度取本节对应试件的现浇混凝土和废弃混凝土组合强度 $f'_{cu}$，且 $f_c=0.88\times0.76f'_{cu}$，纵筋屈服强度取 375MPa。

表 10 所示为钢筋混凝土短柱的抗压承载力计算结果与本节试件相应试验结果的对比。从表中可以看出，薄壁钢管再生混合短柱的抗压承载力明显高于具有相同横截面积和用钢量的钢筋混凝土短柱。此外，从图 9 可以看出，薄壁钢管再生混合短柱的下降段相比于钢筋混凝土短柱明显趋缓，即前者延性更好。

**计算结果与试验结果的比较** **表 10**

| 试件编号 | 试件型式 | 混凝土组合强度 $f'_{cu}$（MPa） | 混合比 $\eta$ | 钢筋混凝土短柱抗压承载力计算值与本节试件抗压承载力实测值之比 |
|---|---|---|---|---|
| C1-A | 全现浇 | 36.3 | 0 | 0.867 |
| C1-B | 全现浇 | 36.3 | 0 | 0.753 |
| K1-C1-A | 块体型 | 36.5 | 25.0% | 0.803 |
| K1-C1-B | 块体型 | 36.5 | 24.8% | 0.781 |
| K2-C1-A | 块体型 | 36.6 | 33.2% | 0.884 |
| K2-C1-B | 块体型 | 36.6 | 35.0% | 0.810 |
| J2-C1-A | 节段型 | 36.6 | 35.3% | 0.766 |
| J2-C1-B | 节段型 | 36.6 | 33.6% | 0.716 |
| J3-C1-A | 节段型 | 36.7 | 48.2% | 0.748 |
| J3-C1-B | 节段型 | 36.7 | 49.4% | 0.724 |
| C2-A | 全现浇 | 50.3 | 0 | 0.811 |
| C2-B | 全现浇 | 50.3 | 0 | 0.778 |

续表

| 试件编号 | 试件型式 | 混凝土组合强度 $f'_{cu}$（MPa） | 混合比 $\eta$ | 钢筋混凝土短柱抗压承载力计算值与本节试件抗压承载力实测值之比 |
|---|---|---|---|---|
| K1-C2-A | 块体型 | 47.0 | 24.8% | 0.831 |
| K1-C2-B | 块体型 | 47.0 | 24.6% | 0.808 |
| K2-C2-A | 块体型 | 45.7 | 34.6% | 0.786 |
| K2-C2-B | 块体型 | 46.1 | 31.8% | 0.811 |
| J2-C2-A | 节段型 | 45.7 | 34.5% | 0.719 |
| J2-C2-B | 节段型 | 45.7 | 34.7% | 0.749 |
| J3-C2-A | 节段型 | 44.3 | 45.3% | 0.715 |
| J3-C2-B | 节段型 | 43.8 | 49.5% | 0.719 |

### 15.2.3　薄壁圆钢管再生混合柱的受剪性能

（1）试验方案

27 个试件均采用外径 200mm、壁厚 1mm（径厚比 200）的直缝焊接钢管，对应不同剪跨比的钢管高度分别为 280mm、340mm 和 420mm，根据标准试验方法测得钢材屈服强度 266.7MPa。其中 22 个试件的钢管内部填充不同型式（节段或块体）的废弃混凝土，余下 5 个试件的钢管内部全部填充现浇混凝土以作对比。这样，所有试件可分为节段型钢管再生混合柱、块体型钢管再生混合柱、全现浇钢管混凝土柱三类。各试件的基本参数见表 11，试件编号的具体含义见图 10。

1–42.0MPa
2–47.5MPa
现浇混凝土立方体抗压强度
轴压比
1–0.2
2–0.4
3–0.6
K1 – C1 – 1 – 1
剪跨比
1–0.15
2–0.3
3–0.5
K–块体型
J–节段型
混合比
1–25%
2–35%

图 10　试件编号的含义

现浇混凝土考虑 C40 和 C45 两个强度等级，前者采用矿渣硅酸盐水泥（P. S. A32.5），后者采用普通硅酸盐水泥（P. O42.5），二者均采用中砂和最大粒径 20mm 碎石。废弃混凝土取自几年前华南理工大学本科教学试验多余的钢筋混凝土梁，将去除了保护层、纵筋、箍筋之后的核心部分，制备成所需节段或块体备用。试件制作过程与前面中厚壁圆钢管再生混合短柱类似，在此不再赘述。

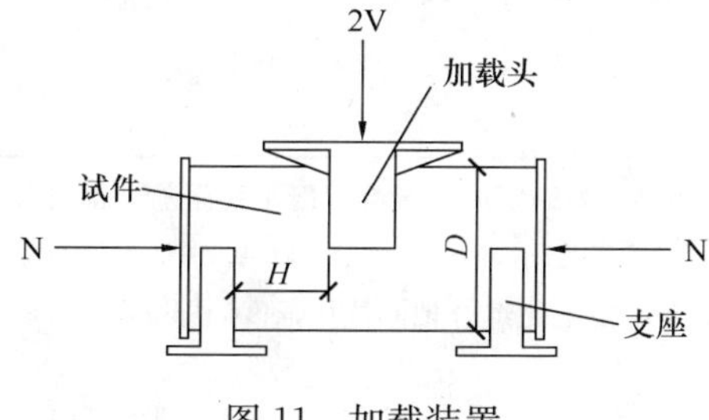


图 11　加载装置

图 11 所示为加载装置示意图，图中 $H$ 为剪跨长度。试件水平放置，两端支承在支座上，支座与钢管接触部分的宽度 40mm。横向力采用 2000kN 油压千斤顶施加在试件跨中，加载头与钢管接触部分的宽度 80mm。轴向压力通过自平衡式水平反力架支承，采用 1000kN 油压千斤顶施加。支座和加载头与钢管的接触部分均加工成圆弧形状。

**试 件 参 数** **表 11**

| 序号 | 试件编号 | $D\times t\times L$ (mm×mm×mm) | $f_{c1u}$ (MPa) | $f_{c2u}$ (MPa) | 试件型式 | 混合比 $\eta$ | 轴压比 $n$ | 剪跨比 $\lambda$ |
|---|---|---|---|---|---|---|---|---|
| 1 | K1-C2-1-1 | 200×1×280 | 47.5 | | 块体型 | 26% | 0.2 | 0.15 |
| 2 | C1-2-1 | 200×1×280 | 42.0 | | 全现浇 | 0 | 0.4 | 0.15 |
| 3 | K1-C1-2-1 | 200×1×280 | 42.0 | | 块体型 | 26% | 0.4 | 0.15 |
| 4 | K2-C1-2-1 | 200×1×280 | 42.0 | | 块体型 | 34% | 0.4 | 0.15 |
| 5 | C2-2-1 | 200×1×280 | 47.5 | | 全现浇 | 0 | 0.4 | 0.15 |
| 6 | K1-C2-2-1 | 200×1×280 | 47.5 | | 块体型 | 26% | 0.4 | 0.15 |
| 7 | K2-C2-2-1 | 200×1×280 | 47.5 | | 块体型 | 35% | 0.4 | 0.15 |
| 8 | J2-C2-2-1 | 200×1×280 | 47.5 | | 节段型 | 35% | 0.4 | 0.15 |
| 9 | K1-C2-3-1 | 200×1×280 | 47.5 | | 块体型 | 26% | 0.6 | 0.15 |
| 10 | K1-C1-1-2 | 200×1×340 | 42.0 | | 块体型 | 26% | 0.2 | 0.3 |
| 11 | K1-C2-1-2 | 200×1×340 | 47.5 | | 块体型 | 26% | 0.2 | 0.3 |
| 12 | C1-2-2 | 200×1×340 | 42.0 | | 全现浇 | 0 | 0.4 | 0.3 |
| 13 | K1-C1-2-2 | 200×1×340 | 42.0 | 37.1 | 块体型 | 26% | 0.4 | 0.3 |
| 14 | K2-C1-2-2 | 200×1×340 | 42.0 | | 块体型 | 33% | 0.4 | 0.3 |
| 15 | C2-2-2 | 200×1×340 | 47.5 | | 全现浇 | 0 | 0.4 | 0.3 |
| 16 | K1-C2-2-2 | 200×1×340 | 47.5 | | 块体型 | 26% | 0.4 | 0.3 |
| 17 | J1-C2-2-2 | 200×1×340 | 47.5 | | 节段型 | 25% | 0.4 | 0.3 |
| 18 | K2-C2-2-2 | 200×1×340 | 47.5 | | 块体型 | 34% | 0.4 | 0.3 |
| 19 | K1-C1-3-2 | 200×1×340 | 42.0 | | 块体型 | 26% | 0.6 | 0.3 |
| 20 | C2-3-2 | 200×1×340 | 47.5 | | 全现浇 | 0 | 0.6 | 0.3 |
| 21 | K1-C2-3-2 | 200×1×340 | 47.5 | | 块体型 | 26% | 0.6 | 0.3 |
| 22 | K2-C2-3-2 | 200×1×340 | 47.5 | | 块体型 | 36% | 0.6 | 0.3 |
| 23 | K1-C2-1-3 | 200×1×420 | 47.5 | | 块体型 | 26% | 0.2 | 0.5 |
| 24 | K1-C1-2-3 | 200×1×420 | 42.0 | | 块体型 | 26% | 0.4 | 0.5 |
| 25 | K1-C2-2-3 | 200×1×420 | 47.5 | | 块体型 | 26% | 0.4 | 0.5 |
| 26 | J1-C2-2-3 | 200×1×420 | 47.5 | | 节段型 | 26% | 0.4 | 0.5 |
| 27 | K1-C2-3-3 | 200×1×420 | 47.5 | | 块体型 | 26% | 0.6 | 0.5 |

注：$f_{c1u}$和$f_{c2u}$分别为试验当天现浇和废弃混凝土的立方体抗压强度；$\eta$为钢管内部废弃混凝土质量与全部混凝土质量之比；$\lambda$为剪跨长度与钢管外径之比；$n$为轴向压力与$f_cA_c+f_sA_s$之比，其中$A_c$和$A_s$分别为混凝土和钢管的横截面积，$f_s$为钢材屈服强度，$f_c$取现浇混凝土和废弃混凝土的组合轴心抗压强度，即$f_c=0.88\times0.76\times[f_{c1u}\cdot(1-\eta)+f_{c2u}\cdot\eta]$。

首先以50kN的荷载分别进行轴向和横向预压并卸载，随后将轴向压力加至预定值并在试验过程中基本保持恒定；分级施加横向荷载，每级荷载持荷数分钟。以事先预估的极限抗剪承载力的80%为界，之前每级加载50kN，之后每级加载25kN，试件临近破坏时缓慢连续加载直至最终破坏。

加载头与 2000kN 油压千斤顶之间设置力传感器，加载头上表面左右两侧各设置一个量程 50mm 的百分表，两个百分表所测数据的平均值作为试件的横向变形。

(2) 试验结果与分析

①破坏形态及荷载—变形曲线

大部分试件的荷载—变形曲线的上升段略呈 S 形，其原因在于：壁厚 1mm 的薄壁钢管在加工卷制过程中不可避免地产生焊接变形，使得钢管与支座及加载头之间无法做到完全圆弧面接触而局部存有间隙，加载时该类间隙导致试件荷载—变形曲线的初始刚度略低。

剪跨比＝0.5 的试件，以 K1-C2-2-3 为例，其荷载—变形曲线及最终破坏形态见图 12。从图中可以看出：由于剪跨比相对较大，试件在极限抗剪承载力之后的破坏过程呈现一定延性；试件与加载头接触一侧凹进，另一侧鼓出，加载头边缘处钢管发生环向和斜向破裂（注：斜向破裂自加载头向支座方向开展），内部混凝土外露，但支座边缘处钢管未出现破裂。

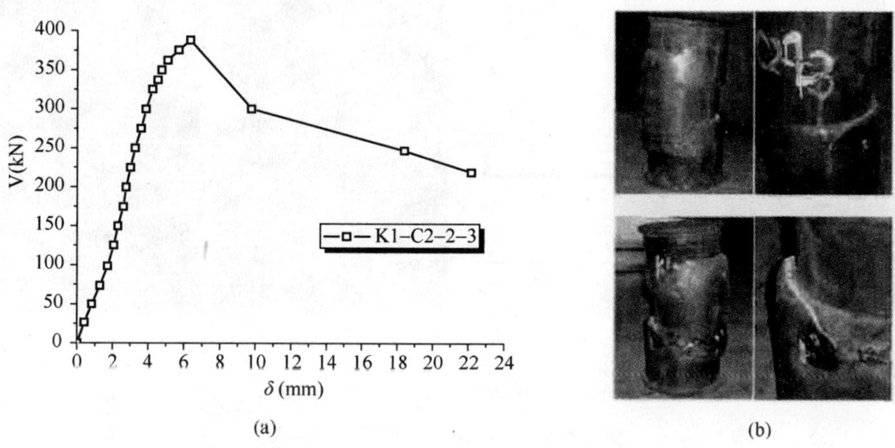


图 12　试件 K1-C2-2-3 的试验结果

(a) 荷载—变形曲线；(b) 破坏形态

剪跨比＝0.15 的试件，以 K2-C1-2-1 为例，其荷载—变形曲线及最终破坏形态见图 13。从图中可以看出：由于剪跨比很小，试件在极限抗剪承载力之后的破坏十分突然，受试验设备限制未能采集到试件的下降段曲线；试件在剪跨区被完全剪断，断裂面总体沿加载头—支座走向；断裂面上无法分辨新、旧混凝土及其界面，表明二者整体性较好。

剪跨比＝0.3 的试件，以 K2-C2-2-2 为例，其荷载—变形曲线及最终破坏形态见图 14。

从图中可以看出：加载至极限抗剪承载力的 85％之前，试件呈现出较好的弹性，在此之后曲线斜率逐渐减小，横向变形加快；极限抗剪承载力之后试件的承载能力出现陡降，受试验设备限制未能采集到试件的下降段曲线；试件与加载头接触一侧凹进，另一侧鼓出，加载头和支座边缘处均发生钢管的环向和斜向破裂，内部混凝土外露。

图 15 所示为全部试件的最终破坏形态。

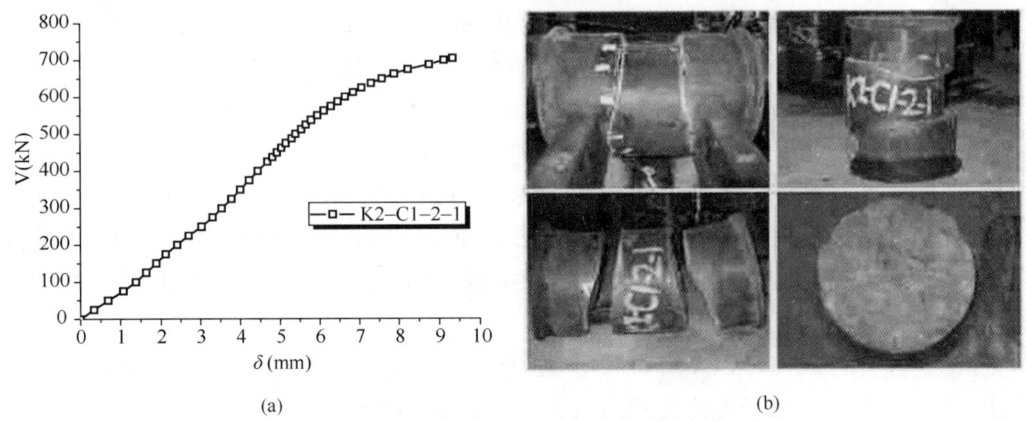


(a) (b)

图 13 试件 K2-C1-2-1 的试验结果

(a) 荷载—变形曲线；(b) 破坏形态

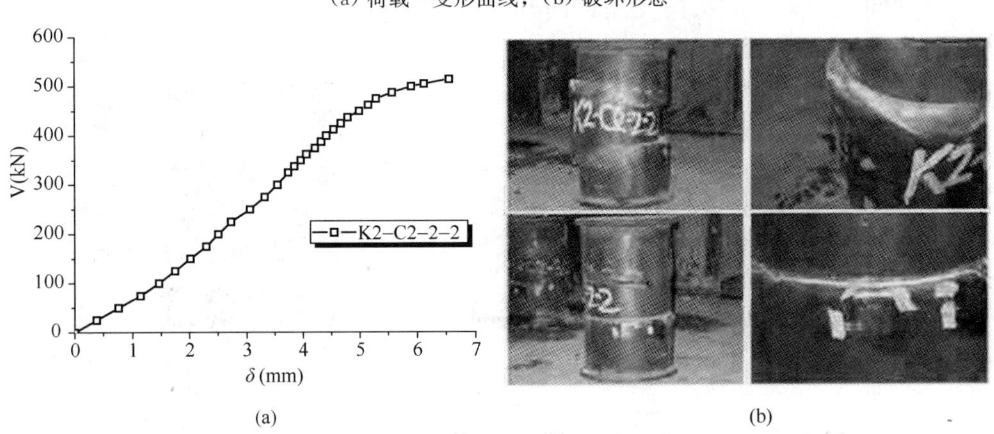


(a) (b)

图 14 试件 K2-C2-2-2 的试验结果

(a) 荷载—变形曲线；(b) 破坏形态

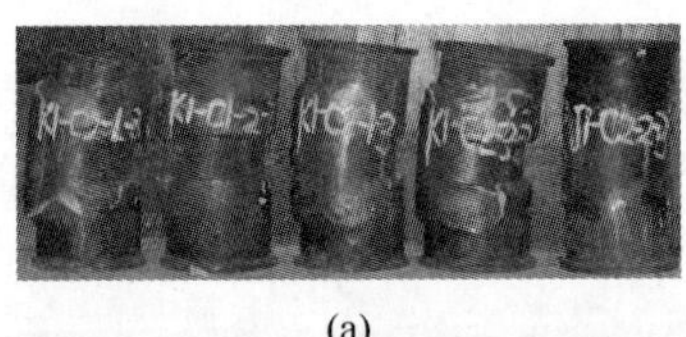

(a)

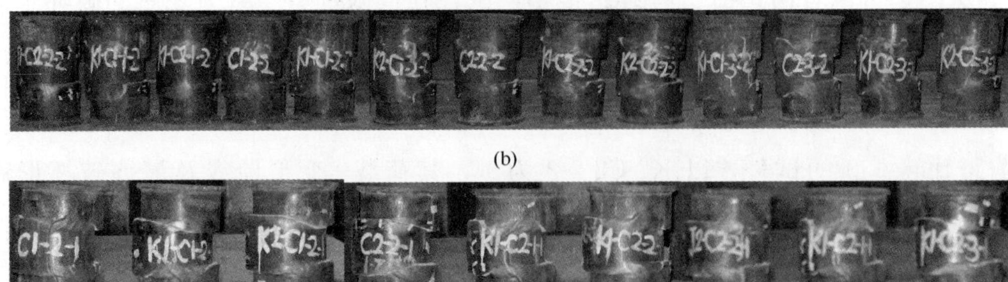

(b)

(c)

图 15 试件破坏形态

(a) 剪跨比＝0.5；(b) 剪跨比＝0.3；(c) 剪跨比＝0.15

②各因素对极限抗剪承载力的影响

A. 现浇混凝土强度

图 16 所示为现浇混凝土强度对试件极限抗剪承载力的影响情况。从图中可以看出，当剪跨比很小或轴压比很大时［见图 16（a）和（b）］，现浇混凝土强度对极限抗剪承载力影响较小；除此之外，极限抗剪承载力均随着现浇混凝土强度的增加而明显增大。需要指出的是，上述现象是在现浇混凝土与废弃混凝土的立方体抗压强度差最大约 10MPa（见表 11）的前提下获得的。

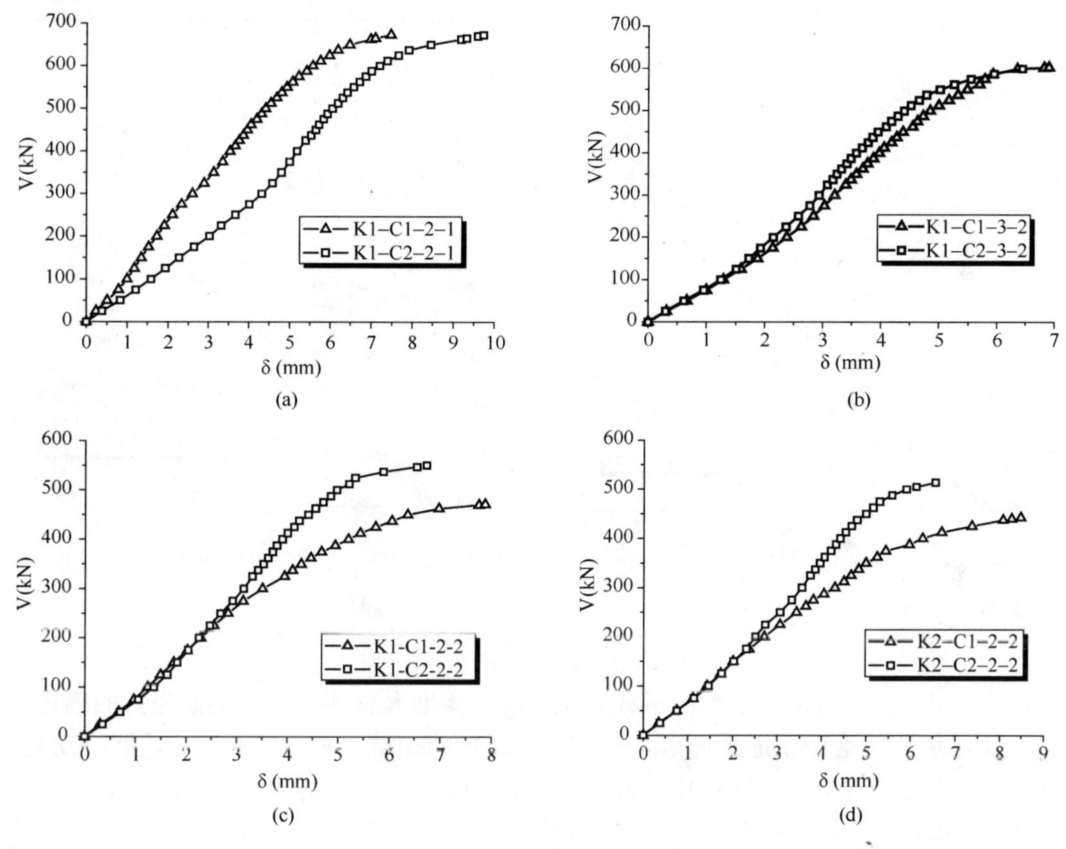


图 16　现浇混凝土强度的影响

B. 轴压比

图 17 所示为轴压比对试件极限抗剪承载力的影响情况。从图中可以看出，在其他条件相同的情况下，试件的极限抗剪承载力随着轴压比增大明显提高。这一趋势与全现浇钢管混凝土柱的相关研究结果是一致的。

C. 剪跨比

图 18 所示为剪跨比对试件极限抗剪承载力的影响情况。从图中可以看出，在其他条件相同的情况下，试件的极限抗剪承载力随着剪跨比增大显著降低。这一趋势也与全现浇钢管混凝土柱的相关研究结果一致。

D. 混合比

图 19 所示为混合比对试件极限抗剪承载力的影响情况。从图中可以看出，在其他条

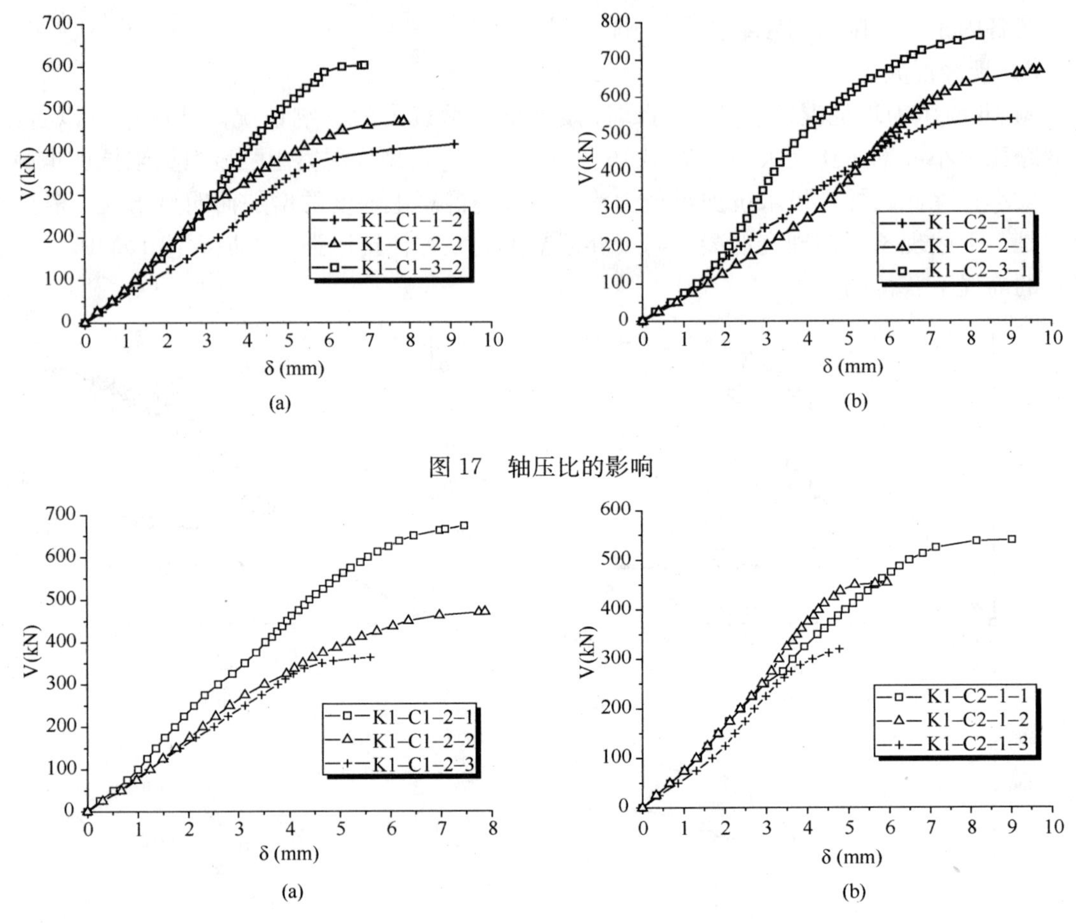


图 17 轴压比的影响

图 18 剪跨比的影响

件相同的情况下，试件的极限抗剪承载力随着混合比增大总体呈现出逐渐降低的趋势。这主要是因为废弃混凝土的抗压强度小于现浇混凝土的相应强度（见表 11），混合比越大，废弃混凝土在全部混凝土中所占比例就越大，导致新旧混凝土的组合抗压强度减小，进而使得试件的极限抗剪承载力降低。

E. 废弃混凝土类型

图 20 所示为废弃混凝土类型对试件极限抗剪承载力的影响情况。从图中可以看出，在其他条件相同的情况下，节段型钢管再生混合柱相比于块体型钢管再生混合柱具有更高的极限抗剪承载力和更大的弹性刚度。这可能是由于节段型废弃混凝土的表面积小于相同质量的块体型废弃混凝土，前者与现浇混凝土之间出现界面缺陷的可能性小于后者的相应可能性，从而使得节段型钢管再生混合柱的极限抗剪承载力和弹性刚度偏高。

（3）计算分析

关于中厚壁钢管混凝土柱的受剪承载力计算问题，CECS 28：90 未做统一规定，该规程主编蔡绍怀通过对国内外弯剪试验结果的分析，认为剪跨比大于 2 时钢管混凝土柱都是弯曲破坏，提出以剪跨 2D 时抗弯强度与剪跨的比值作为受剪承载力的下限值[37]。CECS 159：2004[38]则不考虑核心混凝土的作用，假定剪力完全由钢管承担。DL/T 5085—1999 通

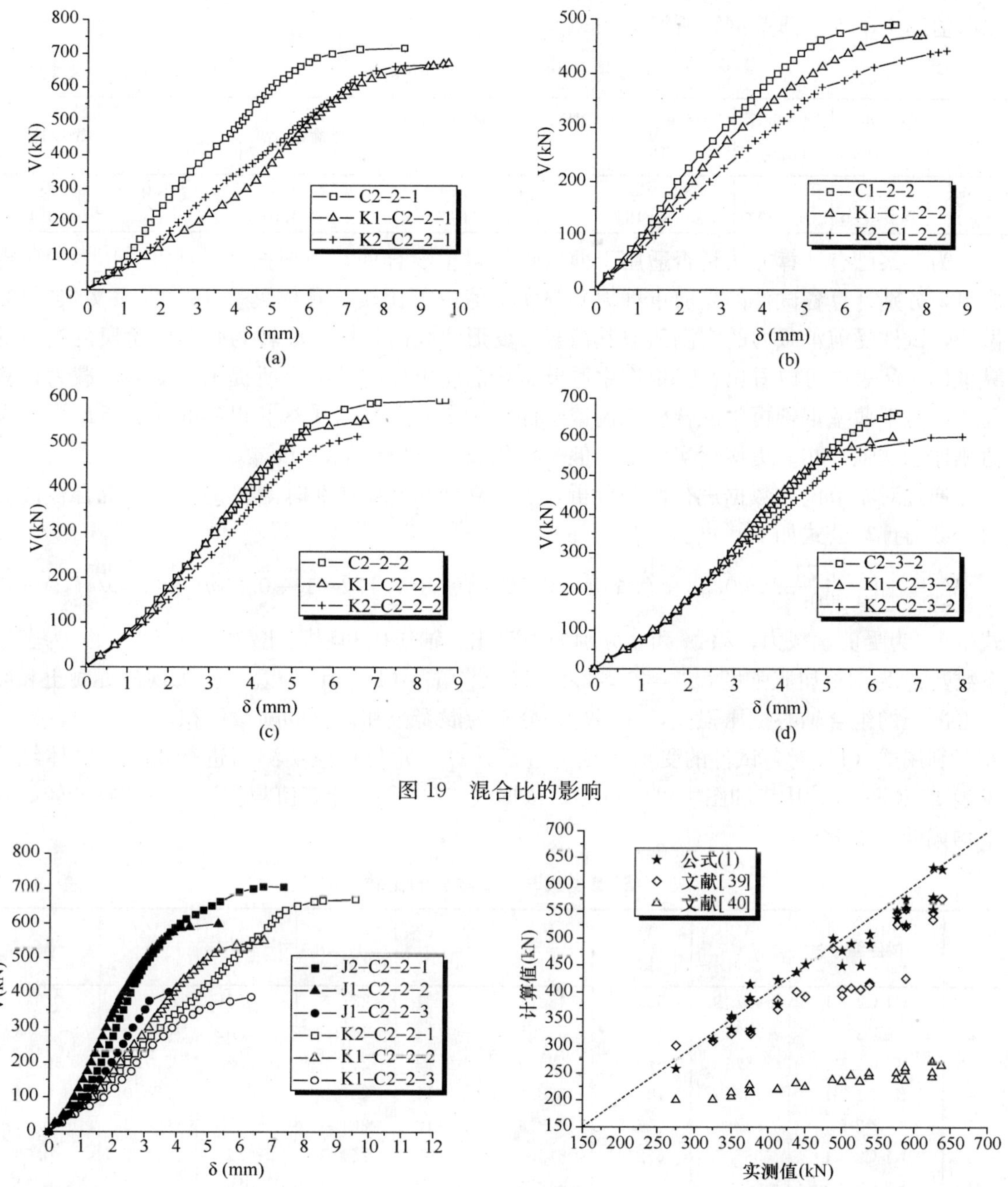


图 19　混合比的影响

图 20　废弃混凝土类型的影响

图 21　受剪承载力计算值与实测值的比较

过定义钢管混凝土组合抗剪强度，建立了相应的受剪承载力计算公式。美国 ANSI/AISC（2005）和日本 AIJ（1997）也只考虑钢管的抗剪贡献。欧洲 EC4（2004）分别演算钢管和混凝土的抗剪能力，但其简化算法同 ANSI/AISC（2005）和 AIJ（1997）一样也假定剪力完全由钢管承担。

文献［39，40］以试验为基础，对钢管混凝土柱的受剪性能进行了分析，提出了相应的受剪承载力计算公式，具体试验参数见表 12。从表中可以看出，两篇文献中钢管的径厚比取值远小于本节薄壁钢管所对应的200。

**文献［39，40］的试验参数** **表12**

| 文献 | 钢管外径(mm) | 钢管壁厚(mm) | 钢管屈服强度(MPa) | 混凝土立方体抗压强度(MPa) | 套箍系数 | 剪跨比 | 轴压比 |
|---|---|---|---|---|---|---|---|
| ［39］ | 194 | 5.5，7.5 | 330，422 | 54，71，84 | 0.61～1.94 | 0.1～1.0 | 0～0.5 |
| ［40］ | 160～166 | 3.0～6.5 | 345～445 | 30～50 | 0.90～4.06 | 0.14～1.0 | 0～0.4 |

为考察已有计算方法是否适用于薄壁钢管再生混合柱，现根据文献［39，40］的研究成果，分别计算各试件的受剪承载力并与实测值进行比较，具体见表13。参照文献［40］做法，试件受剪承载力的实测值取其荷载—变形曲线由弹性阶段转为弹塑性阶段的拐点的纵坐标。从表中可以看出：①根据中厚壁钢管混凝土柱试验结果所提出的受剪承载力计算公式，对于薄壁钢管再生混合柱来说偏于保守。②依据文献［39］得到的受剪承载力计算值相比于文献［40］更接近实测值，但平均仍较实测值偏低约11%。

通过对本节试验数据进行回归分析，可建立剪跨比≤0.5时薄壁钢管再生混合柱的受剪承载力计算公式如下：

$$V_s=\left[\left(\frac{0.74}{\lambda+0.75}-0.52+0.44n-0.16n^2\right)(1+1.17\theta)(1-0.23\eta)+0.58\theta\right]f_cA_c \quad (1)$$

式中$V_s$为受剪承载力；$\lambda$、$n$和$\eta$分别为剪跨比、轴压比和混合比；$\theta=f_sA_s/f_cA_c$为套箍系数，$f_s$为钢材屈服强度，$f_c=0.88\times0.76\times[f_{c1u}\cdot(1-\eta)+f_{c2u}\cdot\eta]$为现浇混凝土和废弃混凝土的组合轴心抗压强度，$A_c$和$A_s$分别为混凝土和钢管的横截面积。

利用式（1）对各试件的受剪承载力进行计算，并与相应实测值进行比较，具体结果见表13和图21。从表和图中可以看出，依据式（1）所得计算值与实测值总体吻合较好且平均偏低约4%。

**计算结果与试验结果的比较** **表13**

| 序号 | 试件编号 | $V_s$ (kN) | $V_u$ (kN) | $V_1$ (kN) | $\frac{V_1-V_s}{V_s}$ | $V_2$ (kN) | $\frac{V_2-V_s}{V_s}$ | $V_3$ (kN) | $\frac{V_3-V_s}{V_s}$ |
|---|---|---|---|---|---|---|---|---|---|
| 1 | K1-C2-1-1 | 487.5 | 539.6 | 482.3 | −0.01 | 237.6 | −0.51 | 500.3 | 0.03 |
| 2 | C1-2-1 | 625 | 688.8 | 535.1 | −0.14 | 243.7 | −0.61 | 576.9 | −0.08 |
| 3 | K1-C1-2-1 | 575 | 673.2 | 525.9 | −0.09 | 238.9 | −0.58 | 537.0 | −0.07 |
| 4 | K2-C1-2-1 | 587.5 | 704.9 | 523.1 | −0.11 | 237.3 | −0.60 | 525.0 | −0.11 |
| 5 | C2-2-1 | 637.5 | 716.8 | 573.9 | −0.10 | 264.8 | −0.58 | 628.2 | −0.01 |
| 6 | K1-C2-2-1 | 587.5 | 672.4 | 555.0 | −0.06 | 254.5 | −0.57 | 572.7 | −0.03 |
| 7 | K2-C2-2-1 | 625 | 669.3 | 548.4 | −0.12 | 250.9 | −0.60 | 554.2 | −0.11 |
| 8 | J2-C2-2-1 | 575 | 704.5 | 547.6 | −0.05 | 250.5 | −0.56 | 552.2 | −0.04 |
| 9 | K1-C2-3-1 | 625 | 763.4 | 571.5 | −0.09 | 271.4 | −0.57 | 631.6 | 0.01 |
| 10 | K1-C1-1-2 | 350 | 417.0 | 352.0 | 0.01 | 206.1 | −0.41 | 355.8 | 0.02 |
| 11 | K1-C2-1-2 | 412.5 | 454.0 | 368.2 | −0.11 | 219.3 | −0.47 | 376.8 | −0.09 |
| 12 | C1-2-2 | 450 | 490.8 | 391.6 | −0.13 | 225.0 | −0.50 | 452.3 | 0.01 |
| 13 | K1-C1-2-2 | 412.5 | 471.1 | 385.6 | −0.07 | 220.5 | −0.47 | 422.7 | 0.02 |
| 14 | K2-C1-2-2 | 375 | 442.6 | 384.0 | 0.02 | 219.3 | −0.42 | 414.9 | 0.11 |
| 15 | C2-2-2 | 537.5 | 595.5 | 416.9 | −0.22 | 244.5 | −0.55 | 490.3 | −0.09 |
| 16 | K1-C2-2-2 | 500 | 550.4 | 404.6 | −0.19 | 234.9 | −0.53 | 449.2 | −0.10 |

续表

| 序号 | 试件编号 | $V_s$ (kN) | $V_u$ (kN) | $V_1$ (kN) | $\frac{V_1-V_s}{V_s}$ | $V_2$ (kN) | $\frac{V_2-V_s}{V_s}$ | $V_3$ (kN) | $\frac{V_3-V_s}{V_s}$ |
|---|---|---|---|---|---|---|---|---|---|
| 17 | J1-C2-2-2 | 525 | 598.4 | 404.6 | −0.23 | 234.9 | −0.55 | 449.2 | −0.14 |
| 18 | K2-C2-2-2 | 437.5 | 514.5 | 400.7 | −0.08 | 231.9 | −0.47 | 437.0 | 0.00 |
| 19 | K1-C1-3-2 | 500 | 603.5 | 393.1 | −0.21 | 234.9 | −0.53 | 477.2 | −0.05 |
| 20 | C2-3-2 | 587.5 | 662.5 | 425.7 | −0.28 | 260.9 | −0.56 | 556.0 | −0.05 |
| 21 | K1-C2-3-2 | 537.5 | 600.5 | 412.8 | −0.23 | 250.5 | −0.53 | 508.1 | −0.05 |
| 22 | K2-C2-3-2 | 512.5 | 602.5 | 407.9 | −0.20 | 246.5 | −0.52 | 490.3 | −0.04 |
| 23 | K1-C2-1-3 | 275 | 320.0 | 301.1 | 0.09 | 199.7 | −0.27 | 258.2 | −0.06 |
| 24 | K1-C1-2-3 | 325 | 363.8 | 309.5 | −0.05 | 200.8 | −0.38 | 313.0 | −0.04 |
| 25 | K1-C2-2-3 | 350 | 388.0 | 322.9 | −0.08 | 213.9 | −0.39 | 330.6 | −0.06 |
| 26 | J1-C2-2-3 | 376.5 | 401.5 | 322.9 | −0.14 | 213.9 | −0.43 | 330.6 | −0.12 |
| 27 | K1-C2-3-3 | 375 | 420.0 | 327.9 | −0.13 | 228.1 | −0.39 | 389.5 | 0.04 |
| 平均 | — | — | — | — | −0.11 | — | −0.50 |  | −0.04 |

注：$V_s$ 和 $V_u$ 分别为试件受剪承载力和极限抗剪承载力的实测值；$V_1$、$V_2$ 和 $V_3$ 分别为依据文献［39］、［40］和式（1）所得受剪承载力计算值；为与各自受剪承载力计算公式建立时混凝土轴心抗压强度的取法一致，$V_1$ 和 $V_2$ 计算过程中分别取 $f_c=0.76\times[f_{c1u}\cdot(1-\eta)+f_{c2u}\cdot\eta]$（对应文献［39］）和 $f_c=0.88\times0.76\times[f_{c1u}\cdot(1-\eta)+f_{c2u}\cdot\eta]$（对应文献［40］）。

## 15.3　再生混合钢筋混凝土构件的基本力学性能

下面主要介绍再生混合钢筋混凝土短柱的轴压性能，以及再生混合钢筋混凝土梁的受弯和受剪性能。

### 15.3.1　再生混合钢筋混凝土短柱的轴压性能

（1）试验方案

8 个方形截面钢筋混凝土短柱的高宽比均等于 3。其中 2 个为全现浇混凝土短柱以作对比，2 个为包含有废弃混凝土块体的再生混合短柱（以下简称块体型再生混合短柱），4 个为包含有废弃混凝土节段的再生混合短柱（以下简称节段型再生混合短柱）。各试件的具体尺寸和配筋情况见图 22，其他参数见表 14。表中混合比 $\eta$ 为试件中废弃混凝土质量与全部混凝土质量之比。

现浇混凝土采用粤秀牌普通硅酸盐水泥 P. O42.5、中砂、最大粒径 20mm 的碎石，每立方米现浇混凝土的材料用量为石子：砂：水泥：水＝1150：595：455：220。自然条件下成型养护，28d 实测 150mm×150mm×150mm 立方体抗压强度 $f_{c1u}$＝58.9MPa，相应的轴心抗压强度 $f_{c1}$ 取为 0.67$f_{c1u}$。

废弃混凝土采用华南理工大学几年前本科教学试验多余的钢筋混凝土梁，将去除了保护层、纵筋、箍筋之后的核心部分制成节段和块体备用。废弃混凝土的当前强度采用与钢筋混凝土梁同时浇注的剩余立方体试块测得，实测值为 $f_{c2u}$＝37.1MPa，相应的轴心抗压强度 $f_{c2}$ 取为 0.67$f_{c2u}$。

纵筋和箍筋分别采用 HRB335 螺纹钢筋和 HPB235 光圆钢筋，纵筋直径 12mm，箍筋直径 6mm。纵筋的实测屈服强度和极限强度分别为 363.4MPa 和 516.4MPa。

试　件　参　数　　表 14

| 序号 | 试件编号 | 混合比 $\eta$ | $f_{c1}$ (MPa) | $f_{c2}$ (MPa) | $f_y$ (MPa) | 废弃混凝土类型 | 极限抗压承载力 (kN) | |
|---|---|---|---|---|---|---|---|---|
| | | | | | | | 实测值 | 均　值 |
| 1 | RC-A | 0 | 39.5 | 24.9 | 363.4 | 全现浇 | 1452 | 1413 |
| 2 | RC-B | 0 | | | | 全现浇 | 1373 | |
| 3 | RC-K1-A | 24.9% | | | | 块 体 | 1189 | 1187 |
| 4 | RC-K1-B | 24.9% | | | | 块 体 | 1184 | |
| 5 | RC-J1-A | 25.1% | | | | 节 段 | 1319 | 1373 |
| 6 | RC-J1-B | 22.6% | | | | 节 段 | 1427 | |
| 7 | RC-J2-A | 32.8% | | | | 节 段 | 1331 | 1210 |
| 8 | RC-J2-B | 32.4% | | | | 节 段 | 1089 | |

注：$f_y$ 为纵筋屈服强度。

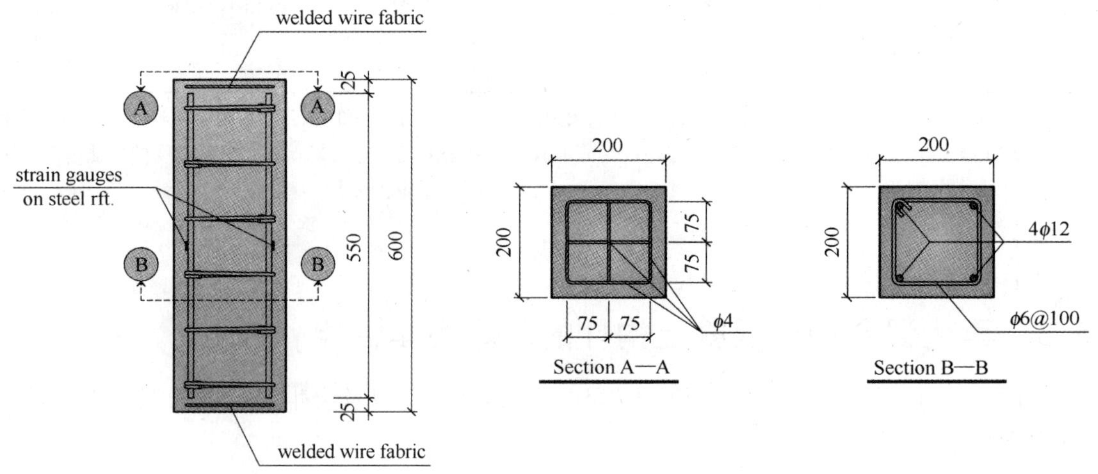


图 22　试件尺寸及配筋情况（单位：mm）

制作块体型再生混合短柱时，首先将绑扎好的钢筋笼放入模板内，然后在模板底部浇筑一层现浇混凝土，随后将废弃混凝土块体和现浇混凝土交替置入模板内并充分振捣，直至模板填满。制作节段型再生混合短柱时，首先将废弃混凝土节段置入钢筋笼内，然后在模板底部浇筑一层现浇混凝土，随后将钢筋笼放入模板内，最后在节段与模板内壁之间灌入现浇混凝土并充分振捣，直至模板填满。图 23 是试件制作过程的典型照片。

(a)

(b)

图 23　试件制作过程典型照片

(*a*) 节段型再生混合短柱；(*b*) 块体型再生混合短柱

需要指出的是，对于实际工程中的柱式构件，由于模板通常采用竖向架设方式，制作节段型再生混合柱时，可先将绑扎好的钢筋笼放入模板内，然后在模板底部浇筑一层现浇混凝土，随后将废弃混凝土节段从上至下吊入钢筋笼内部，最后在节段与模板内壁之间灌入现浇混凝土并充分振捣，直至模板填满。

试验在华南理工大学结构实验室的 1500t 长柱压力机上进行。在每个试件表面半高处对称布设四个应变花，用以量测受力过程中混凝土的应变；同时在四根纵筋的半高处也各粘贴一个应变片，用以量测纵筋的应变。

加载时先采用力控制策略，加载速率 2kN/s；纵筋应变 1400$\mu\varepsilon$ 后改用位移控制策略（注：纵筋的实测屈服应变约 1800$\mu\varepsilon$，由于纵筋屈服后不久试件就发生最终破坏，试验过程中纵筋应变 1400$\mu\varepsilon$ 后即采用位移控制），相应的加载速率为 0.005mm/s。图 24 所示为实际加载装置。

图 24　加载装置

（2）试验结果与分析

图 25 所示为试验后各试件的宏观破坏形态。试验过程中，节段型/块体型再生混合短柱与全现浇混凝土短柱呈现出相似的破坏过程。随着竖向荷载的增加，试件顶端或低端首先出现纵向小裂缝，随后柱身开始出现微细裂缝且宽度逐渐增加，最后柱表面出现明显的纵向裂缝且承载力不再增加，试件随即发生破坏。

图 26 所示为试件的实测荷载—变形曲线，图中各曲线的纵坐标都是该曲线所对应工况下两个试件（注：表 14 中“试件编号”的最后一个字母 A 和 B 分别表示相同或相近工况下的两个试件）的平均值。从图 26 和表 14 中可以看出：

①加载之初试件的竖向刚度偏低，这主要是因为试件上、下表面略有不平使得初始荷载作用下加载装置上、下压头之间的竖向位移相对较大所致。

②除混合比约 25％的节段型再生混合短柱（RC-J1）以外，其他三类短柱的荷载－变形曲线大部分几乎重合，表明在混合比为 25％～33％且新、旧混凝土轴心抗压强度相差约 15MPa 的情况下，块体型和节段型再生混合短柱的轴向刚度不低于全现浇混凝土短柱。

③相比于全现浇混凝土短柱的抗压承载力，混合比约 25％的块体型再生混合短柱（RC-K1）和节段型再生混合短柱（RC-J1），以及混合比约 33％的节段型再生混合短柱（RC-J2）分别降低约 16.0％、2.8％和 14.4％。这表明混合比相近时，节段型再生混合短柱的抗压承载力大于块体型再生混合短柱；而废弃混凝土类型一定时，再生混合短柱的抗

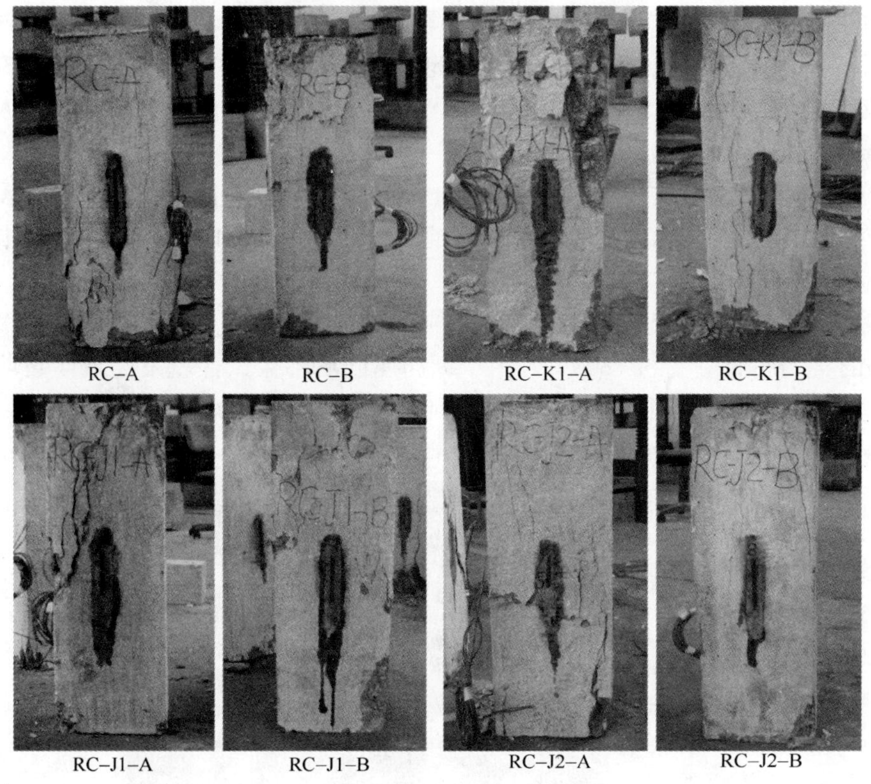


图 25　试件破坏形态

压承载力随混合比增大而降低。这是因为在废弃混凝土用量相近的情况下，节段型废弃混凝土的表面积小于块体型废弃混凝土，前者与现浇混凝土之间出现界面缺陷的可能性小于后者的相应可能性，从而使得节段型再生混合短柱的抗压承载力偏高。由于本节试验中废弃混凝土的轴心抗压强度低于现浇混凝土约15MPa，显然混合比越大，再生混合短柱的抗压承载力就越小。

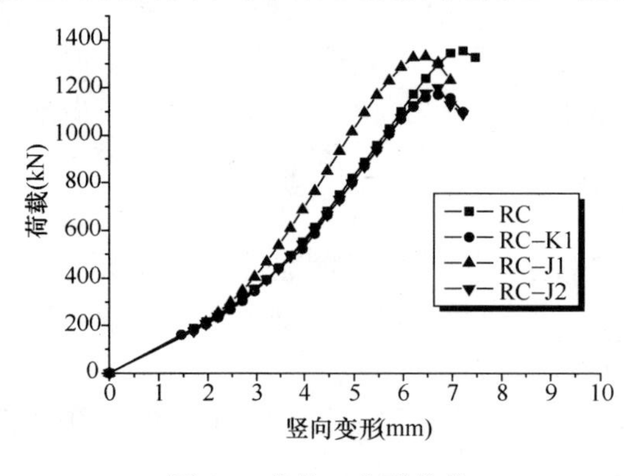


图 26　荷载—变形曲线

基于上述试验结果，建议实际工程中应尽可能优先采用废弃混凝土节段，并适当控制混合比的大小。我们在试件制作过程中发现，混合比越大施工越困难。为此，实际工程中为保证施工质量，建议混合比的取值范围为1/5～1/3。

（3）计算比较

按照《混凝土结构设计规范》（GB 50010—2002）中的相关公式，对各试件的抗压承载力进行计算，计算结果与试验实测结果的比较见表 15。计算过程中，混凝土轴心抗压强度取现浇混凝土和废弃混凝土的组合强度 $f'_{c}=f_{c1}\times(1-\eta)+f_{c2}\times\eta$，其中 $f_{c1}$、$f_{c2}$ 和 $\eta$ 的具体取值见表 14。从表 15 中可以看出，块体型和节段型再生混合短柱的抗压承载力计算结果分别比实测结果平均偏大约 21.5%和 11.3%，与此同时全现浇混凝土短柱的抗压承载力计算结果比实测结果平均偏大约 11%。这表明采用现行规范计算节段型再生混合短柱的抗压承载力具有与全现浇混凝土短柱相近的安全性，而计算块体型再生混合短柱的抗压承载力则需再乘以 0.9 左右的折减系数，才能具有与全现浇混凝土短柱相近的安全性。

**计算结果与试验结果的比较　表 15**

| 序号 | 试件编号 | $f'_c$ (MPa) | $f_y$ (MPa) | 抗压承载力 | | 计算/实测 |
|---|---|---|---|---|---|---|
| | | | | 计算值（kN） | 实测值（kN） | |
| 1 | RC-A | 39.50 | 363.4 | 1570 | 1452 | 1.08 |
| 2 | RC-B | 39.50 | | 1570 | 1373 | 1.14 |
| 3 | RC-K1-A | 35.87 | | 1439 | 1189 | 1.21 |
| 4 | RC-K1-B | 35.87 | | 1439 | 1184 | 1.22 |
| 5 | RC-J1-A | 35.83 | | 1438 | 1319 | 1.09 |
| 6 | RC-J1-B | 36.19 | | 1451 | 1427 | 1.02 |
| 7 | RC-J2-A | 34.72 | | 1398 | 1331 | 1.05 |
| 8 | RC-J2-B | 34.76 | | 1399 | 1089 | 1.29 |

### 15.3.2　再生混合钢筋混凝土梁的受弯性能

（1）试验方案

共制作 4 根截面尺寸 150mm×300mm 的矩形梁，其中 3 根为再生混合钢筋混凝土梁，1 根为常规混凝土梁，梁长 1800mm，净跨 1500mm（见图 27）。下部受拉钢筋分别采用两根直径 10mm、14mm 和 18mm 的 HRB335 级钢筋，根据标准试验方法测得其屈服强度依次为 356MPa、363MPa 和 384MPa，纵筋保护层厚度 25mm。各试件的具体尺寸和基本参数见表 16。表中 $f_{c1u}$和 $f_{c2u}$分别为试验当天现浇和废弃混凝土的实测立方体抗压强度；混合比 $\eta$ 为试件中废弃混凝土质量与全部混凝土质量之比；试件编号 RB 表示再生混合钢筋混凝土梁，B 表示常规混凝土梁，其后数字表示下部受拉钢筋的直径。

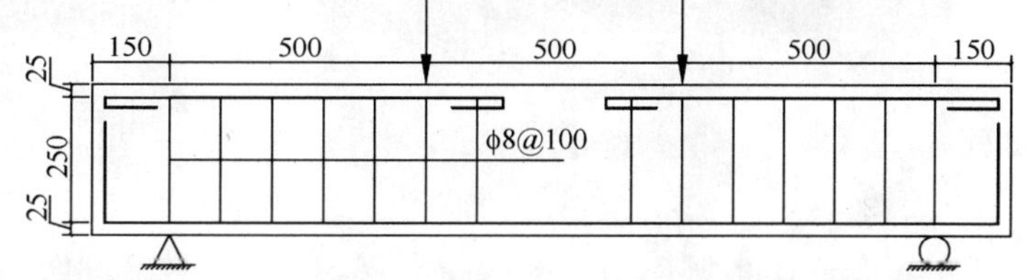


图 27　试件示意图（单位：mm）

现浇混凝土采用矿渣硅酸盐水泥 P. S. A32.5、中砂和最大粒径 20mm 的碎石。废弃混凝土采用华南理工大学几年前本科教学试验多余的钢筋混凝土梁，将去除了保护层、纵筋、箍筋之后的核心部分制备成块体备用。废弃混凝土的当前强度采用与本科教学试验钢筋混凝土梁同时浇注的剩余立方体试块测得。试件制作过程与前面块体型再生混合短柱类似，在此不再赘述。

**试　件　参　数　表 16**

| 试件编号 | $b$ (mm) | $h_0$ (mm) | $f_{c1u}$ (MPa) | $f_{c2u}$ (MPa) | $f_{cu}$ (MPa) | $f_c$ (MPa) | $f_t$ (MPa) | 配筋率 | 试件类型 | 混合比 $\eta$ |
|---|---|---|---|---|---|---|---|---|---|---|
| RB-10 | 146 | 270 | 45.2 | 32.9 | 42.7 | 28.6 | 2.74 | 0.4% | 混合型 | 20% |
| B-14 | 146 | 268 | 44.7 | | 44.7 | 29.9 | 2.81 | 0.8% | 全现浇 | 0 |
| RB-14 | 154 | 268 | 44.7 | | 42.3 | 28.3 | 2.73 | 0.7% | 混合型 | 20% |
| RB-18 | 150 | 266 | 45.2 | | 42.7 | 28.6 | 2.74 | 1.3% | 混合型 | 20% |

注：$f_{cu}=[f_{c1u}\cdot(1-\eta)+f_{c2u}\cdot\eta]$ 为现浇混凝土和废弃混凝土的组合立方体抗压强度；

$f_c=0.88\times0.76f_{cu}$为组合轴心抗压强度；$f_t=0.88\times0.395f_{cu}^{0.55}$为组合轴心抗拉强度。

图 28 加载装置

试验采用两点加载，利用分配梁在试件中部形成 500mm 的纯弯段，加载装置见图 28。试验过程中量测跨中不同高度处的混凝土应变、下部受拉钢筋应变、跨中挠度等。竖向千斤顶与分配梁之间设置力传感器，跨中及两支座处各设置一个量程 50mm 的百分表。

正式加载时每级荷载约为预估极限荷载的 10%，持荷 3min 后采集相关数据。接近预估开裂荷载时，适当降低每级荷载增量以期较准确地获取实际开裂荷载。

（2）试验结果与分析

开始加载时，各试件基本处于弹性阶段，受拉钢筋应变与同一高度处的混凝土应变相近且数值很小。随着荷载的增加，受拉区混凝土边缘开裂，裂缝一出现即具有一定高度，同时可量测到受拉钢筋应变跳跃性增大。荷载继续增加，跨中附近新裂缝陆续出现且向上延伸，受压区高度渐小，梁的跨中挠度增长比弹性阶段略快。随着受拉纵筋屈服，跨中挠度急剧增加，但荷载增加不多，最终各试件均呈现出适筋梁的破坏特征。总体上看，再生混合钢筋混凝土梁的破坏过程与常规混凝土梁类似。部分试件的破坏形态见图 29。

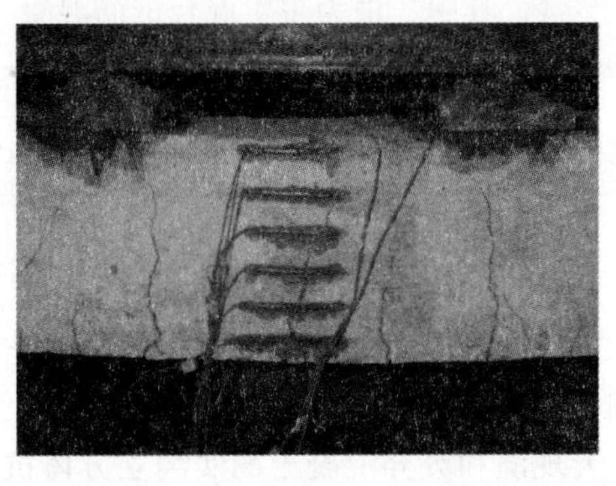

(a)

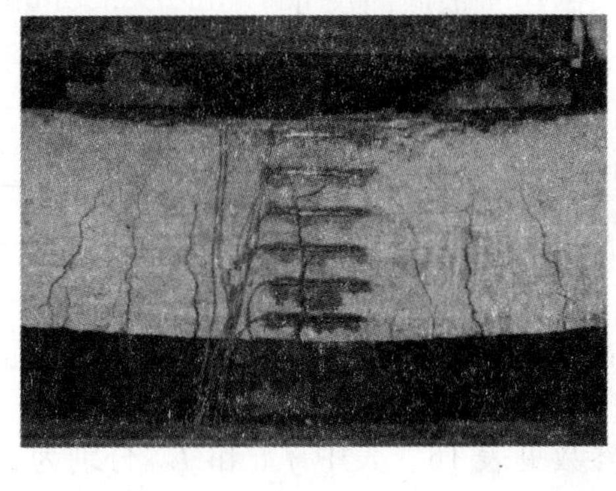

(b)

图 29 部分试件的破坏形态

（a）B-14；（b）RB-14

扣除微小支座位移之后，各试件的实测荷载—跨中挠度曲线见图 30。从图中可以看出：

①再生混合钢筋混凝土梁的荷载—跨中挠度曲线的总体趋势与常规混凝土梁相似。

②对于再生混合钢筋混凝土梁，在其他参数相近的情况下，随着受拉钢筋直径增大，初始刚度略有增加，同时加载中后期相同跨中挠度所对应的荷载显著增大。

③对比试件 RB-14 与 B-14 的初始刚度基本相同。

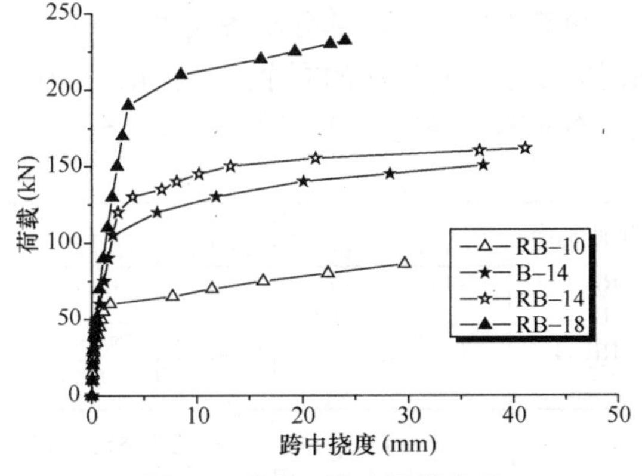


图 30 荷载—跨中挠度曲线

④对比试件 RB-14 与 B-14 的荷载—跨

中挠度曲线总体较为接近，表明在其他参数相同的情况下，再生混合钢筋混凝土梁具有与常规混凝土梁相近的受弯性能。

各试件的主要试验结果汇总于表 17。表中开裂荷载对应跨中竖向裂缝刚出现时的荷载值，屈服荷载对应受拉钢筋屈服时的荷载值，极限荷载对应试件最终破坏时的荷载值。开裂弯矩、屈服弯矩和抗弯承载力则是分别与开裂荷载、屈服荷载和极限荷载对应的梁跨中弯矩。

**试验结果**　　**表 17**

| 试件编号 | 开裂荷载 (kN) | 开裂弯矩 (kN·m) | 屈服荷载 (kN) | 屈服弯矩 (kN·m) | 极限荷载 (kN) | 抗弯承载力 (kN·m) |
|---|---|---|---|---|---|---|
| RB-10 | 32.5 | 8.13 | 58 | 14.5 | 85.9 | 21.5 |
| B-14 | 42.5 | 10.63 | 98 | 24.5 | 150.4 | 37.6 |
| RB-14 | 37.5 | 9.38 | 107 | 26.8 | 161.4 | 40.4 |
| RB-18 | 35.0 | 8.75 | 190 | 47.5 | 232.1 | 58.0 |

利用《混凝土结构设计规范》(GB 50010—2002) 给出的相关公式，对各试件的抗弯承载力进行计算，并与相应实测值进行比较，具体见表 18。计算过程中混凝土的轴心抗压强度取表 16 中的组合轴心抗压强度。从表 18 中可以看出：

A. 再生混合钢筋混凝土梁的计算结果比试验结果平均偏低 26%，而常规混凝土梁则偏低 24%，这表明采用现行规范计算得到的再生混合钢筋混凝土梁的抗弯承载力具有与常规混凝土梁相近的安全性。

B. 对比试件 RB-14 与 B-14 的抗弯承载力较为相近。这是因为影响受弯构件抗弯承载力的主要因素包括受拉钢筋的抗拉性能以及受压区混凝土的抗压性能，虽然再生混合钢筋混凝土梁内部存在大量新旧混凝土界面，但这对混凝土的受压行为影响相对较小。

**抗弯承载力计算值与实测值的对比**　　**表 18**

| 试件编号 | 计算值 (kN·m) | 实测值 (kN·m) | (计算值−实测值) / 实测值 |
|---|---|---|---|
| RB-10 | 14.7 | 21.5 | −31% |
| B-14 | 28.5 | 37.6 | −24% |
| RB-14 | 28.5 | 40.4 | −29% |
| RB-18 | 47.6 | 58.0 | −18% |

利用《混凝土结构设计规范》(GB 50010—2002) 给出的相关公式，对各试件的开裂弯矩进行计算，并与相应实测值进行比较，具体见表 19。计算过程中混凝土的轴心抗拉强度取表 16 中的组合轴心抗拉强度。从表 19 中可以看出：

A. 当采用新、旧混凝土的组合轴心抗拉强度时，再生混合钢筋混凝土梁的开裂弯矩计算值较实测值平均偏高 21%，而常规混凝土梁则基本接近。这表明采用现行规范计算得到的再生混合钢筋混凝土梁的开裂弯矩偏于不安全，有必要进行修正。根据本次试验结果，我们建议在再生混合钢筋混凝土梁的计算结果基础上乘以调整系数 0.8 以近似考虑其特殊性。表中数据显示，修正后开裂弯矩的计算值较实测值平均偏低 3%。

B. 在其他参数相近的情况下，试件 RB-14 的开裂弯矩实测值较对比试件 B-14 偏低约

12%。这是因为再生混合钢筋混凝土梁内部大量存在的新、旧混凝土界面在处于受拉状态时可能成为薄弱区域，致使跨中附近的竖向裂缝更易产生和发展。

**开裂弯矩计算值与实测值的对比** **表 19**

| 试件编号 | 计算值 (kN·m) | 实测值 (kN·m) | (计算值－实测值)/实测值 | 修正值 (kN·m) | (修正值－实测值)/实测值 |
|---|---|---|---|---|---|
| RB-10 | 9.85 | 8.13 | 21% | 7.88 | －3% |
| B-14 | 10.59 | 10.63 | －0.4% | — | — |
| RB-14 | 10.80 | 9.38 | 15% | 8.64 | －8% |
| RB-18 | 11.21 | 8.75 | 28% | 8.97 | 3% |

### 15.3.3 再生混合钢筋混凝土梁的受剪性能

（1）试验方案

共制作 14 根矩形截面梁，其中 7 根为无腹筋梁，7 根为有腹筋梁。有腹筋梁的上部架力筋和箍筋分别采用直径 8mm 和 6.5mm 的 HPB235 级钢筋，有腹筋梁和无腹筋梁的下部受拉钢筋采用直径 25mm 的 HRB400 级钢筋。纵筋保护层厚度 25mm。下部受拉钢筋和箍筋的实测屈服强度分别为 414.4MPa 和 346MPa。各试件的具体尺寸和基本参数见表 20。表中 $f_{c1u}$ 和 $f_{c2u}$ 分别为试验当天现浇和废弃混凝土的实测立方体抗压强度，混合比 $\eta$ 为试件中废弃混凝土质量与全部混凝土质量之比，剪跨比 $\lambda$ 为剪跨长度与截面有效高度之比。试件编号的具体含义见图 31，其中 RB 表示再生混合钢筋混凝土梁，B 表示常规混凝土梁。废弃混凝土块体和现浇混凝土的制备，以及试件制作过程与第 3.2 节相同，在此不再赘述。

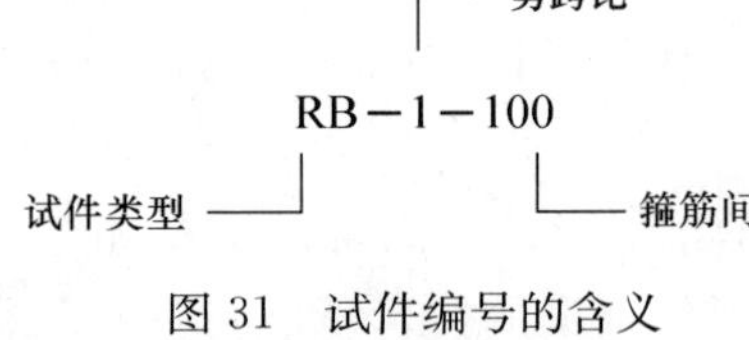

图 31 试件编号的含义

试验采用两点加载，加载装置见图 28。加载过程与第 3.2 节相同。纵筋和箍筋应变采用电阻式应变片量测。

**试 件 参 数** **表 20**

| 序号 | 试件编号 | 截面宽度 (mm) | 截面有效高度 (mm) | 梁长 (mm) | 净跨 (mm) | $f_{c1u}$ (MPa) | $f_{c2u}$ (MPa) | 试件型式 | 混合比 $\eta$ | 剪跨比 $\lambda$ |
|---|---|---|---|---|---|---|---|---|---|---|
| 1 | RB-1 | 149 | | 1800 | 1500 | 48.8 | | 混合型 | 20% | 1 |
| 2 | RB-1.5 | 147 | | 1800 | 1500 | 48.8 | | 混合型 | 20% | 1.5 |
| 3 | B-2 | 146 | | 1800 | 1500 | 49.0 | | 全现浇 | 0 | 2 |
| 4 | RB-2 | 148 | | 1800 | 1500 | 49.0 | | 混合型 | 20% | 2 |
| 5 | RB-2-150 | 154 | | 1800 | 1500 | 47.0 | | 混合型 | 20% | 2 |
| 6 | RB-2-100 | 149 | | 1800 | 1500 | 47.0 | | 混合型 | 20% | 2 |
| 7 | RB-2.5 | 144 | | 2300 | 2000 | 53.4 | | 混合型 | 20% | 2.5 |
| 8 | B-3-150 | 152 | 262.5 | 2300 | 2000 | 47.1 | 32.9 | 全现浇 | 0 | 3 |
| 9 | RB-3 | 147 | | 2300 | 2000 | 53.4 | | 混合型 | 20% | 3 |
| 10 | RB-3-150 | 156 | | 2300 | 2000 | 47.1 | | 混合型 | 20% | 3 |
| 11 | RB-3-100 | 145 | | 2300 | 2000 | 50.4 | | 混合型 | 20% | 3 |
| 12 | RB-4 | 150 | | 2900 | 2600 | 51.5 | | 混合型 | 20% | 4 |
| 13 | RB-4-150 | 147 | | 2900 | 2600 | 53.0 | | 混合型 | 20% | 4 |
| 14 | RB-4-100 | 150 | | 2900 | 2600 | 53.0 | | 混合型 | 20% | 4 |

(2) 试验结果与分析

①破坏形态

试验过程中，大部分梁发生受剪破坏，依剪跨比不同具体可分为斜压破坏、剪压破坏和斜拉破坏三种情况，另有少量梁因剪跨比较大而发生弯曲破坏。剪跨比 $\lambda=1$ 的无腹筋梁（RB-1）发生斜压破坏。剪跨比 $1.5\leqslant\lambda\leqslant2.5$ 的无腹筋梁（RB-1.5、B-2、RB-2、RB-2.5）、剪跨比 $\lambda=2$ 的有腹筋梁（RB-2-150、RB-2-100），以及剪跨比 $\lambda=3$ 且箍筋间距相对较大的有腹筋梁（B-3-150、RB-3-150）发生剪压破坏。剪跨比 $\lambda\geqslant3$ 的无腹筋梁 RB-3 和 RB-4 发生斜拉破坏。剪跨比 $\lambda=3$ 且箍筋间距相对较小的有腹筋梁（RB-3-100），以及剪跨比 $\lambda=4$ 的有腹筋梁（RB-4-150、RB-4-100）发生受弯破坏。部分试件的破坏形态见图 32。

图 32　部分试件的破坏形态

(a) RB-1；(b) RB-2；(c) RB-3-150；(d) RB-3；(e) RB-4；(f) RB-4-100

②荷载—跨中挠度曲线

部分试件的荷载—跨中挠度曲线见图 33。从图中可以看出：

A. 对于发生斜压破坏和斜拉破坏的无腹筋梁，在 80%破坏荷载之前，试件的荷载—跨中挠度曲线呈现出较明显的线性特性，随后曲线斜率有所减小。试件 RB-3 因斜拉破坏前临界斜裂缝开展较宽，曲线末段挠度增长相对较快；而试件 RB-4 的临界斜裂缝出现十分突然，破坏过程迅速且无任何征兆，曲线末段斜率几乎没有减小。

B. 对于发生剪压破坏的无腹筋梁和有腹筋梁，随着荷载增加，试件的荷载—跨中挠度曲线的斜率逐渐变小，呈现出较明显的渐变非线性特征。

C. 对于发生受弯破坏的有腹筋梁，试件的荷载—跨中挠度曲线起初近似呈直线发展，后期曲线斜率逐渐变小，最终破坏时曲线几乎水平，相应的跨中挠度明显大于斜压、斜拉

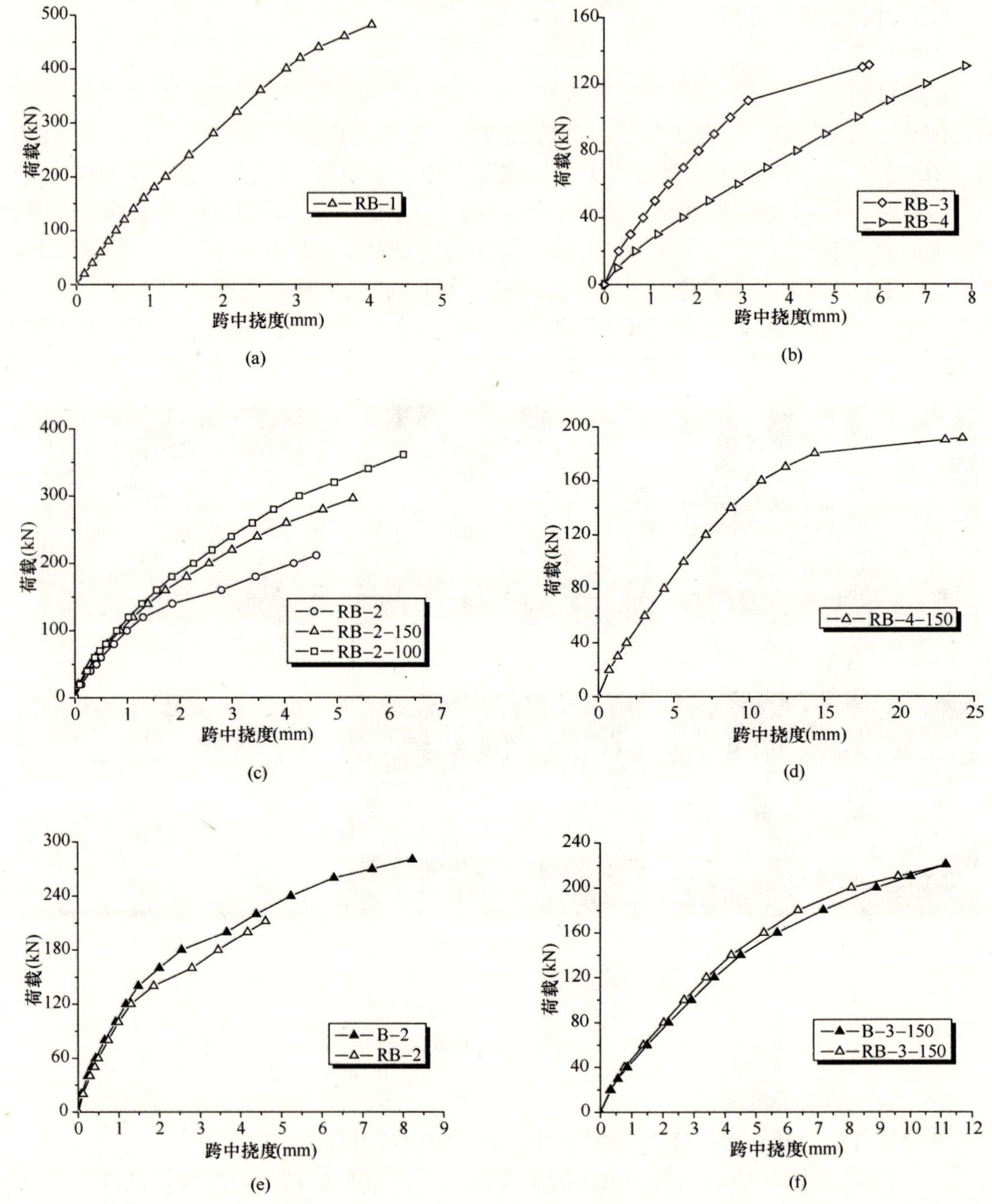


图 33 荷载—跨中挠度曲线

(a) 斜压破坏；(b) 斜拉破坏；(c) 剪压破坏；(d) 受弯破坏；(e) 无腹筋梁；(f) 有腹筋梁

和剪压破坏情况。

D. 在其他条件相同的情况下，无论是无腹筋还是有腹筋，再生混合钢筋混凝土梁的荷载—跨中挠度曲线与常规混凝土梁的相应曲线均较为接近，表明二者具有几乎相同的初始刚度，但前者的破坏荷载小于后者且无腹筋时偏小更为明显。

③受剪承载力

各试件的试验结果汇总于表 21。表中开裂荷载对应斜裂缝刚出现时的荷载值，极限荷载则根据文献［41］确定，即受剪破坏试件在加载或持载过程中出现下列现象之一时所对应的荷载：①斜裂缝端部受压区混凝土剪压破坏；②沿斜截面混凝土斜向受压破坏；③沿斜截面撕裂形成斜拉破坏；④箍筋与斜裂缝交会处的裂缝宽度达到 1.5mm。表中受剪承载力取为极限荷载的一半。

**试验结果**　　**表 21**

| 序号 | 试件编号 | $f_{cu}$ (MPa) | $f_t$ (MPa) | 开裂荷载 (kN) | 极限荷载 (kN) | 破坏荷载 (kN) | 受剪承载力 (kN) | 破坏形态 |
|---|---|---|---|---|---|---|---|---|
| 1 | RB-1 | 45.6 | 2.84 | 150 | 481.4 | 481.4 | 240.7 | 斜压 |
| 2 | RB-1.5 | 45.6 | 2.84 | 130 | 320 | 341.6 | 160 | 剪压 |
| 3 | B-2 | 49.0 | 2.96 | 120 | 210 | 280.6 | 105 | 剪压 |
| 4 | RB-2 | 45.8 | 2.85 | 90 | 180 | 212 | 90 | 剪压 |
| 5 | RB-2-150 | 44.2 | 2.79 | 90 | 270 | 296.1 | 135 | 剪压 |
| 6 | RB-2-100 | 44.2 | 2.79 | 120 | 350 | 360.9 | 175 | 剪压 |
| 7 | RB-2.5 | 49.3 | 2.97 | 75 | 150 | 195.1 | 75 | 剪压 |
| 8 | B-3-150 | 47.1 | 2.89 | 70 | 190 | 220.3 | 95 | 剪压 |
| 9 | RB-3 | 49.3 | 2.97 | 85 | 120 | 131.2 | 60 | 斜拉 |
| 10 | RB-3-150 | 44.3 | 2.80 | 70 | 190 | 218.8 | 95 | 剪压 |
| 11 | RB-3-100 | 46.9 | 2.89 | 90 | — | 240.6 | — | 受弯 |
| 12 | RB-4 | 47.8 | 2.92 | 55 | 130.6 | 130.6 | 65.3 | 斜拉 |
| 13 | RB-4-150 | 49.0 | 2.96 | 60 | — | 191.3 | — | 受弯 |
| 14 | RB-4-100 | 49.0 | 2.96 | 70 | — | 190.8 | — | 受弯 |

注：$f_{cu}=[f_{c1u}\cdot(1-\eta)+f_{c2u}\cdot\eta]$ 为现浇混凝土和废弃混凝土的组合立方体抗压强度；
$f_t=0.88\times0.395f_{cu}^{0.55}$ 为现浇混凝土和废弃混凝土的组合轴心抗拉强度。

对比表 21 中的试件 RB-2 和 B-2 可以看出，在其他条件相同的情况下，无腹筋再生混合钢筋混凝土梁 RB-2 的开裂荷载、受剪承载力和破坏荷载分别比无腹筋常规混凝土梁 B-2 降低 25%、14.3%和 24.4%。这一方面是因为前者内部废弃混凝土和现浇混凝土的组合强度略低于后者全现浇混凝土的缘故（见表 20 和表 21），另一方面则是因为前者内部大量存在的新、旧混凝土界面可能成为薄弱区域，致使斜裂缝更易产生和开展，并最终削弱无腹筋试件的抗剪能力。

对比表 21 中的试件 RB-3-150 和 B-3-150 可以看出，在其他条件相同的情况下，有腹筋再生混合钢筋混凝土梁 RB-3-150 的开裂荷载、受剪承载力和破坏荷载几乎与有腹筋常规混凝土梁 B-3-150 完全相同，表明上述两方面因素在有腹筋时影响十分有限。这一方面是因为箍筋提供的抗剪能力对于前者和后者是相同的，使得新、旧混凝土组合强度略低于全现浇混凝土所导致的不利效应对前者总抗剪能力的影响有所减弱，另一方面则可能源于箍筋对前者更易产生和开展的斜裂缝具有一定抑制作用。

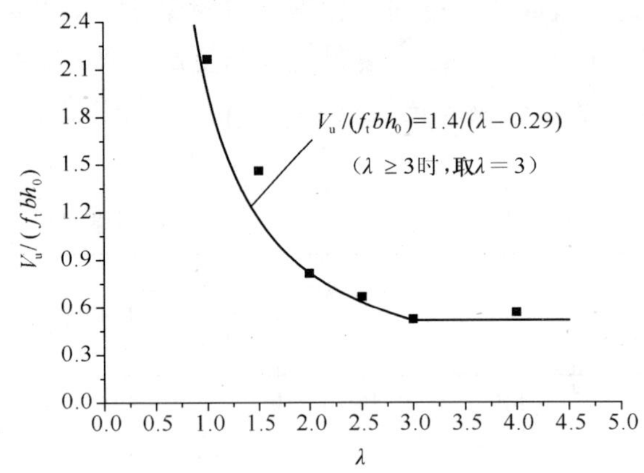


图 34 受剪承载力—剪跨比关系曲线

图 34 所示为无腹筋再生混合钢筋混凝土梁 RB-1、RB-1.5、RB-2、RB-2.5、RB-3 和 RB-4 的受剪承载力—剪跨比关系曲线。从图中可以看出，与无腹筋常规混凝土梁相似，无腹筋再生混合钢筋混凝土梁的受剪承载力随剪跨比增大逐渐减小，但当剪跨比增至 3 以后，受剪承载力变化不大。

通过对图 34 所示数据点进行回归分析，可建立剪跨比 $1 \leqslant \lambda \leqslant 4$ 时无腹筋再生混合钢筋混凝土梁在集中荷载作用下的如下受剪承载力计算公式：

$$V_u = \frac{1.4}{\lambda - 0.29} f_t b h_0 \quad (\lambda \geqslant 3 \text{时，取} \lambda = 3) \tag{2}$$

对于有腹筋再生混合钢筋混凝土梁，其受剪承载力可在无腹筋梁基础上增加箍筋的贡献。为此，剪跨比 $1 \leqslant \lambda \leqslant 4$ 时有腹筋再生混合钢筋混凝土梁在集中荷载作用下的受剪承载力可按下式计算：

$$V_u = \frac{1.4}{\lambda - 0.29} f_t b h_0 + f_{yv} \frac{A_{sv}}{s} h_0 \quad (\lambda \geqslant 3 \text{时，取} \lambda = 3) \tag{3}$$

式中 $f_{yv}$ 为箍筋屈服强度；$A_{sv}$ 为配置在同一截面内箍筋各肢的全部横截面积；$s$ 为箍筋间距。

利用式（2）和式（3）以及《混凝土结构设计规范》（GB 50010—2002）给出的相应公式，对各试件的受剪承载力进行计算，并与对应实测值进行比较，具体如图 35。从图中可以看出：

A. 针对无腹筋再生混合钢筋混凝土梁，现行规范计算结果比试验结果偏低 17%～68%，且总体上剪跨比越小偏低越显著，平均偏低约 35.5%。这是因为规范所给计算公式一般对应大量试验点的偏下值。

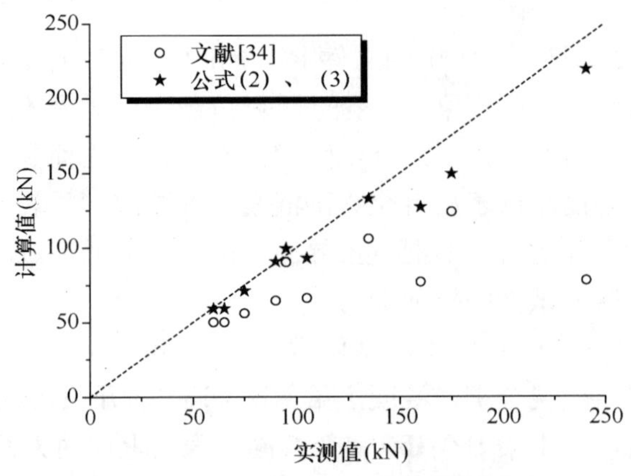


图 35 受剪承载力计算值与实测值的比较

B. 针对有腹筋再生混合钢筋混凝土梁，现行规范计算结果与试验结果相比偏低 5%～29%；对比试件 RB-3-150 和 B-3-150 发现，二者的现行规范计算结果均比试验结果偏低 5%，表明在其他因素相同的情况下，采用现行规范计算有腹筋再生混合钢筋混凝土梁的受剪承载力具有与常规混凝土梁相近的安全性。

C. 针对无腹筋再生混合钢筋混凝土梁，式（2）计算结果与试验结果总体吻合较好，平均偏低 7.3%。

D. 针对有腹筋再生混合钢筋混凝土梁，式（3）计算结果与试验结果总体吻合较好，平均偏低 4.3%。

## 15.4　工程应用

近两年来，本课题组与深圳市建筑设计研究总院合作，对混凝土再生混合构件的应用技术进行了有益的探索，2010 年 5 月 18 日已正式将钢管再生混合柱应用于实际工程，迈出了混凝土再生混合构件工程应用的第一步。

### 15.4.1　工程背景

该工程位于广东省紫金县县城中心区，工程计划将原建于上世纪 70 年代的紫金县影剧院拆除，并在原址建设新的紫金县文化活动中心。工程为一公共建筑，具有观影（1200 座）、办公、商业等多项功能。建筑主体平面为矩形，平面尺寸 36.6m×72m，总建筑面积 6759m$^2$。建筑总层数 4 层（舞台区设有地下室），总高度（舞台区）27.3m。建筑采用框架结构，局部设有剪力墙。乙类抗震设防，设防烈度 7 度。图 36 和图 37 所示分别为已拆除的紫金县影剧院原貌，以及修建中的紫金县文化活动中心效果图。

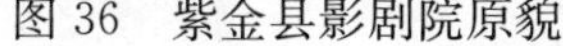

图 36　紫金县影剧院原貌

图 37　紫金县文化活动中心建筑效果图

作为初次应用，该建筑舞台区附近的每层 6 根柱子被设计为钢管再生混合柱。圆钢管外径 600mm、壁厚 10mm，废弃混凝土采用块体型式，混合比取为 25%。建筑中钢管再生混合柱的平面位置见图 38。

钢管再生混合柱使用的废弃混凝土直接来源于紫金县影剧院拆除现场。由于建设年代久远，废弃混凝土强度等级普遍偏低。废弃混凝土的现场抽芯检测结果见表 22。图 39 所示为紫金县影剧院的拆除现场，拆除的同时挑选出约 30～40t 废弃混凝土存贮堆放在施工现场附近（见图 40）。

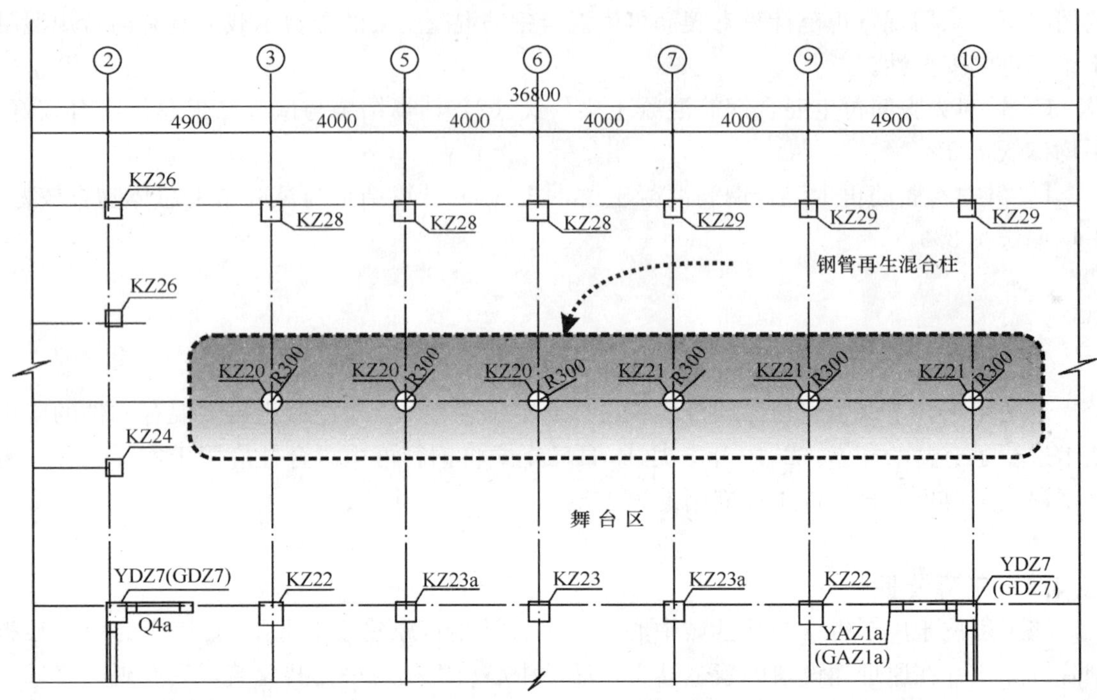


图 38 钢管再生混合柱的平面位置

图 39 紫金县影剧院拆除现场

图 40 废弃混凝土现场

废弃混凝土的强度检测结果 表 22

| 组号 | 换算所得立方体强度 (MPa) | | | 每组立方体强度平均值 (MPa) | 废弃混凝土立方体强度 $f_{c2u}$ (MPa) |
|---|---|---|---|---|---|
| 1 | 13.6 | 13.0 | 11.6 | 12.7 | 15.43 |
| 2 | 19.6 | 17.4 | 15.3 | 17.4 | |
| 3 | 16.4 | 15.5 | 16.7 | 16.2 | |

## 15.4.2 足尺试验

(1) 试验方案

为考察现场浇注钢管再生混合柱的可行性并验证设计方法的准确性，在工程正式施工前进行了 2 根钢管再生混合短柱的足尺试验。试件采用 Q235B 直缝焊接钢管，外径 600mm，壁厚 10mm，高度 1800mm。废弃混凝土由紫金县影剧院拆除而得。试件的基本参数和实测抗压承载力见表 23。

试件基本参数和实测抗压承载力　　表 23

| 试件编号 | $f_{c1u}$ (MPa) | $f_{c2u}$ (MPa) | 钢管屈服强度 $f_s$ (MPa) | 混合比 $\eta$ | 实测抗压承载力 (kN) |
|---|---|---|---|---|---|
| Z1 | 31.24 | 15.43 | 305 | 25% | 14500 |
| Z2 | | | | | |

注：$f_{c1u}$和 $f_{c2u}$分别为轴压试验当天现浇混凝土和废弃混凝土的立方体抗压强度。

现浇混凝土现场拌制，每立方米现浇混凝土的材料用量为石子：砂：水泥：水＝1031：747：426：196。利用简易工具将足量废弃混凝土破碎成块体，特征尺寸控制在 100～250mm。一切准备就绪后，施工单位安排工人顺利完成了两个试件的现场浇捣工作。从施工情况看，工人很快就可掌握钢管再生混合柱的施工要领，同时混合比为 25%时废弃混凝土块体与新混凝土的混合浇捣也较为容易。图 41 所示为试件的现场浇注情况。现场养护两天后，试件被运回华南理工大学结构实验室继续养护。

试件浇注 36d 后在 1500t 长柱压力机上进行轴压试验（见图 42）。由于预估的试件抗压承载力相当大，已接近试验设备的承受极限，试验时采用力控制加载方案，每级荷载 500kN。当加载至 14500kN 时，两个试件的压力机油压表指针均突然迅速下掉，随即停止试验。

图 41　现场浇注情况

（2）试验结果与分析

图 43 所示为试件破坏形态。当荷载达到极限荷载的 95%时，试件上端钢管壁出现局部屈曲，其中 Z2 更为明显。图 44 所示为试件的荷载－竖向变形曲线。从图中可以看到，

图 42 试件加压前照片

两个试件显示了相似的屈服前和屈服后力学行为。若以钢管纵向应变达到 2000$\mu\varepsilon$ 作为试件的屈服应变[42]，则两个试件对应的屈服荷载分别为 10000kN 和 10500kN，而紫金县文化活动中心的钢管再生混合柱的设计轴力仅为 2800kN，可见设计是足够安全的。当然本次足尺试验采用的是短柱，忽略了弯矩和长细比等因素对承载力的影响，这是需要额外考虑的。

图 45 所示为试验后割开钢管后显示的核心混凝土外表面，除试件上端混凝土轻微压碎外，核心混凝土外表面基本光洁。采用气刨破开核心混凝土后的内部结构见图 46。从图中可以看出，新、旧混凝土界面清晰可辨，试件浇捣情况良好，内部结构密实。从废弃混凝土块体在试件内的分布情况看，其沿试件高度方向的分布基本均匀，没有因自重影响而出现上部块体少、下部块体多的现象。

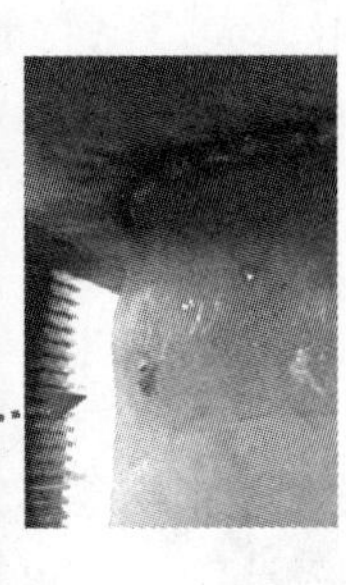

图 43 试件破坏形态

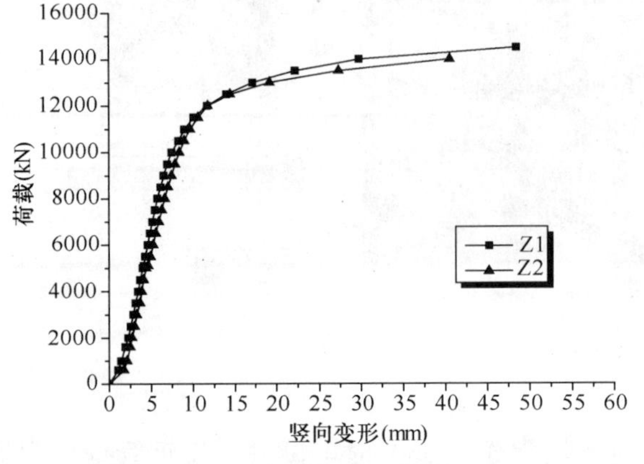


图 44 荷载—变形曲线

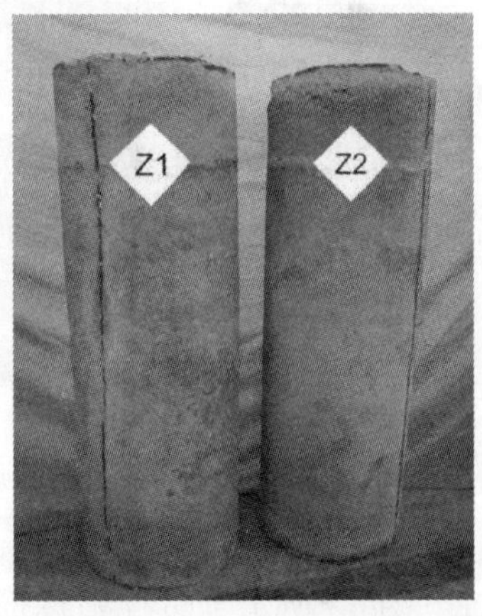


图 45 核心混凝土表面

图 46　核心混凝土内部结构

（3）计算分析

为考察第 2.1 节的研究结论对实际工程的适用性，对两个足尺试件采用 JCJ 01—89、CECS 28 ：90、DL/T 5085—1999、DBJ 13-51-2003、ACI（2005）、AISC（2005）、EC4（2004）和 AIJ（1997）共 8 部设计标准进行了计算。计算过程中，混凝土抗压强度按现浇和废弃混凝土的组合强度 $f'_{cu}=f_{c1u}\times(1-\eta)+f_{c2u}\times\eta$ 取值。计算结果见表 24。

**计算结果与试验结果的比较　表 24**

| 设计规范 | JCJ | CECS | DL/T | DBJ | ACI | AISC | EC4 | AIJ |
|---|---|---|---|---|---|---|---|---|
| $N_{uc}$（kN） | 14945 | 15697 | 13137 | 12054 | 10557 | 11066 | 14383 | 12267 |
| $N_{ue}/N_{uc}$ | 0.970 | 0.924 | 1.104 | 1.203 | 1.373 | 1.310 | 1.008 | 1.182 |

注：$N_{uc}$和 $N_{ue}$分别为计算值和实测值。

从表 24 中可以看出，相对于其他 6 部设计标准，依据 JCJ 01—89 和 EC4（2004）给出的抗压承载力计算结果与试验实测值最为接近，这与第 2.1 节的研究结论是一致的。

### 15.4.3　现场施工

2010 年 5 月 18 日开始施工首层 6 根钢管再生混合柱。施工前采用简易工具破碎制取首层浇注所需的废弃混凝土块体共 4.8t（两个工人一个工作日即可完成）。施工时首先将废弃混凝土块体浇水侵润，再用小斗车将其运至柱旁，然后利用小型龙门架将废弃混凝土块体和新混凝土向上提送到钢管再生混合柱管口进行投放和混合浇捣。工程施工现场见图 47，钢管再生混合柱的浇捣情况见图 48。

图 47　施工现场

图 48　浇捣钢管再生混合柱

该工程现已顺利封顶。从现场施工情况看，钢管再生混合柱的浇捣过程较为顺利，实现了设计意图，表明其在多层建筑中应用是完全可行的。

## 15.5 施工工艺

下面对节段型钢管再生混合柱和块体型钢管再生混合柱的施工工艺做一简单介绍。

对于节段型钢管再生混合柱，可首先在钢管底部浇注一定厚度的新混凝土，然后根据预定方案在钢管内放置一根较长的或沿竖向依次放置 2～3 根较短的废弃混凝土节段，最后浇注新混凝土填充废弃混凝土节段与钢管之间的横向空隙。当钢管内沿竖向依次放置有 2～3 根较短的废弃混凝土节段时，相邻节段之间应间隔一定厚度的新混凝土。施工过程中，为避免废弃混凝土节段与钢管内壁接触，吊装时可在节段两端各固定一个钢筋箍，每个钢筋箍上焊接 3～4 根水平短钢棒以隔离节段和钢管内壁。图 49 所示为节段型钢管再生混合柱的施工工艺示意图。

对于块体型钢管再生混合柱，可首先在钢管底部浇注一定厚度的新混凝土，然后将装有废弃混凝土块体的钢槽箱起吊到钢管上部，利用钢槽箱的底部插板控制块体的分层投放，最后浇注新混凝土与废弃混凝土块体进行混合浇捣。钢槽箱内的横隔板是活动的，可根据废弃混凝土块体的一次投放量进行灵活布置。图 50 所示为块体型钢管再生混合柱的施工工艺示意图。

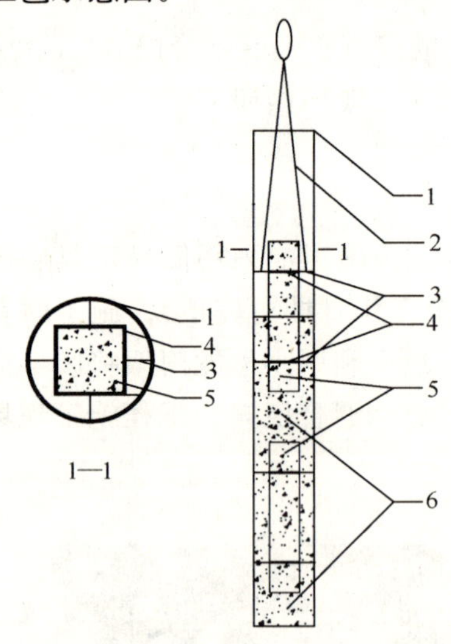


图 49 节段型钢管再生混合柱的施工工艺
1—钢管；2—吊钩；3—短钢棒；4—钢筋箍；
5—废弃混凝土节段；6—新混凝土

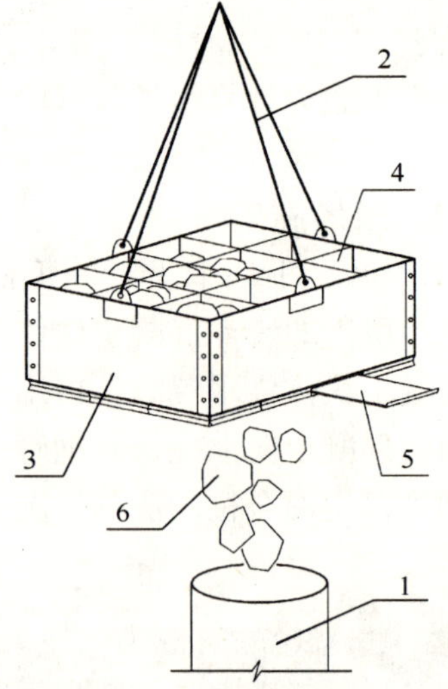


图 50 块体型钢管再生混合柱的施工工艺
1—钢管；2—吊钩；3—钢槽箱；4—活动隔板；
5—底部插板；6—废弃混凝土块体

## 15.6 结论与展望

通过本文的研究，可以得到如下初步结论：

(1) 废弃混凝土大尺度回收利用省去了再生骨料的繁琐生产过程及其大量能源消耗，以及再生骨料混凝土配制所耗费的水、水泥和能源，经济性优于再生骨料层次的回收利用。

(2) 混凝土再生混合构件的力学性能总体上接近或略低于全现浇构件，现行规范有关后者的计算公式可直接或略作修正后应用于前者。

(3) 新、旧混凝土的混合强度可在二者各自强度的基础上进行初步估算。

(4) 采用废弃混凝土节段的再生混合构件的力学性能总体上优于采用废弃混凝土块体的再生混合构件。

(5) 在用钢量和横截面积相同的情况下，薄壁钢管再生混合柱的抗压承载力明显高于钢筋混凝土柱。

(6) 从确保施工质量和提高节能减排环保效益两方面综合考虑，再生混合构件中废弃混凝土块体的混合比宜取为 20%～33%，废弃混凝土节段的混合比可适当提高。

虽然本课题组已经并正在对混凝土再生混合构件开展系列研究，但仍有许多问题有待进一步探讨，例如：

A. 混凝土再生混合构件的长期受力性能；

B. 薄壁钢管再生混合柱的有效节点形式及性能；

C. 外置型钢再生混合构件的抗震性能；

D. 外置型钢再生混合构件的抗火性能；

E. 新、旧混凝土的界面行为；

F. 混凝土再生混合构件的耐久性能；

G. 混凝土再生混合构件的高效施工方法；

H. 混凝土再生混合构件的技术标准；

I. 废弃混凝土回收管理的有效机制。

总之，混凝土再生混合构件是废弃混凝土循环利用的一种新的有效途径，具有广阔应用前景，同时又是一项新生事物，还有许多工作有待开展。进一步发展和应用混凝土再生混合构件是一项系统工程，涉及行政管理、科研、设计、施工、监理等各个方面，有赖全社会有识之士的广泛关注和积极参与。

## 参考文献

[1] 2006 年中国经济年鉴. 北京：中国经济年鉴社，2006.

[2] 王罗春，赵由才. 建筑垃圾处理与资源化. 北京：化学工业出版社，2004.

[3] 朱红兵，王顺林，熊汉林，李秀. 混凝土废弃物的再生利用. 公路交通技术，2005，(1)：52-53，76.

[4] Hansen T C. Recycling of demolished concrete and masonry. London：E&FN SPON，1992.

[5] Buck A D. Recycled concrete as a source of aggregate. ACI Journal，Proceedings，1977，(74-22)：212-219.

[6] Topcu I B. Physical and mechanical properties of concretes produced with waste concrete. Cement and Concrete Research，1997，27(12)：1817-1823.

[7] Tam W Y V，Gao X F，Tam C M. Microstructural analysis of recycled aggregate concrete produced

from two-stage mixing approach. Cement and Concrete Research，2005，35(6)：1195-1203.

[8] Xiao J Z，Sun Y D，Falknerb H. Seismic performance of frame structures with recycled aggregate concrete. Engineering Structures，2006，28(3)：1-8.

[9] Eguchi K，Teranishi K，Nakagome A，Kishimoto H. Application of recycled coarse aggregate by mixture to concrete construction. Construction and Building Materials，2007，21(7)：1542-1551.

[10] Limbachiya M C，Marrocchino E，Koulouris A. Chemical-mineralogical characterization of coarse recycled concrete aggregate. Waste Management，2007，27(2)：201-208.

[11] Evangelista L，Brito J. Mechanical behavior of concrete made with fine recycled concrete aggregates. Cement and Concrete Composites，2007，29 (5)：397-401.

[12] Poon C S，Chan D. The use of recycled aggregate in concrete in Hong Kong. Resources，Conservation and Recycling，2007，50(3)：293-305.

[13] Li X. Recycling and reuse of waste concrete in China：part I. material behaviour of recycled aggregate concrete. Resources，Conservation and Recycling，2008，53(1-2)：36-44.

[14] Fathifazl G，Razaqpur A G，Isgor O B，et al. Flexural performance of steel-reinforced recycled concrete beams. ACI Structural Journal，2009，106(6)：858-867.

[15] 肖建庄，李佳彬，兰阳. 再生混凝土技术研究最新进展与评述. 混凝土，2003，(10)：17- 20.

[16] 肖建庄. 再生混凝土. 北京：中国建筑工业出版社，2008.

[17] 万惠文，水中和，林宗寿，肖开涛. 再生混凝土的环境评价. 武汉理工大学学报，2003，25(4)：17-20.

[18] 武洪明. 水泥单位产品能耗电力折算标准煤取值探讨. 中国水泥，2008，(4)：71-73.

[19] 中华人民共和国国家建筑材料工业局标准 JCJ 01-89. 钢管混凝土结构设计与施工规程. 上海：同济大学出版社，1989.

[20] 中国工程建设标准化协会标准 CECS 28：90. 钢管混凝土结构设计与施工规程. 北京：中国计划出版社，1992.

[21] 中华人民共和国电力行业标准 DL/T 5085-1999. 钢-混合体组合结构设计规程. 北京：中国电力出版社，1999.

[22] 福建省工程建设标准 DBJ 13-51-2003. 钢管混凝土结构技术规程. 福州：福建省建设厅，2003.

[23] 天津市工程建设标准 DB 29-57-2003. 天津市钢结构住宅设计规程. 天津，2003.

[24] ACI Committee 318 (ACI 318-05)，2005. Building code requirements for structural concrete and commentary. American Concrete Institute，Detroit，USA.

[25] ANSI/AISC 360-05，2005. Specification for structural steel buildings. American Institute of Steel Construction (AISC)，Chicago，USA.

[26] AIJ，1997. Recommendations for design and construction of concrete filled steel tubular structures. Architectural Institute of Japan (AIJ)，Tokyo，Japan.

[27] British Standards Institutions BS 5400，2005. Steel，concrete and composite bridge，Part 5：Code of practice for design composite bridges. London，UK.

[28] Eurocode 4 (EC 4)，Design of steel and concrete structures-Part 1：General rules and rules for building. EN 1994-1-1：2004，Brussels，European Committee for Standardization.

[29] O'Shea M D，Bridge R Q. Test on circular thin-walled steel tubes filled with medium and high strength concrete. Department of Civil Engineering，Research Report No. R755，Australia：the University of Sydney，1997.

[30] Uy B. Local and post-local buckling of concrete filled steel welded box columns. Journal of Construc-

tional Steel Research，1998，47(1-2)：47-72.

[31] Nishiyama I，Morino S，Sakino K，et al. Summary of research on concrete-filled structural steel column system carried out under the US-Japan cooperative research program on composite and hybrid structures. BRI Research Paper No. 147，Japan：Building Research Institute，2002.

[32] 张耀春，王秋萍，毛小勇，曹宝珠. 薄壁钢管混凝土短柱轴压力学性能试验研究. 建筑结构. 2005，35(1)：22-27.

[33] 占美森. 薄壁钢管混凝土柱轴心受压承载性能研究. 南昌：南昌大学，2007.

[34] 中华人民共和国国家标准(GB 50010—2002). 混凝土结构设计规范. 北京：中国建筑工业出版社，2002.

[35] 哈尔滨工业大学，大连理工大学，北京建筑工程学院，华北水利水电学院合编. 混凝土及砌体结构. 北京：中国建筑工业出版社，2002.

[36] 马欣伯. 圆钢管混凝土构件承载力设计方法比较及新方法初探. 哈尔滨：哈尔滨工业大学，2005.

[37] 蔡绍怀. 现代钢管混凝土结构. 北京：人民交通出版社，2003.

[38] 中国工程建设标准化协会标准 CECS 159：2004. 矩形钢管混凝土结构技术规程. 北京：中国计划出版社，2004.

[39] 钱稼茹，崔瑶，方小丹. 钢管混凝土柱受剪承载力试验. 土木工程学报，2007，40(5)：1-9.

[40] 肖从真，蔡绍怀，徐春丽. 钢管混凝土抗剪性能试验研究. 土木工程学报，2005，38(4)：5-11.

[41] 中华人民共和国国家标准 GB 50152-92. 混凝土结构试验方法. 北京：中国建筑工业出版社，1992.

[42] Schneider S P. Axially loaded concrete-filled steel tubes. Journal of Structural Engineering，1998，124(10)：1125-1138.

# 第 16 章 Chapter 16

# 超高韧性水泥基复合材料与既有混凝土粘结力学性能及耐久性防护加层的试验研究

徐世烺[1]，王　楠[2]
（1. 浙江大学建筑工程学院，浙江 310027；2. 大连理工大学建设工程学部，大连 116024）

**提　要：**本文通过超高韧性水泥基复合材料(UHTCC)与既有混凝土组成粘结试件的抗拉和抗剪试验、后浇UHTCC、既有混凝土复合梁和后浇UHTCC耐久性防护层加固钢筋混凝土梁的弯曲试验以及集中荷载作用下简支双向复合板试验，研究了UHTCC与既有混凝土两者的粘结性能以及不同因素对两者粘结强度的影响，以及加固的复合试件弯曲性能、加固效果、破坏形态和裂缝扩展等。试验结果表明，UHTCC与既有混凝土材料间具有较好的粘结性能，通过改善两者的粘结性状能够提高两者间粘结强度；在复合梁板试件中，UHTCC材料的加入改善了既有混凝土试件的整体韧性，提高既有试件的开裂荷载和正常使用极限荷载；UHTCC后浇层有效限制了受弯底面上裂缝的开裂和扩展。运用解析和数值方法，对复合梁板试件的承载能力等性能进行计算分析，并提出相应的求解过程，所得的计算结果与试验实测值符合较好。

**关键词：**粘结性能；复合试件；弯曲性能；解析计算；数值计算

## 16.1　引言

混凝土材料作为当今世界上用途最广、用量最大的建筑材料，由于其具有抗压强度高、耐久性好、强度等级范围广，以及原料丰富、价格低廉、生产工艺简单等特点，被广泛用于土木、水利建筑工程，海洋及港湾建筑工程，交通运输与铁路工程等与国民经济生产生活息息相关的基础性设施中，取得了良好的经济效益和社会效益[1~3]。作为发展中国家，基础设施的建设与完善一直是我国发展过程中的重要环节，而其中绝大部分都是混凝土结构，因此从20世纪90年代起我国就已经是世界上混凝土生产与应用最多的国家之一，其中混凝土和水泥的产量连续20年居世界第一，占全球总量的40%以上，据统计我国年用量约为25亿$m^3$。近年来，为了加速经济发展和社会进步，我国每年投入到重大工程建设项目(铁路、公路、机场等)、城市基础设施建设、新农村建设等的资金用量一直呈现出递增势头[4,5]。据国家发展改革委统计，2002～2008年，国家仅在西部大开发一项中就新开工重点工程102项，投资总规模达17400多亿元，2009年在4万亿资金拉动内需下，又批准新开工建设18项重点工程，投资总规模达4689亿元。目前多个重点工程已经开始运作或是已经竣工，如西藏铁路、南水北调、西气东输、北煤南运、西油南输、西电东送、西面东调、南菜北运等[6,7]。然而我国自1949年新中国成立后在东部大规模兴建，到目前所开展的西部大开发等项目中所采用的混凝土结构，在广泛使用的同时，还面临着诸多耐久性问题，使混凝土结构承载能力下降、服役寿命衰减，给国家和人民生产生活带来诸多安全隐患和经济损失。特别是海洋与近海水利工程以及西部盐湖地区工程项目等，由于所处区域内普遍存在高浓度的氯、硫酸根、镁等多种有害离子，以及潮湿或者干冷、干热的严酷气候条件，导致当地混凝土结构损伤劣化速度明显高于我国其他地区[10~12]。然而，目前加固方法中较为常见的传统方法依旧采用混凝土材料加固，由于混凝土自身缺点(脆性大、抗拉强度小等)，加固后整体耐久性只能得到短时间提高，之后又会陷入之前的耐久性问题中；而近年来出现的加固技术中大多采用的都是刚性材料，比如侧重于提高结构承载能力的钢板和纤维布等，这些材料由于刚性较大，故无法适应结构承载中出现的较大变形，往往在应用中需要进行严格的设计和施工，而侧重于提高结构耐久性的无机刚性防水材料、聚合物水泥防水砂浆、水泥基渗透结晶防水材料等，主要以防水为主，基本不能承载，对于既有结构由于承载或收缩变形等引起的裂缝，通常无法阻止，进而降低了其耐久性加固后的效果。因此，寻找一种有效易行的加固(或防护)已受损(或未受损)既有混凝土结构的方法，是当前较为迫切的工作。

超高韧性水泥基复合材料(Ultra High Toughness Cementitious Composites，简称UHTCC)是一种借助细观力学和断裂力学基本原理设计而成的纤维增强水泥基复合材料。通过优化基体、纤维和纤维与基体界面的基本性能以及三者之间的相互作用，使得纤维在少掺量(通常小等于2%)下仍能很好地满足应变硬化特性的两个设计准则。与传统的刚性建筑加固材料(纤维布、钢材等)不同，UHTCC弹性模量相对较低，有较强的变形能力，具有类似金属材料拉伸强化的特征，极限拉应变可稳定地达到3%以上，其单轴拉伸应力应变曲线如图1-a所示；与传统水泥基材料在抗拉荷载下单一裂纹的宏观开裂模式也不同，UHTCC开裂为多条细密裂纹的微观开裂模式，具有优越的裂缝分散能力，在极限拉应变时该材料的平均裂缝宽度可控制在100μm以内，图1-b为直接拉伸试验中临近极限应变状态时裂缝发展情况[13~15]。UHTCC还具有较好的阻裂、耐冲击、抗疲劳、高韧性、高耐久性等特点，并且开裂后的UHTCC即使在拉伸应变达到百分之几的情况下，仍能表现出几乎和完好的混凝土材料相同的渗透性[14,16~19]。因此将其用于修补既有混凝土建筑物，通过在结构外部浇筑UHTCC耐久性防护层，进而提高或维持既有结构的力学性能、提高结构整体耐久性，符合目前钢筋混凝土设计理念，即以强度和耐久性并重，并以耐久

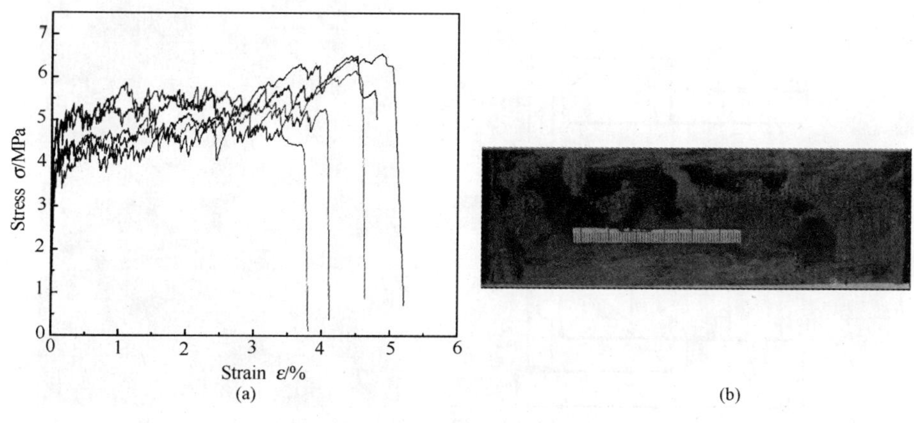


图1 拉伸荷载作用下UHTCC

(a)应力应变曲线；(b)裂缝形态

性为重点。特别是如水坝、跨海大桥等一些对阻裂、限裂要求较为严格的大型混凝土工程中，具有重要的实际意义。这些通常需要严格的配筋设计和混凝土施工养护的建筑物在实际使用中仍难以避免出现开裂、裂缝过大等现象，采用这项新技术后将会大大增强整体结构的裂缝防护能力。在国外部分地区，已尝试将其用于全比例工程试验。例如在日本 Mitaka 大坝受损坝面的维修，日本东海道新干线部分高架桥的耐久性加固，日本滋贺县中心主渠道的表面修补，以及美国密歇根某四跨简支钢梁桥的桥面板修补等[20～22]。但目前相关研究仍主要集中在其自身各项性能以及全比例试验的工程效果等展开，对 UHTCC 加固既有结构后的受力性能研究相对较少。因此，本文针对 UHTCC 与既有混凝土的粘结性能以及将其作为耐久性防护层与既有混凝土构成的复合试件的弯曲性能进行试验研究。

## 16.2 超高韧性水泥基复合材料与既有混凝土粘结性能试验研究

分别采用如图 2，图 3 所示试件形式，研究了不同因素对两者结合面粘结强度的影响，包括既有混凝土界面粗糙度、既有混凝土强度、界面干湿状态以及 UHTCC 的浇筑方位等因素。研究结果分析表明：

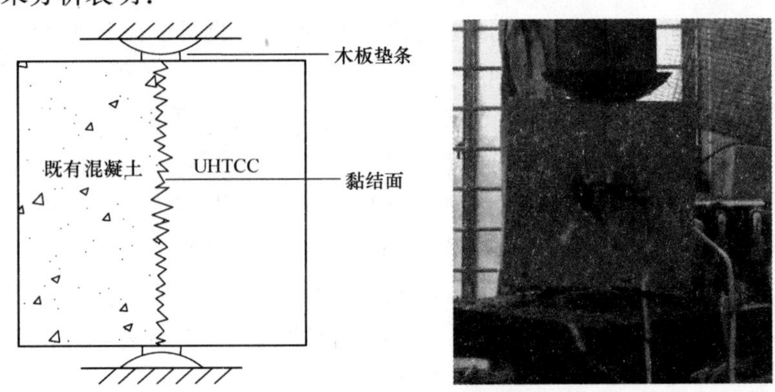


图 2 劈拉试验示意图

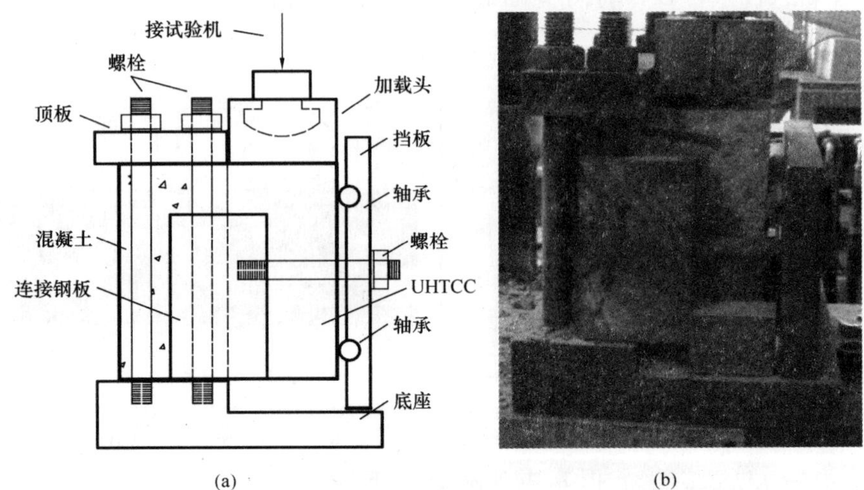


图 3 剪切试验示意图

在无界面剂、锚筋等增强界面粘结状态的措施下，单纯通过改变既有混凝土构件性状可使UHTCC与既有混凝土粘结劈拉强度达到既有混凝土相应强度72.8%以上、粘结剪切强度达到52.9%以上，表明此材料与既有混凝土结构具有较好的粘结性能。在实际工程应用中，如有较高的粘结要求，可通过改善既有混凝土界面微细观结构和增大UHTCC与既有混凝土两者间接触面积等来提高两者间粘结强度。

粘结界面的粗糙度对UHTCC与既有混凝土粘结强度有较大影响。随着粘结面粗糙度的增加，相应粘结试件的粘结强度随之增大，两者近似呈现线性关系；在四种界面处理方法下，各组试件中Ⅳ界面(自然劈拉界面)浇筑而成的复合试件粘结效果最好，粘结劈拉强度可达伴随整体混凝土试件强度的72.8%（A组）、81.6%（B组）、89.2%（C组）和98.9%（D组）；黏结剪切强度可达伴随整体混凝土试件强度的52.9%（A组）、56.0%（B组）、77.2%（C组）和87.5%（D组）。

当粘结面粗糙度相同时，复合试件的粘结强度随着既有混凝土强度的增加而增加；随着粘结面粗糙度的增加，既有混凝土强度对试件的粘结强度的影响也越大，混凝土强度最高的A组与强度最低的D组相比，劈拉强度提高的幅值为60%（Ⅰ界面）、70 %（Ⅱ界面）、87 %（Ⅲ界面)和103%（Ⅳ界面）；剪切强度提高的幅值为27%（Ⅰ界面）、32 %（Ⅱ界面）、46 %（Ⅲ界面)和42%（Ⅳ界面）。

不同的UHTCC浇筑方向对UHTCC与既有混凝土粘结性能也有一定影响，水平向(粘结面)浇筑UHTCC而成的复合试件粘结强度基本高于竖直向(粘结面)浇筑的试件，两者间存在一定的比例关系，竖直向(粘结面)浇筑试件粘结强度约为水平向(粘结面)浇筑试件的：劈拉强度0.75～0.83（B组），0.85～0.93（C组）；剪切强度0.86～1.06（B组），0.86～0.96（C组）。

既有混凝土界面的干湿状态同样影响着UHTCC与既有混凝土的粘结强度，其中界面饱和状态时浇筑的试件粘结强度高于丁燥状态时所浇筑的同组试件。原因是干燥界面会导致UHTCC与既有混凝土交界面附近的UHTCC水化不完全，产生大量气泡，形成新的薄弱区域，进而减小了复合试件的粘结强度。

具体试验结果如表1，表2所示。

**劈拉试验结果均值表** **表1**

<table>
<tr><th rowspan="2" colspan="2">组别</th><th colspan="2">既有混凝土</th><th rowspan="2">粘结面方向</th><th colspan="5">粗糙度(mm)</th><th colspan="6">平均劈拉强度(MPa)</th></tr>
<tr><th>抗压强度(MPa)</th><th>界面状态</th><th colspan="4">实测值</th><th>均值</th><th>伴随试件</th><th colspan="4">实测值</th><th>均值</th></tr>
<tr><td rowspan="4">A</td><td>Ⅰ</td><td rowspan="4">50.7</td><td rowspan="4">饱和</td><td rowspan="4">水平</td><td>1.51</td><td>1.25</td><td>1.32</td><td>1.39</td><td>1.37</td><td rowspan="4">5.11</td><td>2.73</td><td>2.43</td><td>2.56</td><td>2.59</td><td>2.58</td></tr>
<tr><td>Ⅱ</td><td>3.35</td><td>2.71</td><td>3.12</td><td>3.10</td><td>3.06</td><td>3.13</td><td>2.87</td><td>2.65</td><td>3.00</td><td>2.91</td></tr>
<tr><td>Ⅲ</td><td>4.38</td><td>5.10</td><td>4.59</td><td>4.78</td><td>4.71</td><td>2.98</td><td>3.61</td><td>3.32</td><td>3.25</td><td>3.29</td></tr>
<tr><td>Ⅳ</td><td>6.49</td><td>5.73</td><td>6.35</td><td>6.04</td><td>6.15</td><td>3.91</td><td>3.64</td><td>3.49</td><td>3.85</td><td>3.72</td></tr>
</table>

续表

| 组别 | | 既有混凝土 | | 粘结面方向 | 粗糙度(mm) | | | | | | 平均劈拉强度(MPa) | | | | |
|---|---|---|---|---|---|---|---|---|---|---|---|---|---|---|---|
| | | 抗压强度(MPa) | 界面状态 | | 实测值 | | | | 均值 | 伴随试件 | 实测值 | | | | 均值 |
| B | Ⅰ | 42.5 | 饱和 | 水平 | 1.28 | 1.41 | 1.12 | 1.31 | 1.28 | 4.24 | 2.56 | 2.18 | 2.2 | 2.35 | 2.32 |
| | Ⅱ | | | | 2.96 | 3.10 | 2.58 | 2.87 | 2.88 | | 2.62 | 2.79 | 2.65 | 2.59 | 2.66 |
| | Ⅲ | | | | 4.56 | 4.38 | 4.49 | 4.53 | 4.49 | | 3.12 | 3.5 | 2.84 | 3.27 | 3.18 |
| | Ⅳ | | | | 6.59 | 6.25 | 6.35 | 6.18 | 6.34 | | 3.21 | 3.72 | 3.69 | 3.23 | 3.46 |
| | Ⅰ | | | 竖直 | 1.38 | 1.55 | 1.27 | 1.50 | 1.43 | | 1.90 | 1.63 | 1.85 | 1.59 | 1.73 |
| | Ⅱ | | | | 2.93 | 2.56 | 2.65 | 2.80 | 2.74 | | 2.52 | 2.17 | 2.49 | 2.21 | 2.30 |
| | Ⅲ | | | | 4.25 | 4.85 | 4.33 | 4.69 | 4.53 | | 2.72 | 2.47 | 2.37 | 2.68 | 2.56 |
| | Ⅳ | | | | 6.38 | 6.16 | 6.30 | 6.29 | 6.28 | | 2.65 | 3.10 | 2.75 | 2.97 | 2.87 |
| C | Ⅰ | 31.2 | 饱和 | 水平 | 1.59 | 1.35 | 1.42 | 1.49 | 1.46 | 3.34 | 2.00 | 1.89 | 2.08 | 1.97 | 1.99 |
| | Ⅱ | | | | 2.20 | 2.51 | 2.65 | 2.44 | 2.45 | | 2.33 | 2.48 | 2.28 | 2.35 | 2.36 |
| | Ⅲ | | | | 4.25 | 4.36 | 4.78 | 4.45 | 4.46 | | 2.66 | 2.91 | 2.75 | 2.92 | 2.81 |
| | Ⅳ | | | | 5.18 | 5.52 | 5.03 | 5.23 | 5.24 | | 3.11 | 2.96 | 2.86 | 2.99 | 2.98 |
| | Ⅰ | | | 竖直 | 1.41 | 1.32 | 1.27 | 1.39 | 1.35 | | 1.91 | 1.72 | 1.94 | 1.81 | 1.85 |
| | Ⅱ | | | | 2.25 | 2.67 | 2.37 | 2.53 | 2.46 | | 1.91 | 2.25 | 2.15 | 2.31 | 2.16 |
| | Ⅲ | | | | 4.28 | 4.42 | 4.05 | 4.26 | 4.25 | | 2.57 | 2.51 | 2.22 | 2.26 | 2.39 |
| | Ⅳ | | | | 5.05 | 5.13 | 5.38 | 5.21 | 5.19 | | 2.83 | 2.71 | 2.66 | 2.49 | 2.67 |
| | Ⅰ | | 干燥 | 水平 | 1.67 | 1.49 | 1.51 | 1.43 | 1.53 | | 1.42 | 1.23 | 1.36 | 1.16 | 1.29 |
| | Ⅱ | | | | 2.78 | 2.43 | 2.51 | 2.75 | 2.62 | | 1.91 | 1.68 | 1.71 | 1.76 | 1.77 |
| | Ⅲ | | | | 4.62 | 4.16 | 4.34 | 4.27 | 4.35 | | 1.96 | 2.26 | 2.36 | 2.16 | 2.19 |
| | Ⅳ | | | | 5.25 | 5.13 | 5.27 | 5.39 | 5.26 | | 2.63 | 2.35 | 2.19 | 2.49 | 2.42 |
| D | Ⅰ | 19.0 | 饱和 | 水平 | 1.69 | 1.83 | 1.76 | 1.66 | 1.74 | 1.85 | 1.57 | 1.71 | 1.67 | 1.53 | 1.62 |
| | Ⅱ | | | | 3.18 | 3.07 | 2.97 | 3.23 | 3.11 | | 1.58 | 1.87 | 1.76 | 1.64 | 1.71 |
| | Ⅲ | | | | 4.56 | 4.32 | 4.26 | 4.19 | 4.33 | | 1.85 | 1.62 | 1.83 | 1.73 | 1.76 |
| | Ⅳ | | | | 5.13 | 4.71 | 4.82 | 5.06 | 4.93 | | 1.97 | 1.78 | 1.82 | 1.73 | 1.83 |
| | Ⅰ | | | | 1.41 | 1.6 | 1.53 | 1.49 | 1.51 | | 1.58 | 1.41 | 1.46 | 1.67 | 1.53 |
| | Ⅱ | | | | 2.97 | 3.32 | 3.17 | 3.27 | 3.18 | | 1.62 | 1.83 | 1.70 | 1.58 | 1.68 |
| | Ⅲ | | | | 3.96 | 4.49 | 4.00 | 4.56 | 4.25 | | 1.85 | 1.61 | 1.72 | 1.91 | 1.77 |
| | Ⅳ | | | | 4.94 | 4.62 | 4.74 | 5.01 | 4.83 | | 2.01 | 1.75 | 1.93 | 1.76 | 1.83 |

**剪切试验结果均值表** 表2

| 组别 | | 既有混凝土 | | 粘结面方向 | 粗糙度(mm) | | | | | | 平均劈拉强度(MPa) | | | | |
|---|---|---|---|---|---|---|---|---|---|---|---|---|---|---|---|
| | | 抗压强度(MPa) | 界面状态 | | 实测值 | | | | 均值 | 伴随试件 | 实测值 | | | | 均值 |
| A | Ⅰ | 50.7 | 饱和 | 水平 | 1.42 | 1.11 | 1.35 | 1.24 | 1.28 | 7.31 | 2.43 | 2.71 | 2.49 | 2.58 | 2.55 |
| | Ⅱ | | | | 2.98 | 3.05 | 2.76 | 2.79 | 2.90 | | 3.24 | 3.09 | 2.97 | 3.37 | 3.17 |
| | Ⅲ | | | | 3.93 | 4.6 | 4.38 | 4.52 | 4.36 | | 3.93 | 3.37 | 3.86 | 3.55 | 3.68 |
| | Ⅳ | | | | 5.96 | 6.51 | 6.38 | 6.61 | 6.37 | | 3.56 | 4.05 | 3.76 | 4.12 | 3.87 |
| B | Ⅰ | 42.5 | 饱和 | 水平 | 1.58 | 1.3 | 1.2 | 1.37 | 1.36 | 6.64 | 1.93 | 2.26 | 2.39 | 2.19 | 2.19 |
| | Ⅱ | | | | 3.25 | 2.77 | 2.91 | 3.11 | 3.01 | | 3.12 | 2.75 | 3.07 | 2.85 | 2.95 |
| | Ⅲ | | | | 4.67 | 5.02 | 4.95 | 4.91 | 4.89 | | 3.63 | 3.72 | 3.21 | 3.28 | 3.45 |
| | Ⅳ | | | | 6.62 | 6.21 | 6.73 | 6.23 | 6.45 | | 3.74 | 4.05 | 3.61 | 3.49 | 3.72 |
| | Ⅰ | | | 竖直 | 1.41 | 1.23 | 1.18 | 1.35 | 1.29 | | 2.09 | 2.22 | 2.12 | 2.05 | 2.32 |
| | Ⅱ | | | | 3.4 | 2.95 | 3.13 | 3.24 | 3.18 | | 3.03 | 2.66 | 2.8 | 2.67 | 2.79 |
| | Ⅲ | | | | 5.23 | 4.68 | 5.02 | 4.75 | 4.92 | | 3.29 | 2.93 | 2.81 | 3.16 | 3.05 |
| | Ⅳ | | | | 6.32 | 5.88 | 6.13 | 6.12 | 6.11 | | 2.97 | 3.18 | 3.49 | 3.21 | 3.21 |
| C | Ⅰ | 31.2 | 饱和 | 水平 | 1.45 | 1.59 | 1.35 | 1.46 | 1.46 | 4.43 | 2.15 | 1.88 | 2.09 | 2.05 | 2.04 |
| | Ⅱ | | | | 2.9 | 2.55 | 2.85 | 2.46 | 2.69 | | 2.72 | 2.29 | 2.4 | 2.52 | 2.48 |
| | Ⅲ | | | | 4.46 | 4.72 | 4.74 | 4.71 | 4.66 | | 3.3 | 3.21 | 3.22 | 3.23 | 3.24 |
| | Ⅳ | | | | 6.17 | 5.78 | 6.01 | 5.98 | 5.99 | | 3.36 | 3.67 | 3.24 | 3.42 | 3.42 |
| | Ⅰ | | | 竖直 | 1.31 | 1.19 | 1.27 | 1.35 | 1.28 | | 2.12 | 1.81 | 1.89 | 2.03 | 1.96 |
| | Ⅱ | | | | 2.57 | 2.36 | 2.28 | 2.34 | 2.39 | | 2.52 | 2.28 | 2.44 | 2.4 | 2.41 |
| | Ⅲ | | | | 3.95 | 4.76 | 4.49 | 4.33 | 4.38 | | 3.05 | 2.76 | 2.81 | 2.95 | 2.89 |
| | Ⅳ | | | | 5.88 | 5.67 | 5.53 | 5.41 | 5.62 | | 3.12 | 3.21 | 2.76 | 2.95 | 3.01 |
| | Ⅰ | | 干燥 | 水平 | 1.41 | 1.55 | 1.33 | 1.39 | 1.42 | | 1.86 | 1.72 | 1.91 | 1.93 | 1.86 |
| | Ⅱ | | | | 2.62 | 2.92 | 2.75 | 2.63 | 2.73 | | 2.5 | 2.15 | 2.27 | 2.45 | 2.34 |
| | Ⅲ | | | | 4.74 | 4.51 | 4.35 | 4.61 | 4.55 | | 3.23 | 2.79 | 2.87 | 3.02 | 2.98 |
| | Ⅳ | | | | 5.86 | 5.5 | 5.22 | 5.39 | 5.49 | | 3.34 | 2.91 | 3.15 | 3.25 | 3.16 |
| D | Ⅰ | 19.0 | 饱和 | 水平 | 1.32 | 1.19 | 1.2 | 1.36 | 1.27 | 3.11 | 1.98 | 1.85 | 2.17 | 2.03 | 2.01 |
| | Ⅱ | | | | 2.21 | 1.93 | 2.18 | 2.11 | 2.11 | | 2.14 | 2.38 | 2.02 | 2.31 | 2.41 |
| | Ⅲ | | | | 4.05 | 3.72 | 3.96 | 3.92 | 3.91 | | 2.6 | 2.47 | 2.5 | 2.52 | 2.52 |
| | Ⅳ | | | | 5.37 | 4.96 | 5.21 | 5.16 | 5.18 | | 2.93 | 2.67 | 2.57 | 2.72 | 2.72 |
| | Ⅰ | | 干燥 | | 1.25 | 1.42 | 1.23 | 1.39 | 1.32 | | 1.69 | 1.41 | 1.52 | 1.66 | 1.57 |
| | Ⅱ | | | | 2.22 | 2.07 | 2.36 | 2.39 | 2.26 | | 1.86 | 2.21 | 2.05 | 2.17 | 2.07 |
| | Ⅲ | | | | 4.46 | 3.75 | 3.93 | 4.25 | 4.10 | | 2.26 | 2.51 | 2.15 | 2.33 | 2.31 |
| | Ⅳ | | | | 5.94 | 5.19 | 5.36 | 5.75 | 5.56 | | 2.77 | 2.46 | 2.62 | 2.51 | 2.59 |

## 16.3 后浇超高韧性水泥基复合材料\既有混凝土复合梁弯曲控裂性能研究

由于实际混凝土构件承载力主要取决于受拉钢筋，而钢筋的位置和强度对混凝土的开裂以及裂缝发展都有较大影响，因此为了能准确获得混凝土和UHTCC材料的相互影响关系，本次试验在混凝土和UHTCC材料中均没有进行配筋。在不同厚度的素混凝土梁上浇筑相应厚度的UHTCC材料组成定截面复合梁，通过四点弯曲试验来研究这两种材料各自的受力状态，裂缝发展以及整体的工作性能。研究结果分析表明：

各组伴随混凝土梁达到开裂应变时，混凝土内部裂缝贯通，短时间内即发生脆性破坏，跨中挠度几乎没有变化，约0.1～0.2mm。

由不同厚度的UHTCC与既有混凝土(A组)构成的复合梁试件中，都具有明显的延性破坏特征，随着UHTCC在截面中所占比例的增大，对混凝土的限制能力也越强，极限荷载和所对应的跨中挠度也相应提高；但当UHTCC层厚度超过40mm时，随着荷载的增加，复合梁中弯剪区出现斜裂缝，表明梁内产生应力重分布，破坏形式从弯曲破坏向剪切破坏转化。

在相同UHTCC厚度下，当既有混凝土强度不同时，随着混凝土强度的提高，复合梁的极限承载力提高幅度相对减小，但挠度仍提高了约8～31倍；由于混凝土开裂后，UHTCC承受了复合梁中绝大部分的拉应力，因此各复合梁后期承载曲线基本一致；但当混凝土强度较高时，其开裂所释放的能量不能短时间内被UHTCC层完全吸收，多余的能量会通过UHTCC中部分纤维被拔出或拔断来抵消，因此造成UHTCC层有效截面的减小。

当既有混凝土粘结面粗糙度(均为人工凿毛)不同时，随着粗糙度的提高，复合梁的承载能力并无明显差别；而既有混凝土粘结面为自然浇筑界面的复合试件中，其承载能力出现一定的随机性，个别UHTCC层厚度小的复合梁承载力比厚度大的还要高，或者同样厚度的复合梁开裂强度和极限强度差别较大的情况。这是由于上层混凝土界面没有经过任何处理，UHTCC与混凝土表面浆体粘结强度较低，两者界面上容易发生滑移。若对于UHTCC厚度相对较小的复合梁，两者界面出现滑移的距离足够长，裂缝发展的范围就更广，UHTCC层的作用越突出，就会出现这样的情况。说明当界面粗糙度较小时，UHTCC和既有混凝土在受力时会出现滑移现象，使UHTCC受力区域增大，有益于复合结构承载能力提高，但全界面随机性较大，使用时还须进行计算设计滑移带；当界面粗糙度达到一定程度后，由于UHTCC和既有混凝土两者界面上的约束作用，界面不会出现滑移，UHTCC层裂缝扩展区域较为一致，因此粗糙度的增大对整体性能的影响不大。其加载形式如图4所示，各组试件设计和试验结果如表3、表4所示。另以UH10-A试件加载过程为例，其裂缝扩展形式如图5所示。

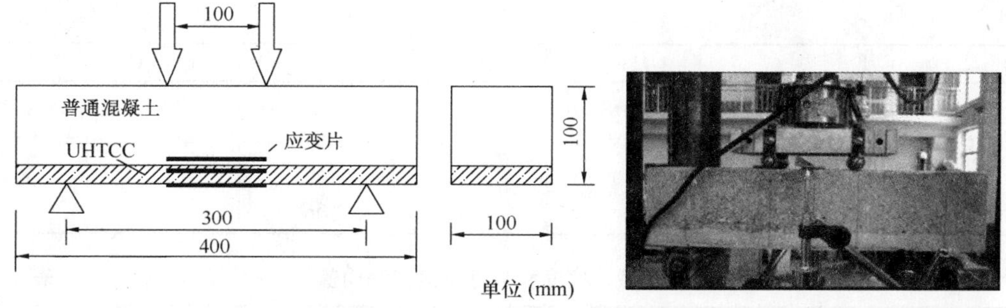


图 4 试件加载示意图和实图

图 5
(a)复合梁加载过程侧面图；(b)放大图；(c)裂缝观测图

**各组试件设计** **表 3**

| 类型 | | CON-A | UH10-A | UH20-A | UH30-A | UH40-A |
|---|---|---|---|---|---|---|
| 既有混凝土 | 伴随试件强度(MPa) | 18.0 | 18.0 | 18.0 | 18.0 | 18.0 |
| | 界面粗糙度(mm) | — | 1.4 | 1.4 | 1.3 | 1.5 |
| | 厚度(mm) | — | 90 | 80 | 70 | 60 |
| UHTCC | 厚度(mm) | — | 10 | 20 | 30 | 40 |
| 类型 | | UH50-A | CON-B1 | UH20-B1 | CON-B2 | UH-A-B2 |
| 既有混凝土 | 伴随试件强度(MPa) | 18.0 | 37.1 | 37.1 | 42.1 | 42.1 |
| | 界面粗糙度(mm) | 1.6 | — | 1.5 | — | 0.6 |
| | 厚度(mm) | 50 | — | 80 | — | 75 |
| UHTCC | 厚度(mm) | 50 | — | 20 | — | 25 |
| 类型 | | UH-B-B2 | UH-C-B2 | UH-D-B2 | CON-B3 | UH20-B3 |
| 既有混凝土 | 伴随试件强度(MPa) | 42.1 | 42.1 | 42.1 | 34.3 | 34.3 |
| | 界面粗糙度(mm) | 1.6 | 2.8 | 4.1 | — | 0.5 |
| | 厚度(mm) | 75 | 75 | 75 | — | 80 |
| UHTCC | 厚度(mm) | 25 | 25 | 25 | — | 20 |

续表

| 类　型 | | UH30-B3 | UH40-B3 | UH50-B3 | CON-C | UH20-C |
|---|---|---|---|---|---|---|
| 既有混凝土 | 伴随试件强度(MPa) | 34.3 | 34.3 | 34.3 | 48.9 | 48.9 |
| | 界面粗糙度(mm) | 0.3 | 0.4 | 0.5 | — | 1.7 |
| | 厚度(mm) | 80 | 80 | 80 | — | 80 |
| UHTCC | 厚度(mm) | 20 | 20 | 20 | — | 20 |

**复合梁荷载、挠度和裂缝宽度实测均值** **表 4**

| 试件编号 | 开裂 | | | 底面 $w_{max}=0.2$mm | | 极限 | | | |
|---|---|---|---|---|---|---|---|---|---|
| | 荷载(kN) | 挠度(mm) | $w_{max}$(mm) | 荷载(kN) | 挠度(mm) | 荷载(kN) | 提高量(%) | 挠度(mm) | 提高量(%) |
| CON-A-1 | 8.78 | 0.12 | >>0.2 | — | — | 8.78 | — | 0.12 | — |
| CON-A-2 | 9.70 | 0.10 | >>0.2 | — | — | 9.70 | — | 0.10 | — |
| CON-A-3 | 9.70 | 0.10 | >>0.2 | — | — | 9.70 | — | 0.10 | — |
| 均值 | 9.39 | 0.11 | >>0.2 | — | — | 9.39 | — | 0.11 | — |
| UH10-A-1 | 14.74 | 0.15 | 0.01 | 13.96 | 0.78 | 14.74 | 56.98 | 1.04 | 845.45 |
| UH10-A-2 | 14.41 | 0.18 | 0.02 | 13.64 | 0.71 | 14.41 | 53.46 | 1.06 | 863.64 |
| UH10-A-3 | 13.69 | 0.19 | 0.02 | 12.40 | 0.65 | 13.69 | 45.79 | 1.08 | 881.82 |
| 均值 | 14.28 | 0.17 | 0.02 | 13.24 | 0.74 | 14.28 | 52.08 | 1.06 | 863.64 |
| UH20-A-1 | 17.25 | 0.31 | 0.01 | 21.45 | 2.22 | 21.92 | 133.44 | 4.33 | 3836.36 |
| UH20-A-2 | 15.53 | 0.27 | 0.01 | 17.77 | 2.27 | 19.38 | 106.39 | 2.75 | 2400.00 |
| UH20-A-3 | 15.77 | 0.20 | 0.01 | 20.65 | 2.70 | 21.71 | 131.20 | 3.45 | 3036.36 |
| 均值 | 16.18 | 0.26 | 0.01 | 19.96 | 2.40 | 21.00 | 123.64 | 3.51 | 3090.91 |
| UH30-A-1 | 18.87 | 0.29 | 0.01 | 22.76 | 2.95 | 25.63 | 172.95 | 4.40 | 3900.00 |
| UH30-A-2 | 17.07 | 0.23 | 0.02 | 25.31 | 3.15 | 26.36 | 180.72 | 4.19 | 3709.09 |
| UH30-A-3 | 17.59 | 0.30 | 0.01 | 22.01 | 3.03 | 24.20 | 157.72 | 4.12 | 3645.45 |
| 均值 | 17.84 | 0.27 | 0.01 | 23.36 | 3.04 | 25.40 | 170.50 | 4.24 | 3754.55 |
| UH40-A-1 | 22.15 | 0.46 | 0.01 | 29.80 | 3.08 | 30.64 | 226.30 | 4.23 | 3745.45 |
| UH40-A-2 | 21.69 | 0.36 | 0.01 | 29.18 | 3.49 | 30.34 | 223.11 | 4.33 | 3836.36 |
| UH40-A-3 | 21.33 | 0.31 | 0.02 | 28.77 | 3.41 | 31.60 | 236.53 | 6.07 | 5418.18 |
| 均值 | 21.72 | 0.38 | 0.01 | 29.25 | 3.33 | 30.86 | 228.65 | 4.91 | 4363.64 |
| UH50-A-1 | 22.90 | 0.36 | 0.01 | 31.70 | 4.37 | 32.78 | 249.09 | 6.01 | 5363.64 |
| UH50-A-2 | 22.32 | 0.39 | 0.02 | 30.22 | 4.15 | 32.30 | 243.98 | 5.50 | 4900.00 |
| UH50-A-3 | 23.67 | 0.34 | 0.02 | 31.14 | 3.95 | 32.72 | 248.46 | 4.70 | 4172.73 |
| 均值 | 22.96 | 0.36 | 0.02 | 31.02 | 4.16 | 32.60 | 247.18 | 5.40 | 4809.09 |
| CON-B1-1 | 17.17 | 0.13 | >>0.2 | — | — | 17.17 | — | 0.13 | — |
| CON-B1-2 | 18.71 | 0.16 | >>0.2 | — | — | 18.71 | — | 0.16 | — |
| CON-B1-3 | 16.12 | 0.12 | >>0.2 | — | — | 16.12 | — | 0.12 | — |
| 均值 | 17.33 | 0.14 | >>0.2 | — | — | 17.33 | — | 0.14 | — |

续表

| 试件编号 | 开裂 | | | 底面 $w_{max}$=0.2mm | | 极限 | | | |
|---|---|---|---|---|---|---|---|---|---|
| | 荷载（kN） | 挠度（mm） | $w_{max}$（mm） | 荷载（kN） | 挠度（mm） | 荷载（kN） | 提高量（%） | 挠度（mm） | 提高量（%） |
| UH20-B1-1 | 18.79 | 0.20 | 0.02 | 20.87 | 1.17 | 21.88 | 26.26 | 1.67 | 1092.86 |
| UH20-B1-2 | 16.18 | 0.27 | 0.02 | 20.01 | 1.15 | 21.70 | 25.22 | 1.50 | 971.43 |
| UH20-B1-3 | 19.31 | 0.22 | 0.02 | 21.35 | 1.20 | 22.01 | 27.01 | 1.65 | 1078.57 |
| 均值 | 18.09 | 0.23 | 0.02 | 20.74 | 1.17 | 21.86 | 26.14 | 1.61 | 1050.00 |
| CON-B2-1 | 19.11 | 0.16 | >>0.2 | — | — | 19.11 | — | 0.16 | — |
| CON-B2-2 | 20.02 | 0.17 | >>0.2 | — | — | 20.02 | — | 0.17 | — |
| CON-B2-3 | 20.68 | 0.18 | >>0.2 | — | — | 20.68 | — | 0.18 | — |
| 均值 | 19.94 | 0.17 | >>0.2 | — | — | 19.94 | — | 0.17 | — |
| UH-A-B2-1 | 24.67 | 0.31 | 0.01 | 24.59 | 1.90 | 26.02 | 30.49 | 2.29 | 1247.06 |
| UH-A-B2-2 | 25.72 | 0.25 | 0.02 | 26.69 | 1.79 | 28.22 | 41.52 | 3.14 | 1747.06 |
| UH-A-B2-3 | 26.71 | 0.22 | 0.02 | 28.06 | 1.92 | 30.55 | 53.21 | 2.42 | 1323.53 |
| 均值 | 25.70 | 0.26 | 0.02 | 26.45 | 1.87 | 28.26 | 41.73 | 2.62 | 1441.18 |
| UH-B-B2-1 | 23.41 | 0.30 | 0.01 | 26.00 | 1.52 | 26.62 | 33.50 | 1.90 | 1017.65 |
| UH-B-B2-2 | 25.35 | 0.26 | 0.01 | 24.91 | 1.73 | 25.35 | 27.13 | 1.97 | 1058.82 |
| UH-B-B2-3 | 23.30 | 0.29 | 0.02 | 27.61 | 1.93 | 29.02 | 45.54 | 2.37 | 1294.12 |
| 均值 | 24.02 | 0.28 | 0.01 | 26.17 | 1.73 | 27.00 | 35.41 | 2.08 | 1123.53 |
| UH-C-B2-1 | 23.33 | 0.14 | 0.02 | 24.87 | 2.66 | 27.74 | 39.12 | 3.35 | 1870.59 |
| UH-C-B2-2 | 25.97 | 0.23 | 0.02 | 27.01 | 1.21 | 27.91 | 39.97 | 2.33 | 1270.59 |
| UH-C-B2-3 | 23.52 | 0.24 | 0.02 | 25.04 | 1.12 | 26.42 | 32.50 | 1.32 | 676.47 |
| 均值 | 24.27 | 0.20 | 0.02 | 25.64 | 1.66 | 27.36 | 37.21 | 2.33 | 1270.59 |
| UH-D-B2-1 | 26.47 | 0.17 | 0.02 | 26.33 | 2.53 | 27.53 | 38.06 | 2.98 | 1652.94 |
| UH-D-B2-2 | 25.39 | 0.23 | 0.01 | 25.74 | 1.78 | 28.48 | 42.83 | 2.13 | 1152.94 |
| UH-D-B2-3 | 26.00 | 0.25 | 0.02 | 26.66 | 1.52 | 27.94 | 40.12 | 1.94 | 1041.18 |
| 均值 | 25.95 | 0.22 | 0.02 | 26.24 | 1.94 | 27.98 | 40.32 | 2.35 | 1282.35 |
| CON-B3-1 | 18.49 | 0.16 | >>0.2 | — | — | 18.49 | — | 0.16 | — |
| CON-B3-2 | 19.21 | 0.19 | >>0.2 | — | — | 19.21 | — | 0.19 | — |
| CON-B3-3 | 18.01 | 0.07 | >>0.2 | — | — | 18.01 | — | 0.07 | — |
| 均值 | 18.57 | 0.14 | >>0.2 | — | — | 18.57 | — | 0.14 | — |
| UH20-B3-1 | 20.38 | 0.23 | 0.02 | 22.12 | 1.24 | 22.13 | 19.17 | 1.49 | 964.29 |
| UH20-B3-2 | 20.18 | 0.15 | 0.02 | 21.59 | 2.67 | 23.17 | 24.77 | 3.29 | 2250.00 |
| UH20-B3-3 | 19.15 | 0.19 | 0.01 | 18.44 | 1.07 | 21.15 | 13.89 | 1.39 | 892.86 |
| 均值 | 19.90 | 0.19 | 0.02 | 20.72 | 1.66 | 22.15 | 19.28 | 2.06 | 1371.43 |
| UH30-B3-1 | 21.94 | 0.26 | 0.01 | 26.98 | 2.63 | 27.77 | 49.54 | 3.05 | 2078.57 |
| UH30-B3-2 | 21.67 | 0.44 | 0.01 | 22.23 | 2.63 | 23.98 | 29.13 | 3.10 | 2114.29 |

续表

| 试件编号 | 开裂 | | | 底面 $w_{max}$=0.2mm | | 极限 | | | |
|---|---|---|---|---|---|---|---|---|---|
| | 荷载(kN) | 挠度(mm) | $w_{max}$(mm) | 荷载(kN) | 挠度(mm) | 荷载(kN) | 提高量(%) | 挠度(mm) | 提高量(%) |
| UH30-B3-3 | 20.76 | 0.34 | 0.01 | 25.99 | 4.26 | 27.11 | 45.99 | 4.96 | 3442.86 |
| 均值 | 21.46 | 0.35 | 0.01 | 25.07 | 3.17 | 26.29 | 41.57 | 3.70 | 2542.86 |
| UH40-B3-1 | 20.01 | 0.31 | 0.02 | 25.40 | 3.75 | 28.24 | 52.07 | 4.52 | 3128.57 |
| UH40-B3-2 | 24.02 | 0.36 | 0.01 | 26.50 | 2.95 | 27.78 | 49.60 | 3.48 | 2385.71 |
| UH40-B3-3 | 23.25 | 0.22 | 0.01 | 26.18 | 2.69 | 27.17 | 46.31 | 4.62 | 3200.00 |
| 均值 | 22.43 | 0.30 | 0.01 | 26.03 | 3.13 | 27.73 | 49.33 | 4.21 | 2907.14 |
| UH50-B3-1 | 23.31 | 0.38 | 0.02 | 28.46 | 2.97 | 29.22 | 57.35 | 3.45 | 2364.29 |
| UH50-B3-2 | 23.98 | 0.40 | 0.01 | 31.18 | 3.66 | 31.54 | 69.84 | 4.43 | 3064.29 |
| UH50-B3-3 | 25.30 | 0.51 | 0.02 | 30.94 | 4.07 | 32.34 | 74.15 | 4.86 | 3371.43 |
| 均值 | 24.20 | 0.43 | 0.02 | 30.19 | 3.57 | 31.03 | 67.10 | 4.25 | 2935.71 |
| CON-C-1 | 21.28 | 0.15 | >>0.2 | — | — | 21.28 | — | 0.15 | — |
| CON-C-2 | 21.73 | 0.17 | >>0.2 | — | — | 21.73 | — | 0.17 | — |
| CON-C-3 | 20.17 | 0.19 | >>0.2 | — | — | 20.17 | — | 0.19 | — |
| 均值 | 21.06 | 0.17 | >>0.2 | — | — | 21.06 | — | 0.17 | — |
| UH20-C-1 | 20.92 | 0.19 | 0.02 | 16.30 | 1.38 | 20.92 | −0.66 | 1.51 | 788.24 |
| UH20-C-2 | 21.24 | 0.18 | 0.02 | 14.88 | 1.19 | 21.24 | 0.85 | 1.69 | 894.12 |
| UH20-C-3 | 20.83 | 0.18 | 0.02 | 14.33 | 1.33 | 20.83 | −1.09 | 1.49 | 776.47 |
| 均值 | 21.00 | 0.18 | 0.02 | 15.17 | 1.30 | 21.00 | −0.28 | 1.56 | 817.65 |

为了能够合理分析复合梁的弯曲受力状态，采用正截面承载力计算方法来进行分析计算。考虑混凝土材料所受 UHTCC 层约束作用，选取混凝土非脆性开裂受拉本构模型[23]，即混凝土达到抗拉强度后不立即下降为零，而是有一个逐渐下降的过程。UHTCC 材料模型通过试验测得，并简化为双斜线模型[24~26]。各模型如图 6 所示。通过求解平衡方程，考虑各阶段受力状态(如图 7 所示)，获得其加载曲线图(图 8 所示，以 UH10-A，UH20-A 和 UH30-A 为例)，计算结果与试验数据符合较好，另外还计算了复合梁 UH20-B1 和 UH20-C 的加载曲线，结果表明在开裂前计算曲线与试件结果符合较好，随着上层既有混凝土强度的提高，后期曲线有一定的差别。这是由于上层混凝土强度较高，因此开裂时所释放的能量较大，造成了 UHTCC 层有效截面面积的减小。

比较上层混凝土开裂时受拉区混凝土和 UHTCC 中所产生应变能的比值与试验测量荷载-应变曲线。应变能计算公式如式 1，其中 $\int_{\varepsilon}\sigma d\varepsilon$ 为单位体积应变能，$\sigma$ 为广义应力，$\varepsilon$ 为对应的应变，$V$ 为体积。

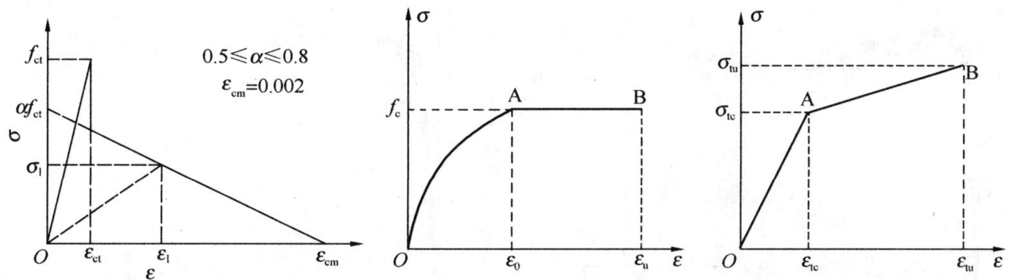


图 6　复合梁中混凝土受拉、受压和 UHTCC 的受拉本构模型

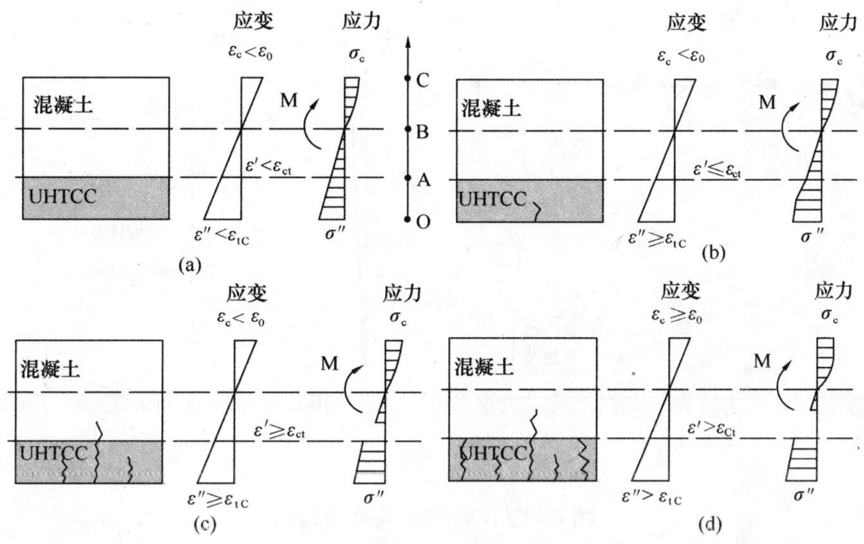


图 7　正截面应力应变分布图

$$U=\int_V\left(\int_\varepsilon\sigma\mathrm{d}\varepsilon\right)\mathrm{d}V \tag{1}$$

以 UH20-B 为例，当混凝土和 UHTCC 界面应变 $\varepsilon'$ 达到 $\varepsilon_{ct}$ 时，代入图 6 中材料模型，可得

$$\begin{aligned}N=&\int_0^{\varepsilon_c}f_c\left[2\frac{\varepsilon(x)}{\varepsilon_0}-\left(\frac{\varepsilon(x)}{\varepsilon_0}\right)^2\right]\mathrm{d}\varepsilon(x)-\int_0^{\varepsilon_{ct}}\frac{f_{ct}}{\varepsilon_{ct}}\varepsilon(x)\mathrm{d}\varepsilon(x)\\&-\int_{\varepsilon_{ct}}^{\varepsilon_{tc}}\frac{\sigma_{tc}}{\varepsilon_{tc}}\varepsilon(x)\mathrm{d}\varepsilon(x)-\int_{\varepsilon_{tc}}^{\varepsilon''}\left(\sigma_{tc}+\frac{\sigma_{tu}-\sigma_{tc}}{\varepsilon_{tu}-\varepsilon_{tc}}(\varepsilon(x)-\varepsilon_{tc})\right)\mathrm{d}\varepsilon(x)=0\end{aligned} \tag{2}$$

$$\begin{aligned}N=&-\frac{1}{3}\frac{f_c(h-c)^3\varepsilon''^3}{\varepsilon_0^2c^3}+\frac{f_c(h-c)^2\varepsilon''^2}{\varepsilon_0c^2}-\frac{1}{2}f_{ct}\varepsilon_{ct}-\frac{1}{2}\frac{\sigma_{tc}\left(\varepsilon_{tc}^2-\frac{(c-t)^2\varepsilon''^2}{c^2}\right)}{\varepsilon_{tc}}\\&-\frac{1}{2}\frac{(\sigma_{tu}-\sigma_{tc})(\varepsilon''^2-\varepsilon_{tc}^2)}{\varepsilon_{tu}-\varepsilon_{tc}}-\sigma_{tc}(\varepsilon''-\varepsilon_{tc})+\frac{(\sigma_{tu}-\sigma_{tc})\varepsilon_{tc}(\varepsilon''-\varepsilon_{tc})}{\varepsilon_{tu}-\varepsilon_{tc}}=0\end{aligned} \tag{3}$$

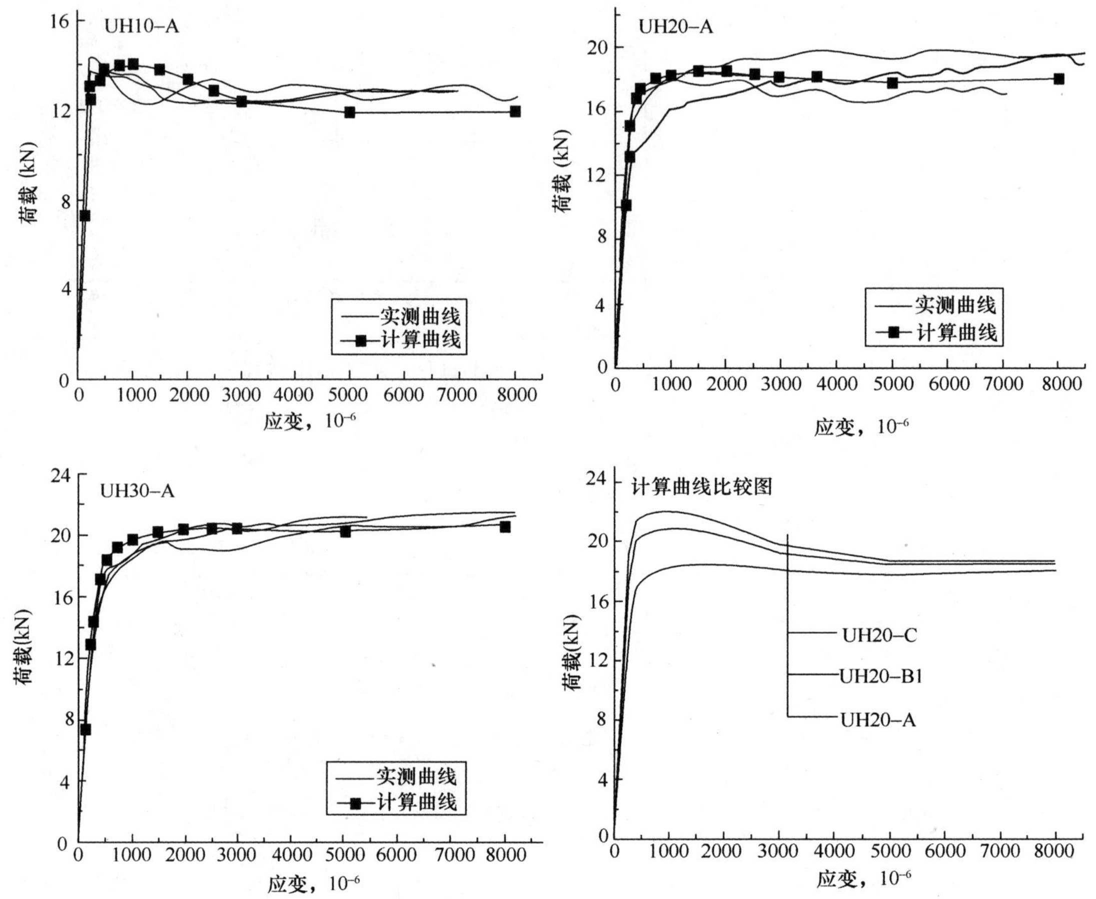


图 8 复合梁荷载-应变曲线图

求解式(3)，可求得中性轴 c，结合式(1)可求出受拉区混凝土和 UHTCC 应变能比值：

$$\frac{U_{con}}{U_{uh}}=\frac{\int_{V_{con}}\left(\int_{\varepsilon_{con}}\sigma_{con}\mathrm{d}\varepsilon_{con}\right)\mathrm{d}V_{con}}{\int_{V_{uh}}\left(\int_{\varepsilon_{uh}}\sigma_{uh}\mathrm{d}\varepsilon_{uh}\right)\mathrm{d}V_{uh}}$$

$$=\frac{\int_0^{\varepsilon_{ct}}\left(\int_0^{\varepsilon_{ct}}\frac{f_{ct}}{\varepsilon_{ct}}\varepsilon(x)\mathrm{d}\varepsilon(x)\right)\mathrm{d}\varepsilon(x)}{\int_{\varepsilon_{ct}}^{\varepsilon_{tc}}\left(\int_{\varepsilon_{ct}}^{\varepsilon_{tc}}\frac{\sigma_{tc}}{\varepsilon_{tc}}\varepsilon(x)\mathrm{d}\varepsilon(x)\right)\mathrm{d}\varepsilon(x)+\int_{\varepsilon_{tc}}^{\frac{c\varepsilon_{ct}}{c-t}}\left(\int_{\varepsilon_{tc}}^{\frac{c\varepsilon_{ct}}{c-t}}\left(\sigma_{tc}+\frac{\sigma_{tu}-\sigma_{tc}}{\varepsilon_{tu}-\varepsilon_{tc}}(\varepsilon(x)-\varepsilon_{tc})\right)\mathrm{d}\varepsilon(x)\right)\mathrm{d}\varepsilon(x)} \tag{4}$$

通过求解各组复合梁试件的应变能比值，与试验曲线比较可得，当 $\frac{U_{con}}{U_{uh}}\leqslant 2.9$ 时，复合梁试件中 UHTCC 层截面受损程度较小，对复合梁后期持载能力影响不大，可保证整体梁试件后期持载能力。

## 16.4 后浇 UHTCC 耐久性防护层加固钢筋混凝土梁弯曲试验研究

在既有钢筋混凝土梁受弯拉面浇筑 UHTCC 防护层构成整体加固和局部加固的复合试件，通过四点弯曲试验，考虑不同 UHTCC 加固厚度、加入纵向钢筋、端部植筋对复合梁弯曲性能的影响，并与相应的高强度混凝土等级加固试件相比较，对加固梁的破坏形式、裂缝发展以及承载能力等进行研究分析。试件尺寸和具体设计见图 9 和表 5，试验结果见表 6 所列。试验结果分析表明：

相比对比梁 Con，加固的复合梁(包括 UHTCC 加固梁和高强度等级混凝土加固梁)的开裂荷载和极限荷载均得到一定提高。

其中使用高强度等级混凝土加固的复合梁，开裂前刚度提高显著，并随着随着梁试件截面有效面积的增加而增大。但由于混凝土材料自身的缺点，复合梁 CON-CB1 和 CON-CB3(试件均有混凝土材料加固，内部无纵向钢筋)承载性能基本一致，在试件开裂后即表现出部分脆性破坏特征，开裂后裂缝迅速扩展并成为构件中主裂缝，当其扩展到既有钢筋混凝土纵向钢筋位置时，裂缝宽度受到一定限制，但仍作为构件中主裂缝；由于在原结构受拉表面上又浇筑了一层混凝土，造成既有结构中纵向钢筋的保护层厚度过大，不能有效控制裂缝宽度，开裂后裂缝宽度可达 4mm，相应的挠度不超过对比混凝土梁的开裂挠度。而后浇 UHTCC 加固试件中 UH-CB1、UH-CB2 和 UH-CB3 仍以弯曲破坏为主，且随着荷载的增加，跨中受拉区混凝土首先出现可视的竖向裂缝，宽度逐渐增大，裂缝条数增多。当上层混凝土梁中产生的单条裂缝扩展到 UHTCC 层中，即被分散成多条细密的微观裂缝。当上层混凝土裂缝宽度达到 0.7～1.5mm 时，底面的 UHTCC 裂缝仍保持在 0.07mm 左右。受拉钢筋接近屈服时，UHTCC 裂缝宽度快速增长，直至破坏。随着加固层厚度的增加，复合梁的开裂和极限承载能力均得到提高，相对于极限荷载提高幅度较小，约为 10%～17%；但相同裂缝宽度(0.2mm)下所承受的荷载最大能提高了 98%，显示出优异的裂缝控制和分散能力。

UH-CB4、UH-CB5 和 UH-CB6 是以 UHTCC 为基体加入纵向受力钢筋对原有结构进行局部加固而成的复合梁试件。由于防护层中加入了纵向受力钢筋，增强了加固层刚度和复合梁配筋率，其整体加固效果同刚性加固材料(如纤维布、板，钢板以及高强砂浆等)相似，无论是承载能力，还是对上层混凝土中裂缝发展的控制效果更加明显；承载能力也大幅提高，混凝土中裂缝宽度和裂缝间距较同截面尺寸的 UH-CB2 明显减小。与同截面尺寸和加入同直径纵向钢筋的高强度等级混凝土局部加固的复合梁 CON-CB3 相比，两者的承载能力较为接近，UHTCC 加固复合梁略高，说明构件中配置的钢筋仍是主要的承载体。但 UHTCC 加固复合梁受拉面裂缝宽度随荷载的增加，扩展更为缓慢，其最大裂缝宽度较混凝土加固复合梁低 1～2 个数量级，且受拉面裂缝分布更为均匀。同时由于纵向钢筋的加入，加固层刚度相应变大，结构先弯曲持载，跨中出现多条竖向裂缝；当荷载达到 43.9～55.4kN 时，加固层端部发生开裂，斜裂缝向加载点延伸，发展到既有混凝土保护层时横向裂缝出现，并导致既有混凝土保护层脱开，发生剥离破坏。UH-CB5 和 UH-CB4 在各材料性能和设计参数均相同，唯一不同的是 UH-CB5 加固层中距端部 60mm 处各植入两根直径为 6.5mm 的 L 型锚筋，从试验中发现，锚筋的植入改善了 UHTCC 与混凝土间的界面应力分布状态，减小了加固层端部集中应力幅值，进而推迟了端部开裂，使得 UH-CB5 中斜裂缝出现时的荷载比 UH-CB4 提高了 17.6%。而 UH-CB6 虽然加固层中受拉钢筋直径较 UH-CB4 大，但加固层刚度增加幅度也相应提高，使得交界面上应力幅值也随之增大，导致其界面破坏过早，破坏荷载较 UH-CB4 还要低 7.3%。与 UH-CB4 相比，同截面尺寸和配筋设计的整体加固复合梁 UH-CB8 前期加载过程基本一致，但由于其交界面上无截面突变，且整体加固在提高弯曲承载力的同时，也提高了构件的剪切承载力，因此复合梁仍为弯曲破坏，加载过程中没有斜裂缝产生。

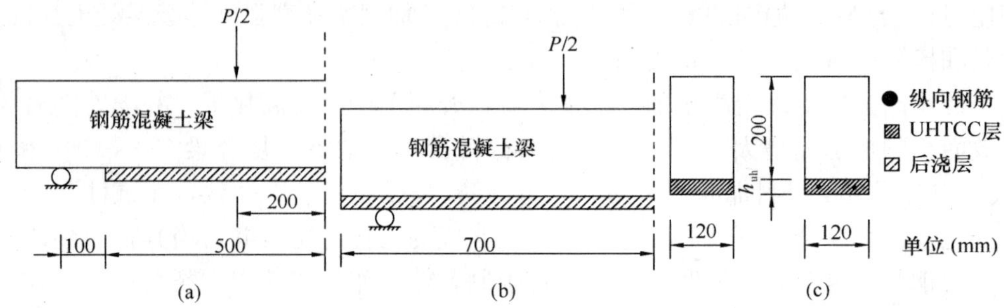


图 9 加固方法和加载方式示意图

(a)局部加固梁；(b)整体加固梁；(c)加固后复合梁截面图

**试件编号和加固方式** **表 5**

| 编号 | 加固方案 | | | | |
|---|---|---|---|---|---|
| | 材料 | 长度(mm) | 厚度(mm) | 纵筋直径(mm) | 植筋 |
| Con | — | — | — | — | — |
| CON-CB1 | 混凝土 | 1000 | 30 | — | — |
| CON-CB2 | 混凝土 | 1400 | 30 | — | — |
| CON-CB3 | 混凝土+纵筋 | 1000 | 30 | 6 | — |
| CON-CB4 | 混凝土+纵筋 | 1400 | 30 | 6 | — |
| UH-CB1 | UHTCC | 1000 | 15 | — | — |
| UH-CB2 | UHTCC | 1000 | 30 | — | — |
| UH-CB3 | UHTCC | 1000 | 45 | — | — |
| UH-CB4 | UHTCC+纵筋 | 1000 | 30 | 6 | — |
| UH-CB5 | UHTCC+纵筋 | 1000 | 30 | 6 | 有 |
| UH-CB6 | UHTCC+纵筋 | 1000 | 30 | 8 | — |
| UH-CB7 | UHTCC | 1400 | 30 | — | — |
| UH-CB8 | UHTCC+纵筋 | 1400 | 30 | 6 | — |

**复合梁承载能力与裂缝宽度实测值**　　**表 6**

| 编　号 | 荷　载(kN) | | | | |
|---|---|---|---|---|---|
| | 开裂时 | $w_{max-C}$=0.1mm | $w_{max-C}$=0.2mm | $w_{max-P}$=0.2mm | 极限 |
| Con | 12.7 | 18.0 | 26.9 | — | 50.0 |
| CON-CB1 | 22.7 | 22.7 | 22.7 | 22.7 | 22.7 |
| CON-CB2 | 20.3 | 20.3 | 20.3 | 20.3 | 20.3 |
| CON-CB3 | 20.3 | 34.6 | 34.6 | — | 69.4 |
| CON-CB4 | 21.5 | 36.5 | 49.2 | 49.2 | 72.3 |
| UH-CB1 | 16.1 | 27.8 | 36.0 | 50.4 | 54.9 |
| UH-CB2 | 23.5 | 33.5 | 44.6 | 52.5 | 56.5 |
| UH-CB3 | 28.2 | 40.1 | 52.4 | 53.2 | 58.3 |
| UH-CB4 | 30.0 | 47.1 | 47.1 | — | 73.3 |
| UH-CB5 | 31.4 | 55.4 | 55.4 | — | 70.8 |
| UH-CB6 | 32.7 | 43.9 | 43.9 | — | 72.6 |
| UH-CB7 | 24.2 | 36.3 | 44.2 | 51.7 | 56.0 |
| UH-CB8 | 32.1 | 44.0 | 66.5 | 80.1 | 88.9 |

由于梁试件在承载过程中发生了弯曲破坏和剥离破坏两种破坏方式，因此在计算分析时也需要分别考虑两种情况下的破坏承载力，计算结果与试验结果相比，符合较好，相应计算方法如下。复合梁的正截面承载力可通过分析，取 UHTCC 层达到自身极限应变或混凝土中钢筋达到屈服应变时的荷载作为加固梁失效荷载。在计算正截面失效荷载前，可判断出既有梁体中受拉钢筋应变在屈服应变 $\varepsilon_y$ 以内，UHTCC 层已经超过其 $\varepsilon_{tc}$，在计算时可先取 UHTCC 加固层底部应变等于 $\varepsilon_{tu}$来进行计算，并与钢筋应变达到 $\varepsilon_y$ 时所得比较，较小值作为失效荷载。计算中采用混凝土和 UHTCC 模型同第三节，钢筋采用理想弹塑性模型。参考公式 4 可计算出本次试验中 UHTCC 加固梁体均满足应变能比值要求，因此后续计算中后浇层有效截面面积保持不变。由于梁体截面尺寸发生变化，因此中性轴以上混凝土的受力状态需要假设成两种状态，即 1)$\varepsilon_c \leqslant \varepsilon_{c0}$ 和 2)$\varepsilon_c \geqslant \varepsilon_{c0}$，如图 10 所示。根据以上分析，下面分为两种情况讨论。

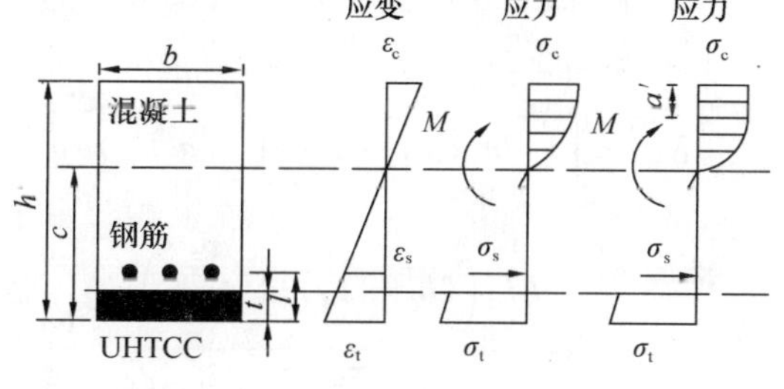


图 10　正截面应力应变分布图

设 UHTCC 底部受拉应变为 $\varepsilon_t$，受压区混凝土的合力可由积分获得：

$$\frac{\varepsilon_t}{c}=\frac{\varepsilon_c}{h'-c}=\frac{\varepsilon_s}{c-l'} \tag{5}$$

$$F_c=\int_0^{h'-c} b\sigma_c(x)\mathrm{d}x=\int_0^{h'-c} bf_c\left[2\left(\frac{\varepsilon_c}{\varepsilon_{c0}}\right)-\left(\frac{\varepsilon_c}{\varepsilon_{c0}}\right)^2\right]\mathrm{d}x=\frac{f_c\varepsilon_t b(h'-c)^2}{c\varepsilon_{c0}}-\frac{f_c\varepsilon_t^2 b(h'-c)^3}{3c^2\varepsilon_{c0}^2} \tag{6}$$

或

$$F_c=\int_0^{h'-c} b\sigma_c(x)\mathrm{d}x=\int_0^{h'-c-a'} bf_c\left[2\left(\frac{\varepsilon_c}{\varepsilon_{c0}}\right)-\left(\frac{\varepsilon_c}{\varepsilon_{c0}}\right)^2\right]\mathrm{d}x+\int_{h'-c-a'}^{h'-c} bf_c\mathrm{d}x$$
$$=f_cb\left(h'-c-\frac{\varepsilon_{cu}c}{\varepsilon_t}\right)+\frac{f_cc\varepsilon_{cu}b}{\varepsilon_t}\left(\frac{2\varepsilon_{cu}}{\varepsilon_{co}}-\left(\frac{\varepsilon_{cu}}{\varepsilon_{co}}\right)^2\right) \tag{7}$$

钢筋合力为

$$F_s=A_sf_s=E_sA_s\varepsilon_s \tag{8}$$

UHTCC 加固层合力

$$F_t=\int_{c-t}^{c} b\sigma_t(x)\mathrm{d}x=\int_{c-t}^{c} b\ (\sigma_{tc}+K_2\ (\varepsilon_t-\varepsilon_{tc}))\mathrm{d}x=\frac{(\sigma_{tu}-\sigma_{tc})\varepsilon_tb\ (c^2-(c-t)^2)}{2\ (\varepsilon_{tu}-\varepsilon_{tc})c}$$
$$+\left(\sigma_{tc}-\frac{(\sigma_{tu}-\sigma_{tc})\varepsilon_{tc}}{\varepsilon_{tu}-\varepsilon_{tc}}\right)bt \tag{9}$$

正截面受弯承载力通过求解以下算式来获得。

$$\Sigma N=F_c-F_s-F_t=0 \tag{10}$$

通过力的平衡可以求出 $c$ 值，即确定中性轴的位置。由式(7)对中性轴取距，

$$M_u=\int_0^{h'-c} b\sigma_c(x)x\mathrm{d}x+\int_{c-t}^{c} b\sigma_t(x)x\mathrm{d}x+A_sf_s(c-l')=\frac{2f_c\varepsilon_tb(h'-c)^3}{3c\varepsilon_{c0}}$$
$$-\frac{f_c\varepsilon_t^2b(h'-c)^4}{4c^2\varepsilon_{c0}^2}+\frac{(\sigma_{tu}-\sigma_{tc})\varepsilon_tb(c^3-(c-t)^3)}{3(\varepsilon_{tu}-\varepsilon_{tc})c}$$
$$+\frac{1}{2}\left(\sigma_{tc}-\frac{(\sigma_{tu}-\sigma_{tc})\varepsilon_{tc}}{\varepsilon_{tu}-\varepsilon_{tc}}\right)b(c^2-(c-t)^2+A_sf_s(c-l')) \tag{11}$$

或

$$M_u=\int_0^{h'-c} b\sigma_c(x)x\mathrm{d}x+\int_{c-t}^{c} b\sigma_t(x)xg\,\mathrm{d}x+A_sf_s(c-l')=\frac{bf_cc^2\varepsilon_{cu}^2\left(\frac{2\varepsilon_{cu}}{\varepsilon_{c0}}-\frac{\varepsilon_{cu}^2}{\varepsilon_{c0}^2}\right)}{2\varepsilon_t^2}$$
$$+\frac{bf_c\left((h'-c)^2-\frac{c^2\varepsilon_{cu}^2}{\varepsilon_t^2}\right)}{2}\times\frac{(\sigma_{tu}-\sigma_{tc})\varepsilon_tb(c^3-(c-t)^3)}{3(\varepsilon_{tu}-\varepsilon_{tc})c}$$
$$+\frac{1}{2}\left(\sigma_{tc}-\frac{(\sigma_{tu}-\sigma_{tc})\varepsilon_{tc}}{\varepsilon_{tu}-\varepsilon_{tc}}\right)b(c^2-(c-t)^2+A_sf_s(c-l')) \tag{12}$$

将式(10)中所求得中性轴值代入式(11)或(12)，即可求得加载值。用钢筋应变进行验算时，方法相同。

端部破坏的计算方法如下，参考 Vilnaly，Roberts 和 Malek 等[27]的研究，在弹性梁理论的基础上，结合部分组合截面假定，推导了有关公式用以推算粘结界面的剪应力和正应力，由于本次试验中端部开裂是在混凝土中出现，因此将混凝土破坏准则代入计算分析，以求得复合梁极限破坏荷载。其中基本假定如下：假定钢筋混凝土梁、粘结层和加固层均为各向同性的弹性材料；钢筋混凝土梁和加固层粘结可靠，共同受力，试验结果也证实，沿梁截面高度应变符合平截面假定，呈线性分布；粘结层为钢筋混凝土受拉面凿毛处的混凝土与 UHTCC 共同组成，厚度取为常数(平均凿毛深度)，且剪应力在粘结胶层的厚度方向不变化，忽略其在垂直方向的变化影响。加固长度为 $l$，加载跨度为 $l_1$，高为 $h$，

宽为 $b$。粘结层厚 $t_a$，宽 $b$，加固层厚度为 $h_{uh}$。

沿梁纵向取一微元体，如图 11 所示。由加固层 $x$ 方向平衡可得：

$$\tau(x)=\frac{G_a}{t_a}u(x,y) \tag{13}$$

式中 $G_a$ 为粘结层的剪切模量，由粘结试件剪切试验测得。$u(x,y)$为混凝土梁底面 $u_c(x)$加固层顶面 $u_{uh}(x)$的纵向位移差，可以表示为：

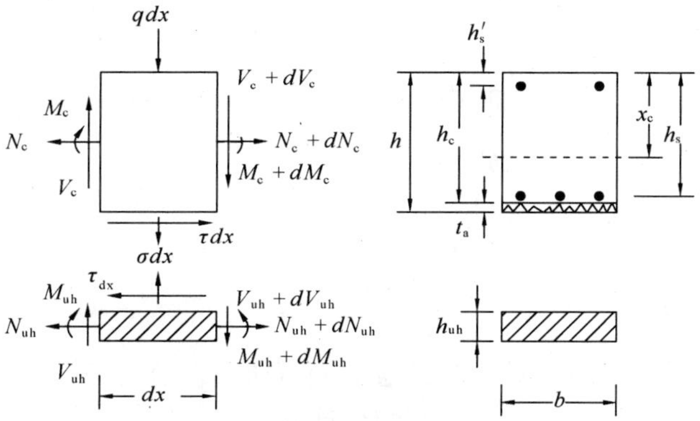

$$u(x,y)=u_{uh}(x)-u_c(x) \tag{14}$$

图 11　单元体受力状体及尺寸

对式(13)求导二次，可得：

$$\frac{d^2\tau(x)}{dx^2}=\frac{G_a}{t_a}\left[\frac{d^2u_{uh}(x)}{dx^2}-\frac{d^2u_c(x)}{dx^2}\right] \tag{15}$$

忽略 UHTCC 层的弯曲变形和钢筋混凝土梁的轴向变形，由图 11 中梁微元的平衡可得

$$\frac{d^2u_{uh}(x)}{dx^2}=\left(\frac{N_{uh}(x)}{E_{uh}A_{uh}}\right)'=\frac{\tau(x)}{E_{uh}A_{uh}} \tag{16}$$

$$\frac{d^2u_c(x)}{dx^2}=\left(\frac{M_c(x)(h_c-x_c)}{E_cI_0}\right)'=\frac{(h_c-x_c)}{E_cI_0}V_c(x)-\tau(x)(h_c-x_c) \tag{17}$$

上式中 $t_a$ 为粘结层厚度，$E_{uh}$和 $A_{uh}$分别为加固层的等效弹性模量和面积，$E_c$ 为混凝土弹性模量，$I_0$ 为加固后梁的界面惯性矩。其中加固后梁体的受压区高度 $x_c$ 可按截面平衡方程求出，

$$\frac{1}{2}bx_c^2+\left(\frac{E_sA_s}{E_c}x_c+\frac{E'_sA'_s}{E_c}x_c\right)+\frac{E_{uh}A_{uh}}{E_c}x_c-\frac{E_sA_s}{E_c}h_s-\frac{E_sA_s}{E_c}h'_s-\frac{E_{uh}A_{uh}}{E_c}\left(h+\frac{h_{uh}}{2}\right)=0 \tag{18}$$

加固后梁的截面惯性矩为：

$$I_0=\frac{1}{3}bx_c^3+\frac{E_s}{E_c}I_s+\frac{E_s}{E_c}A_s(h_s-x_c)^2+\frac{E'_s}{E_c}I'_s+\left(\frac{E'_s}{E_c}-1\right)$$
$$A'_s(x_c-h'_s)^2+\frac{E_{uh}}{E_c}I_{uh}+\frac{E_{uh}}{E_c}A_{uh}\left(h+\frac{h_{uh}}{2}-x_c\right)^2 \tag{19}$$

式中 $E_s$、$A_s$ 和 $I_s$ 为钢筋混凝土受拉纵向钢筋的弹性模量、面积和惯性矩，$E'_s$、$A'_s$ 和 $I'_s$ 为钢筋混凝土受压区钢筋的弹性模量、面积和惯性矩。

将式(13)、(14)代入式(12)，整理可得关于 $\tau(x)$的界面剪应力的控制微分方程。

$$\frac{d^2\tau(x)}{dx^2}-\lambda^2\tau(x)=-m_1\lambda^2V_c(x) \tag{20}$$

其中

$$\lambda^2=\frac{G_ab}{t_a}\left(\frac{1}{E_{uh}h_{uh}b}+\frac{(h_c-x_c)^2}{E_cI_0}\right) \tag{21}$$

$$m_1 = \frac{G_a(h_c - x_c)}{E_c I_0 t_a \lambda^2} \tag{22}$$

结合实际四点弯曲试验的边界条件，可得粘结剪应力的解为：

当 $0 \leqslant x \leqslant l_1/2$

$$\tau(x) = C_1 e^{\lambda x} + C_2 e^{-\lambda x} \tag{23}$$

当 $l_1/2 \leqslant x \leqslant l/2$

$$\tau(x) = C_3 e^{\lambda x} + C_4 e^{-\lambda x} - m_1 \frac{P}{2} \tag{24}$$

设 $v_F(x,\ y)$为粘结层下表面和上表面的竖向位移差，即 UHTCC 层上表面 $v_{uh}$和钢筋混凝土梁下表面 $v_c$ 的竖向位移差，则界面正应力可表示为：

$$\sigma(x) = \frac{E_a}{t_a} v(x, y) \tag{25}$$

$$v(x,y) = v_{uh}(x) - v_c(x) \tag{26}$$

对式(22)求四次导数可得：

$$\frac{d^4\sigma(x)}{dx^4} = \frac{E_a}{t_a}\left[\frac{d^4 v_{uh}(x)}{dx^4} - \frac{d^4 v_c(x)}{dx^4}\right] \tag{27}$$

考虑混凝土的剪切变形，可得梁挠度的微分方程如下：

$$\frac{d^2 v_c(x)}{dx^2} = -\frac{M_c(x)}{E_c I_c} - \frac{\alpha_s q(x)}{G_c A_c} \tag{28}$$

$$\frac{d^2 v_{uh}(x)}{dx^2} = -\frac{M_{uh}(x)}{E_{uh} I_{uh}} \tag{29}$$

上式中，$G_c$ 和 $A_c$ 为混凝土的弹性模量和面积。$I_c$ 为钢筋混凝土梁的惯性矩，可按下式求出：

$$I_c = \frac{1}{3}bx_c^3 + \frac{E_s}{E_c}I_s + \frac{E_s}{E_c}A_s(h_s - x_c)^2 + \left(\frac{E'_s}{E_c} - 1\right)I'_s + \left(\frac{E'_s}{E_c} - 1\right)A'_s(x_c - h_s)^2 \tag{30}$$

由图 11 中的平衡关系可求得：

$$\frac{d^4 v_c(x)}{dx^4} = -\frac{q(x) - b\sigma(x)}{E_c I_c} \tag{31}$$

$$\frac{d^4 v_{uh}(x)}{dx^4} = -\frac{b\sigma(x)}{E_{uh} I_{uh}} \tag{32}$$

将式(28)、式(29)代入式(24)，可得粘结正应力的微分控制方程：

$$\frac{d^4\sigma(x)}{dx^4} + 4\beta^4\sigma(x) = n_1\frac{d\tau(x)}{dx} - n_2 q(x) \tag{33}$$

其中

$$4\beta^4 = \frac{E_a b}{t_a}\left(\frac{1}{E_{uh} I_{uh}} + \frac{1}{E_c I_c}\right) \tag{34}$$

$$n_1 = \frac{E_a b}{t_a}\left(\frac{h_{uh}}{2E_{uh} I_{uh}} - \frac{h_c - x_c}{E_c I_c}\right) \tag{35}$$

$$n_2 = \frac{E_a}{E_c I_c t_a} \tag{36}$$

结合实际四点弯曲试验的边界条件，可得粘结正应力的解为：

当 $0 \leqslant x \leqslant l_1/2$

$$\sigma(x) = e^{-\beta x}\left[D_1\cos(\beta x) + D_2\sin(\beta x)\right] + \frac{n_1\lambda}{4\beta^4}\left(C_1 e^{\lambda x} - C_2 e^{-\lambda x}\right) \tag{37}$$

当 $l_1/2 \leqslant x \leqslant l/2$

$$\sigma(x) = e^{-\beta x}\left[D_1\cos(\beta x) + D_2\sin(\beta x)\right] + \frac{n_1\lambda}{4\beta^4}\left(C_3 e^{\lambda x} - C_4 e^{-\lambda x}\right) \tag{38}$$

加固梁试验中的剥离破坏主要发生在加固层端部，由端部混凝土开裂引起，此时加固区内混凝土裂缝较小，且无界面剥离发生，因此可按下列边界条件，求得参数解。

(a) $x=0$，$\tau=0$；

(b) $x=l_1/2$，$\tau$ 连续；

(c) $x=l_1/2$，$\dfrac{\mathrm{d}\tau}{\mathrm{d}x}$ 连续；

(d) $x=l/2$，$\dfrac{\mathrm{d}\tau}{\mathrm{d}x}=m_2=-\dfrac{G_a(h_c-x_c)aP}{2E_cI_0t_a}$；

(e) $x=l/2$，$\dfrac{\mathrm{d}^2\sigma(x)}{\mathrm{d}x^2}=\dfrac{E_a aP}{2E_cI_ct_a}$；

(f) $x=l/2$

$$\frac{\mathrm{d}^3\sigma(x)}{\mathrm{d}x^3} = \frac{E_a}{t_a}\left(\frac{h_{uh}}{2E_{uh}I_{uh}} - \frac{h_c-x_c}{E_cI_c}\right)\tau(x) - \frac{E_aP}{2t_aE_cI_c}\text{；}$$

(g) $x=l_1/2$，$\sigma(x)$ 连续；

(h) $x=l_1/2$，$\dfrac{\mathrm{d}^3\sigma(x)_-}{\mathrm{d}x^3}=\dfrac{\mathrm{d}^3\sigma(x)_+}{\mathrm{d}x^3}=\dfrac{E_aP}{2E_cI_ct_a}$。

解得：

$$C_1 = \frac{4m_2 - m_1P\lambda e^{-\frac{\lambda l}{2}}e^{\frac{\lambda l_1}{2}} + m_1P\lambda e^{\frac{\lambda}{2}(l-l_1)}}{4\lambda\left(e^{-\frac{l\lambda}{2}} + e^{\frac{l\lambda}{2}}\right)}$$

$$C_2 = \frac{-4m_2 + m_1P\lambda e^{-\frac{\lambda l}{2}}e^{\frac{\lambda l_1}{2}} - m_1P\lambda e^{\frac{\lambda}{2}(l-l_1)}}{4\lambda\left(e^{-\frac{l\lambda}{2}} + e^{\frac{l\lambda}{2}}\right)}$$

$$C_3 = \frac{4m_2 - m_1P\lambda e^{-\frac{\lambda(l-l_1)}{2}} - m_1P\lambda e^{-\frac{\lambda(l+l_1)}{2}}}{4\lambda\left(e^{-\frac{l\lambda}{2}} + e^{\frac{l\lambda}{2}}\right)}$$

$$C_4 = -\frac{4m_2 + m_1P\lambda e^{\frac{\lambda(l-l_1)}{2}} + m_1P\lambda e^{\frac{\lambda(l+l_1)}{2}}}{4\lambda\left(e^{-\frac{l\lambda}{2}} + e^{\frac{l\lambda}{2}}\right)}$$

其中 $D_1$ 和 $D_2$ 由于符号运算公式较长，文中暂不描述，直接编程代入求解方程中。

当混凝土处于双向拉、压状态下的破坏准则为：

$$\sigma_1 = \left(1 - 0.8\frac{\sigma_2}{f_c}\right)f_t \tag{39}$$

式中，$f_c$、$f_t$ 为混凝土轴心抗压强度和抗拉强度。由于式(36)中 $\sigma_2/f_c$ 的值较小，所以将其近似写为：

$$\sigma_1 = 0.95f_t \tag{40}$$

则梁底混凝土在粘结剪应力、剥离正应力的作用下，主应力可按下式计算，并引入影响因子 $\varphi$，计算结果与试验结果相比较，进行回归计算，取 $\varphi=0.53$

$$\varphi\sigma_1=\frac{\sigma_{\max}}{2}\pm\sqrt{\left(\frac{\sigma_{\max}}{2}\right)^2+\tau_{\max}^2} \tag{41}$$

将上述计算结果，即在 $x=l/2$ 时，可计算出界面剪应力和正应力的最大值，可求出加载值 $P$ 的关系式：

$$P=Kf_t \tag{42}$$

根据上述分析方法，分别计算了试验梁的正常使用极限荷载和端部开裂荷载。从表 7 中可以看出，计算值与试验值吻合较好，有一定的计算精度。剥离破坏通常在加固层端部发生，因此在设计应用时，可将端部开裂荷载作为初步判断出加固梁的剥离是否发生的依据。由于端部开裂荷载较剥离荷载要小，以端部开裂荷载为判据，计算出的剥离荷载偏小，因此可以提高结构整体安全性。端部植入锚筋可以改善 UHTCC 和既有混凝土两者界面间的粘结状态，故试件 UH-CB5 的计算值和试验值会有较大的偏差。

**加固梁试验值与计算值** **表 7**

| 试件编号 | 试验值(kN) | | 计算值(kN) | | 误差(%) | |
|---|---|---|---|---|---|---|
| | 端部开裂 | 极限荷载 | 端部开裂 | 极限荷载 | 端部开裂 | 极限荷载 |
| UH-CB1 | — | 47.4 | — | 44.2 | — | 6.8 |
| UH-CB2 | — | 52.6 | — | 52.3 | — | 0.6 |
| UH-CB3 | 40.1 | 58.3 | 43.8 | 59.6 | 9.2 | 2.2 |
| UH-CB4 | 47.1 | 73.3 | 43.5 | — | 7.6 | — |
| UH-CB5 | 55.4 | 70.8 | 43.5 | — | 21.8 | — |
| UH-CB6 | 43.9 | 72.6 | 47.3 | — | 7.7 | — |
| UH-CB7 | — | 56.0 | — | 52.3 | — | 6.7 |
| UH-CB8 | — | 88.9 | — | 91.7 | — | 3.1 |

## 16.5 集中荷载作用下后浇 UHTCC 简支双向混凝土板的试验研究

通过集中荷载作用下的 7 块简支混凝土双向板试验结果和分析，表明使用后浇 UHTCC 材料作为混凝土构件的防护层有较高的可行性和有效性。既有混凝土尺寸和界面处理图 12、图 13，加固方法和加载方式见表 8 和图 14。

试验结果分析表明，对比既有混凝土双向板，各复合试件的开裂荷载和极限荷载普遍提高。选取底面最大裂缝宽度为 0.2mm 时对比板件的荷载值，所对应的各复合试件裂缝宽度值，UHTCC 复合板底面最大裂缝宽度均保持在 0.02mm 以下；当各试件底面最大裂缝宽度均达到 0.2mm 时，相比既有混凝土板 B2-SL1，复合板 B2-SL3～5 以及 B2-SL7 相应承载力分别提高了 165%，443%，576%和 419%。在端部植筋的局部加固试件(B2-SL6～7)也表现出同整体加固试件相近的承载效果。

各试件加载过程中所测得数据结果见表 9。后浇高强度混凝土加固复合板虽然开裂荷

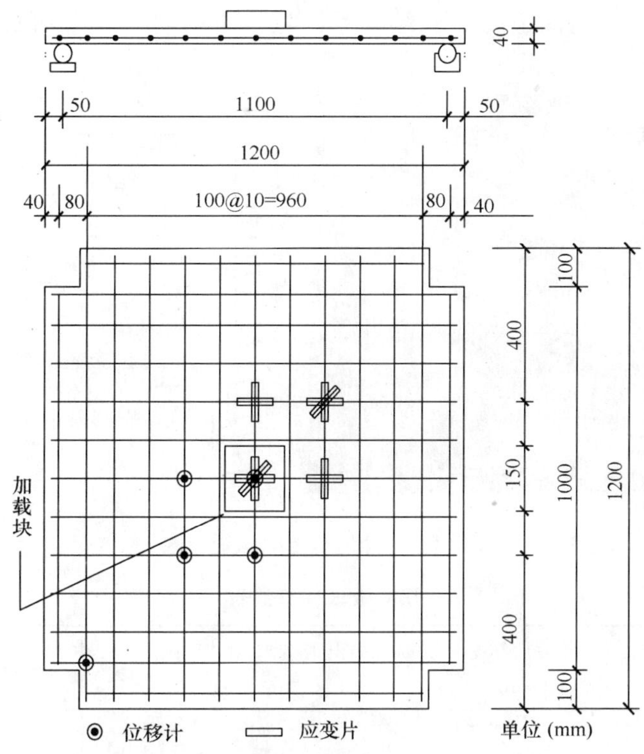


图 12　混凝土板示意图

图 13　混凝土板受拉面界面处理

载和极限承载能力得到提高，但是由于开裂后裂缝扩展迅速，造成后浇材料过早退出工作，仅在加载前期共同承载，后期只通过既有混凝土板承载，整体承载能力并未改善，承载能力远低于同厚度 UHTCC 加固复合板。当复合试件底面最大裂缝宽度达到 0.2mm 时，B2-SL4 以及 B2-SL7 相应荷载值分别较同截面高强度等级混凝土加固试件的提高了 57%和 124%。

为了观察复合板底裂缝的扩展，选取后浇 UHTCC 的复合板 B2-SL4 和 B2-SL5，加载至可视裂缝出现即停。此时在中心位置已经有裂缝宽度超过 0.02～0.04mm，裂缝观测仪画面如图 15(a)，但在其余部位没有发现可视裂缝。将显示液涂抹于复合板底面，如图 15(b)所示，才能够看到有许多细小致密裂缝的扩展和分散路径。通过涂抹显示液对裂缝进行描绘，得到复合板受弯面的整体裂缝分布图[图 15(c)和图 15(d)]。从图中可见，两块复合板底面裂缝已扩展到四边的加载端位置，且分布均匀，较为细密，细小裂缝条数较多，表现出优异的控裂和分散裂缝的能力。

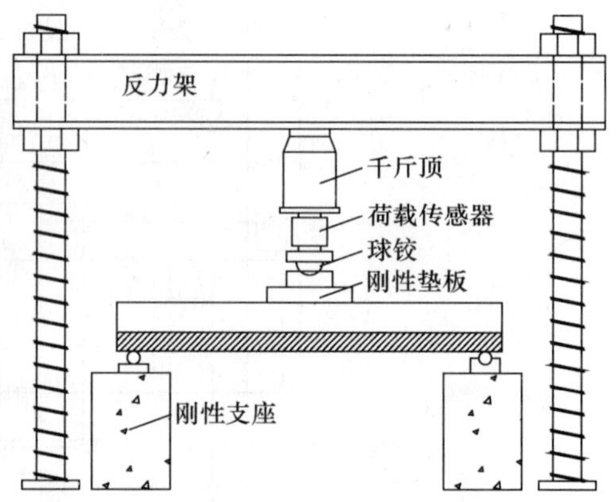


图 14 试件加载实图和示意图

双向板试件参数表 表 8

| 试件编号 | 整体厚度(mm) | 加固层厚度(mm) | 加固面积比(%) | 加固面积(mm×mm) | 后浇材料 | 植筋 |
|---|---|---|---|---|---|---|
| B2-SL1 | 40 | — | — | — | — | — |
| B2-SL2 | 70 | 30 | 100 | 1200×1200 | 混凝土 | — |
| B2-SL3 | 55 | 15 | 100 | 1200×1200 | UHTCC | — |
| B2-SL4 | 70 | 30 | 100 | 1200×1200 | UHTCC | — |
| B2-SL5 | 85 | 45 | 100 | 1200×1200 | UHTCC | — |
| B2-SL6 | 70 | 30 | 60 | 930×930 | 混凝土 | 是 |
| B2-SL7 | 70 | 30 | 60 | 930×930 | UHTCC | 是 |

荷载、中心挠度、裂缝宽度测量值 表 9

| 编号 | 后浇材料 | 开裂 | | | | 当受拉面最大裂缝宽度达到0.2mm时 | |
|---|---|---|---|---|---|---|---|
| | | 荷载(kN) | 相对增量(%) | 中心挠度(mm) | 最大裂缝宽度(mm) | 荷载(kN) | 中心挠度(mm) |
| B2-SL1 | — | 8.3 | — | 1.17 | 0.2 | 8.3 | 1.17 |
| B2-SL2 | 混凝土 | 28.7 | 246 | 1.12 | >0.2 | 28.7 | 1.12 |
| B2-SL3 | UHTCC | 15.0 | 81 | 1.73 | 0.03 | 22.1 | 4.01 |
| B2-SL4 | UHTCC | 30.4 | 267 | 2.75 | 0.02 | 45.2 | 8.63 |
| B2-SL5 | UHTCC | 42.5 | 413 | 2.56 | 0.02 | 56.0 | 4.35 |
| B2-SL6 | 混凝土 | 20.9 | 152 | 0.88 | >0.2 | 20.9 | 0.88 |
| B2-SL7 | UHTCC | 36.9 | 345 | 3.43 | 0.03 | 46.7 | 8.68 |

由于构件受力状态较为复杂，故使用非线性有限元方法对简支双向复合板的加载过程进行分析[28~30]。复合板模型如图 16(a)、(c)所示。为了提高计算效率，取双向板的四分之一进行建模，如图 16(b)、(d)所示。建模中的加载方式和边界条件与实际相同，如图

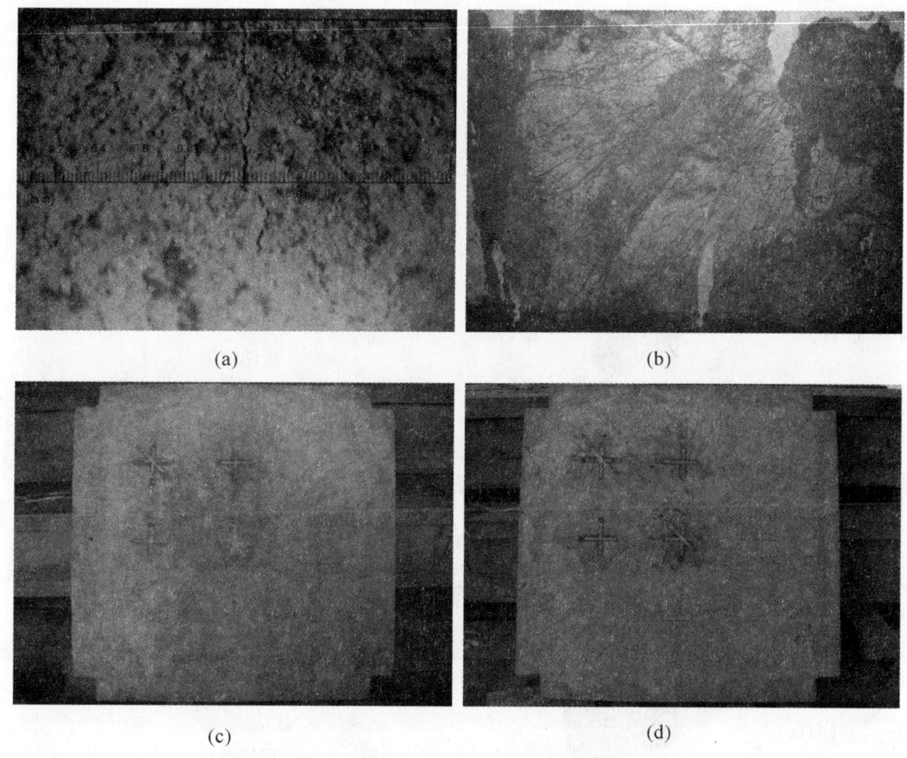

(a)　(b)　(c)　(d)

图 15　UHTCC 底面裂缝图

15(b)、(d)所示，其中加载钢板与简支板、滚轴与简支板之间的法向接触采用硬接触，即垂直于接触面的压力可以完全在界面间传递，界面切向力较小，故假定为无摩擦状态，以此来模拟接触面分离及相互摩擦作用。由于 UHTCC 后浇层采用全加固、和端部植筋的局部加固方式，因此加载过程中不考虑既有混凝土和 UHTCC 界面间两者的滑移，将界面定义为绑定约束。取试件的四分之一进行分析，分别在两个对称面上施加对称边界条件，在滚轴底面施加位移约束。

混凝土材料采用混凝土塑性损伤模型（Concrete Damage Plasticity）进行定义，此模型时依据 Lubliner，Lee 和 Fenves（1998）提出的损伤塑性模型确定的，它考虑了材料拉压性能的差异，主要用于模拟低静水压力下由损伤引起的不可恢复的材料退化。这种退化主要表现在材料宏观属性的拉压屈服强度不同、拉伸屈服后材料表现为软化及压缩屈服后材料先硬化后软化等。由于本章试验中集中加载方式为单调加载，且主要针对构件屈服前计算分析，因此分析主要利用其塑性模型部分，文中对损伤所带来的影响不考虑。混凝土的单轴受拉受压过程应力-应变曲线分别采用双线性简化受拉曲线和 Carreria 与 Chu 提出的混凝土受压计算应力-应变全曲线的经验公式。由于 UHTCC 材料在受拉和受压时均会表现出应力应变硬化特征，因此选用铸铁塑性模型（Cast Iron Plasticity）来进行计算。此模型在受到拉伸和压缩时，其弹塑性状态会有不同的屈服强度和硬化，拉伸服从 Rankine 屈服准则，压缩服从 Mises 屈服准则。UHTCC 单轴受拉和受压 UHTCC 采用简化受压双折线和单线性受拉硬化模型，弹性模量使用实测初裂点应力和应变的比值来确定。两种材料的单元均采用八节点减缩积分格式的三维实体单元（C3D8R）。钢筋使用满足 Von

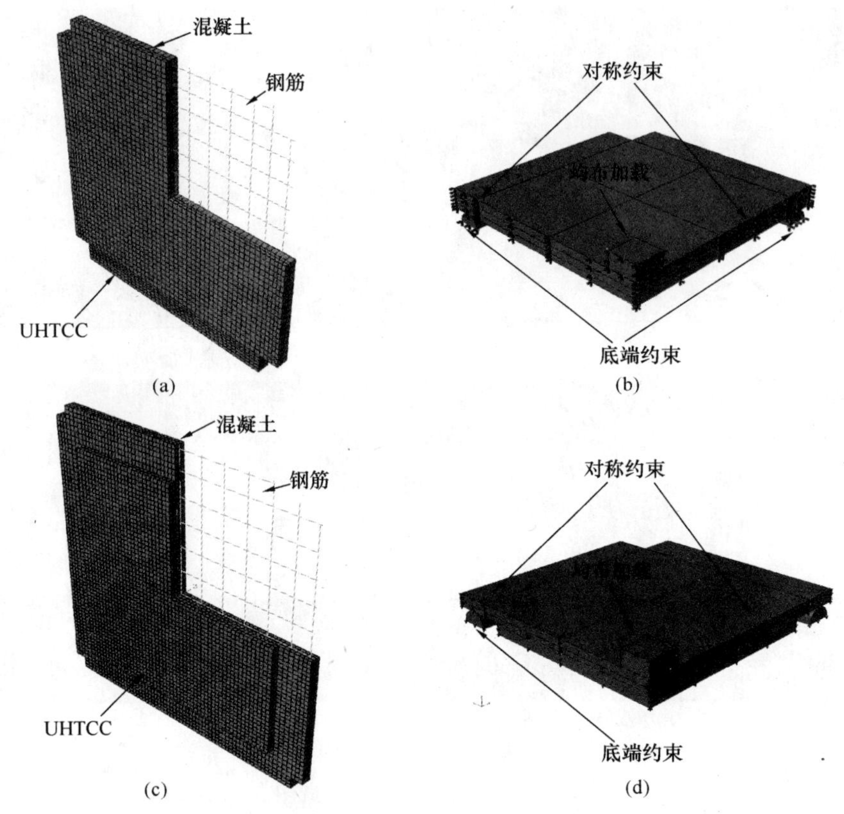


图 16 有限元模型示意图和实际建模图，整体（a）和（b），局部（c）和（d）

Mises 屈服准则的等向弹塑形模型，单元选择三维桁架单元，即两节点线性位移杆单元，采用嵌入单元的方法来进行模拟。分别针对荷载-挠度曲线、沿跨度方向距离的挠度和应变分布进行计算，见图 17，图 18 和图 19 所示，相应计算结果与试验结果相比较，两者符合较好，具有一定精度。

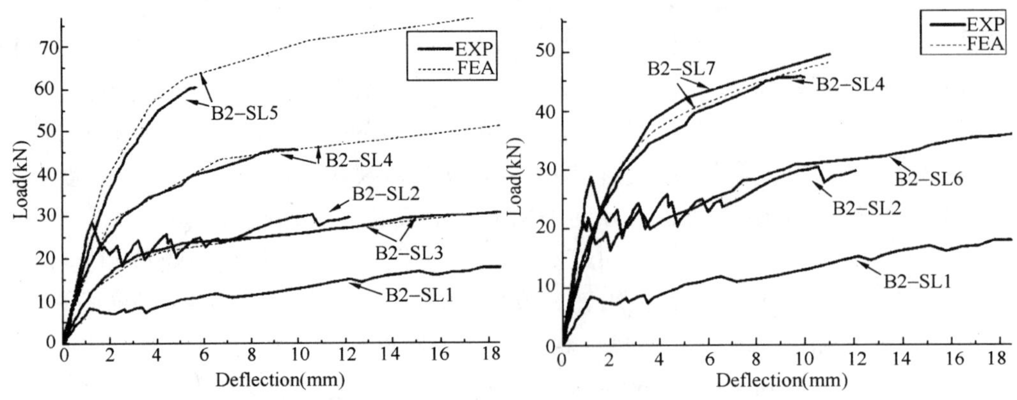


图 17 试件荷载-挠度曲线（试验和有限元）

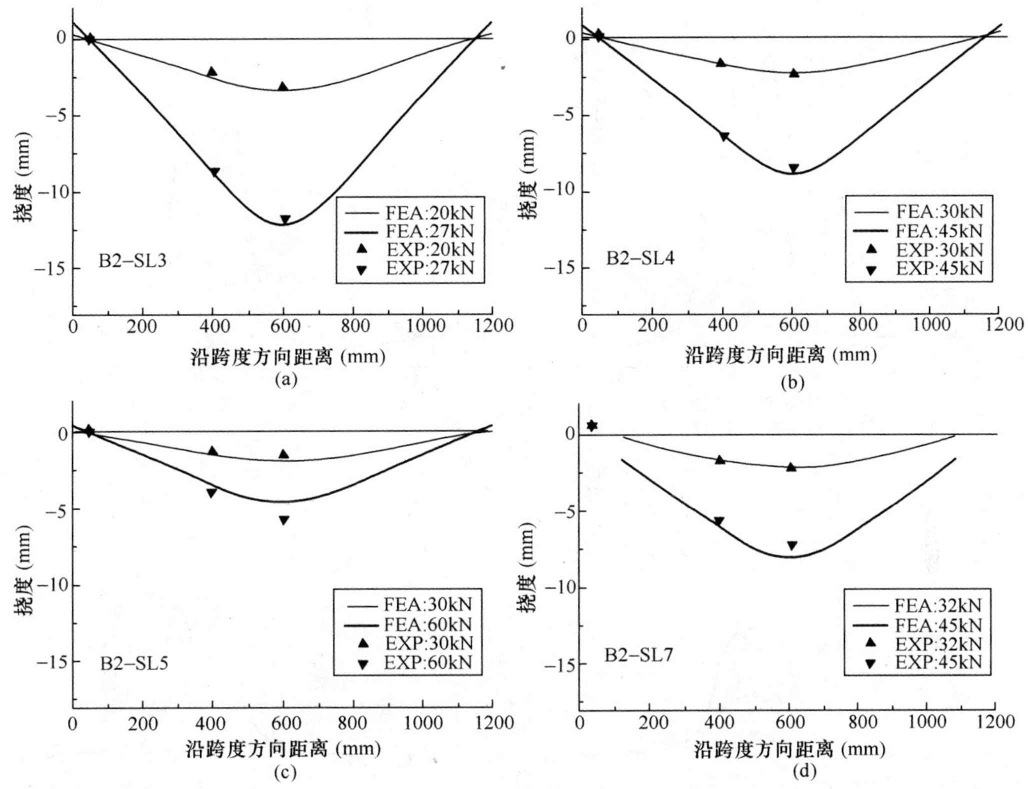


图 18　中心剖面的挠度分布

## 16.6　结论

（1）通过 UHTCC 与既有混凝土组成的粘结试件抗拉和抗剪试验，表明在无界面剂、锚筋等增强界面粘结状态的措施下，单纯通过改变既有混凝土构件性状可使 UHTCC 与既有混凝土粘结劈拉强度达到既有混凝土相应强度 72.8%以上、粘结剪切强度达到 52.9%以上，表明此材料与既有混凝土结构具有较好的粘结性能。

（2）相比既有混凝土梁、钢筋混凝土梁和板试件，后浇 UHTCC 组成的复合试件的弯曲承载能力和韧性都得到了显著改善，呈现出延性破坏特征；同时同级荷载作用下，同截面配置的 UHTCC 加固试件中裂缝宽度较高强度混凝土加固试件能小 1～2 个数量级，在加载过程中裂缝分布均匀，且较为细密，在试件临近屈服前，能够将最大裂缝宽度值限制在 0.06mm 左右，表现出优异的控裂和分散裂缝的能力。

（3）基于弹塑性截面分析法，对加固梁正常使用状态下的弯曲承载力进行了计算。通过平截面假定、力学平衡和几何方程，采用相应的混凝土和 UHTCC 计算模型，对复合梁承载力进行分析；并根据弹性梁理论和组合截面假定对复合梁中 UHTCC 和既有混凝土间界面的应力状态进行分析，推导了界面应力的解析解，结合加载时构件实际中性轴位置，推算此时构件等效截面惯性矩，近似求出后浇层端部开裂时所对应的荷载值，计算结果与实测值符合较好。

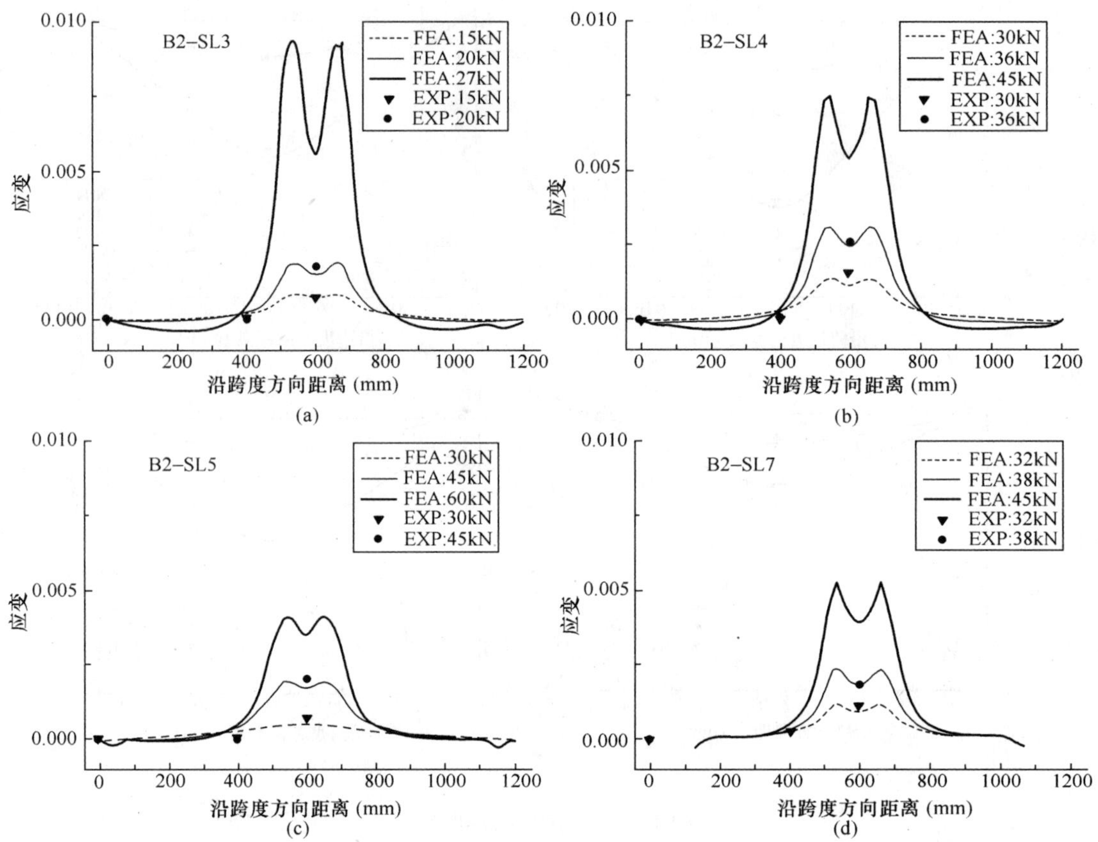


图 19 中心剖面的应变分布

(4) 由于构件受力状态较为复杂，采用非线性有限元计算软件，按照试验中实际的材料性能和加载方式等条件，来定义各材料参数模型、复合试件几何模型和约束条件等，对集中荷载作用下简支双向板试验进行数值分析。计算所得到的荷载-中心挠度曲线、受弯底面上挠度和应变分布与试验中实测值符合较好，具有一定精度，表明有限元计算中所采用的材料模型和分析方法较为合理。

(5) 通过以上研究分析说明 UHTCC 作为结构耐久性防护加层是一种较为有效和实用的措施，文中所涉及的试验研究和分析方法能为实际工程设计应用提供一定的实际和理论依据。

## 参考文献

[1] Metha P K, Monteiro P J M. Concrete: Microstructure, properties, and materials (third edition). New York: McGraw-Hill, 2006.

[2] 叶列平. 混凝土结构. 北京：清华大学出版社，2005.

[3] 沈蒲生，梁兴文. 混凝土结构设计原理. 北京：高等教育出版社，2005.

[4] 吴兴祖，韩素芳，路来军. 竞争与发展同在，挑战和机遇并存——回顾我国混凝土产业的发展历程. 混凝土世界，2009，9(3)：20-24.

[5] 李明顺. 我国混凝土结构技术五十年的发展与展望. 建筑结构，1999，29(10)：3-8.

[6] 国家发展改革委，国务院西部开发办. 西部大开发“十一五”规划. 西部大开发，2007，(4)：13-27.

[7] 刘卫东，刘毅，秦玉才等. 2009中国区域发展报告—西部开发的走向. 北京：商务印书馆，2010.

[8] 李治国. 病害隧道调查及裂缝整治技术. 现代隧道技术(增刊)，2002：15-21.

[9] 陈肇元. 土建结构工程的安全性与耐久性. 北京：建筑工业出版社，2003：174-187.

[10] Maher A B. Performance of concrete in a coastal environment. Cement & Concrete Composites. 2003，(25)：539-548.

[11] 金祖权. 西部地区严酷环境下混凝土的耐久性与寿命预测.（博士学位论文）. 南京：东南大学，2006.

[12] 刘连新. 察尔汗盐湖及超盐渍上地区混凝土腐蚀及预防初探. 建筑材料学报，2001，4(4)：395-400.

[13] Li V C, Wang S, Wu H C. Tensile strain-hardening behavior of PVA-ECC. ACI Materials Journal, 2001, 98(6)：483-492.

[14] 徐世烺，李贺东. 超高韧性水泥基复合材料研究进展及其工程应用. 土木工程学报，2008，41(6)：45-60.

[15] Li V C. Reflections on the research and development of engineered cementitious composites (ECC). Proceedings of the JCI International Workshop on Ductile Fiber Reinforced Cementitious Composites (DFRCC). Tokyo：Japan Concrete Institute, 2002：1-21.

[16] 徐世烺，蔡新华，李贺东. 超高韧纤维增强水泥基复合材料抗冻耐久性能试验研究. 土木工程学报，2009，42(9)：42-46.

[17] Ahmed S F U, Mihashi. A review on durability properties of strain hardening fibre reinforced cementitious composites (SHFRCC). Cement and Concrete Composites, 2007, 29：365-376.

[18] Lepech M, Li V C. Water permeability of cracked cementitious composites. ICF 11, Turin, Italy, 2005：4539 4544.

[19] Li V C. Reflections on the research and development of engineered cementitious composites (ECC). Proceedings of the JCI International Workshop on Ductile Fiber Reinforced Cementitious Composites (DFRCC). Tokyo：Japan Concrete Institute, 2002：1-21.

[20] Minoru K, Keitetsu R. Recent progress on HPFRCC in Japan：Required performance and applications. Journal of Advanced Concrete Technology, 2006, 4(1)：19-33.

[21] Kim Y Y, Fischer G & Li V C. Performance of Bridge Deck Link Slabs Designed with Ductile ECC. ACI Structural J, 2004, 101(6)：792-801.

[22] Kojima S, Sakat N, Kanda T & Hiraishi T. Application of Direct Sprayer ECC for Retrofitting Dam Structure Surface - Application for Mitaka-Dam. Concrete Journal, 2004, 42(5)：35-39.

[23] 朱昌明，金永杰. 钢筋混凝土梁的非线性分析. 工程力学，1990，7(1)：50-56.

[24] Xu S L, Li Q H. Theoretical analysis on bending behavior of functionally graded composite beam crack-controlled by ultra high toughness cementitious composites. Science in China Series E：Technological Sciences, 2009, 52(2)：363-378.

[25] Li V C, Wang S, Wu H C. Tensile strain-hardening behavior of PVA-ECC. ACI Materials Journal, 2001, 98(6)：483-492.

[26] 李贺东. 超高韧性水泥基复合材料试验研究. 大连：大连理工大学，2008.

[27] Malek A M, Saadatmanesh H, and Ehsani M R. Prediction of failure load of R/C beams strength-

ened with FRP plate due to stress concentration at the plate end. ACI Structural Journal. 1998，95 (1)：142-153.

[28] 王勖成. 有限单元法. 清华大学出版社，2003.

[29] 吕西林. 钢筋混凝土结构非线性有限元分析与应用. 同济大学出版社，1997.

[30] Hibbitte，Karlsson，Sorenson，INC. ABAQUS/Standard User's Manual. 2002.

# 第 17 章 Chapter 17

# 城市轻轨混凝土梁长期性能试验研究和时随分析*

薛伟辰[1]，刘　婷[1]，王　巍[2]

（1. 同济大学，土木工程学院，上海 200092；

2. 同济大学，上海 200092）

**摘　要**：城市轻轨是目前国内外城市中普遍应用的中等容量轻轨交通，其轨道梁的主要结构形式是预应力混凝土梁。本文通过 4 根 1：5 城市轻轨预应力混凝土轨道梁模型的 1500d 长期试验，对其时随性能进行较系统的研究与分析，重点考察预应力筋张拉方式和截面上下缘应力差等因素对轨道梁的长期挠曲变形、徐变应变、截面曲率、预应力筋应变和钢筋应变等的影响。试验结果表明，采用 2 次张拉方式和降低截面上、下缘应力差均能减少混凝土梁的徐变变形和徐变应变。基于龄期调整有效模量法，编制步进法时随有限元分析程序，实现了城市轻轨混凝土轨道梁的长期挠曲变形的时随全过程模拟分析。应用该程序对本文试验结果进行模拟分析，计算值与试验结果吻合良好。

**关键词**：城市轻轨；混凝土轨道梁；长期性能；时随全过程分析；长期挠曲变形

# EXPERIMENTAL STUDY ON LONG-TERM BEHAVIORS OF URBAN LIGHT RAIL CONCRETE TRACK GIRDERS

W. C. Xue[1]，T. Liu[1]，W. Wang[2]

（1. School of Civil Engineering，Tongji University，Shanghai 200092，China；

2. Tongji University，Shanghai 200092，China.）

**Abstract**：Urban light rail is a kind of rail transit system，which is used more and more widely. And the Urban light rail in the existing field applications is to consist primarily of prestressed concrete girders. Experimental and analytical studies are conducted to investigate time-dependent behaviors of 4 concrete track girder models（1：5）of urban light rail for about 1500 days. The research is carried out with an emphasis on the influence of methods of tensioning，stress differences between upper and lower fibers of midspan sections on

* 基金项目：建设部研究开发项目（04-2-026）和交通部西部交通建设科技项目（200631882244）.

long-term deflections, creep strains, section curvatures, strains of reinforcements and prestressing tendons. In the test, the prestressed concrete girders with twice prestress stretching showed a better creep behavior in comparison to those with once prestress stretching and better creep behavior of prestressed concrete girders is found with lower stress difference between upper and lower fiber of the section at midspan. Moreover, step-by-step finite element analysis program based on Age-adjusted Effective Modulus Method (AEMM) is developed for the time-dependent full-range analysis of creep deformation of track girders. Calculations by using this program agree well with the test results.

**Keywords**: urban light rail; concrete track girder; long-term behavior; time-dependent analysis; long-term deflection

## 17.1　引言

为克服道路拥挤、减少环境污染，德国、英国、美国等国采用了全新的现代化轻轨交通系统。轻轨交通系统覆盖面较广、灵活性较大，具有建设投资少、运能大、乘坐方便等优点。自 2000 年我国建设首批轻轨城市以来，城市轻轨日益成为现代化城市的重要交通通道。目前，城市轻轨线路以无碴轨道为主，其主要结构形式是预应力混凝土梁。而由于混凝土徐变的发展，轨道梁梁体长期挠曲变形随时间持续增长[1-7]，成桥后的不良线型将影响乘车的舒适性，甚至危及行车安全。因此，为了满足城市轻轨混凝土轨道梁对变形的高精度要求并保证安全性和长期稳定性，对其长期性能进行深入的试验研究和时随分析是十分必要的。

近年来，我国已对混凝土梁[8-13]的长期性能进行了一些试验研究和有限元分析。但总的来看，在已有的研究中，专门针对城市轻轨混凝土轨道梁长期性能的试验和有限元分析很少。鉴于此，本文基于城市轻轨混凝土轨道梁 1500d 的长期性能试验研究和时随全过程分析，对其长期挠曲变形、徐变应变、截面曲率、预应力筋应变和钢筋应变等长期性能进行了研究。

## 17.2　试验设计

### 17.2.1　试件设计

以实际工程梁为原型，根据梁跨中截面应力等效和施工工艺相似的原则，共设计了 4 根 1∶5 大尺度轨道梁模型，编号分别为 RC-1、PC-1、PC-2 和 PC-4 。其中 RC-1 为混凝土基准梁，用于测试混凝土收缩对模型梁的影响。模型梁的具体参数见表 1。4 根模型梁的截面形状与尺寸相同，梁长为 5240mm，计算跨径为 5000mm，梁高 500mm，高跨比均为 1∶10。模型梁和预应力筋布置如图 1 所示。试件与原型梁的结构尺寸对比见表 2。4 根预应力混凝土梁均采用高性能混凝土，掺加了 15%的粉煤灰，强度等级为 C60。普通钢筋的配置均相同，预应力筋采用高强低松弛钢绞线。

模型梁参数 表1

| 模型梁 | | PC-1 | PC-2 | PC-4 | RC-1 |
|---|---|---|---|---|---|
| 预应力筋 | | 6 $\phi^{s}$ 13 | 6 $\phi^{s}$ 13 | 6 $\phi^{s}$ 13 | — |
| 第1次张拉 | 张拉控制应力 | 0.40 $f_{ptk}$ | 0.60 $f_{ptk}$ | 0.40 $f_{ptk}$ | — |
| | 张拉根数 | 4 | 6 | 4 | — |
| 第2次张拉 | 张拉控制应力 | 0.60 $f_{ptk}$ | — | 0.65 $f_{ptk}$ | — |
| | 张拉根数 | 6 | — | 6 | — |

原型梁与模型梁的尺寸对比（m） 表2

| 参数 | 计算跨度 | 全长 | 梁高 | 顶板宽 | 顶板高 | 跨中腹板宽 | 底板宽 | 底板高 |
|---|---|---|---|---|---|---|---|---|
| 原型 | 24000 | 24600 | 2000 | 12800 | 300 | 450 | 6620 | 300 |
| 试件 | 5000 | 5240 | 500 | 600 | 100 | 150 | 250 | 100 |

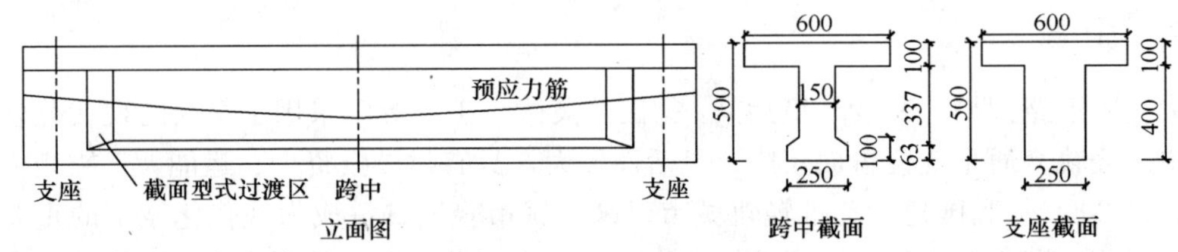


图1 模型梁（单位：mm）

### 17.2.2 预应力张拉与加载方案

预应力筋采用2次张拉方式（表1），即在梁体养护3d时第1次张拉预应力筋，第10d时进行2次张拉。为了模拟轨道梁的实际施工过程，在张拉完成80d后施加二期恒载，见表3。二期恒载按照结构静力试验中长期堆载的方法，采用素混凝土加载条和铁块进行加载，以保证长期加载的稳定性。RC-1不进行加载，用于测试混凝土收缩对模型梁的影响。

模型梁二期恒载（kN/m） 表3

| 试件编号 | PC-1 | PC-2 | PC-4 | RC-1 |
|---|---|---|---|---|
| 二期恒载 | 8.97 | 8.97 | 7.28 | — |

### 17.2.3 量测内容

分别在梁端和跨中布置百分表，测量梁跨中变形；在跨中截面沿梁高布置3个千分表，测量截面曲率；混凝土收缩由RC-1两侧的各1个拉线式位移计测量；混凝土应变、钢筋应变和预应力筋应变则分别在梁跨中截面处混凝土、钢筋和预应力筋上布置测点测量；试验中同时测量环境温度和相对湿度的时随变化。

## 17.3 长期试验结果与分析

预应力筋张拉完成并施加二期恒载后，梁体上下缘的应力差值见表4。加载产生的实测变形值与现行公路规范计算值的误差在 8%以内。以第二次张拉预应力筋完成为起点，在 1500d 的持续时段内，研究了环境温度、湿度、梁跨中长期变形、跨中距梁顶面 190mm 处的徐变应变的时随变化规律，并以 PC-2 为例，给出了典型的梁跨中截面曲率、下缘钢筋应力和预应力筋应变的时程曲线图，如图 2～图 8 所示。

**梁跨中截面应力**（MPa）　　**表 4**

| 试件编号 | 预应力筋张拉完成后 | | | 施加二期恒载后 | | |
|---|---|---|---|---|---|---|
| | 上缘 | 下缘 | 差值 | 上缘 | 下缘 | 差值 |
| PC-1 | −0.6 | −9.05 | −8.45 | −2.25 | −6.36 | −4.11 |
| PC-2 | −0.6 | −9.05 | −8.45 | −2.25 | −6.36 | −4.11 |
| PC-4 | −0.6 | −10.32 | −9.72 | −1.93 | −8.14 | −6.21 |

（1）徐变的时随发展规律。以预应力筋第 2 次张拉完成后的数值作为观测初值。在预应力筋张拉之前，由于试验梁是平放在水平地面上的，因此并不产生长期挠曲变形和徐变应变；在预应力筋张拉完成后初期，长期挠曲变形和徐变应变都有较大幅度的增长，分别达到二期恒载施加之前最大值的 87%和 92%；二期恒载施加时，长期挠曲变形立即发生突变，下降的幅度即为二期恒载产生的瞬时弹性变形值；徐变应变在二期恒载施加之后的突变没有长期挠曲变形那么明显，只有 $20\times10^{-6}$左右，而在此之后，由于受到温度和湿度等时随因素的综合影响，徐变应变产生了较大的波动；在二期恒载施加之后的 40d 内，长期挠曲变形有较大幅度的变化，而在次之后就趋向于平缓；3 根预应力梁在 1500d 时的长期挠曲变形和徐变应变终值分别在 2.20mm 和 $777\times10^{-6}$以内，并且 3 根梁变形和应变值最大的为 PC-2，其次依次为 PC-4，PC-1 最小；梁跨中截面曲率的变化规律与长期挠曲变形基本相同，1500d 内 PC-2 的曲率终值约为 $0.715\times10^{-6}\text{mm}^{-1}$。

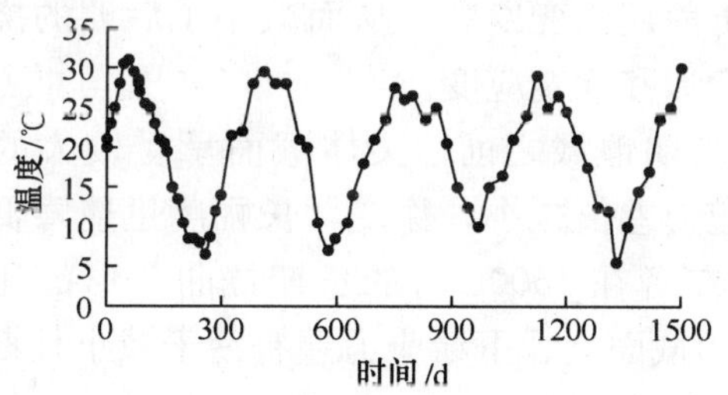


图 2　环境温度时程曲线

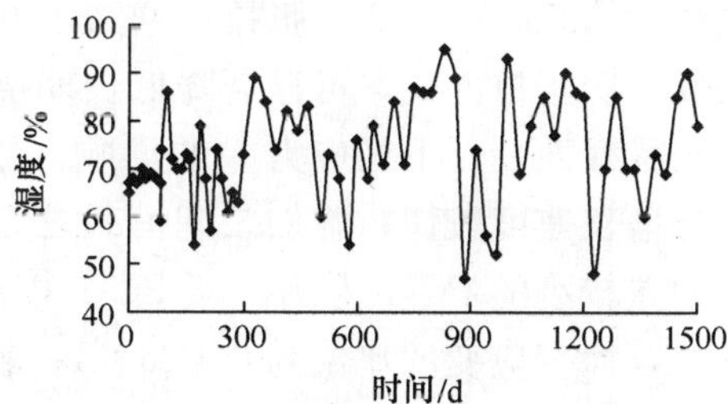


图 3　环境湿度时程曲线

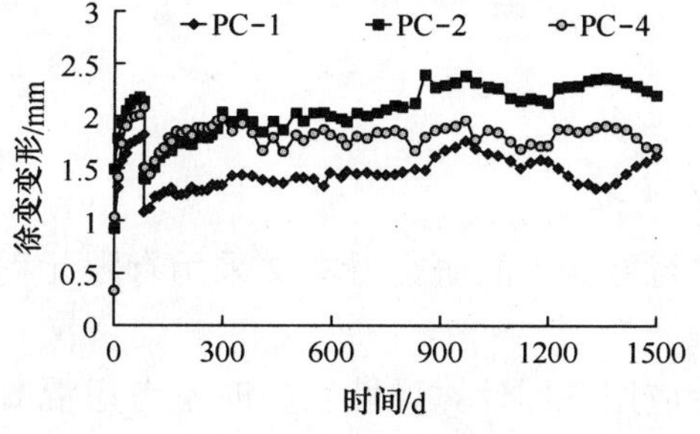


图 4　梁跨中长期变形时程曲线

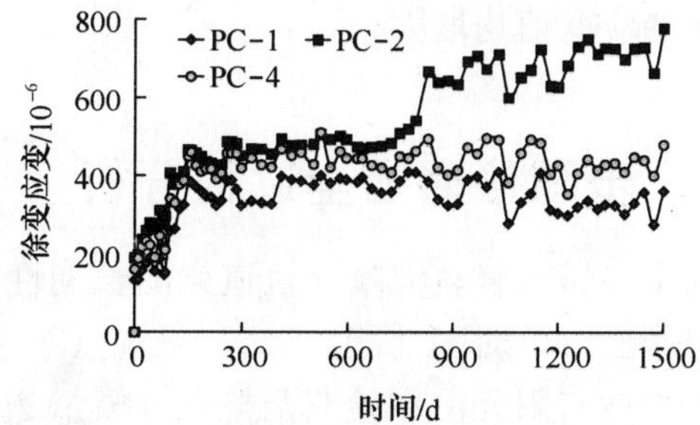


图 5　梁跨中截面混凝土徐变应变时程曲线

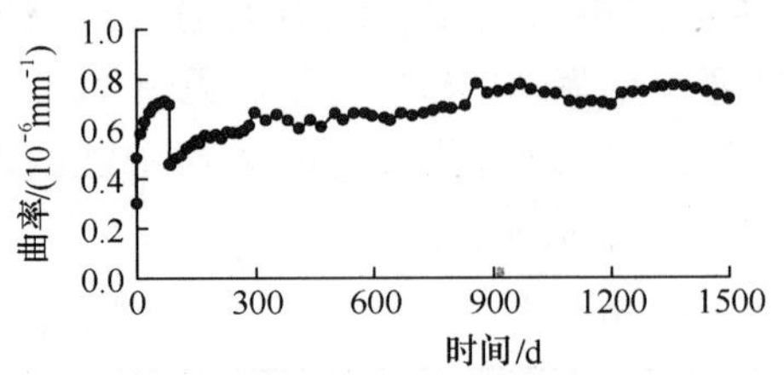


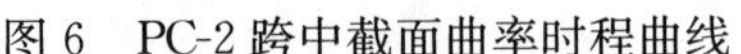
图 6　PC-2 跨中截面曲率时程曲线

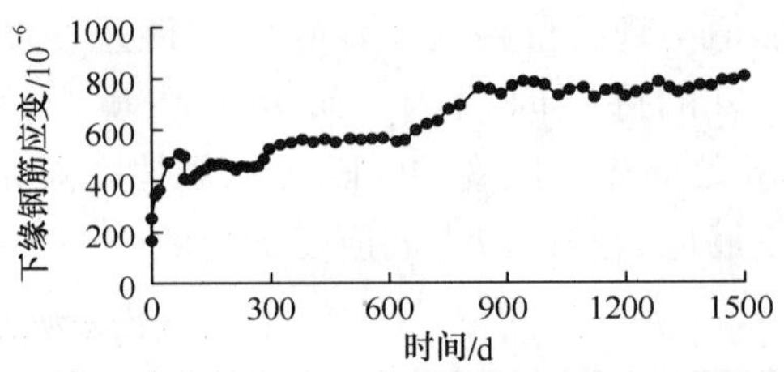


图 7　PC-2 跨中截面下缘钢筋应力时程曲线

（2）钢筋应变的时随发展规律。由于混凝土收缩和徐变的综合影响，梁体内钢筋的应变变化规律很复杂，波动性较大。如图 7 所示，以 PC-2 跨中截面下缘钢筋为例，初期增长迅速，300d 时达到 1500d 内最大值的 65.2%左右，之后逐渐趋于平缓；预应力筋张拉完成 1500d 时的 PC-2 跨中截面下缘钢筋应变值约为 807$\mu\varepsilon$。

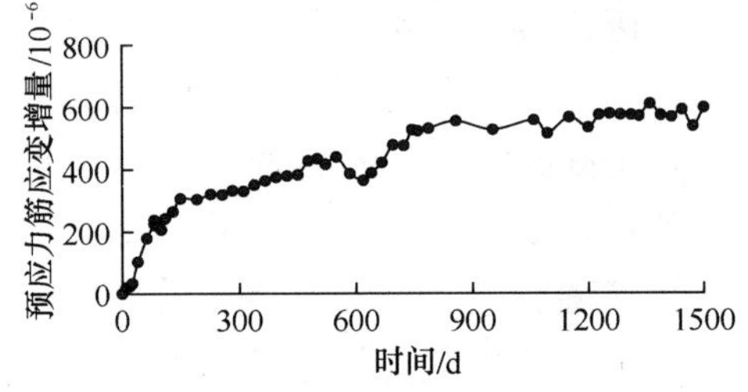


图 8　PC-2 跨中截面预应力筋应变增量时程曲线

（3）预应力筋应变的时随发展规律。3 根预应力混凝土试验梁的预应力筋应变变化规律基本相同；如图 8 所示，以 PC-2 为例，在施加二期恒载以前，预应力筋应变增长幅度较大，80d 时达到 1500d 应变增量的 39.8%；施加二期恒载后，虽有较大的波动，但总体增长趋势较为平缓，PC-2 预应力筋 1500d 的总应变增量值约为 $598\times10^{-6}$。

（4）预应力筋张拉方式的影响。采用 2 次张拉方式的 PC-1 与采用 1 次张拉方式的 PC-2 相比，1500d 长期挠曲变形减少 26.4%，徐变应变减少了 53.6%；2 次张拉，增加了混凝土前期徐变发展，从而减小了后期的挠曲变形；采取 2 次张拉方式，可显著降低长期挠曲变形和徐变应变。

（5）梁截面上、下缘应力差的影响。施加二期恒载之前，梁体截面承受较大的应力差，徐变增长速度较快；施加二期恒载之后，应力差值减小，徐变增长幅度迅速降低；以截面应力差较小的 PC-4 为例，与 PC-1 相比，二者在 1500d 时的长期挠曲变形的比值为 0.768∶1，徐变应变的比值为 0.616∶1；较小的截面上、下缘应力差有助于减小长期挠曲变形和徐变应变。

（6）对温湿度变化的敏感度。随着温度的升高和相对湿度的降低，试件的长期挠曲变形和徐变应变值均增大。

## 17.4　步随法时随全过程分析

为了对城市轻轨混凝土轨道梁的长期性能进行更深入的研究，本文采用有限元程序来模拟混凝土梁的时随性能。

将弹性有限元法和龄期调整有效模量法结合应用，即计算弹性效应时不考虑混凝土的收缩、徐变，将时间全过程按施工和运营阶段的关键时间点划分时段，将每个时段按前密后疏原则再划分为若干个小时段。在每个小时段内对结构的徐变性能进行详细分析，求出各节点的位移增量和节点力增量，并与该小时段初始时刻的位移和节点力进行叠加，即可得到结构在施工、运营全过程中的内力和变形值。

采用 Trost-Bažant 龄期调整有效模量法[14]来编制徐变计算程序。该法引入老化系数来考虑混凝土老化对最终徐变值的影响，实质上就是应用积分中值定理，使过去求解徐变问题的积分方法转变为代数方法，计算简易，精度较高[7]。

表 5 列出了梁跨中长期挠曲变形的程序计算结果与试验值的对比情况。从表中可见：各个不同时间段内长期挠曲变形和 1500d 时的长期挠曲变形终值均吻合较好。

**长期挠曲变形对比**　　**表 5**

| 模型 | 10d | 50d | 100d | 500d | 800d | 1095d | 1500d |
|---|---|---|---|---|---|---|---|
| PC-1 | 0.96 | 0.86 | 0.87 | 1.06 | 1.14 | 1.09 | 1.08 |
| PC-2 | 0.95 | 0.92 | 0.89 | 1.04 | 1.03 | 1.01 | 1.01 |
| PC-4 | 0.89 | 0.93 | 0.87 | 1.05 | 1.05 | 1.12 | 1.17 |

注：表中数值为程序计算结果与试验值之比。

## 17.5　结论

（1）在预应力筋张拉完成后初期即前 30d 内，试验梁跨中截面的长期挠曲变形和徐变应变增长较快，1500d 终值分别在 2.20mm 和 $777\times10^{-6}$以内，跨中截面曲率的变化规律与跨中变形相似。

（2）钢筋应变的时随变化规律和梁跨中长期挠曲变形的变化规律相似，1500d 终值约为 $807\times10^{-6}$；预应力筋应变在二期恒载施加以前的变化速度较大，施加二期恒载后，变化平缓，1500d 的总应变增量约为 $598\times10^{-6}$。

（3）采用 2 次张拉工艺增加了混凝土前期徐变发展，并减小了后期的徐变发展。

（4）环境温度和湿度对预应力混凝土梁徐变的影响规律基本相似，当温度升高或相对湿度降低时，长期挠曲变形和徐变应变均增大。

（5）编制了基于按龄期调整有效模量法的时随有限元全过程分析程序，实现了对预应力混凝土梁长期挠曲变形的时随全过程分析，程序计算值与试验值吻合良好。

## 参考文献

[1] Paul A, Franz-Josef Ulm. Creep and shrinkage of concrete: physical origins and practical measurements. Nuclear Engineering and Design, 2001(203): 143-158.

[2] Bažant Z P. Prediction of concrete creep and shrinkage: past, present and future. Nuclear Engineering and Design. 2001(203): 27-38.

[3] McHenry D. A new aspect of creep in concrete and its applications to design. Proc. ASTM. 1943, 43: 1069-1084.

[4] Neville A M, Dilger W H, Brooks J J. Creep of plain and structural concrete. London and New York: Construction Press, 1983.

[5] Bažant Z P, Panula L. Practical prediction of time-dependent deformations of concrete. Materials and

Structures. 1978，11(65)：307-328.

[6] 劳远昌. 劳远昌桥梁论文选集．北京：中国铁道出版社，1992.

[7] 周履，陈永春．收缩徐变．北京：中国铁道出版社，1994.

[8] 薛伟辰，王巍. 轨道交通预应力混凝土梁施工阶段徐变性能研究. 铁道学报，2008，30(1)：53-59.

[9] 曹国辉，方志，孟新田．开裂混凝土梁长期挠曲变形研究．四川建筑科学研究，2008，34(1)：69-72.

[10] 潘立本，钟文乐，庞春龙．预应力混凝土梁长期荷载下的挠度．扬州大学学报(自然科学版)，1998，1(4)：71-74.

[11] 丁大钧，黄德富，兰宗建．预应力粉煤灰陶粒混凝土梁长期荷载试验与刚度计算．工业建筑，1985，(6)：8-13.

[12] 胡狄，陈政清．预应力混凝土桥梁收缩与长期变形试验研究. 土木工程学报，2003，36(8)：79-85.

[13] 潘祫棣．长期可变荷载作用下预应力混凝土梁的刚度试验研究．工业建筑，2008，38(5)：50-53.

[14] Bazant Z P. Prediction of concrete creep effect using age-adjusted effective modulus method. ACI Material Journal，1972，69(2)：212-217.

# 第 18 章 Chapter 18

# 丙类与乙类设防 RC 框架结构抗地震倒塌能力对比*

**叶列平，陆新征，汤保新**

（清华大学土木工程系，清华大学结构工程与振动教育部重点实验室，北京 100084）

**摘　要：**按 2001 版《建筑抗震设计规范》6～8 度丙类与乙类设防建筑设计了 10 个典型钢筋混凝土框架结构算例，基于 IDA 结构地震倒塌易损性分析方法进行了抗地震倒塌能力评价，并按近似对数正态分布拟合了地震倒塌易损性曲线，确定了各算例结构的倒塌概率。基于概率论方法指出，结构抗地震倒塌能力需从绝对抗地震倒塌能力和离散程度两方面进行全面评价。结合汶川地震建筑震害统计数据，参照美国 ATC 建议和本文算例的分析结果，建议了我国框架结构抗地震倒塌概率的目标值，并依据该目标值对 10 算例的抗地震倒塌能力进行评价。结果表明，对于 6 度和 7 度设防算例，因绝对抗地震倒塌能力不足，仅由丙类提高到乙类，遭遇大震和特大地震时倒塌率明显高于 8 度设防算例。据此，本文建议采取适当提高设防烈度方法来增强 6 度和 7 度较高设防要求建筑的抗地震倒塌能力，以减小大震和特大地震下 6 度和 7 度较高设防要求建筑的倒塌概率。

**关键词：**框架结构；动力增量时程分析；抗地震倒塌能力；抗倒塌可靠指标；倒塌概率

# COMPARISON IN THE COLLAPSE-RESISTANT CAPACITY OF RC FRAMES WITH DIFFERENT SEISMIC FORTIFICATION CATEGORY OF B AND C

L. P. Ye，X. Z. Lu，B. X. Tang

(1. Department of Civil Engineering，Tsinghua University，Key Laboratory of Structural Engineering and Vibration of China Education Ministry，Beijing 100084. )

**Abstract**：Based on "Code for Seismic Design of Buildings" （GB50011-2001），10 Classic examples of reinforced concrete frame structures are designed with different levels of seismic fortification intensity and seismic fortification category for structures. And the seismic

* 基金项目：中国工程院咨询研究项目（2010-ZD-4）；国家自然科学基金重大研究计划重点项目资助（90815025）；国家科技支撑计划项目（2009BAJ28B01）。

collapse-resistant capacities of these structures are evaluated quantitatively by use of IDA method. Because the collapse margin ratio (CMR) proposed by the U. S. ATC can't reflect the discrete of collapse fragility curve, this paper calculates the structural collapse probability by means of the approximate logarithmic normal distribution curve of seismic collapse fragility, and on probability theory, proposes that the collapse-resistant capacity of a structure should be evaluated in two aspects of it's absolute capacity and it's relative discrete. Referring to the American ATC suggestion and analysis result of the 10 examples , this paper propose structural target values of anti-seismic collapse probability for our country, and evaluates the collapse-resistant capacities of the 10 examples according to the target values. The results show that, it is not enough to achieve the target value of 8 degree for the frame structures of 6 degree and 7 degree only to be upgraded the seismic fortification category from B to C, because their absolute capacities are insufficient. Therefore the other effective measures must be taken to enhance further the collapse-resistant capacity of frame structures.

**Keywords**: frame structure; incremental dynamic analysis; seismic collapse-resistant capacity; reliable index of collapse-resistant capacity, Collapse probability

## 18.1 引言

2008年5月12日汶川特大地震的震害调查表明，按2001版《建筑抗震设计规范》GB 50011—2010[1]（以下简称《抗规》）设计的房屋建筑，总体上基本能实现预期的抗震设防目标[2]。尽管如此，由于部分地区遭遇地震烈度超过设防罕遇地震烈度、甚至达到特大地震，导致很多按《抗规》设计建造的房屋建筑发生了较严重震害，甚至倒塌[3]，其中校舍建筑倒塌尤为受到关注。汶川地震后，提高校舍建筑的抗震能力、尤其是抗地震倒塌能力，成为社会各界的共识，但具体提高幅度多少合适，缺乏定量评价目标。2008年6月，中国工程院土木、水利与建筑学部《汶川地震建筑震害分析与重建研讨会》建议书[4]提出："对灾害发生时逃生能力差的人员的建筑，包括如幼儿园、中小学、老人院等，应达到大震可修的抗震设防目标。"这相当于要求将抗震设防烈度提高1度。汶川地震后修订的《建筑工程抗震设防分类标准》GB 50223—2008[5]（以下简称《设防标准》）调整了校舍建筑的抗震设防标准，要求"教育建筑中，幼儿园、小学、中学的教学用房以及学生宿舍和食堂，抗震设防类别应不低于重点设防类（乙类）"。可见工程院的建议书[4]和《设防标准》[5]对校舍建筑抗震能力的提高程度是有所差别的。比如，位于7度区（100Gal）的框架结构，按《设防标准》[5]要求将丙类提高到乙类，抗震设防烈度不变，即地震作用不变，抗震等级提高一级；而按工程院建议书的要求，抗震设防烈度相当于7度大震（220Gal），比8度设防地震（200Gal）略高。这两种提高校舍建筑抗震能力的措施有多大差别，在遭遇罕遇地震烈度和特大地震烈度下的倒塌情况如何，目前未见有研究分析。

避免强震下倒塌是建筑抗震的核心性能目标[6]。本文以强震下建筑结构的倒塌率作为评价指标，采用基于动力增量时程分析（Incremental Dynamic Analysis，简称IDA）的地震倒塌率作为评价结构抗地震倒塌能力的指标[7]，通过对一组不同抗震设防烈度的丙类和乙类钢筋混凝土框架结构教学楼算例分析，比较不同抗震能力增强措施对提高学校建筑抗地震倒塌能力的改善程度，并根据分析结果，给出了罕遇地震和特大地震下校舍建筑倒塌率合理目标值，并据此提出了增强校舍建筑抗震能力措施的建议。

## 18.2 算例结构设计

算例为5层钢筋混凝土框架结构教学楼，按照Ⅱ类场地、第1设计地震分组进行设计。抗震设防烈度分别取6度（0.05$g$）、7度（0.10$g$）、7度（0.15$g$）、8度（0.20$g$）和8度（0.30$g$）的丙类设防（代号B）和乙类设防（代号Y），共10个结构模型。抗震设防烈度为9度的5层建筑，不再适合设计为纯框架结构，未考虑。各算例结构按《抗规》[1]的最低要求进行设计，未进行人为调整，以检验仅满足《抗规》建筑的抗震能力。各算例结构布置相同，结构平面见图1（a），结构立面见图1（b）。荷载标准值取值如下：

（1）楼面：恒载4kN/m²，活载2kN/m²，楼梯间恒载8.0kN/m²，楼梯间活载3.5kN/m²，走廊活载2.5kN/m²；

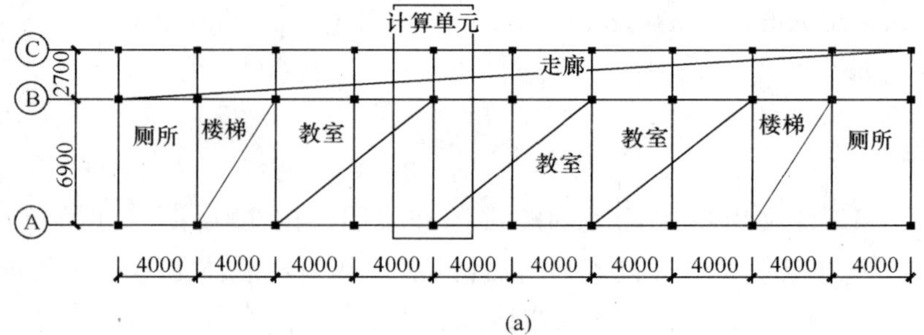


(a)

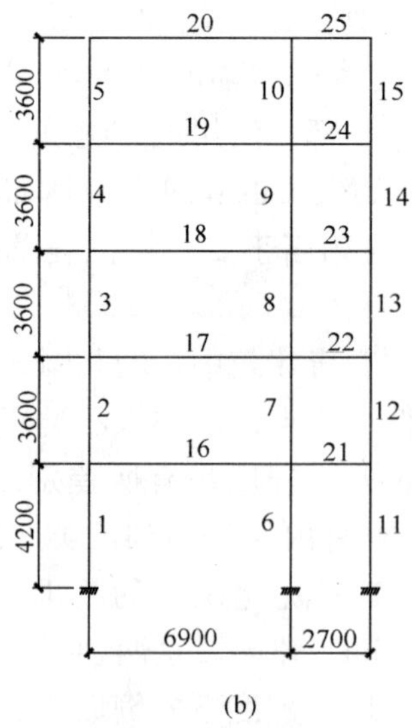


(b)

图1 结构布置

(a) 平面布置；(b) 立面布置

(2) 屋面：恒载 $7kN/m^2$，活载 $0.5kN/m^2$，雪载 $0.5kN/m^2$；

(3) 墙体：横墙荷载 9kN/m，纵墙荷载 6kN/m，走廊栏板、女儿墙荷载 2kN/m。截面尺寸、材料与抗震等级等信息见表 1。

**算例的截面尺寸、材料与抗震等级**　　**表 1**

| 结构编号 | 6B | 6Y | 7B | 7Y | 7.5B | 7.5Y | 8B | 8Y | 8.5B | 8.5Y |
|---|---|---|---|---|---|---|---|---|---|---|
| A、B 轴柱 (mm) | 400×400 | | 400×400 | | | | 550×550 | | 600×600 | |
| C 轴柱 (mm) | 300×300 | | 350×350 | | | | 400×400 | | 400×400 | |
| 6.9m 跨梁 (mm) | 250×600 | | 250×600 | | | | 300×600 | | 300×600 | |
| 2.7m 跨梁 (mm) | 250×400 | | 250×400 | | | | 300×450 | | 300×450 | |
| 抗震等级 | 四级 | 三级 | 三级 | 二级 | 三级 | 二级 | 二级 | 一级 | 二级 | 一级 |
| 混凝土 | C20 | C20 | C20 | C25 | C20 | C25 | C20 | C20 | C20 | C20 |
| 钢筋种类 | 纵筋 HRB400，箍筋 HRB335， | | | | | | | | | |

**算例的弹性抗震计算结果**　　**表 2**

| 结构编号 | | 6B，6Y | 7B | 7Y | 7.5B | 7.5Y | 8B，8Y | 8.5B，8.5Y |
|---|---|---|---|---|---|---|---|---|
| 结构总质量 (t) | | 2193 | 2212 | 2212 | 2212 | 2212 | 2467 | 2532 |
| 一阶周期 (s) | X 向 | 1.2586 | 1.2124 | 1.1570 | 1.2124 | 1.1570 | 0.8700 | 0.8382 |
| | Y 向 | 1.1957 | 1.1610 | 1.1079 | 1.1610 | 1.1079 | 0.8751 | 0.8422 |
| | 扭转 | 1.1493 | 1.1098 | 1.0591 | 1.1098 | 1.0591 | 0.8339 | 0.8025 |
| 基底剪力 (kN) | X 向 | 344 | 707 | 736 | 1001 | 1041 | 2078 | 2185 |
| | Y 向 | 360 | 744 | 775 | 1053 | 1096 | 2067 | 2171 |
| 最大层间位移 | X 向 | 1/1535 | 1/789 | 1/831 | 1/560 | 1/589 | 1/560 | 1/584 |
| | Y 向 | 1/1675 | 1/861 | 1/907 | 1/610 | 1/643 | 1/570 | 1/594 |
| 最大轴压比 | 边柱 | 0.76 | 0.79 | 0.64 | 0.81 | 0.66 | 0.51 | 0.44 |
| | 中柱 | 0.84 | 0.85 | 0.69 | 0.86 | 0.70 | 0.53 | 0.45 |
| 主要控制指标 | | 重力 | 轴压比 | | 轴压比、层间位移 | | 层间位移、节点抗剪 | |

结构设计计算采用 PKPM 系列软件，各算例结构的主要性能指标见表 2。设计计算结果表明：

(1) 对于 6 度和 7 度设防结构，结构设计主要控制指标为重力和轴压比，即竖向荷载起控制作用；而 7.5～8.5 度设防结构，主要控制指标为层间位移和节点抗剪，即水平地震作用起控制作用。

(2) 设防烈度相同时，乙类与丙类结构的基底剪力基本相近，抗震等级提高一级，主要体现在框架柱轴压比有所降低、加密区配箍率略有增大，这将增强框架柱的塑性变形能力。箍筋对混凝土的约束作用体现在，一是混凝土峰值强度和峰值应变有所增加，二是混凝土极限应变有所增加，即会增加结构的塑性变形能力。本文以下分析中按文献 [10]、[11] 的方法考虑箍筋的影响。结构由丙类设防提高到乙类设防，其抗地震倒塌能力是否有所改善，需通过以下 IDA 地震倒塌率分析来进一步研究。

## 18.3　基于 IDA 的地震倒塌率分析

本文仅以各算例中间短轴框架方向建模进行 IDA 地震倒塌率分析。分析模型采用

PKPM 系列软件计算的实配钢筋，竖向荷载取图 1（a）阴影范围内所有恒载和 50%活载。各算例配筋见图 2～图 6。

结构弹塑性地震倒塌分析采用基于 MSC. MARC 分析软件开发的钢筋混凝土杆系结构纤维模型 ThuFiber 程序计算。文献［8］、［9］对 ThuFiber 程序的适用性和准确性进行了验证，对分析结构倒塌过程也具有很好的收敛性。

地震倒塌分析用地震动数据库采用美国 ATC-63 报告[11]建议的 22 条远场地震动记录，另外增加常用的 El-Centro 地震动记录，共 23 条。按设计截面尺寸、实际配筋和材料强度标准值，用 ThuFiber 程序建立了上述 10 个算例结构的计算模型，用各条地震动记录分别进行 IDA 弹塑性时程倒塌分析，得到各算例的地震倒塌率，具体计算步骤如下：

（1）分别对每一条地震动记录，采用逐步增大峰值加速度 PGA，当 PGA 达到某一临界值时，结构刚好倒塌。该临界值用 $R_{PGA}$ 表示，它代表结构在该地震动输入下的抗地震倒塌能力。并根据该地震动的弹性反应谱，计算结构基本周期 $T_1$ 在临界地震动峰值 $R_{PGA}$ 时对应的谱加速度，该临界谱加速度用 $S_a(T_1)$ 表示[12]，它是地震动强度指标另一种表示，也可作为结构在该地震动作用下的抗地震倒塌能力指标。文献[11]认为，$S_a(T_1)$ 有助于减小不同地震动引起弹塑性分析结果的离散性。

（2）总共输入 23 条地震动记录，得到 23 组临界值 $R_{PGA}$ 或 $R_{Sa}(T_1)$。

（3）对于给定的地震动强度 PGA 或 $S_a(T_1)$，如果它大于上述临界值，则表示结构会倒塌，记有 $n$ 个，则（$n$/23）为该地震动强度下的倒塌率。

显然，随 PGA 或 $S_a(T_1)$ 的逐渐增大，倒塌率也逐渐增大。计算结果见图 2～图 7 中的数据点。经检验，每个算例的 23 个 PGA 和 $S_a(T_1)$ 值均符合对数正态分布。于是按美国 ATC-63[11]建议，以 $S_a(T_1)$ 作为统计对象，按照对数正态分布进行拟合，得到各算例结构的四种倒塌率曲线，见图 2～图 7。

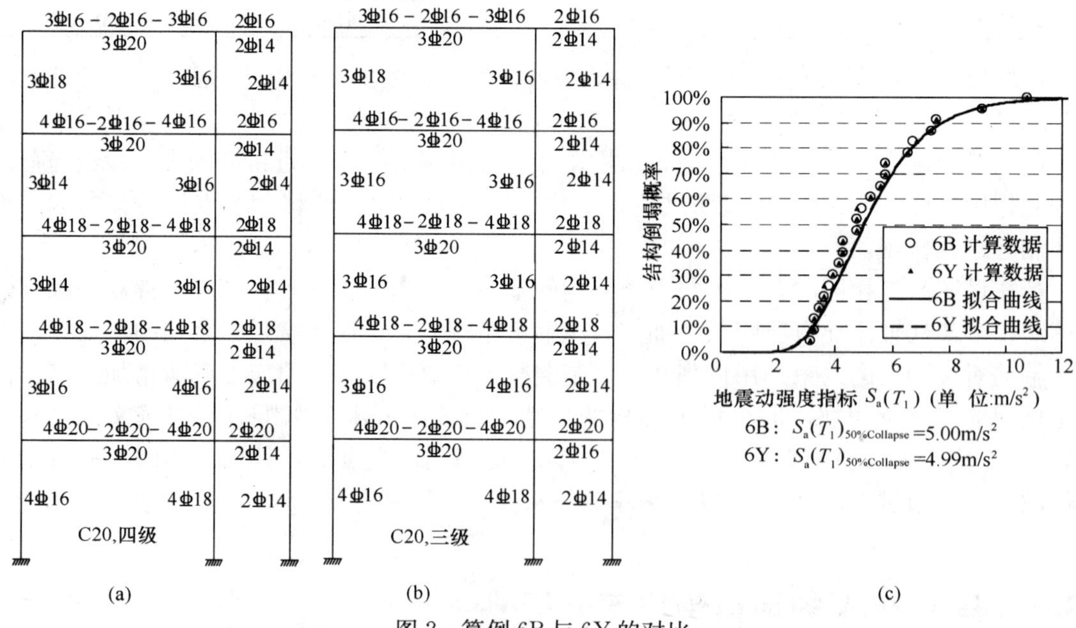


图 2 算例 6B 与 6Y 的对比

（a）6B 配筋；（b）6Y 配筋；（c）6B 与 6Y 的倒塌概率曲线对比

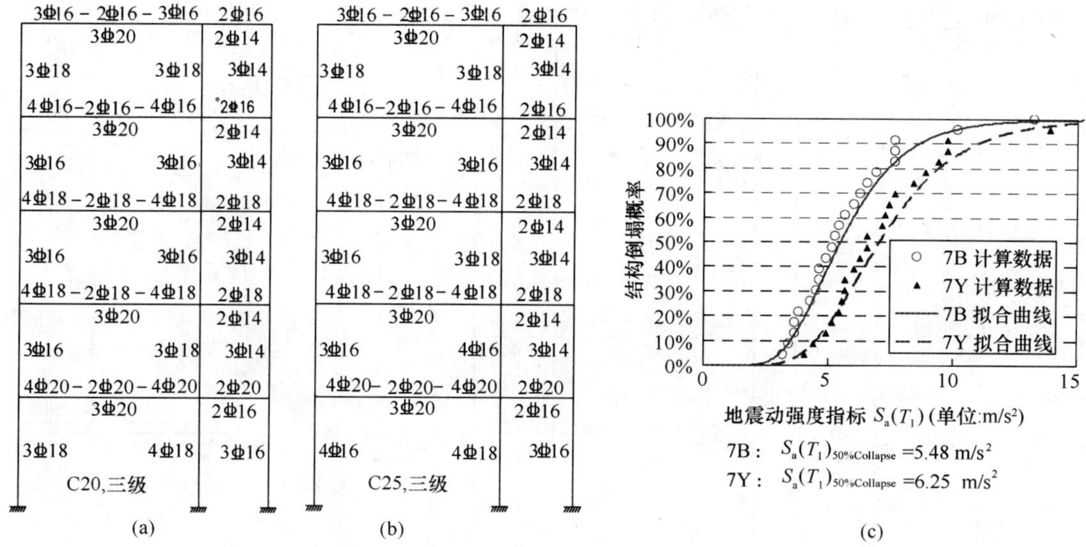


图 3　算例 7B 与 7Y 的对比

(a) 7B 配筋；(b) 7Y 配筋；(c) 7B 与 7Y 的倒塌概率曲线对比

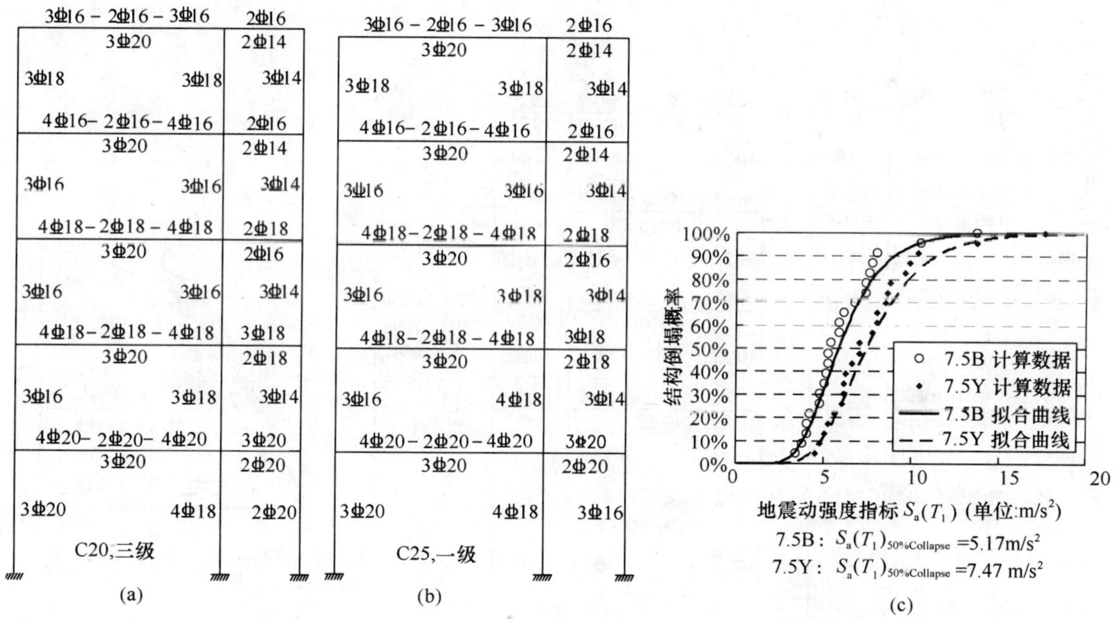

图 4　算例 7.5B 与 7.5Y 的对比

(a) 7.5B 配筋；(b) 7.5Y 配筋；(c) 7.5B 与 7.5Y 的倒塌概率曲线对比

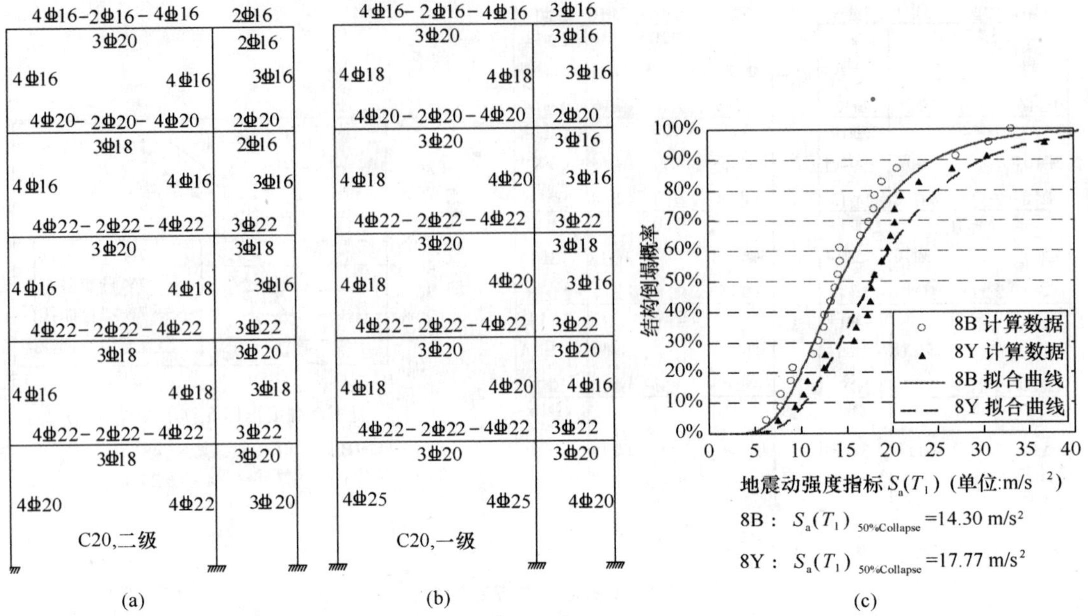

图5 算例8B与8Y的对比

(a) 8B配筋；(b) 8Y配筋；(c) 8B与8Y的倒塌概率曲线对比

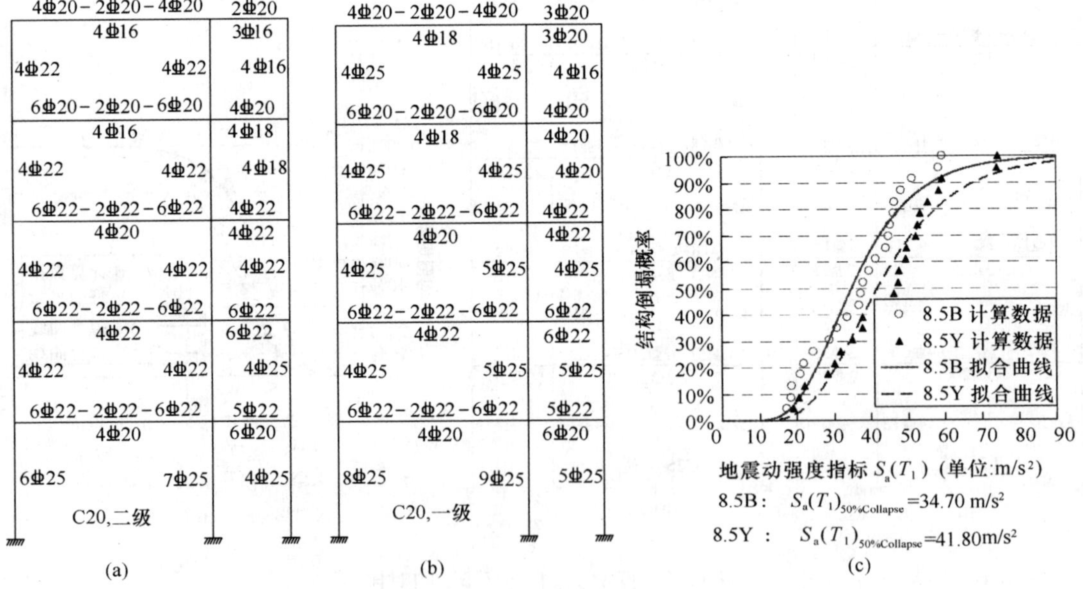

图6 算例8.5B与8.5Y的对比

(a) 8.5B配筋；(b) 8.5Y配筋；(c) 8.5B与8.5Y的倒塌概率曲线对比

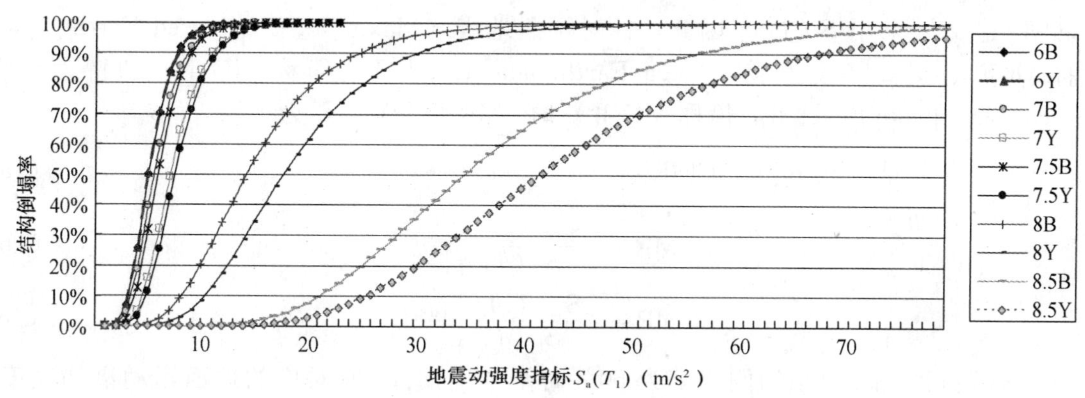


图 7　各算例的倒塌概率曲线对比

## 18.4 抗地震倒塌能力评估

基于 IDA 结构抗地震倒塌能力的定量评估，主要是给出罕遇地震和特大地震下结构的倒塌概率。由于地震的复杂性，我国发生的破坏性地震中超过地震区划罕遇地震（大震）水平的特大地震时有发生，如 2008 年汶川地震，设防烈度为 7～8 度，遭遇烈度 9～11 度；又如 1966 年邢台地震，设防 7 度，遭遇烈度 10 度；1975 年海城地震，设防 6 度，遭遇烈度 9～11 度；1976 年唐山地震，设防 6 度，遭遇烈度 10～11 度。本文依据课题组所承担国家科技支撑计划项目（2009BAJ28B01）制定的目标，分别以"提高一度大震"和"特大地震"（Mega Earthquake）下的倒塌率来综合评价结构的抗地震倒塌能力。提高一度大震，是指遭遇烈度比地震区划烈度增加一度所对应的大震；特大地震烈度，抗震设防烈度为 6、7、8 度时，遭遇烈度分别取 9、10、10 度。根据《抗规》5.1.2 条表 5.1.2-2 的规定，本文建议遭遇烈度地震动加速度最大值见表 3。

**抗倒塌分析用地震加速度时程曲线的最大值**（PGA）　　**表 3**

| 设防烈度 | 6(0.05$g$) | 7(0.10$g$) | 7(0.15$g$) | 8(0.20$g$) | 8(0.30$g$) | 9(0.40$g$) |
|---|---|---|---|---|---|---|
| 设防地震（Gal） | 50 | 100 | 150 | 200 | 300 | 400 |
| 罕遇地震（Gal） | 110* | 220 | 310 | 400 | 510 | 620 |
| 提高一度大震（Gal） | 220 | 400 | 510 | 620 | 730* | 930* |
| 特大地震（Gal） | 400 | 620 | 620 | 930* | 930* | — |

注：*者为本文设定值。

### 18.4.1 美国 ATC-63 建议方法（以 $S_a(T_1)$ 作为地震动强度指标，CMR 作为评价指标）

为了比较不同结构抗地震倒塌能力的差异，美国 ATC 建议采用结构抗地震倒塌储备系数 CMR（Collapse Margin Ratio）作为评价指标[11]。CMR 系数的定义为：结构抗地震倒塌能力与结构设防抗震能力之比，其概念相当于安全系数。结构的抗地震倒塌能力和结构设防抗震能力采用相应的地震动强度指标表示，如 PGA 或 $S_a(T_1)$。由于结构的弹塑性地震响应与谱加速度 $S_a(T_1)$ 的线性相关性较好，适用的基本周期 $T_1$ 范围较大[11],[12]，故

ATC-63 建议[11]采用 $S_a(T_1)$。因此，针对罕遇地震（Maximum Considered Earthquake，简称 MCE）、提高一度大震（Mega Earthquake Level Ⅰ，简称 ME-Ⅰ）和特大地震（Mega Earthquake Level Ⅱ，简称 ME-Ⅱ）的 CMR 可分别表示为：

$$CMR_{MCE}=\frac{S_a(T_1)_{50\%collapse}}{S_a(T_1)_{MCE}} \tag{1a}$$

$$CMR_{ME\text{-}I}\ \frac{S_a(T_1)_{50\%collapse}}{S_a(T_1)_{ME\text{-}I}} \tag{1b}$$

$$CMR_{ME\text{-}II}\ \frac{S_a(T_1)_{50\%collapse}}{S_a(T_1)_{ME\text{-}II}} \tag{1c}$$

式中，$S_a(T_1)_{50\%collapse}$是结构倒塌率为 50％时基本周期 $T_1$ 所对应的地面运动谱加速度；$S_a(T_1)_{MCE}$ 为相应设防地震的罕遇地震下基本周期 $T_1$ 所对应的地面运动谱加速度；$S_a(T_1)_{ME\text{-}Ⅰ}$ 和 $S_a(T_1)_{ME\text{-}Ⅱ}$ 分别为相应设防地震“提高一度大震”（ME-Ⅰ）和“特大地震”（ME-Ⅱ）下基本周期 $T_1$ 所对应的地面运动谱加速度。$S_a(T_1)_{MCE}$ 和 $S_a(T_1)_{ME}$ 可根据结构基本周期 $T_1$，由 23 条地震记录的罕遇地震或特大地震弹性反应谱平均值确定。

按上述方法分析得到的各算例的 $S_a(T_1)_{50\%collapse}$ 以及在罕遇地震、提高一度大震和特大震下的倒塌率及其相应的 CMR 计算结果见表 4 和图 8。$S_a(T_1)_{50\%collapse}$ 越大，或 CMR 越大，则结构的抗倒塌能力就越大。但 $S_a(T_1)_{50\%collapse}$ 反映的是结构抗倒塌能力的绝对量，其值随设防烈度的提高而增加，见图 8(a)。由图 8(a) 可见，相对于 6 度设防，7 度及 7.5 度的 $S_a(T_1)_{50\%collapse}$ 无明显提高，如 7B 相对于 6B 仅提高 10％，7.5B 相对于 6B 仅提高

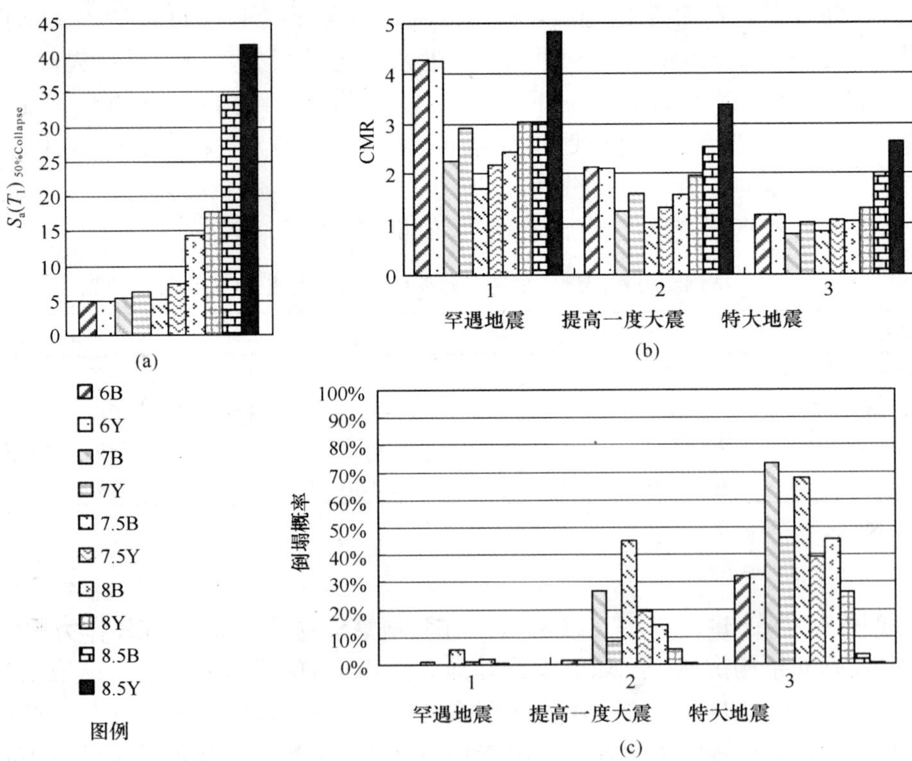


图 8 各算例的 $S_a(T_1)_{50\%Collapse}$ 和 CMR 及倒塌率对比

(a) $S_a(T_1)_{50\%Collapse}$；(b) CMR；(c) 倒塌率

17%；而相对于 7 度设防，8 度的 $S_a(T_1)_{50\%collapse}$ 有明显提高，如 8B 相对于 7B 提高 261%，相对于 7.5B 提高 244%；相对于 8 度设防，8.5 度的 $S_a(T_1)_{50\%collapse}$ 的提高十分显著，为 242%～235%。

CMR 反映的是结构抗倒塌能力的相对量，即相对于设防烈度的结构抗地震倒塌能力，故称为"结构抗地震倒塌储备系数"。由表 4 和图 8(*b*) 可见，我国不同设防烈度结构的 CMR 并不一致，7～7.5 度设防的 CMR 普遍偏小，而由前述结果控制，7～7.5 度设防的绝对量 $S_a(T_1)_{50\%collapse}$ 比 6 度设防也没有明显增加，因此 7 度和 7.5 度设防的框架结构遭遇提高一度大震和特大地震时倒塌率较大，分别达到 8.62%～45.2%到 39.2%～67.6%，甚至 7 度丙类在罕遇地震下的倒塌率也分别达到 5.34%。汶川地震震害调查也表明 7 度设防区的建筑震害相对更为严重，倒塌率较高[3],[16]。

不同设防标准框架结构的抗倒塌能力指标　表 4

| 算例结构 | | 6B | 6Y | 7B | 7Y | 7.5B | 7.5Y | 8B | 8Y | 8.5B | 8.5Y |
|---|---|---|---|---|---|---|---|---|---|---|---|
| $S_a(T_1)_{50\%Collapse}$ | | 5.00 | 4.99 | 5.48 | 6.25 | 5.17 | 7.47 | 14.30 | 17.77 | 34.70 | 41.80 |
| 罕遇地震 | $S_a(T_1)_{MCE}$ | 1.18 | 1.18 | 2.42 | 2.42 | 3.41 | 3.41 | 5.86 | 5.86 | 9.50 | 8.67 |
| | CMR | 4.28 | 4.24 | 2.27 | 2.91 | 1.72 | 2.19 | 2.44 | 3.03 | 3.04 | 4.82 |
| | 倒塌率（%） | 0.001 | 0.001 | 1.03 | 0.10 | **5.34** | 0.89 | 1.89 | 0.38 | 0.032 | 0.002 |
| 提高一度大震 | $S_a(T_1)_{ME\text{-}I}$ | 2.35 | 2.35 | 4.40 | 4.40 | 5.61 | 5.61 | 9.08 | 9.08 | 13.60 | 12.42 |
| | CMR | 2.13 | 2.12 | 1.25 | 1.60 | 1.04 | 1.33 | 1.58 | 1.96 | 2.55 | 3.37 |
| | 倒塌率（%） | 1.29 | 1.38 | **26.7** | **8.62** | **45.2** | **19.4** | **14.5** | 5.30 | 0.683 | 0.07 |
| 特大地震 | $S_a(T_1)_{ME\text{-}II}$ | 4.27 | 4.27 | 6.82 | 6.82 | 6.82 | 6.82 | 13.62 | 13.62 | 17.33 | 15.82 |
| | CMR | 1.17 | 1.17 | 0.80 | 1.03 | 0.86 | 1.09 | 1.05 | 1.30 | 2.00 | 2.64 |
| | 倒塌率（%） | **32.1** | **32.5** | **73.2** | **46.2** | **67.6** | **39.2** | **45.5** | **26.1** | 3.37 | 0.53 |

注：(1) 表中数值的单位为加速度 m/s²；
(2) 表中带下划线粗体数据表示倒塌率偏大的结果。

## 18.4.2　本文建议方法

由 (1) 式可知，如果将 CMR 视为安全系数概念，则 CMR 值相当于安全系数平均值。尽管 CMR 越大，结构的相对抗地震倒塌能力越大，但 CMR 没有体现结构倒塌率曲线的离散性，可能出现 CMR 相同、而倒塌率曲线的离散程度不同情况，因此仅用 CMR 评价结构的抗倒塌能力还不全面。为此，本文将结构倒塌率曲线近似按对数正态分布拟合，来计算结构的倒塌概率，并以大震和特大地震下的倒塌概率作为结构抗倒塌能力的评价指标。

为与目前结构可靠度符号表达形式一致，将结构倒塌时的 PGA 或谱加速度 $S_a(T_1)$ 用符号 $R_{PGA}$ 表示，代表结构抗地震倒塌能力的绝对量，属于抗力项，为随机变量；将大震或特大震时的 PGA 或谱加速度 $S_a(T_1)$ 用符号 $E_{PGA}$ 表示，属于荷载项，为确定量。这样，根据可靠度理论引入结构抗地震倒塌极限状态功能函数：

$$Z = R_{PGA} - E_{PGA} \tag{2}$$

根据上述计算结果检验，$R_{PGA}$一般服从对数正态分布。于是（2）式极限状态功能函数可表示为：

$$Z=\ln R_{PGA}-\ln E_{PGA} \tag{3}$$

于是，结构抗倒塌可靠指标 $\beta_{collapse}$ 可表示为：

$$\beta_{collapse}=\frac{E(\ln R_{PGA})-\ln E_{PGA}}{\sigma_{\ln R_{PGA}}} \tag{4}$$

式中，$E(\ln R_{PGA})$ 为结构抗地震倒塌能力对数均值；$\sigma_{\ln R_{PGA}}$ 为结构抗地震倒塌能力对数标准差。

由可靠指标 $\beta_{collapse}$ 可直接按下式计算结构倒塌概率，

$$P_{collapse}=\Phi(-\beta_{collapse})=\int_{-\infty}^{-\beta_{collapse}}\frac{1}{\sqrt{2\pi}}\exp\left(-\frac{1}{2}x^2\right)\mathrm{d}x \tag{5}$$

当以 PGA 作为地震地面运动强度指标时，按上式计算得到的各算例倒塌概率见表 5。

**不同设防标准框架结构的抗倒塌能力指标** **表 5**

| 算例结构 | | 6B | 6Y | 7B | 7Y | 7.5B | 7.5Y | 8B | 8Y | 8.5B | 8.5Y |
|---|---|---|---|---|---|---|---|---|---|---|---|
| $E(\ln R_{PGA})$ | | 1.615 | 1.613 | 1.686 | 1.938 | 1.751 | 1.995 | 2.317 | 2.534 | 3.021 | 3.253 |
| $\sigma_{\ln R_{PGA}}$ | | 0.454 | 0.453 | 0.456 | 0.453 | 0.439 | 0.440 | 0.480 | 0.460 | 0.527 | 0.460 |
| 罕遇地震 | $\ln E_{PGA}$ | 0.075 | 0.075 | 0.768 | 0.768 | 1.111 | 1.111 | 1.366 | 1.366 | 1.609 | 1.609 |
| | $\beta_{collapse}$ | 3.391 | 3.392 | 2.014 | 2.581 | 1.459 | 2.005 | 1.981 | 2.54 | 2.676 | 3.572 |
| | 倒塌概率（%） | 0.035 | 0.035 | 2.20 | 0.49 | **7.23** | 2.25 | 2.38 | 0.56 | 0.99 | 0.074 |
| 提高一度大震 | $\ln E_{PGA}$ | 0.768 | 0.768 | 1.366 | 1.366 | 1.609 | 1.609 | 1.804 | 1.804 | 1.968 | 1.968 |
| | $\beta_{collapse}$ | 1.865 | 1.863 | 0.701 | 1.261 | 0.32 | 0.875 | 1.068 | 1.584 | 2.00 | 2.792 |
| | 倒塌概率（%） | 3.11 | 3.12 | **24.1** | **10.4** | **37.3** | **19.1** | **14.3** | 5.66 | 2.29 | 0.26 |
| 特大地震 | $\ln E_{PGA}$ | 1.366 | 1.366 | 1.804 | 1.804 | 1.804 | 1.804 | 2.210 | 2.210 | 2.210 | 2.210 |
| | $\beta_{collapse}$ | 0.548 | 0.544 | −0.26 | 0.294 | −0.12 | 0.432 | 0.223 | 0.704 | 1.537 | 2.267 |
| | 倒塌概率（%） | **29.2** | **29.3** | **60.3** | **38.4** | **54.8** | **33.3** | **41.2** | **24.1** | 6.21 | 1.17 |

注：本表是以 PGA 作为地震地面运动强度指标计算的。

需说明的是，表 4 是采用 $S_a(T_1)$ 作为统计指标，直接由倒塌分析结果给出的倒塌率；表 5 是采用 PGA 作为统计指标，按式(3)～式(5) 近似概率模型得到的倒塌概率，两者都是根据有限样本的统计特征。这两种表示方法在统计对象上有差别，PGA 是地震动强烈程度的一个客观特征，与结构无关，直观反映了地震动峰值的随机性；$S_a(T_1)$ 除与地震的地震动强烈程度有关外，还与地震动的频谱特征和结构的一阶弹性周期 $T_1$ 有关，对不同的结构，即使地震输入相同，$S_a(T_1)$ 也不相同，所以对不同结构，地震动的随机性的影响情况也有所不同，即 $S_a(T_1)$ 指标已不单纯反映地震动的随机性。两种表示方法中，地震的频谱特征是通过抽样来体现的，它依赖于抽样的数量和代表性。所以，在比较结构的抗地震倒塌能力时，两种表示方法会存在数值上的差别，但在数量级上是基本一致的。当样本数量足够多时，两者在数值上将各自趋于稳定，但不会完全相同。从工程实用角度，采用 PGA 更客观一些，也更方便。因此，在概率意义上，本文以下采用式(3)～式(5) 近似概率模型，通过倒塌概率来反映结构的抗地震倒塌能力。

### 18.4.3 结构抗倒塌能力的评估

如前所述，$R_{PGA}$为结构的抗力项，是代表结构抗倒塌能力的一个绝对量，为随机变量；$S_a(T_1)_{MCE}$、$S_a(T_1)_{ME\text{-}Ⅰ}$ 和 $S_a(T_1)_{ME\text{-}Ⅱ}$ 或相应的 PGA 为给定荷载项，为确定量；而结构抗地震倒塌储备系数 CMR 和结构抗地震倒塌可靠指标 $\beta_{collapse}$ 是一个相对量。当抗力项 $R_{PGA}$减小（相应设防烈度低），或荷载项 PGA 或 $S_a(T_1)$ 增大（相应遭遇烈度增大），都会使得结构的抗倒塌安全储备系数 CMR 或抗倒塌可靠指标 $\beta_{collapse}$ 减小，倒塌概率增大。所以，结构抗地震倒塌能力评估应同时考虑其绝对抗力和相对安全度。由表 5 和图 9 的不同设防烈度丙类和乙类框架结构抗倒塌能力的分析计算结果可知：

(1) 6 度丙类和乙类框架的绝对抗力最小，但由于其设防地震作用也相对较小，故其抗地震倒塌储备系数 CMR 和可靠指标 $\beta_{collapse}$ 反而相对较高。需注意的是，相对 6 度丙类，6 度乙类框架的抗震能力几乎没有提高，这是由于 6 度框架设计是由重力荷载控制。在 6 度大震下的倒塌率为 0.035%，提高一度大震下的倒塌率为 3.11%，还可接受，但特大地震下倒塌率为 29.3%，难以接受。

(2) 7 度丙类框架的绝对抗力较低，在 7 度大震下的倒塌率为 2.20%，勉强还可接受；提高一度大震下的倒塌率为 24.1%，已不可接受；特大地震下倒塌率为 60.3%，已完全不可接受。同时，需注意的是，相对于 7 度丙类框架，7 度乙类框架实配钢筋几乎不变，而轴压比有所降低，但倒塌率降低不显著。7 度乙类框架在大震下的倒塌率为

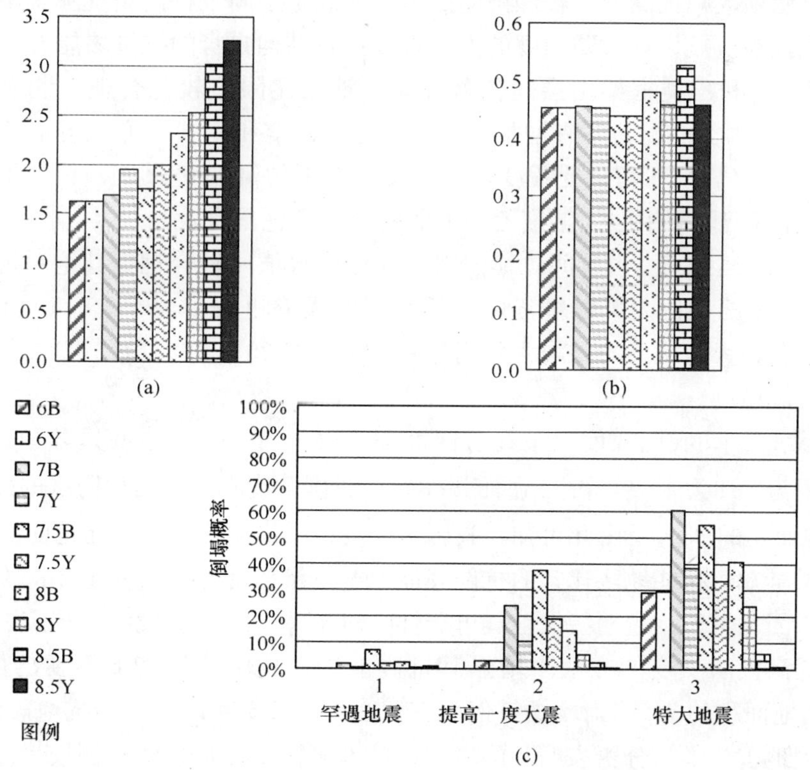


图 9 不同设防标准框架结构的倒塌概率对比

(a) $R_{PGA}$对数的均值；(b) $R_{PGA}$对数的标准差；(c) 倒塌概率

0.49%，可以接受；提高一度大震下的倒塌率为 10.4%，基本不可接受；在特大地震下倒塌率为 38.4%，难以接受。

（3）7.5 度丙类框架的绝对抗力也较低，这与框架柱轴压比较高有关[14]；同时，其设计地震下层间位移较大，X 向 1/560，Y 向 1/610，已接近允许最大值 1/550，所以其抗倒塌安全储备严重不足。在 7.5 度大震下的倒塌率为 7.23%，不可接受；在提高一度大震下的倒塌率为 37.3%，在特大地震下倒塌率为 54.8%，均属于难以接受的程度。相对于 7.5 度丙类，7.5 度乙类框架实配钢筋几乎不变，但轴压比有所降低，其倒塌率有所降低。7.5 度乙类框架在 7.5 度大震下的倒塌率为 2.25%，勉强可以接受；在提高一度大震下的倒塌率为 19.1%，已不可接受；在特大地震下倒塌率为 33.3%，虽然相对于 7.5 度丙类框架有明显降低，但仍难以接受。

（4）总体而言，7 度和 7.5 度设防框架结构的倒塌率较高，将丙类提高到乙类后，倒塌率虽有降低，但仍不可接受，尤其是遭遇特大地震时，倒塌率难以接受。文献［2］调查的汶川特大地震中的 123 个框架结构，大多属于 7 度设防区，其中需停止使用的有 8 个，需立即拆除（含倒塌）的有 9 个，二者占 13.8%，从汶川地震后的社会反映来看，这一比例已难以接受。本文计算结果也表明，相对于其他设防烈度情况，7 度设防框架的震害较为严重。玉树地震区为 7.5 设防区，震害也同样较为严重。

（5）对本文算例而言，7 度及 7.5 度丙类框架的绝对抗力和 6 度及 6 度乙类的绝对抗力相差很小，亟须大幅提高。比较经济的措施是在满足层间位移的情况下大幅降低轴压比限值，或采取提高抗震设计设防烈度的方法直接提高结构自身的抗倒塌能力绝对量。

（6）8 度丙类和乙类框架由层间位移控制，轴压比限值较为合理，且绝对抗力也较高，8 度乙类比 8 丙类的配筋略有增加。8 度乙类在大震和提高一度大震下的倒塌率分别为 0.56%和 5.66%，属于可接受程度；但在特大震下的倒塌率为 24.1%，虽偏高，但仍基本可接受。相对于 8 度丙类，8 度乙类采取的抗震措施是有效的。

（7）8.5 度丙类和乙类由层间位移控制，轴压比限值比较合理，其绝对抗力高，倒塌率均较小。8.5 度乙类比丙类的配筋略有增加，倒塌率进一步减小。8.5 度丙类在大震、提高一度大震和特大地震下的倒塌率分别为 0.37%、2.29%和 6.21%，均属可接受范围，可不再提高抗震措施。

根据文献［15］，我国 6 度以下设防区面积为 201 万 $km^2$，占我国陆地面积 21%，6 度设防区面积为 361 万 $km^2$，占陆地面积 38%；我国 7 度和 7.5 度设防区的面积为 320 万 $km^2$，占陆地面积的 33%。由此可知，我国 7.5 度以下设防区面积占陆地面积达 92%。由本文分析结果可知，这些地区建筑的抗倒塌能力绝对量偏低，且历史上多次发生超设防烈度特大地震，因此考虑到遭遇特大地震可能性，对于 7.5 度以下设防区抗震要求较高的建筑，宜采取提高设防烈度的方法来增强其抗震能力。我国 8 度和 8.5 度的面积为 68 万 $km^2$，仅占陆地面积的 7%。对本文算例而言，8 度和 8.5 度设防区的抗地震倒塌能力强，即使遭遇特大地震，本文分析表明其倒塌率也基本可接受，且总体比例较少。我国 9 度及以上地区的面积为 10 万 $km^2$，占陆地面积的 1%，即使遭遇烈度达 11 度，仅高出设防烈度 2 度，根据本文分析结果推断，其倒塌率应在可接受范围。

## 18.5 结构地震抗倒塌能力的合理目标值

在“小震不坏、中震可修、大震不倒”三水准要求中，保证“大震不倒”是核心目标，但我国目前尚没有明确大震和特大震的倒塌率目标，为此需要采用一个合理的倒塌率指标来统一不同设防烈度和设防类别结构的抗地震倒塌能力，以合理评价《抗规》中的各种抗震措施对增强结构抗倒塌能力的综合效果。

美国 ATC-63 报告建议：“在设防大震下倒塌概率小于 10%即认为达到大震性能的要求”。但需注意的是，美国抗震设防标准较高，如美国加州抗震设防基本相当于我国 9 度设防，大震（MCE）反应谱峰值超过 2G，所以倒塌率 10%是可接受的结果。

根据《汶川地震震害地图集》[16]粗略估计，汶川地震部分地区房屋倒塌率见表 6。由表 6 可知：

(1) 城乡地区房屋倒塌率差别明显：广大农村房屋，绝大部分没有按国家抗震标准进行设防，普遍倒塌，灾情很严重；乡镇房屋有许多也没有按国家抗震标准进行设防，倒塌严重；县城新建房屋大部分按国家抗震标准进行设防，也有许多房屋达不到现行的国家抗震标准，总体倒塌率偏高；市区房屋大部分按国家抗震标准进行设防，倒塌率相对较低。

(2) 除损毁的房屋外，当房屋平均倒塌率大于 40%时，灾情极为严重，社会完全不能接受；当房屋平均倒塌率在 20%～ 40%时，灾情很严重，社会不能接受；当房屋平均倒塌率在 20%左右时，灾情严重，社会也难以接受。

(3) 遭遇烈度远超过设防烈度。以都江堰市区为例，原设防烈度为 7 度，遭遇烈度为 9 度，城镇非住宅用房中倒塌或损毁的房屋超过 400 万平方米，而灾情最严重的北川县城区倒塌或损毁的房屋不到 200 万平方米[14]，即都江堰市区绝对损失比北川县城区还要大，社会难以接受。如提高一度设防（8 度），则遭遇 9 度地震的倒塌率将小于 20%，所以震后提高了设防标准。

**汶川地震部分地区房屋倒塌率的估计** **表 6**

| 序号 | 地点 | 设防烈度 | 遭遇烈度 | 房屋倒塌率估计 | 震后修改设防烈度 |
|---|---|---|---|---|---|
| 1 | 汶川县城区 | 7 度(0.1$g$) | 9 度 | 大部分倒塌率＜20%，<br>少部分倒塌率 20%～40%， | 8 度(0.2$g$) |
| 2 | 汶川县映秀镇 | 7 度(0.1$g$) | 11 度 | 断裂带上倒塌率＞80%<br>断裂带附近倒塌率 60～80% | 8 度(0.2$g$) |
| 3 | 北川县城区 | 7 度(0.1$g$) | 11 度 | 倒塌率＞80% | 8 度(0.2$g$) |
| 4 | 青川县城区 | 7 度(0.1$g$) | 9 度 | 部分区域倒塌率 40%～80% | 7 度(0.15$g$) |
| 5 | 茂县城区 | 7 度(0.15$g$) | 9 度 | 倒塌率 40%～60% | 8 度(0.2$g$) |
| 6 | 安县城区 | 7 度(0.1$g$) | 8 度 | 倒塌率 60%～80% | 7 度(0.15$g$) |
| 7 | 安县安昌镇 | 7 度(0.1$g$) | 8 度 | 倒塌率＞80% | 7 度(0.15$g$) |
| 8 | 都江堰市区 | 7 度(0.1$g$) | 9 度 | 倒塌率 5%～30%，<br>平均 15%～20% | 8 度(0.2$g$) |

续表

| 序号 | 地点 | 设防烈度 | 遭遇烈度 | 房屋倒塌率估计 | 震后修改设防烈度 |
|---|---|---|---|---|---|
| 9 | 平武县城区 | 7度(0.15$g$) | 8度 | 倒塌率20%～80% | 8度(0.2$g$) |
| 10 | 平武县南坝镇 | 7度(0.15$g$) | 8度 | 倒塌率60%～80% | 8度(0.2$g$) |
| 11 | 绵竹市区 | 7度(0.1$g$) | 9度 | 少部分区域20%～40%<br>大部分区域<20% | 7度(0.15$g$) |
| 12 | 绵竹市汉旺镇 | 7度(0.1$g$) | 9度 | 一部分区域60%～80%<br>一部分区域>80% | 7度(0.15$g$) |
| 13 | 罗江县城区* | 6度(0.05$g$) | 7度 | 倒塌率<20% | 7度(0.1$g$) |

注：*者为重灾区，其余均为极重灾区。

另外，汶川地震灾区除房屋倒塌造成人员伤亡外，还存在大量虽未倒但已严重损坏、不具有修复价值或修复可能的房屋，也存在大量虽可修但修复代价较高的房屋。所以，为了尽量减少地震造成的人员伤亡、减少经济损失和灾后重建代价，以及尽量减少因停工停产等原因造成的间接生产损失，在我国综合财力增长较快的经济背景下，抗震设计标准应以人文本，超前规划，适当提高，以逐步达到发达国家的水准。

综合考虑本文倒塌率分析结果、特大地震下的各种地震损失和我国未来发展，本文建议允许倒塌率见表7。

**建议允许倒塌率** **表7**

| 设防类别 | 丙类 | 乙 类 |
|---|---|---|
| 罕遇烈度 | 2% | 1% |
| 提高一度大震 | 10% | 5% |
| 特大地震 | 20% | 10% |

根据表7允许倒塌率，按2001版《抗规》设计的8.5B与8.5Y符合上述要求，而6B与6Y、7B与7Y、7.5B与7.5Y、8B与8Y均不同程度地不满足上述抗倒塌目标要求，建议抗震规范做必要的修改，增强抗震能力。在《抗规》修订时，应明确满足规范最低要求并不能保证结构具有足够的安全度，同时应明确抗地震倒塌是抗震工程的核心任务，否则就会出现按规范最低要求进行所谓“优化设计”而结构抗地震倒塌能力实际很差这种本末倒置的问题。

汶川地震后，提高中小学校舍建筑抗震能力成为社会共识，尤其应保证抗地震倒塌能力。根据本文分析结果，相对于同一地区的其他普通建筑，对于7度以下设防区，建议采取提高一度设防的方法。对本文前述原6度与7度框架结构按提高一度重新设计，如6Y*按7度(0.10$g$)设计，7Y*、7.5Y*按8度(0.20$g$)设计，结构在罕遇地震下的倒塌概率降到0.1%以下，效果很显著，可以满足大震不倒的目标，而且在提高一度大震下的倒塌概率降到2.22%以下，在特大大震下的倒塌概率降到10.4%以下，基本满足表7的允许倒塌率的要求。具体结果见表8。

原 6 度与 7 度框架提高一度重新设计后的抗倒塌能力指标　**表 8**

| 算例结构 | | 6Y* | 7Y* | 7.5Y* |
|---|---|---|---|---|
| 计算设防烈度 | | 7 度(0.10$g$) | 8 度(0.20$g$) | 8 度(0.20$g$) |
| $E(\ln R_{PGA})$ | | 1.938 | 2.534 | 2.534 |
| $\sigma_{\ln R_{PGA}}$ | | 0.453 | 0.460 | 0.460 |
| 罕遇地震 | $\ln E_{PGA}$ | 0.075 | 0.768 | 1.111 |
| | $\beta_{collapse}$ | 4.111 | 3.834 | 3.092 |
| | 倒塌概率(%) | 0.002 | 0.006 | 0.10 |
| 提高度大震 | $\ln E_{PGA}$ | 0.768 | 1.366 | 1.609 |
| | $\beta_{collapse}$ | 2.581 | 2.536 | 2.010 |
| | 倒塌概率(%) | 0.49 | 0.56 | 2.22 |
| 特大地震 | $\ln E_{PGA}$ | 1.366 | 1.804 | 1.804 |
| | $\beta_{collapse}$ | 1.261 | 1.584 | 1.584 |
| | 倒塌概率(%) | 10.4 | 5.66 | 5.66 |

注：表中 * 表示按提高一度重新设计。

## 18.6　结论及需进一步研究的问题

本文按照我国 2001 版《规范》设计了不同设防烈度的丙类和乙类设防的 10 个钢筋混凝土框架结构教学楼，基于 IDA 倒塌率分析进行了抗地震倒塌能力评价，得到以下结论：

(1)相同设防烈度下，丙类设防和乙类设防的框架，由重力荷载起控制作用的 6 度框架的抗地震倒塌能力基本没有差异，其他设防框架的抗地震倒塌能力有一定差异，但差异不显著，即由丙类设防提高到乙类设防，不能明显提高框架结构的抗地震倒塌能力。

(2)低设防烈度、尤其是 7 度(0.10$g$，0.15$g$)设防框架，抗地震倒塌能力需要提高，否则如果轴压比和层间位移都按现行规范的最低要求设计，可能导致抗地震倒塌能力严重不足。

(3)对低设防烈度区的框架，提高设防烈度比提高设防类别能更有效增强结构的抗地震倒塌能力，对 8 设防烈度地区的结构，提高结构的抗震措施基本适宜。

(4)采用 IDA 地震倒塌率分析，应注意 CMR 和 $\beta$ 的相对意义以及设防烈度与实际遭遇烈度的差异性。因基于当前的科技水平，对未来可能遭遇的实际地震动强度的预测还有较大的不确定性，尤其是许多低设防烈度区易发生超设防烈度的特大地震，因此低设防烈度区的重要建筑宜采取提高设防烈度的措施保证其抗地震能力。

影响结构抗倒塌能力的因素很多，比如结构体系与结构布置的差异，主要结构控制参数间耦合，等等。要总结出更普遍适用的结论，还需进一步的深入分析研究，尤其应以抗地震倒塌能力为核心目标，重新对现有的抗震设计技术标准进行系统研究，如：各设防烈度下的各类结构的抗地震倒塌储备系数 CMR 和允许倒塌率(或可靠指标 $\beta$)的可接受水准；进一步加强震害调查分析研究，尤其是对按现行规范设计结构的震害研究。

## 参考文献

[1] 中华人民共和国标准，建筑抗震设计规范(GB 50011—2001)，2001.7.

[2] 王亚勇. 汶川地震建筑震害启示——三水准设防和抗震设计基本要求. 建筑结构学报，2008，29(4)：26-33.

[3] 清华大学土木结构组，西南交通大学土木结构组，北京交通大学土木结构组. 汶川地震建筑震害分析. 建筑结构学报，2008，29(4)：1-9.

[4] 中国工程院土木、水利与建筑学部等. 汶川地震建筑震害调查与灾后重建分析报告. 北京：中国建筑工业出版社，2008.

[5] 中华人民共和国标准. 建筑工程抗震设防分类标准(GB 50023—2008). 2008.

[6] 叶列平，曲哲，陆新征，冯鹏. 提高建筑结构的抗地震倒塌能力的设计思想与方法. 建筑结构学报，Vol. 29，No. 4，42-50，2008.

[7] 叶列平，陆新征，赵世春，李易. 框架结构抗地震倒塌能力的研究. 建筑结构学报，2009，30(6)：67-76.

[8] 汪训流，陆新征，叶列平. 往复荷载下钢筋混凝土柱受力性能的数值模拟. 工程力学，2007，24(12)：76-81.

[9] 叶列平，陆新征，马千里，汪训流，缪志伟. 混凝土结构抗震非线性分析模型、方法及算例. 工程力学，2006，23(sup. 2)：131-140.

[10] Légeron F，Paultre P. Uniaxial confinement model for normal and high-strength concrete columns. Journal of Structural Engineering，ASCE，2005，129(2)：241-252.

[11] 陆新征，叶列平，缪志伟. 建筑抗震弹塑性分析. 北京：中国建筑工业出版社，2009.

[12] Applied Technology Council，Federal Emergency Management Agency. Quantification of Building Seismic Performance Factors. America，FEMA，2008.

[13] 叶列平，马千里，缪志伟. 结构抗震分析用地震动强度指标的研究. 地震工程与工程振动，2009，29(4)：9-22.

[14] 唐代远，陆新征，叶列平，施炜. 柱轴压比对我国RC框架结构抗地震倒塌能力的影响. 北京：第十二届高层建筑抗震技术交流会论文集，2009，10：35-45.

[15] 陶夏新. 我国新的地震区划编图和中国地震烈度区划图(1990). 自然灾害学报，1992，1(1)：99-109.

[16] 汶川地震震害地图集.

# 第 19 章 Chapter 19

# 公路大跨径预应力混凝土桥梁设计指南成果简介

**赵君黎，张喜刚，冯　苠，翟慧娜**

（中交公路规划设计院有限公司，北京　100088）

**提　要**：近几十年来，世界上修建了两百多座跨径 100m 以上的大跨径预应力混凝土梁桥，取得了辉煌的建设成就，为混凝土梁桥跻身四大桥型奠定了坚实的基础。但随着桥梁使用期的增长、桥区自然环境的恶化和汽车荷载的发展增大，大跨径预应力混凝土梁桥上部结构主梁开裂、下挠现象相当普遍，引起了社会各界的高度关注，有关科研项目纷纷立项，众多专家学者潜心研究，从开裂下挠的形成原因、发展机理入手，开展了多项调研、试验和理论分析工作，在研究的基础上提出了多种解决问题的思路和具体措施。2007 年，交通部在项目招标的基础上，委托中交公路规划设计院有限公司组织编写《大跨径预应力混凝土梁桥设计施工技术指南》，要求在和现有的技术标准、设计、施工规范体系相协调的基础上，吸纳最新科研成果，以期能为更好地解决实际问题、建设大跨径梁桥提供集成化、体系化、规范化的实用指南工具。该指南的编制工作目前已经完成征求意见稿和送审稿，本文介绍了指南设计部分的编制背景、章节编排、技术措施和主要成果，请会议代表和各位专家不吝赐教，也可作为指南的一次更广泛地征求意见过程，在此表示感谢。

**关键词**：大跨径；梁桥；设计；指南；规范

## 19.1　指南编制背景

预应力混凝土连续梁、连续刚构、连续梁与刚构组合体系桥梁（以下简称“预应力混凝土梁桥”）是我国桥梁建设中的常见桥型。预应力混凝土梁桥对建筑环境的要求不高，适用范围很广，桥梁跨径在 100～300m 时更是最具竞争力的桥型。据不完全统计，目前我国已建和在建的跨径在 100m 以上的预应力混凝土连续刚构桥已达 90 多座，跨径在 100m 以上的预应力混凝土连续梁桥已有 100 多座。世界范围内共有跨径超过 240m 的特大跨径连续刚构桥 18 座，其中 13 座在中国，占世界总量的 72%。大跨径预应力混凝土梁桥在我国交通建设中发挥着非常重要的作用。

然而，随着预应力混凝土连续结构梁桥，特别是大跨径连续梁桥的大量修建，出现问题的桥梁也越来越多。现役大跨径预应力混凝连续梁、连续刚构桥均是近十几年来才投入运营，总体运营状况良好，但部分桥梁出现了腹板、底板和顶板开裂、跨中下挠等病害，不但在视觉上给人一种不安全感，而且影响了结构的耐久性和适用性。

随着跨径的增大，大跨径预应力混凝土梁桥跨中下挠过大成为一个比较突出的问题。如主跨 270m 的虎门大桥辅航道桥，建成 7 年后实测跨中下挠 260mm；湖北黄石大桥，主跨 245m，建成后 10 年实测下挠 335mm；广东金沙大桥，主跨 120m，建成后 6 年实测下

图 1　苏通大桥辅航道桥（连续刚构桥，主跨 268m）

挠 220mm；三门峡黄河大桥，主跨 140m，建成 10 年跨中下挠 220mm 等。

大跨径预应力混凝土梁桥跨中下挠的问题不仅出现在我国，在国外相当多同类型桥梁中也出现这一病害。如挪威主跨 301m 的 Stolma 桥（连续梁桥），建成 3 年跨中下挠 92mm；主跨 220m 的 Stovset 桥，建成 8 年下挠 220mm；西太平洋帕劳共和国主跨 240.8m 的科偌尔-巴比拉达奥比桥，建成 18 年后下挠 1200mm。

大跨径预应力混凝土梁桥普遍出现的裂缝和跨中下挠病害表明，虽然近年我国预应力混凝土桥梁的建设取得了很大的进步和成绩，但在设计、施工等方面的整体技术水平还有待进一步提高。并且目前我国还没有专门针对大跨径梁桥的规范。在已有设计规范、实践经验和科研成果的基础上，及时总结归纳，提出进一步改进大跨径预应力混凝土梁桥设计及施工的技术措施，编制行业推荐性指南非常必要和紧迫。

## 19.2　章节编排

《大跨径预应力混凝土梁桥设计施工技术指南》分为四个篇幅：总则、设计、施工、附录。

### 19.2.1　总则篇

总则篇包括目的和范围、术语和符号，主要对指南的适用范围、采用的基础设计理论、总体要求、主要术语、符号等作出规定和说明。

### 19.2.2　设计篇

设计篇对大跨径预应力混凝土梁桥的设计作出详细规定。分为以下各章：设计原则、主要技术标准、设计作用和作用效应组合、主要材料、桥梁平纵横设计、结构设计、结构计算分析、耐久性设计、箱梁抗裂设计、箱梁下挠控制、抗震设计、抗风设计。

### 19.2.3 施工篇

施工篇主要对大跨径预应力混凝土梁桥的施工要求作出规定，按照施工顺序分为以下各章：总则、墩身施工、墩顶0号梁段施工、悬浇梁段施工、边跨现浇段施工、合龙段施工、体系转换、施工控制。

### 19.2.4 附录篇

本指南包括如下附录：大跨径预应力混凝土梁桥主要开裂形式、大跨径预应力混凝土梁桥主要参数表、混凝土抗裂性试验方法、混凝土抗渗性快速测定方法、混凝土施工试件耐久性能检验要求、实体混凝土质量检验要求、成桥荷载试验、运营期桥梁结构安全监测。

下面主要介绍指南中有关大跨径预应力混凝土桥梁设计方面的主要规定。

## 19.3 主要内容简介

### 19.3.1 指南采用的设计理论

目前大跨径预应力混凝土梁桥的设计大多参照《公路钢筋混凝土及预应力混凝土桥涵设计规范》(JTG D62—2004)，该规范对公路桥涵钢筋混凝土及预应力混凝土结构构件的设计计算作出规定。本指南的基本设计理论与《公路钢筋混凝土及预应力混凝土桥涵设计规范》保持一致，采用以概率理论为基础的极限状态设计方法，确定汽车活载的设计基准期为100年，并提出设计使用年限，作为开展耐久性设计的基础。

### 19.3.2 结构设计

本指南对大跨径预应力混凝土梁桥的桥跨布置、结构体系选择、主梁设计、预应力布置、桥墩、基础总体设计等作出了指导性规定；从设计角度提出了对施工的要求。

对于主梁设计，根据已有的科研成果和成熟经验，指南中给出了边中跨比例、主梁梁高、顶底板厚度、腹板厚度等设计参数和截面尺寸的建议值。例如，指南6.1节规定，“边中跨的比例宜控制在0.52～0.65。当出现负反力时，应采用配重措施或在过渡墩上设置拉力支座”；6.3.1条中对主梁梁高作出如下规定，“大跨径预应力混凝土梁桥主梁梁高沿纵向采用变高度，并应适当增加1/8～1/3跨区域梁高或腹板厚度。根部梁高控制在主跨跨径的1/15～1/20；跨中（或边跨端部）梁高控制在主跨跨径的1/40～1/60，跨径越大，跨中（或边跨端部）梁高与主跨跨径的比值越小，必要时还应考虑与引桥梁高相协调的要求。”

对于桥墩，指南分别给出了主墩和边墩设计时应考虑的主要因素、截面形式选择的建议等。关于主墩，指南6.4.1条规定，“当采用墩梁固结方式时，通常采用双薄壁墩；当墩高较高时，也采用独柱墩。一方面要满足在各种荷载作用下其强度、刚度和稳定性的要求，另一方面其柔度要能适应由于混凝土收缩、徐变和温度变化等纵向水平荷载引起的效应”；关于边墩，指南 6.5.3 条规定，“边墩通常采用空心截面，根据受力需要，边墩墩身可采用变壁厚，在变壁厚区段应设置一定高度的过渡段。边墩墩底设 2.5～3m 高度的实心段”。

### 19.3.3　结构计算分析

指南规定了大跨径梁桥整体计算分析应采用的计算模型、计算程序的选取、验算内容等；明确了主梁应进行的持久状况承载能力极限状态计算、持久状况正常使用极限状态计算、持久状况主梁应力验算、短暂状况主梁应力验算的具体内容；另外，对墩身、基础设计应进行的计算也做了详细规定。

指南 7.2 节规定：持久状况承载能力极限状态的计算为“预应力混凝土主梁应进行正截面抗弯承载力和斜截面抗剪承载力验算；对锚下等受力复杂部位应进行局部受压区截面尺寸、局部抗压承载力验算及端部锚固区段内的局部应力分析”；持久状况正常使用极限状态的计算为“主梁应进行正截面和斜截面抗裂验算，计算正截面拉应力和斜截面主拉应力；另外，主梁还应进行挠度验算”；持久状况主梁的应力验算为“计算主梁使用阶段正截面混凝土的法向压应力；受拉区预应力钢筋的拉应力；斜截面混凝土的主压应力；结构及受力复杂区域如锚下、0 号梁段等部位局部应力分析”；短暂状况主梁的应力计算为“根据主梁施工过程验算各施工阶段由自重、施工荷载等引起的正截面和斜截面的应力，尤其对主要控制阶段（最大双悬臂、边跨合龙、中跨合龙、合龙后二期恒载）进行分析”。

### 19.3.4　耐久性设计

本指南规定，公路大跨径预应力混凝土梁桥主体结构的设计使用年限宜为 100 年，并对耐久性设计提出了要求。指南中，耐久性设计的基本要求与《公路钢筋混凝土及预应力混凝土桥涵设计规范》（JTG D62—2004）保持一致，对耐久性设计的内容、构造措施、附加防腐蚀措施及设计对检测、养护、维修的要求等作出了详细规定。例如，在构造措施方面，指南 8.2.3 条规定：“结构的形状、布置和构造应有利于阻挡或减轻环境对结构的作用，有利于避免水、水汽和有害物质在混凝土表面的积聚，便于施工时混凝土的捣固和养护，并应有利于保证施工质量”；关于设计对检测、养护、维修的要求，指南 8.4.2 条规定，“不得任意改变桥梁结构的受力状态和提高桥梁的荷载等级”，8.4.3 条规定，“对于处于严重腐蚀环境中的混凝土结构，除了对结构进行定期常规检测外，还应对结构的环境条件、混凝土的性能以及耐久性状况进行跟踪调查和检测”。

### 19.3.5　箱梁抗裂设计

近 20 年来，已建成的大跨径预应力混凝土梁桥普遍存在主梁开裂、下挠等现象。针对这一问题，指南编写组查阅了国内外已有的研究成果，并开展了专题研究。在此基础上，本指南对箱梁抗裂设计作出专门规定，包括设计基本要求、施工基本要求、预应力设计、结构计算及构造与抗裂措施等内容，下面简要列出部分条文的规定。

在设计基本要求中，指南 9.1.1 条规定，“结构设计应在保证结构强度的前提下适当提高结构刚度，控制恒载变形，恒载弹性挠度不宜大于计算跨径的 1/3000～1/4000，汽车荷载作用（含冲击）下的挠度不应超过计算跨径的 1/1000”；在施工基本要求中，指南 9.1.2 条规定，“预应力钢筋的张拉控制应力应符合设计的要求。为防止混凝土早期裂缝，预应力钢筋的张拉可分阶段进行，初次张拉应力宜控制在最终张拉力的 35%，且混凝土强度应大于设计强度的 60%”。

对于箱梁抗裂设计的结构计算，指南 9.3 节规定，“宜采用空间有限元计算方法进行三向预应力混凝土箱梁桥的计算”，“一个箱梁截面可以分为顶板、底板和若干块腹板组成”，如图 2 所示，空间应力状态下应验算的主要部位应力如表 1 所示。

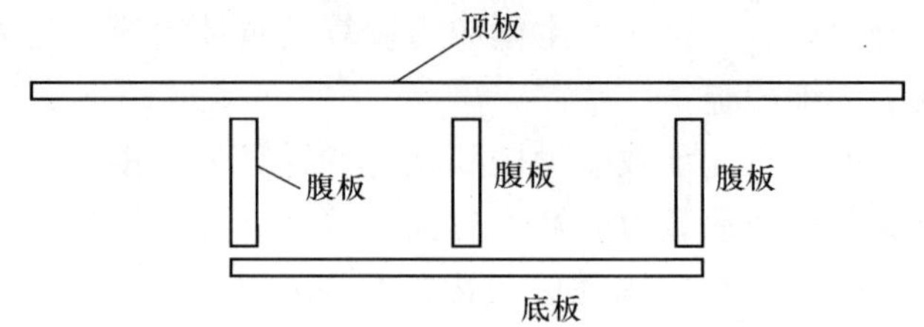


图 2　由“板”表达的单箱双室箱梁截面

**单箱单室箱梁截面应验算的主要部位应力**　　**表 1**

| 构件/受力方向 | 部位 | 验算应力 |
|---|---|---|
| 箱梁顶板纵向面外 | 上缘 | 正应力（面外） |
| 箱梁顶板横向面外 | 上缘<br>下缘 | 正应力（面外） |
| 箱梁顶板面内 | 中间层 | 主应力（面内） |
| 箱梁底板纵向面外 | 下缘 | 正应力（面外） |
| 箱梁底板横向面外 | 上缘<br>下缘 | 正应力（面外） |
| 箱梁底板面内 | 中间层 | 主应力（面内） |
| 箱梁腹板面外 | 中间层 | 主应力（面内） |

在构造与抗裂措施一节中，指南规定了箱梁各板件普通钢筋的最小含筋率，见表 2～表 4。

**箱梁腹板、横隔板的双向钢筋最小含筋率**　　**表 2**

| 混凝土等级 | C40 | C45 | C50 | C55 | C60 | C65 | C70 | C75 | C80 |
|---|---|---|---|---|---|---|---|---|---|
| 最小含筋率 $u_{xymin}$（%） | 1.007 | 1.062 | 1.118 | 1.154 | 1.198 | 1.235 | 1.266 | 1.284 | 1.309 |

**箱梁顶板、底板的横向钢筋最小含筋率**　　**表 3**

| 混凝土等级 | C40 | C45 | C50 | C55 | C60 | C65 | C70 | C75 | C80 |
|---|---|---|---|---|---|---|---|---|---|
| 最小含筋率 $u_{min}$（%） | 0.217 | 0.229 | 0.240 | 0.248 | 0.257 | 0.265 | 0.272 | 0.276 | 0.281 |

**箱梁顶板、底板的纵向钢筋最小含筋率**　　**表 4**

| 混凝土等级 | C40 | C45 | C50 | C55 | C60 | C65 | C70 | C75 | C80 |
|---|---|---|---|---|---|---|---|---|---|
| 最小含筋率 $u_{min}$（%） | 0.51 | 0.54 | 0.56 | 0.58 | 0.605 | 0.62 | 0.64 | 0.65 | 0.66 |

### 19.3.6　箱梁下挠控制

为解决大跨径预应力混凝土梁桥下挠问题，本指南通过调研和专题研究，对箱梁下挠控制作出具体规定。对于桥梁挠度，指南规定，“大跨径预应力混凝土梁桥设计应给出正常使用状态下影响结构下挠的设计主要参数、上部结构施工预抛高值、成桥状态的线形高程数据、汽车荷载及其最大挠度值；宜给出梁桥正常下挠和非正常下挠的限值区间以及开裂后非正常下挠的控制和应对措施”，“设计应根据桥梁建设期间施工和监理反馈的短期材料试验结果修正计算分析模型中的材料参数，降低设计计算和实际结果的差异”；针对箱梁结构剪切开裂产生的非正常下挠，指南规定，“箱梁顶板、腹板和底板均为抗剪构件，须按计算配置抗剪钢筋”，并分别给出了正常使用极限状态和承载能力极限状态时的抗剪配筋要求及最小抗剪钢筋配筋率。

### 19.3.7　大跨径梁桥主要裂缝形式

根据大量样本调查的统计结果，通过从裂缝对结构安全的影响度和出现的频度上进行分析，指南在附录 A 中列出了预应力箱梁常见裂缝的形态和位置，供专业人员参考，以尽可能在设计与施工中避免出现这些裂缝。附录 A 中列出的预应力箱梁常见裂缝包括：底板横向裂缝、腹板下缘竖向裂缝，顶板横向裂缝、腹板上缘竖向裂缝，腹板中部斜裂缝，与底板横向裂缝贯通的腹板裂缝，贯通腹板、底板的螺旋状裂缝，顶、底板纵向裂缝，齿板局部区域裂缝，底板层间横向裂缝，锚下发散裂缝，沿预应力管道裂缝，横隔板裂缝等。

### 19.3.8　大跨径梁桥主要参数表

通过对大量桥梁数据的调研和统计，本指南附录 B 列出了近 20 座典型大跨径预应力混凝土梁桥的主要参数表，供设计人员参考。主要参数包括跨径布置、主梁宽度、梁高、顶底板厚度、混凝土用量、预应力钢筋含量、普通钢筋含量等。

## 19.4　结论

《大跨径预应力混凝土梁桥设计施工技术指南》的编制，使我国近年来公路大跨径预应力混凝土梁桥建设中的很多先进技术和成熟经验得到了进一步总结、提升，并将通过指南的形式迅速在我国公路建设中得以广泛的推广和应用，进一步指导工程实践，为工程质量提供保障。指南颁布实施后，将能有效解决目前大跨径预应力混凝土梁桥存在的问题，提高大跨径梁桥的设计水平，提高桥梁的耐久性，延长桥梁使用寿命。

## 参考文献

[1]　大跨径预应力混凝土梁桥设计施工技术指南. 中交公路规划设计院有限公司主编.

[2]　中交公路规划设计院主编. 公路钢筋混凝土及预应力混凝土桥涵设计规范(JTG D62—2004). 北京：人民交通出版社，2004.

# 第 20 章 Chapter 20

# 活性粉末混凝土梁受力性能*

**郑文忠，卢姗姗，李　莉**

（哈尔滨工业大学　土木工程学院，黑龙江　哈尔滨　150090）

**提　要**：考察了水胶比、钢纤维、石英砂、矿渣粉等对活性粉末混凝土（Reactive Powder Concrete，简称 RPC）强度及流动度的影响，提出了 RPC 的优选配合比，获得了 RPC 的基本力学性能。开展了 6 根钢筋 RPC 简支梁、5 根钢筋 RPC 连续梁和 8 根 GFRP 筋 RPC 简支梁试验。发现试验梁受压边缘 RPC 极限压应变可取 5500$\mu\varepsilon$，高于普通钢筋混凝土梁受压边缘极限压应变 3300$\mu\varepsilon$。发现试验梁截面抵抗矩塑性影响系数与配筋率成正比，建立了考虑配筋率影响的截面抵抗矩塑性影响系数计算公式，可用于较准确计算 RPC 梁开裂弯矩。发现 RPC 梁不论在使用阶段，还是在正截面承载能力极限状态，截面受拉区 RPC 拉应力的合力对刚度、裂缝和承载力的影响都不容忽视，建立了合理考虑受拉区 RPC 拉应力影响的刚度、裂缝及正截面承载力计算公式。发现试验梁等效塑性铰长度与普通混凝土及预应力混凝土梁相近，由于受压边缘极限压应变增大，调幅幅度相对较大，建立了钢筋 RPC 连续梁弯矩调幅系数计算公式。

**关键词**：活性粉末混凝土；梁；正截面承载力；刚度及裂缝；弯矩调幅系数

# MECHANICAL BEHAVIOR OF REACTIVE POWDER CONCRETE BEAMS

W. Z. Zheng，S. S. Lu，L. Li

(School of Civil Engineering，Harbin Institute of Technology，
Harbin 150090，China)

**Abstract**: By studying the effect of water-cement ratio，steel fiber，quartz sand and slag powder on reactive powder concrete (RPC) strength and fluid，the optimal mix and basic mechanical behavior of RPC are given. The experiments of six steel reinforced RPC simply supported beams，five steel reinforced RPC continuous beams and eight GFRP bars reinforced RPC simply supported beams are completed. It is found that ultimate compressive strain at compressiveedge of these beams is 5500$\mu\varepsilon$，which is higher than 3300$\mu\varepsilon$ for normal concrete beams. The influence coefficient of section plastic resistance moment is pro-

* 基金项目：国家教育部长江学者奖励计划资助(2009-37).

portional to reinforcement ratio. Considering the effect of reinforcement ratio, the formula of the influence coefficient is established to calculate the crack moment of RPC beams with greater accuracy. It is also found that the effect of total tensile stress of RPC in the tension zone on stiffness, crack and carrying capacity can not be ignored whether in the design state or in the ultimate state of flexural carrying capacity, and the formula of stiffness, crack, flexural carrying capacity are established considering the effect of tensile stress of RPC in the tension zone. The length of equivalent plastic hinge of these beams is similar to that of normal concrete beams and prestressed concrete beams, and the redistribution is relatively larger for the larger ultimate compressive strain at compressive edge. The formula of moment redistribution of steel reinforced RPC continuous beams is also established.

**Keywords**: reactive powder concrete (RPC); beam; flexural carrying capacity; stiffness and crack; moment redistribution coefficient

## 20.1 引言

活性粉末混凝土（Reactive Powder Concrete，简称 RPC）是由法国 Bouygues 实验室以 Pierre Richard 为首的研究小组于 20 世纪 90 年代率先研制出来的，是一种具有高强度、高韧性、高耐久性等优良性能的水泥基复合材料[1]。

虽然国内外对 RPC 的配制技术进行了一些研究[2~4]，但对基于水胶比、钢纤维、石英砂、矿渣粉等对 RPC 强度及流动度影响的优化配合比尚需深入研究。虽然对 RPC 结构构件开展了一系列的研究[5~7]，但主要是针对轴向受力构件，对 RPC 受弯构件也开展了一些研究，但未给出考虑配筋率影响的截面抵抗矩塑性影响系数计算公式，未给出考虑受拉区拉应力刚度、裂缝及正截面承载力计算公式。GFRP 筋的应力-应变关系一直保持线弹性，对 GFRP 筋 RPC 梁的受力性能研究尚未见报道。虽然国内外学者对普通混凝土连续梁及预应力混凝土连续梁的弯矩调幅计算研究取得成果，但由于 RPC 极限压应变达到了 5500$\mu\varepsilon$，钢筋 RPC 连续梁塑性发展过程应有其自身特点。

## 20.2 RPC 基本力学性能试验研究

### 20.2.1 RPC 的配制

共进行了 13 组 RPC 配合比试验，考察了水胶比、石英砂掺量、矿渣粉及钢纤维掺量等因素对 RPC 强度及其拌合物流动度的影响规律。综合考虑配合比试验中各组试件的抗压强度、流动度性能指标和经济指标，选取如表 1 所示的配合比作为下一步试验的优选配合比[8]。材料选用哈尔滨亚泰牌水泥、哈尔滨晶华石英砂、唐山硅灰、鞍钢 S95 型矿渣粉、山东莱芜汶河 FDN 减水剂和鞍山直径 0.22mm，长度 13mm 的超细超短镀铜平直钢纤维。

RPC 优选配合比 表 1

| 水胶比 | 水泥 | 硅灰 | 矿渣粉 | 石英砂 | 减水剂 | 水 | 钢纤维 |
|---|---|---|---|---|---|---|---|
| 0.2 | 1 | 0.3 | 0.25 | 1.2 | 4% | 0.31 | 2% |

### 20.2.2 试验结果

轴压试验采用尺寸 100mm×100mm×300mm 棱柱体试件。试验在 MTS 试验机上进行，使用特制的钢箍架设位移引伸计，标距取为 150mm。压应力不大于 $0.8f_c$（$f_c$ 为棱柱体强度）时，加载采用力控制。为了得到应力-应变关系曲线的下降段，压应力大于 $0.8f_c$ 时，采用位移控制加载，速度为 0.00015mm/s。力控制和位移控制的采样频率均为 0.5Hz。

拉伸试验采用直接拉伸试验方法，轴拉试件尺寸为 60mm×60mm×400mm，两端分别配有 $\phi$14 钢筋。锚固长度取为 120mm。试验在普通液压拉伸试验机上进行。

RPC 受压应力-应变关系曲线如图 1 所示。RPC 棱柱体抗压强度为 102.28MPa，峰值压应力对应应变 $\varepsilon_c=3600\mu\varepsilon$，泊松比 $\nu=0.22$。

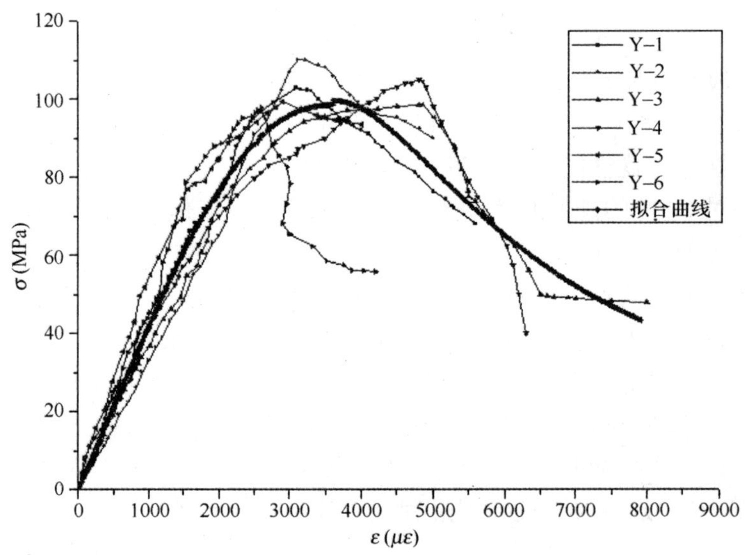


图 1 RPC 受压应力-应变关系曲线

令 $x=\varepsilon/\varepsilon_c$，$y=\sigma/f_c$。经拟合，得到 RPC 受压应力-应变曲线方程为：

$$\begin{cases} y=1.55x-1.20x^4+0.65x^5 & 0\leqslant x\leqslant 1 \\ y=\dfrac{x}{6(x-1)^2+x} & x\geqslant 1 \end{cases} \tag{1}$$

RPC 受拉应力-应变关系曲线如图 2 所示。RPC 抗拉强度为 10.19MPa，峰值拉应力对应应变 $\varepsilon_{t0}=249\mu\varepsilon$。

令 $x=\varepsilon_t/\varepsilon_{t0}$，$y=\sigma_t/f_{ft}$。经拟合，得到 RPC 受拉应力-应变曲线方程为：

$$\begin{cases} y=1.17x+0.65x^2-0.83x^3 & 0\leqslant x<1 \\ y=\dfrac{x}{5.5(x-1)^{2.2}+x} & x\geqslant 1 \end{cases} \tag{2}$$

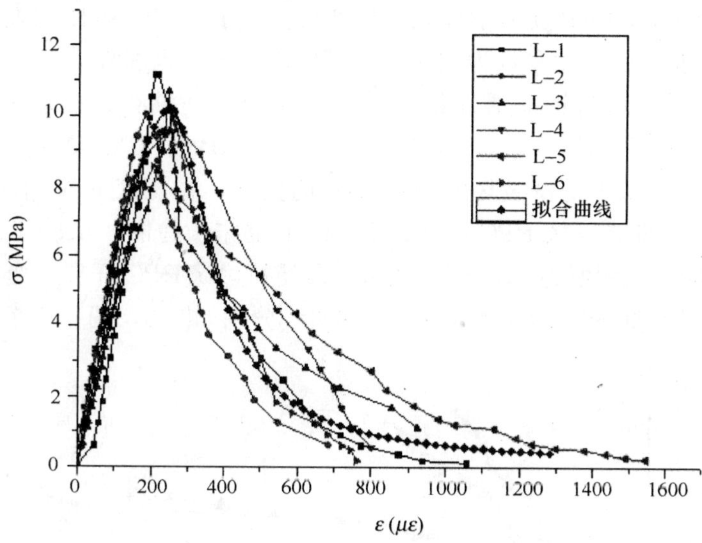


图 2　RPC 受拉应力-应变关系曲线

## 20.3　RPC 简支梁受力性能试验研究

### 20.3.1　试验概况

6 根钢筋 RPC 简支梁的截面尺寸均为 150mm×200mm，计算跨度为 1.2m。纵筋采用 HRB400 级钢筋，箍筋和受压区架立筋采用 HPB235 级钢筋，试验梁配筋情况如表 2 所示，钢筋力学性能指标如表 3 所示。RPC 配比如表 1 所示，力学性能与前述相同。梁的变化参数为配筋率和钢筋直径。试验采用两点对称加载，纯弯区段长度为 0.6m。

试验梁配筋一览表　　表 2

| 梁编号 | L-1 | L-2 | L-3 | L-4 | L-5 | L-6 |
|---|---|---|---|---|---|---|
| 受拉纵筋 | 2A6 | 2C14 | 2C22 | 3C18 | 2C25+1C22 | 2C22+1C40 |
| 架立筋 | 2A6 | 2A6 | 2A 6 | 2A 6 | 2A 6 | 2A6 |
| 箍筋 | A6@150 | A10@80 | A10@80 | A10@80 | C12@60 | C12@60 |

钢筋力学性能指标　　表 3

| 钢　　筋 | A6 | A10 | C12 | C14 | C18 | C25 | C40 | C22 |
|---|---|---|---|---|---|---|---|---|
| 屈服强度 $f_y$（N·mm$^{-2}$） | 275.2 | 275.0 | 480.2 | 476.5 | 467.6 | 460.6 | 465.2 | 478.3 |
| 极限强度 $\sigma_b$（N·mm$^{-2}$） | 310.5 | 310.5 | 619.5 | 615.6 | 624.7 | 660.3 | 621.0 | 613.2 |
| 屈服应变 $\varepsilon_y$（με） | 1310 | 1310 | 2401 | 2382 | 2338 | 2298 | 2301 | 2392 |
| 弹性模量 $E_s$（N·mm$^{-2}$） | 21000 | 21000 | 20000 | 20000 | 20000 | 20000 | 20000 | 20000 |

8 根 GFRP 筋 RPC 简支梁的截面尺寸均为 150mm×280mm，计算跨度为 1.8m。梁底部受拉筋和上部架立筋均选用表面螺旋绕肋 GFRP 筋，箍筋选用 HPB235 级钢筋，试验梁

配筋情况如表 4 所示，GFRP 筋力学性能指标如表 5 所示。RPC 配比如表 1 所示，力学性能与前述相同。试验梁的变化参数为配筋率。试验采用三分点加载。

按照配筋率的大小可将钢筋活性粉末混凝土梁分为适筋梁、超筋梁和少筋梁三类。L-1为少筋梁，L-6 为超筋梁，L-2～L-5 为适筋梁。GFRP 筋 RPC 梁分受拉破坏和受压破坏两类。GL-1～GL-3 属于受拉破坏，即 GFRP 筋拉断，受压区 RPC 未被压碎；GL-4～GL-8属于受压破坏，即受压区 RPC 被压碎，GFRP 筋未被拉断。试验梁的纯弯区段受压边缘 RPC 极限压应变实测值为 5500$\mu\varepsilon$，纯弯区段受拉边缘 RPC 开裂应变实测值为 750$\mu\varepsilon$。钢筋 RPC 梁及 GFRP 筋 RPC 梁跨中荷载-位移曲线分别如图 3，图 4 所示。

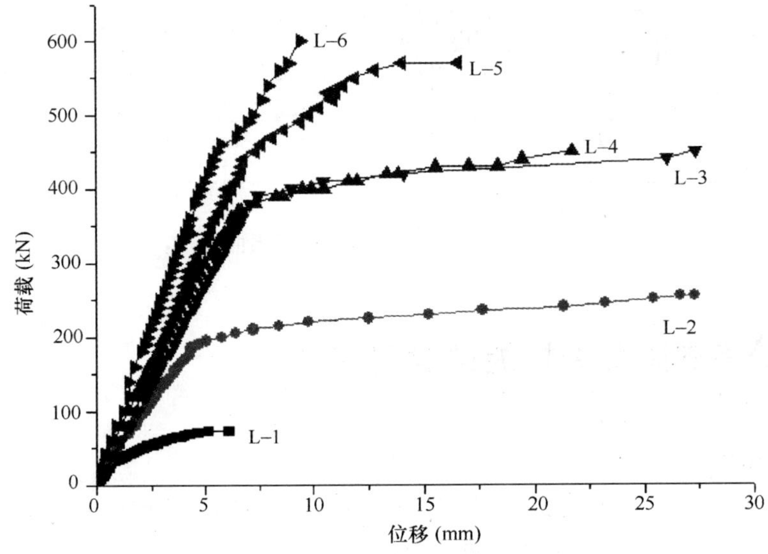


图 3 钢筋 RPC 梁跨中荷载-位移曲线

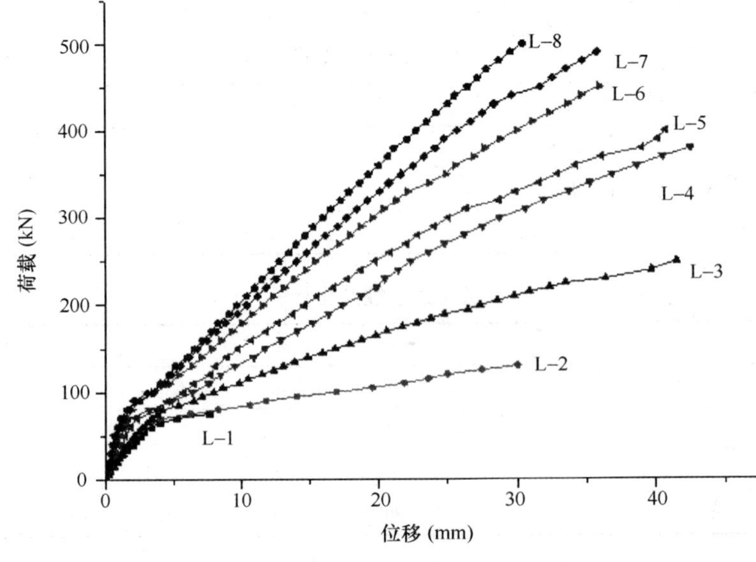


图 4 GFRP 筋 RPC 梁跨中荷载-位移曲线

试验梁配筋一览表　表 4

| 梁编号 | 受拉纵筋 | 架立筋 | 箍　筋 | 受拉纵筋配筋率 |
|---|---|---|---|---|
| GL-1 | $2\phi_f 5.5$ | $2\phi_f 5.5$ | $\phi$12@100 | 0.13% |
| GL-2 | $4\phi_f 5.5$ | $2\phi_f 5.5$ | $\phi$12@100 | 0.25% |
| GL-3 | $2\phi_f 12$ | $2\phi_f 5.5$ | $\phi$12@80 | 0.61% |
| GL-4 | $4\phi_f 12$ | $2\phi_f 5.5$ | $\phi$12@50 | 1.31% |
| GL-5 | $5\phi_f 12$ | $2\phi_f 5.5$ | $\phi$12@50 | 1.61% |
| GL-6 | $6\phi_f 12$ | $2\phi_f 5.5$ | $\phi$12@25 | 1.97% |
| GL-7 | $5\phi_f 14$ | $2\phi_f 5.5$ | $\phi$12@25 | 2.21% |
| GL-8 | $6\phi_f 14$ | $2\phi_f 5.5$ | $\phi$12@25 | 2.69% |

GFRP 筋力学性能指标　表 5

| GFRP 筋 | $\phi_f 5.5$ | $\phi_f 12$ | $\phi_f 14$ |
|---|---|---|---|
| 极限强度 $f_{fu}$（MPa） | 1159 | 990 | 836 |
| 弹性模量 $E_f$（MPa） | $4.94\times10^4$ | $4.76\times10^4$ | $5.00\times10^4$ |

## 20.3.2　开裂荷载及截面抵抗矩塑性影响系数

RPC 梁开裂弯矩 $M_{cr}$ 按下式计算

$$M_{cr}=\gamma_m f_t W_0=f_t W_s \tag{3}$$

$W_0$ 为梁截面对受拉边缘弹性抵抗矩；$W_s$ 为考虑混凝土受拉区塑性变形影响的梁截面对受拉边缘弹塑性抵抗矩；$\gamma_m$ 为梁截面抵抗矩塑性影响系数。由公式（3）得：

$$\gamma_m=W_s/W_0 \tag{4}$$

RPC 梁达到开裂荷载时，受压边缘 RPC 的压应力很小，处于弹性阶段，应力分布近似为三角形。受拉区 RPC 应力分布为曲线形，为了简化计算，近似为三角形分布和梯形分布。图 5 中 $\varepsilon_c$ 为截面开裂时受压边缘应变，$\varepsilon_{t0}$ 为峰值拉应变（$\varepsilon_{t0}=249\mu\varepsilon$），$\varepsilon_s$ 为截面开裂时纵向钢筋拉应变，$\varepsilon_{tu}$ 为 RPC 抗弯初裂拉应变（$\varepsilon_{tu}=750\mu\varepsilon$），$D$ 为截面压区 RPC 合力，$T_1$ 为截面受拉区上三角 RPC 合力，$T_2$ 为截面受拉区梯形 RPC 合力，$T_s$ 为钢筋合力，$x_0$ 为截面受压区高度，$x_t$ 为截面受拉区高度。

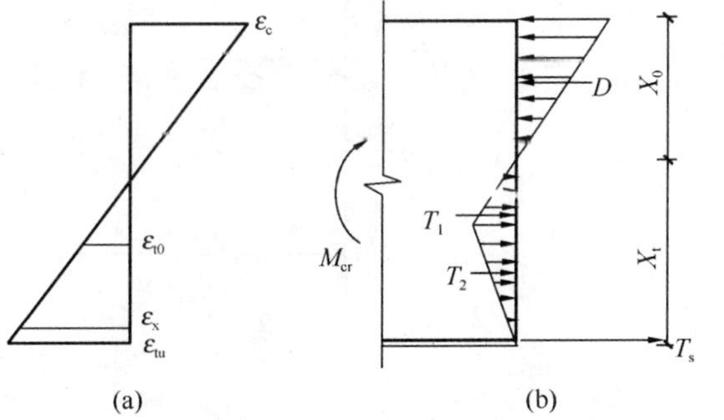


图 5　开裂荷载下截面应力应变分布图
(a) 应变分布；(b) 应力分布

根据几何关系，受拉区 RPC 峰值拉应力对应的应变 $\varepsilon_{t0}$ 与截面受拉边缘 RPC 开裂应变实测值 $\varepsilon_{tu}$ 的比值即受拉区上三角部分与拉区高度的比值为 $\varepsilon_{t0}/\varepsilon_{tu}=249/750=0.33$[10]，因此受拉区上三角部分的高度 $x_1$ 占总拉区高度 $x_t$ 的 33%，受拉区梯形部分的高度 $x_2$ 占总拉区高度 $x_t$ 的 67%。根据试验梁在开裂荷载作用下受压边缘 RPC 压应变和受拉区纵向筋拉应变，可得到此时的截面受压高度 $x_0$，由 RPC 受压应力-应变关系和受拉筋的应力-应变关系可得到RPC压应力合力和受拉筋合力，再根据截面力的平衡条件，可得到受拉边缘RPC的拉应力，取其平均值为$0.1f_t$。由截面力的平衡及弯矩平衡公式，可以计算得到截面弹塑性抵抗矩。再由公式（7）可得到$\gamma_m$的值。

通过与配筋率拟合得到钢筋RPC梁的$\gamma_m$计算公式为

$$\begin{cases}\gamma_m=1.1+18.4\rho & \rho<4.3\% \\ \gamma_m=1.89 & \rho\geqslant4.3\%\end{cases} \tag{5}$$

通过与配筋率拟合得到GFRP筋RPC梁的$\gamma_m$计算公式为

$$\gamma_m=1.1+6\rho\leqslant1.89 \tag{6}$$

开裂荷载的计算公式为

$$M_{cr}=\gamma_m f_t W_0 \tag{7}$$

### 20.3.3 正截面承载力计算

RPC有较高的抗拉强度，开裂部分的RPC由于钢纤维的存在，仍有一定拉应力存在。进行RPC梁正截面承载力计算时，宜考虑RPC受拉的贡献。

（1）截面受压区图形的等效应力系数

首先，根据合力大小和作用点不变的原则、RPC受压和受拉应力-应变关系方程以及试验梁受压边缘的实测极限压应变，将RPC梁正截面压区应力图形等效为矩形应力图形以便进行受弯承载力计算。图6为截面应力分布图和等效后的截面应力分布图。$\alpha$、$\beta$为受压区等效应力图形系数，根据RPC应力-应变关系和截面平衡条件得到$\alpha=0.9$，$\beta=0.77$。等效后RPC压区应力为$\alpha f_c$，压区高度为$x$。

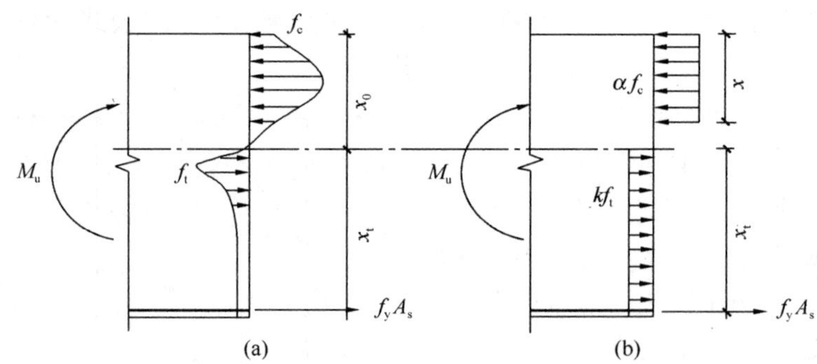


图6 截面应力分布图

(a) 实际应力分布；(b) 等效应力分布

（2）截面受拉区应力图形的等效系数

如图6（a）所示，受拉区实际应力分布为曲线形，为了简化计算，并考虑塑性发展的影响，将曲线拉应力图等效为矩形拉应力图，等效图形的应力高度$kf_t$可通过截面平衡条件和试验结果反推得到。为了简化计算，综合试验梁的计算结果，偏于安全地取$k=0.25$。

（3）正截面受弯承载力计算公式

根据简化的截面应力分布图6（b），由平衡条件得到钢筋RPC适筋梁的正截面受弯承载力计算公式如下

$$\begin{cases}f_c bx=f_y A_s+0.25f_t b\left(h-\dfrac{x}{0.77}\right)\\ M_u=f_c bx\left(h_0-\dfrac{x}{2}\right)-0.25f_t b\left(h-\dfrac{x}{0.77}\right)\left[0.5\left(h-\dfrac{x}{0.77}\right)-a_s\right]\end{cases}\tag{8}$$

对钢筋 RPC 少筋梁，由于达到极限荷载时受拉区裂缝开展宽度较大，因此不考虑受拉区 RPC 作用，其承载力的计算可通过迭代方法得到。虽然超筋梁破坏时裂缝开展宽度不大，但由于 RPC 较高的抗拉强度，计算其承载力时仍考虑拉区 RPC 的作用，近似取 $k=0.25$，根据截面平衡条件，计算公式如下

$$M_{u,max}=\alpha f_c bh_0^2\xi_b\ (1-0.5\xi_b)\ -0.25f_t\left(h-\frac{x}{\beta}\right)\left[\frac{1}{2}\left(h-\frac{x}{\beta}\right)-a_s\right]\tag{9}$$

由于 GFRP 筋的应力-应变关系一直保持线弹性，GFRP 筋 RPC 梁的破坏形式与钢筋混凝土梁有所不同，大致可分为三种：①当受拉 GFRP 筋与受压区边缘 RPC 同时达到极限应变时，发生界限破坏；②当 GFRP 筋配筋率较小，受拉 GFRP 筋达到极限拉应变而被拉断，而受压区边缘 RPC 尚未达到极限压应变，未被压碎；③当 GFRP 筋配筋率较大，受压区边缘 RPC 达到极限压应变而被压碎，而受拉 GFRP 筋未被拉断。图 7 给出了三种破坏形式的截面应力和应变分布图。

界限破坏是区分受拉破坏和受压破坏的临界状态，此时的配筋率称为界限配筋率。如图 7（b）所示，由截面平衡条件及平截面假定推导得到界限配筋率公式如下

$$\rho_{fb}=\frac{\alpha_1\beta_1 f_c}{f_{fu}}\frac{\varepsilon_{cu}}{\varepsilon_{cu}+\varepsilon_{fu}}+\frac{kf_t}{f_{fu}}\left(\frac{\varepsilon_{cu}}{\varepsilon_{cu}+\varepsilon_{fu}}-\frac{h}{h_0}\right)\tag{10}$$

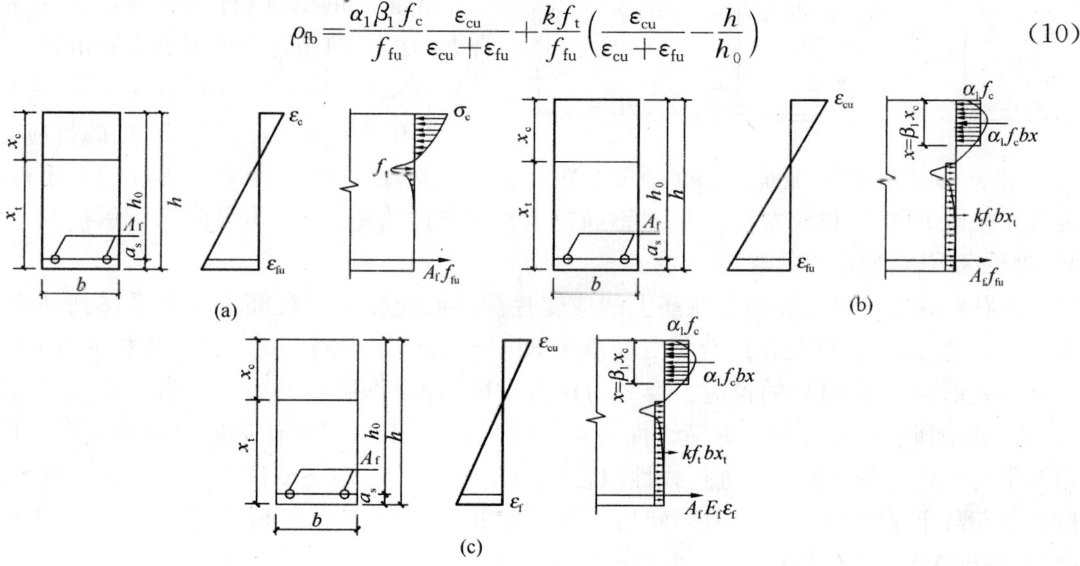


图 7　GFRP 筋 RPC 梁截面应力和应变分布

（a）受拉破坏；（b）界限破坏；（c）受压破坏

受压破坏时，如图 7（c）所示，根据平截面假定及受力平衡得

$$f_f=\frac{1}{2A_f}\left(\sqrt{(kf_t bh-E_f\varepsilon_{cu}A_f)^2+4E_f A_f\varepsilon_{cu}h_0\ (kf_t+\alpha_1\beta_1 f_c b)}-(kf_t bh+E_f\varepsilon_{cu}A_f)\right)\leqslant f_{fu}\tag{11}$$

由此得到受压破坏时正截面承载力公式如下

$$M_u = f_f A_f (h_0 - x/2) + k f_t b x_t (x_t/2 + x/\beta_1 - x/2) \tag{12}$$

受拉破坏时，如图 7（a）所示，GFRP 筋达到极限抗拉强度，而 RPC 受压边缘未达到极限压应变，因此，不能按照等效应力图系数的方法求正截面承载力，可以按以下思路进行计算。首先，先不计受拉区 RPC 的贡献，对受压区 RPC 进行条带积分，使其合力等于 GFRP 筋的合力，结合平截面假定，求得受压区高度 $x'_c$，由此求得受拉区高度 $x'_t$。再由受压区 RPC 合力等于受拉区 RPC 合力与 GFRP 筋合力之和，重新求得 $x'_c$，如此重复数次达到要求的精度，就可得到受压区高度 $x_c$，从而求得正截面承载力。

### 20.3.4 变形计算

由于 RPC 的抗拉强度较高，且掺有钢纤维，所以 RPC 梁截面弯矩由受拉筋和 RPC 共同承担。在使用荷载 $M$ 作用下，梁截面应力分布如图 8 所示，此时的截面拉区 RPC 应力分布情况我们不去考虑，在图中示意为矩形分布。受压区混凝土平均应力为 $\omega\sigma_{ck}$，压区高度 $\zeta h_0$，压区合力作用点到钢筋面积形心的力臂为 $\eta h_0$。

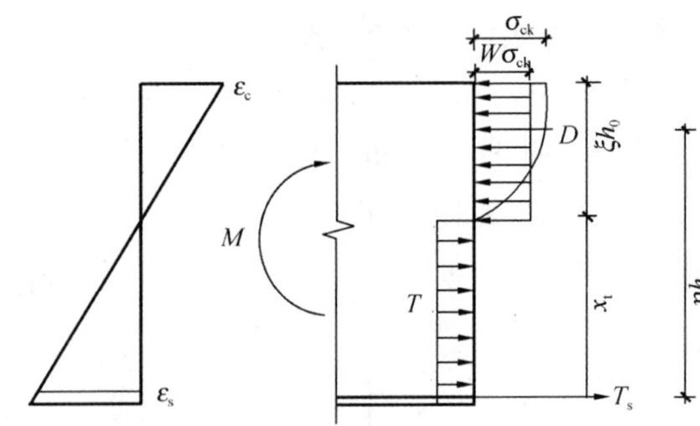

图 8　使用荷载下正截面应力和应变分布图

根据截面力矩平衡条件，拉区 RPC 合力和受拉筋合力向受压区 RPC 合力作用点取矩得到

$$M = M_T + M_s = M_T + \sigma_s A_s \eta h_0 \tag{13}$$

截面压区 RPC 合力和截面拉区 RPC 合力向纵向钢筋合力作用点取矩得到

$$M = M_c - M'_T = \xi \eta h_0 \omega \sigma_c b h_0 - M'_T \tag{14}$$

式中 $M_T$ 为截面拉区 RPC 的拉应力合力 $T$ 对压区 RPC 合力作用点的力矩；$M'_T$ 为截面拉区 RPC 的拉应力合力 $T$ 对钢筋拉力作用点的力矩；$M_s$ 为纵向筋所承担的弯矩；$M_c$ 为压区 RPC 所承担的弯矩。

由粘贴在梁纯弯段纵向受拉筋上的应变片测得的应变及受拉筋本构关系得到其应力，乘以内力臂 $\eta h_0$，得到受拉筋合力对受压区 RPC 合力作用点的力矩，截面承担的总弯矩 $M$ 与其的差值为拉区 RPC 的拉应力合力对压区 RPC 合力作用点的力矩。由试验数据可知，$M_T$ 在 M 中所占比例与配筋率呈线性关系，将各钢筋 RPC 梁在使用阶段各级荷载下的比值取平均值，然后与配筋率进行线性拟合得到公式（15），将各 GFRP 筋 RPC 梁在使用阶段各级荷载下的比值取平均值，然后与配筋率进行线性拟合得到公式（16）。再由公式（13）得到 $M_s$ 的计算公式。

$$M_T/M = 0.228 - 2.05\rho \tag{15}$$

$$M_T/M = 0.36 - 4.3\rho \tag{16}$$

$M'_T$ 为截面拉区 RPC 的拉应力合力 $T$ 对钢筋拉力作用点的力矩。令 $a$ 等于拉区 RPC 合力作用点至筋拉力作用点的力臂，$b$ 等于拉区 RPC 合力作用点至压区 RPC 合力作用点的力臂，则 $M'_T/M_T = a/b$，由粘贴在截面压区的混凝土应变片测得到 RPC 压应变，并根据 RPC 受压本构关系，可计算得到压区 RPC 合力作用点及截面中性轴位置，假定拉区 RPC 拉应力合力点在截面拉区的中央，由此得到使用荷载下的 $a/b$ 值，取平均值，与配筋

率拟合得到钢筋 RPC 梁 $M'_T/M_T$ 计算公式

$$M'_T/M_T=2.85-2.03\,(3.35\times10^{-13})^{\rho} \tag{17}$$

与配筋率拟合得到 GFRP 筋 RPC 梁 $M'_T/M_T$ 计算公式

$$M'_t/M_t=1/\,(1.5+65.6\rho) \tag{18}$$

对于普通钢筋混凝土梁 $\Psi=1.1\,(1-0.8M_{cr}/M)$，对于 RPC 梁暂按此公式进行计算，只是公式中的 $M$ 应减去 $M_t$，即 $\Psi=1.1\,(1-0.8M_{cr}/M_s)$；对于普通钢筋混凝土梁 $\eta=1-0.4\sqrt{\alpha_E\rho}$，对于 RPC 梁暂按此公式进行计算。$\alpha_E\rho/\zeta=0.2+6\alpha_E\rho$，对于 GFRP 筋 RPC 梁暂按此公式进行计算，对钢筋 RPC 梁 $\alpha_E\rho/\zeta=\,(0.2+6\alpha_E\rho)\,/0.62$。

由试验数据，根据以上规定得到钢筋 RPC 梁刚度计算公式如下

$$B_s=\frac{E_sA_sh_0^2}{\frac{\Psi}{\eta}\frac{M_s}{M}+\frac{\alpha_E\rho}{\zeta}\frac{M_c}{M}} \tag{19}$$

由试验数据，根据以上规定得到 GFRP 筋 RPC 梁刚度计算公式如下

$$B_s=\frac{0.83E_sA_sh_0^2}{\frac{\Psi}{\eta}\frac{M_s}{M}+\frac{\alpha_E\rho}{\zeta}\frac{M_c}{M}} \tag{20}$$

### 20.3.5　裂缝开展及分布

平均裂缝间距计算公式可按《混凝土结构设计规范》(GB 50010—2002) 中公式进行计算。

平均裂缝宽度 $w_m$ 等于构件裂缝区段内钢筋的平均伸长与相应水平处构件侧表面混凝土平均伸长的差值，推导得

$$w_m=\alpha_c\Psi\frac{\sigma_s}{E_s}l_m \tag{21}$$

式中，$\alpha_c$ 按照普通混凝土取 0.85，$\sigma_s$ 和 $\Psi$ 可按刚度计算公式中的规定计算得到。

使用荷载作用下的短期最大裂缝宽度 $w_{s,max}$ 可根据平均裂缝宽度乘以考虑裂缝开展不均匀影响的扩大系数 $\tau_s$ 求得，即

$$w_{s,max}=\tau_s w_m \tag{22}$$

短期荷载作用下的裂缝扩大系数 $\tau_s$ 可按裂缝宽度的概率分布规律确定。各试验梁纯弯段各条裂缝宽度 $w_i$ 与对应的平均裂缝宽度 $w_m$ 的比值为 $\tau_i$，其分布规律近似呈正态分布，若按 95% 的保证率考虑，可求得对钢筋 RPC 梁 $\tau_s=1.60$，对 GFRP 筋 RPC 梁 $\tau_s=1.69$，由此可得短期荷载作用下最大裂缝宽度为

$$w_{s,max}=\tau_s\times0.85\Psi\frac{\sigma_s}{E_s}l_m \tag{23}$$

## 20.4　钢筋 RPC 连续梁塑性性能试验研究

### 20.4.1　试验概况

钢筋 RPC 两跨连续梁的试验参数为中支座控制截面弯矩调幅系数。5 根试验梁截面尺

寸均为 180mm×220mm，计算跨度 3.2m。纵筋采用 HRB400 级钢筋，箍筋采用 HRB335 级钢筋。其基本参数如表 6 所示。采用在试验梁两跨中点单点对称施加集中荷载[11]。

**连续梁试件设计参数** **表 6**

| 试件编号 | 试件长度（mm） | 截面尺寸（mm） | 预估调幅系数 | 中支座配筋 | 跨中纵筋 | 箍筋 |
|---|---|---|---|---|---|---|
| LL-1 | 3400 | 180×220 | 0.21 | 2C22+1C14 | 3C22 | B12@80 |
| LL-2 | 3400 | 180×220 | 0.29 | 2C22 | 3C22 | B12@80 |
| LL-3 | 3400 | 180×220 | 0.43 | 2C18 | 3C22 | B12@80 |
| LL-4 | 3400 | 180×220 | 0.57 | 2C14 | 3C22 | B12@80 |
| LL-5 | 3400 | 180×220 | 0.68 | 2B10 | 3C22 | B12@80 |

试验梁达到极限状态有两个标志，分别为设计用承载能力极限状态和真实承载能力极限状态，当中支座控制截面受压边缘混凝土达到极限压应变 $\varepsilon_{cu}=5500\times10^{-6}$ 时为设计用承载能力极限状态，这里称为破坏标志Ⅰ，此时还可以继续承担荷载；当试验梁任一跨出现荷载开始减小，变形持续增大时为真实承载能力极限状态，称为破坏标志Ⅱ。

### 20.4.2 塑性铰长度及塑性转角

（1）基于设计用承载能力极限状态

在各试验梁中支座控制截面 800mm 长范围内的纵向钢筋表面粘贴了间距为 25mm 的应变片，可测得中支座控制截面两侧纵向钢筋应变不小于屈服应变的长度，即中支座两侧的实际塑性铰区长度。$\varphi_d$ 为设计用承载能力极限状态下的极限曲率，$\varphi_y$ 为屈服曲率，以实际塑性铰区内非弹性曲率所包围面积相等为原则，将非弹性曲率等效为矩形分布，由此可确定试验梁在设计用承载能力极限状态下的等效塑性铰区长度。塑性铰的转角为 $\theta_{pd}=(\varphi_d-\varphi_y)\ l_{pd}$。取 5 根试验梁基于设计用承载能力状态时两侧等效塑性铰区长度的平均值，则 $L_{pd}=0.40h_0$。

（2）基于真实承载能力极限状态

$\varphi_u$ 为试验梁真实承载能力状态下极限曲率，按照与实际塑性铰区非弹性曲率所包围面积相等的原则，将曲率等效为矩形分布。从而可得到中支座控制截面两侧等效塑性铰区长度 $l_{pu}$。塑性转角为 $\theta_{pu}=(\varphi_u-\varphi_y)\ l_{pu}$。取 5 根试验梁基于真实承载能力状态时两侧等效塑性铰区长度的平均值，则 $L_{pu}=0.45h_0$。

### 20.4.3 塑性调幅系数的计算

（1）基于设计用承载能力状态下的弯矩调幅系数

以设计用承载能力极限状态弯矩调幅系数 $\beta_d$ 为纵坐标，以达到该状态时相对塑性转角 $\theta_{pd}/h_0$ 为横坐标，得到试验梁试验点分布，数据拟合得到基于设计用承载能力极限状态下以中支座相对塑性转角 $\theta_{pd}/h_0$ 为自变量的弯矩调幅系数的计算公式。

$$\begin{cases}\beta_d=0.145\left(\dfrac{\theta_{pd}}{h_0}\times10^5\right)+0.05 & \dfrac{\theta_{pd}}{h_0}\leqslant5.1\times10^{-5}\\ \beta_d=0.649 & \dfrac{\theta_{pd}}{h_0}>5.1\times10^{-5}\end{cases}\tag{24}$$

以中支座混凝土相对受压区高度 $\xi_i$ 为横坐标，以调幅系数 $\beta_d$ 为纵坐标，可得到 5 根钢筋 RPC 连续梁 $\beta_d$ 与 $\xi_i$ 试验点分布，如图 9 所示。根据图中 $\beta_d$ 与 $\xi_i$ 试验点分布，可得到调幅系数 $\beta_d$ 与中支座混凝土相对受压区高度 $\xi_i$ 关系下包线，并将其绘制于图 9。

由此可得到以中支座混凝土相对受压区高度 $\xi_i$ 为自变量的中支座弯矩调幅系数 $\beta_d$ 的计算公式为：

$$\beta_d=\begin{cases}0.65 & \xi_i<0.038\\0.05 & \xi_i>0.181\end{cases}\tag{25}$$

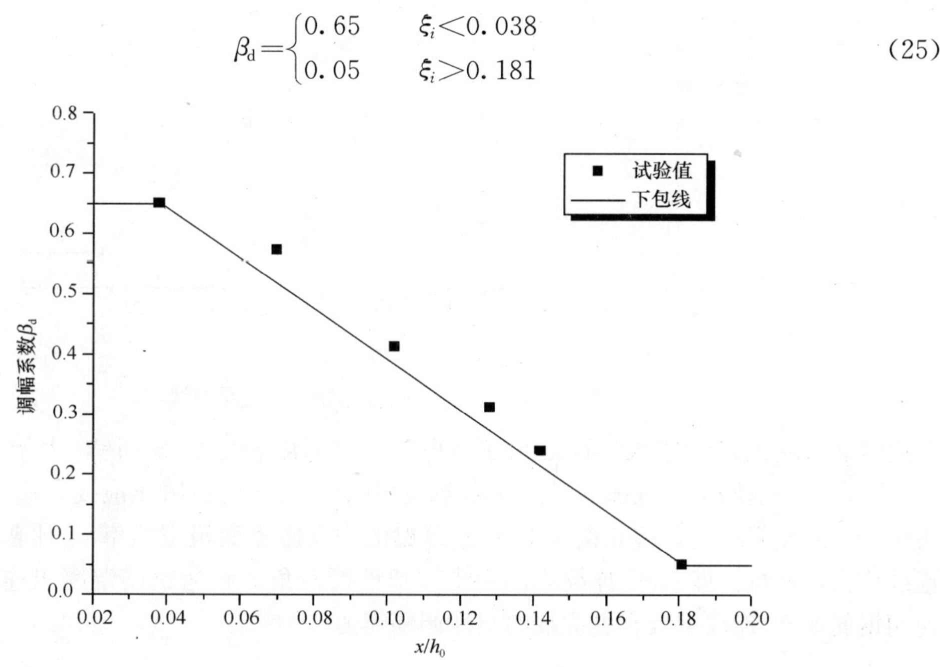


图 9　$\beta_d$ 与 $x/h_0$ 试验点分布及关系曲线

(2) 基于真实承载能力极限状态的弯矩调幅系数

以真实承载能力极限状态弯矩调幅系数 $\beta_u$ 为纵坐标，以达到该状态时相对塑性转角 $\theta_p/h_0$ 为横坐标，得到试验梁试验点分布，数据拟合得到基于真实承载能力极限状态下以中支座相对塑性转角 $\theta_p/h_0$ 为自变量的弯矩调幅系数的计算公式

$$\begin{cases}\beta_u=0.99\left(\dfrac{\theta_p}{h_0}\times10^4\right)+0.05 & \dfrac{\theta_p}{h_0}\leqslant6.06\times10^{-5}\\[2ex]\beta_u=0.652 & \dfrac{\theta_p}{h_0}>6.06\times10^{-5}\end{cases}\tag{26}$$

以中支座混凝土相对受压区高度 $\xi_i$ 为横坐标，以调幅系数 $\beta_u$ 为纵坐标，可得到 5 根钢筋 RPC 连续梁 $\beta_u$ 与 $\xi_i$ 试验点分布，如图 10 所示。根据图中 $\beta_u$ 与 $\xi_i$ 试验点分布，可得到调幅系数 $\beta_u$ 与中支座混凝土相对受压区高度 $\xi_i$ 关系下包线，并将其绘制于图 10。

由此可得到以中支座混凝土相对受压区高度 $\xi_i$ 为自变量的中支座弯矩调幅系数 $\beta_u$ 的计算公式为：

$$\beta_u=\begin{cases}0.653 & \xi_i<0.053\\0.05 & \xi_i>0.216\end{cases}\tag{27}$$

综合以上试验现象及理论分析发现，钢筋 RPC 连续梁的调幅能力比相同截面相同配筋的

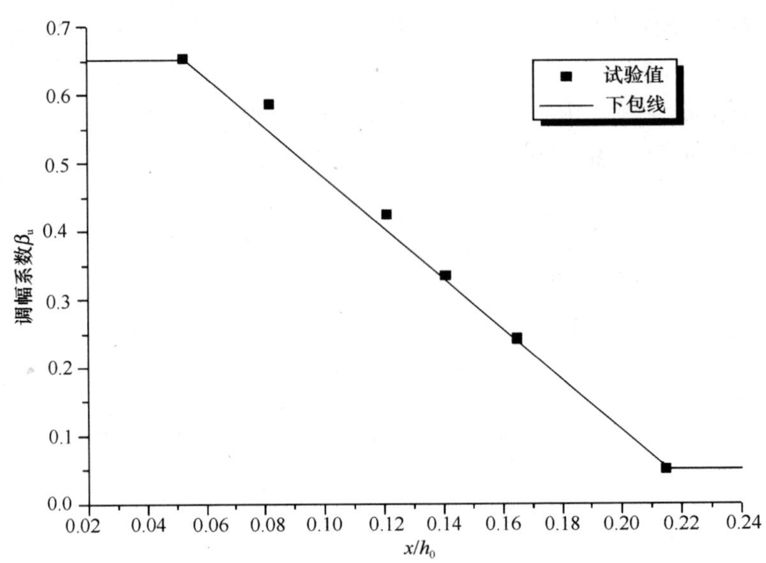


图 10 $\beta_u$ 与 $x/h_0$ 试验点分布及关系曲线

普通钢筋混凝土连续梁大，这是由于 RPC 的极限压应变为 5500$\mu\varepsilon$，大于普通混凝土的 3300$\mu\varepsilon$，对于适筋梁，截面的极限曲率 $\varphi_u$ 取决于混凝土的极限压应变，而屈服曲率 $\varphi_y$ 取决于钢筋的屈服应变，因此钢筋 RPC 连续梁的中支座截面极限曲率 $\varphi_u$ 比普通钢筋混凝土连续梁大，使得钢筋 RPC 连续梁的中支座塑性铰转角比普通钢筋混凝土连续梁大。由此说明钢筋 RPC 连续梁具有更优越的塑性调幅能力。

## 20.5 结论

（1）试验用 RPC 棱柱体达到峰值压应力对应应变 $\varepsilon_0=3600\mu\varepsilon$，高于普通混凝土。试验用 RPC 峰值拉应力对应拉应变 $\varepsilon_{t0}=249\mu\varepsilon$，高于普通混凝土。

（2）试验梁受压边缘 RPC 极限压应变可取 5500$\mu\varepsilon$，高于普通混凝土。矩形截面试验梁截面抵抗矩塑性影响系数介于 1.1 至 1.89，且与受拉纵筋配筋率成正比。

（3）RPC 梁受拉区拉应力对试验梁的刚度、裂缝及正截面承载力的影响应予考虑，建立了考虑这一影响的刚度、裂缝及正截面承载力计算公式。

（4）钢筋 RPC 连续梁等效塑性铰长度与普通混凝土及预应力混凝土梁相近，由于受压边缘极限压应变增大，调幅幅度相对较大，建立了钢筋 RPC 连续梁弯矩调幅系数计算公式。

## 参考文献

[1] Richard P, Cheyrezy M. Composition of reactive powder concrete. Cement and Concrete Research, 1995, 25(11): 1501-1511.

[2] Richard P. Reactive powder concrete: a new ultra-high strength cementitious material. F. de Larrard and R. Lacroix. The 4th International Symposium on Utilization of High Strength/High Performance

Concrete. Presses des Ponts et Chaussees. Paris，France，1996：1343-1349.

[3] 谢友均，刘宝举，龙广成. 掺超细粉煤灰活性粉末混凝土的研究. 建筑材料学报，2001，9：280-284.

[4] Bayard O，Ple O. Fracture mechanics of reactive powder concrete：material modeling and experimental investigations. Engineering Fracture Mechanics，2003，70(7-8)：839-851.

[5] 周文元. 活性粉末混凝土在道路桥梁工程中的应用. 水运工程，2004，(12)：103-105.

[6] Aitcin P C，Richard P and Lachemi M. The sherbrooke reactive powder concrete footbridge. Structural Engineering International. France，1998：47-51.

[7] Behloul B，Lee K C. Ductal seonyu footbridge. Structure Concrete，2003，4(4)：195-201.

[8] 郑文忠，李莉. 活性粉末混凝土配制及其配合比计算方法. 湖南大学学报，2009，35(2)：13-17.

[9] 过镇海. 混凝土的强度和本构关系-原理与应用. 北京：中国建筑工业出版社，2004：33-51.

[10] 卢姗姗，郑文忠. GFRP 筋活性粉末混凝土梁正截面抗裂度计算方法. 哈尔滨工业大学学报，2010，42(4)：536-540.

[11] 李莉，郑文忠. 活性粉末混凝土连续梁塑性性能试验研究. 哈尔滨工业大学学报，2010，42(2)：193-199.

The 4th International Forum on Advances in Structural Engineering（IFASE2010）

# 第四届“结构工程新进展国际论坛”简介

## 论坛主题——混凝土结构与材料新进展

**会议地点：**中国　南京

**会议时间：**2010 年 11 月 5～7 日

**主办单位：**

中国建筑工业出版社

同济大学《建筑钢结构进展》编辑部

香港理工大学《结构工程进展》编委会

**承办单位：**

东南大学土木工程学院
混凝土及预应力混凝土结构教育部重点实验室

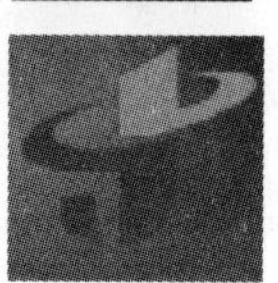

江苏省建筑科学研究院有限公司
高性能土木材料国家重点实验室

# 关 于 论 坛

“结构工程新进展国际论坛”旨在促进我国结构工程界对学术成果和工程经验的总结及交流，汇集国内外结构工程各方面的最新科研信息，提高专业学术水平，推动我国建筑行业科技发展。论坛划分为四个领域，以每次一个主题的形式轮流出现。分别为：

- **新型结构材料与体系（包括组合结构、膜结构等）**
- **钢结构研究和应用的新进展（包括空间结构、大跨度结构）**
- **混凝土结构与材料新进展**
- **结构防灾、监测与控制（包括抗震、抗风、抗火等）**

首届论坛于 2006 年在北京清华大学举办，主题是“新型结构材料与体系”；第二届论坛于 2008 年在大连理工大学举办，主题是“结构防灾、监测与控制”；第三届论坛于 2009 年在上海同济大学举办，主题是“钢结构研究和应用的新进展”。第四届论坛于 2010 年在南京东南大学举办，主题是“混凝土结构与材料新进展”。论坛有两大特点：其一，论坛会议采取特邀报告人专题演讲的形式，其人选在 97 位包括院士、长江学者、国家杰出青年基金获得者及业内专家学者在内所组成的论坛学术委员会推荐的基础上产生，从而保证演讲内容的学术性、代表性及领先水平。其二，“论坛文集”的学术定位：文集按主题每届 1～2 册，于论坛召开之前由中国建筑工业出版社正式出版。文集由两部分组成，第一部分为特邀报告人对演讲内容进行深度展开或延伸而成，充分表达其学术成果；第二部分为参加论坛的代表自由投稿，择优收入文集，并选择部分代表在论坛上交流发言。论坛文集不仅是论坛的重要标志，更重要的是，它将成为结构工程研究领域的风向标，为教学科研人员、在校研究生提供一个既有前瞻性又有可行性的、能引导研究方向的重要文献资料。同时，该书的连续出版，也将成为结构工程科研历程的全记录，对整个研究工作的沿革起着承上启下的作用。

# 组织委员会

顾　　　问：吕志涛　孙　伟（东南大学）

主　　　任：吴智深（东南大学）

副主任委员：吴　刚（东南大学）

缪昌文（江苏省建筑科学研究院有限公司）

委　　　员：陈惠苏　王景全　黄　镇　丁幼亮　梁仁杰　周　臻

冯若强　缪志伟　陆金钰（东南大学）

刘加平（江苏省建筑科学研究院有限公司）

陆　烨　何亚楣（同济大学）

戴建国　余　涛（香港理工大学）

赵梦梅　刘婷婷（中国建筑工业出版社）

秘　书　长：丁幼亮（东南大学）

# 第四届论坛特邀报告人简介（按论坛报告顺序）

**吕志涛** 中国工程院院士，著名结构工程专家，东南大学土木工程学院教授、博导、教授会主席。1965年东南大学结构工程研究生毕业后留校从事土木建筑结构工程领域的教学、科研工作和工程实践，是我国预应力学科的学术带头人，1997年当选为中国工程院院士。现任教育部科学技术委员会委员、江苏省土木建筑学会理事长、东南大学学术委员会主任等职。在国内外已发表学术论文150多篇，出版著作8本，曾获国家级和省部级科技进步奖20项，荣获“国家有突出贡献的中青年专家”（1986）、“全国高校先进科技工作者”（1990）、“全国模范教师”（2001）和“何梁何利基金科学与技术奖”（2000）等荣誉称号。四十多年来，吕志涛院士为我国结构工程科学技术的发展，特别是混凝土及预应力混凝土结构的发展作出了重要贡献。他完善了混凝土结构理论和计算方法，提出了两类斜裂缝理论、抗剪设计方法和双向偏心受拉计算公式等，发展了预应力混凝土结构体系和设计方法，开拓了预应力技术应用的新领域，设计和咨询过上百项预应力工程，如北京西客站、珠海海关大楼、珠海玻璃纤维厂及南京电视塔等重大工程。近10年来，对加气轻质混凝土板（ALC板）技术、预制预应力混凝土建筑（PPB）技术以及为高性能纤维增强复合材料（FRP）在我国土木工程中的应用开展了创新研究。

**Michael P. Collins** 加拿大多伦多大学教授，著名结构工程专家，主要研究领域为混凝土及预应力混凝土结构的设计和分析、核安全壳结构、海上石油平台的设计、结构响应分析的数值技术、结构抗震设计和评价等。他于1964年在新西兰University of Canterbury获学士学位，于1968年在澳大利亚University of New South Wales获博士学位，于1969年至University of Toronto任教。Collins教授在钢筋混凝土结构抗剪方面开展了卓有成效的研究，提出了修正的压力场理论，为结构的抗剪设计提供了理论基础，这一成果已得到世界范围的认可。发表了70余篇论文，并获得了很多奖项。

Hans W. Reinhardt 德国斯图加特大学教授，曾担任斯图加特大学土木和环境工程负责人（主席），建筑材料研究所和材料测试中心主任。Reinhardt 教授于 1968 年在斯图加特大学取得博士学位，后在美国伊利诺伊理工大学从事博士后研究，在荷兰 Delft University of Technology 史蒂文混凝土结构实验室担任教授和负责人，在 Darmstadt University of Technology 担任建筑材料学科负责人，于 1990 年受聘于斯图加特大学。Reinhardt 教授的主要研究领域为混凝土材料的力学特性和耐久性、测试技术等，主要包括混凝土力学性能随时间、温度、湿度的变化、混凝土和木材料的断裂力学、材料的物理特性、无损检测等。他主持和参与编写了 12 本专著，发表了大量学术论文。现担任相关国际期刊编委，RILEM 和美国混凝土协会会员，RILEM 和 CEN 技术委员会主席，在 1995 年至 2004 年间担任 RILEM 德国国家代表，并担任德国结构混凝土协会混凝土方面的技术委员会主席，德国建筑技术研究院学术小组成员。

Koichi Maekawa 东京大学教授，工程学院国际交流与合作办公室主任，主要研究领域为钢筋混凝土非线性力学、钢筋混凝土结构非线性动力分析和抗震设计、多孔介质的固液多相力学、高性能黏性混凝土、混凝土热力学、黏性多孔介质的传质机理、结构混凝土-土-水的多尺度动力分析。于 1980 年和 1982 年在东京大学土木工程系取得学士和硕士学位，于 1985 年在东京大学土木工程系获得博士学位，先后在长冈科技大学、东京大学和泰国亚洲科技研究中心任教，1996 年至今在东京大学土木工程学院任教，曾任土木工程系主任。

黄小坤 研究员，现任中国建筑科学研究院建筑结构研究所副总工程师、建研科技股份有限公司工程咨询事业部副主任。1984 年至今主要从事高层建筑结构、混凝土结构以及建筑幕墙结构的研究、工程技术咨询、规范编制和管理工作。在混凝土方面，完成了钢筋混凝土构件复合受力协调计算、低多层房屋混凝土墙体结构性能、高层混凝土结构楼层刚度比研究等；正在进行“十一五”科技项目课题“500MPa 高强钢筋混凝土构件应用技术研究”、“高强混凝土结构性能研究”以及装配式混凝土结构研究等。完成各类科研项目十余项，获省部级科技进步奖 7 项；主编、参编多项标准规范，如《混凝土结构设计规范》、《高层建筑混凝土结构技术规程》等。兼任建设部工程建设标准强制性条文咨询委员会委员、中国工程建设标准化协会理事兼混凝土结构专业委员会副主任委员、全国混凝土标准化技术委员会(SAC/TC458)副秘书长、中国建筑学会高层建筑结构专业委员会委员、全国工程建设标准设计专家委员会委员等。

**金伟良**　浙江大学求是特聘教授，博士生导师，注册土木工程师。1982年获大连理工大学土木工程系学士学位，1989年获得大连理工大学结构工程博士学位。1991年获德国洪堡（Alexander von Humboldt）基金会资助，在德国Stuttgart大学进行博士后研究工作。1994年获挪威国家研究委员会资助，在挪威科技大学（NTH）从事结构可靠度方面的研究。2003年8月新加坡南洋理工大学OAP基金会客座教授，2008年11月英国女皇大学荣誉教授。从1982年开始从事结构工程的可靠性理论和应用等研究，先后承担了国家自然科学基金、国家863计划等50余项科研项目的研究工作，已在国内外学术期刊上发表论文280余篇，其中被SCI收录20篇，EI收录116篇，出版学术专著3本。以第1完成人获得国家科技进步二等奖1项，浙江省科技奖二等奖4项，教育部科技进步二等奖1项。担任国家自然科学基金会学科评审组成员，中国土木工程学会工程结构可靠性委员会副主任委员，中国建筑学会混凝土结构委员会副主任委员，中国工程建设标准化协会混凝土结构委员会委员等学术团体职务，以及《建筑结构学报》、《水利学报》、《浙江大学学报》（工学版）等10个学术刊物的编委、国际学术刊物《International Journal of Structural Engineering》主编。

**李　兵**　新加坡南洋理工大学副教授。师从于著名地震工程界学者Robert Park（帕克）教授，在香港茂盛（Maunsell）、奥雅纳（Oven Arup）等顶尖国际建筑结构咨询公司从事多年的设计实践工作。1999年赴南洋理工大学任教。长期为新加坡、香港及新西兰的政府部门、建筑业主和国际建筑结构咨询公司和大型保险与再保险公司提供各类咨询，包括高层结构建筑、风险评估、恐怖袭击和地震等人为或自然灾害评估。在钢筋混凝土结构的大型试验和精细分析上拥有丰富的经验。研究方向专注于现役钢筋混凝土结构的抗震抗爆性能评估，重点研究建筑在人为或自然灾害所产生的极端负载下的结构性能表现。目前担任美国土木工程学会（American Society of Civil Engineering，ASCE）属下结构工程期刊的副主编。

李　惠　哈尔滨工业大学教授、博士生导师。长江学者特聘教授、国家杰出青年基金获得者、教育部跨世纪优秀人才计划、新世纪百千万人才（国家级）、享受国务院政府特殊津贴。现任国际结构控制与监测学会理事会理事、国际结构控制与监测学会中国分会秘书长、国际结构健康监测学会规范与标准分委员会委员、泛亚太平洋智能传感与健康监测协作研究中心材料委员会主席、美国土木工程师学会结构控制与监测分委员会委员、中国振动工程学会理事、Structural Control and Monitoring 等期刊编委。曾多次在美国加州理工学院、日本京都大学、意大利罗马大学、香港科技大学等做访问教授和访问学者并讲学。在结构健康监测、结构振动控制、智能材料与结构、地震工程与风工程等方向主持国家自然科学基金重大研究计划重点项目及国家自然科学基金重点项目 3 项，国家 863 项目 4 项、国家科技支撑计划项目 1 项和 973 计划课题 1 项等；获国家科技进步二等奖 2 项（排名均为第二），出版专著 2 部，发表论文 100 余篇。

李　杰　同济大学教授、博士生导师。教育部“长江学者奖励计划”首批特聘教授，国家杰出青年科学基金获得者，国家级“有突出贡献的中青年专家”，上海市科技领军人才。现任同济大学结构工程学科特聘教授，博士研究生导师，上海防灾救灾研究所所长。兼任中国振动工程学会随机振动专业分会理事长，中国建筑学会结构计算理论与工程应用专业委员会主任、混凝土结构理论与工程应用专业委员会副主任，中国城市规划委员会城市安全与防灾专业委员会副主任、“International Journal of Nonlinear Mechanics”编委会顾问编委、“Earthquake Science”编委会委员等 20 余个学术团体及学术期刊的学术职务。长期在结构工程与地震工程领域从事研究工作，在随机动力学、混凝土损伤力学、工程结构可靠度与生命线工程研究中取得了较为深入的研究进展。

**刘伟庆** 南京工业大学教授、博士生导师。现任南京工业大学副校长；江苏省土木工程与防灾减灾重点实验室主任。主要从事结构抗震控制、混凝土结构基本理论和新型复合结构材料等方面的研究。先后承担了国家“八五”科技攻关计划子专题、国家重大基础研究项目前期研究专项、国家自然科学基金项目、中美地震工程科技合作项目、教育部高等学校骨干教师资助计划、江苏省青年基金项目等省部级科研项目十余项，在国内外学术期刊发表论文 120 余篇，出版教材、译著 4 部；指导博士、硕士研究生 30 余名。获江苏省科技进步一等奖 1 项、二等奖 1 项；首批江苏省普通高等院校优秀青年骨干教师、跨世纪学术带头人培养人选；多次被评为江苏省“三育人”先进个人、江苏省优秀教育工作者、江苏省“333”跨世纪学术技术带头人培养工程培养对象。兼任国际减震学会委员，中国建筑学会抗震防灾分会常务理事，中国振动工程学会结构减振控制分会常务理事，中国地震学会地震工程专业委员会委员，江苏省振动工程学会副理事长，江苏省力学学会副理事长，江苏省土木建筑学会常务理事，《地震工程与工程振动》编委会副主任委员，《防灾减灾工程学报》、《工程抗震与加固改造》等期刊编委。

**陆　勇** 爱丁堡大学教授，工程力学系主任。2007 年 3 月进入爱丁堡大学任教，之前在新加坡南洋理工大学从事科研工作。在希腊雅典国立科技大学获博士学位，分别在东南大学和同济大学获硕士学位和学士学位。主要从事地震、冲击和爆炸荷载下的结构反应以及结构参数识别的理论，试验和数值分析。兼任英国土木工程学会理事、美国土木工程师学会结构识别技术委员会委员、ACI 委员会短时动力和振动荷载响应委员会委员（ACI Committee 370），期刊《International Journal of Protective Structures》副主编。先后发表论文 200 多篇，其中被 SCI 收录 60 余篇。

**缪昌文** 江苏省建筑科学研究院院长、教授级高级工程师。被评为第五届南京市十大科技功臣，第九届全国人民代表大会代表，中国混凝土外加剂协会副理事长，中国土木工程学会砼外加剂专业委员会副主任委员。主持或参加重点科研项目 20 余项，获得科技进步奖 14 项，其中部级一等奖 1 项，部省级二等奖 4 项，三等奖 4 项。由他主持完成的“JK 系列混凝土快速修补剂的研制与应用技术”等成果推广至全国 23 个省、市、自治区的公路、市政、桥梁、机场、大坝及水利港口工程，取得了显著的技术经济效益。长期从事混凝 土外加剂、高性能及高强混凝土技术研究，发表论文 40 篇，出版专著 1 本。

**牛荻涛** 1963年10月生，陕西华县人。1983年在原西安冶金建筑学院工业与民用建筑专业获学士学位；1991年在原哈尔滨建筑工程学院获博士学位。现为西安建筑科技大学教授、博士生导师。任西安建筑科技大学土木工程学院院长、结构安全研究所所长。主要从事混凝土结构耐久性、结构抗震与结构可靠度等方面的教学、科研工作，在大气环境混凝土结构耐久性、工程结构抗震、纤维复合材料在土木工程中的应用、结构检测鉴定及评估等领域进行了大量的研究和工程实践。主持完成国家自然科学基金重点项目1项、面上项目2项，参与国家“八五”、“十五”、“十一五”科技攻关与科技支撑计划课题多项，主持编制了我国第一部《混凝土结构耐久性评定标准》(CECS220：2007)。发表学术论文150余篇，出版专著1部。2006年获陕西省科学技术一等奖（排名第一），1998年获陕西省科技进步二等奖（排名第一）、1999年获建设部科技进步一等奖（排名第五），1997年获劳动部技进步二等奖（排名第二）。1999年享受国务院政府特殊津贴，2001年入选陕西省“三五人才”，2006年入选“新世纪百千万人才工程”国家级人选，2007年获国家杰出青年科学基金。兼任美国混凝土协会（ACI）中国分会副理事长，陕西省建筑物鉴定与加固标准技术委员会主任委员，中国土木工程学会结构可靠度委员会委员，全国第四届混凝土耐久性专业委员会委员，全国建筑物鉴定与加固委员会委员。

**吴智深** 东南大学教授，中组部“千人计划”国家特聘专家。分别于1983和1986年获南京工学院（现东南大学）学士和硕士学位，1990年获日本名古屋大学工学博士学位。2002年起任茨城大学教授，成为该国立大学历史上最年轻的教授。研究领域：纤维复合材料、光纤传感技术，智能材料及结构、结构工程、城市安全及节能工程及系统等。曾主持各类重大、国家级项目及人才基金项目40余项，发表学术期刊论文250多篇（其中100篇以上被SCI收录），国际会议论文250多篇（其中主题和特邀报告40余篇），申请及获得国内外专利40余项。2003年教育部“长江学者奖励计划”讲座教授。2002年中科院杰出青年科学基金（海外类）、2003年国家杰出青年科学基金（海外类）、1990年日本土木学会论文奖、2005年度日本复合材料学会技术奖、2007年中冶集团科学技术特等奖、2009年国际结构健康监测年度人物奖（SHM-POY）等。先后被选为IIFC (International Institute for FRP in Construction)，ISHMII (International Society for Structural Health Monitoring of Intelligent Infrastructure)，日本土木学会等学会会士。担任International Journal of Sustainable Materials and Structural Systems主编及其他16种刊物的副主编、客座主编、编委（包括10种SCI收录的学术期刊）及IIFC副会长、ISHMII常务理事以及其他国际学术团体的学术委员会主任、理事等重要职务10余项。

**吴 波** 华南理工大学教授、博士生导师。国家杰出青年科学基金获得者。先后主持各类科研项目51项，其中国家和省部级项目及专题21项、境外合作项目5项；出版专著2部，发表论文235篇（SCI、EI收录近60篇，国内外他引1000余次）；获发明专利授权5项、实用新型专利授权5项；获省部级科学技术一等奖2项、二等奖6项；在国内外学术会议做大会/邀请报告21次。长期致力于混凝土结构抗火和抗震方面的研究工作，并先后涉及新建混凝土结构、加固混凝土结构和循环再利用混凝土结构。主持我国摩擦消能器的首例减震工程应用，以及我国第一部混凝土结构抗火设计标准的编制工作，联合深圳市建筑设计研究总院实现了混凝土再生混合构件的首例工程应用。被评为广东省高等学校特聘教授（珠江学者）、黑龙江省优秀中青年专家、建设部有突出贡献的中青年专家；入选国家"百千万人才工程"第一、二层次；享受国务院政府特殊津贴。担任国家自然科学基金委员会学科评审组成员、中国振动工程学会理事兼随机振动专业委员会副理事长、中国建筑学会结构抗火专业委员会副主任委员等学术兼职。

**徐世烺** 浙江大学求是特聘教授、博士生导师、建筑工程学院院长、高性能建筑结构与材料研究所所长。教育部长江学者奖励计划特聘教授，国家杰出青年科学基金、德国洪堡奖励基金和国务院政府特殊津贴获得者。曾担任大连理工大学科研创新学术团队负责人，辽宁省科研创新团队负责人。先后主持和负责国家重点科技攻关项目、国家自然科学基金重点项目和面上项目、国家973计划子课题、南水北调重大工程关键技术与应用研究项目、德国科学基金项目、国家杰出青年科学基金等专项人才基金项目共十余项。在"International Journal of Fracture"，"Engineering Fracture Mechanics"，《中国科学》、《土木工程学报》、《水利学报》等国内外著名学术刊物及国际会议文集上发表论文200余篇。合作编著3本学术专著。担任美国土木工程师学会会员，德国工程师学会会员和美国混凝土学会会员，中国水力发电工程学会和中国水利学会岩石混凝土断裂力学委员会主任委员。《大连理工大学学报》、《水资源与水工程学报》、《水利水电科技进展》、《Frontiers of Architecture and Civil Engineering in China》等学术期刊编委。1987年和1996年两次获国家教委科技进步二等奖；1990年获第二届霍英东教育基金会高校青年教师奖（研究类）；1991年获得国家教委和国务院学位委员会联合授予的"突出贡献的博士学位获得者"；1991年教育部列为优秀轻年教师重点跟踪支持对象；1992年主持完成的研究成果获得国家科技进步二等奖和能源部科技进步一等奖；1999年获国家科技进步三等奖（第二获奖人）；2006年获教育部自然科学一等奖（第一获奖人）；2006年辽宁省科技进步二等奖（第一获奖人）；2007年获湖北省科技进步二等奖（第一获奖人）。

**薛伟辰** 同济大学教授、博士生导师，现任同济大学建筑工程系现代预应力/预制结构研究室主任。作为第一或第二负责人，主持承担了40余项国家与省部级纵向科研项目，包括国家自然科学基金项目（50008012、50878167、50978193）、国家“十一五”科技支撑计划重点课题（2006BAJ16B07）等。先后入选4项省部级人才计划。此外，还承担了多项重大工程的科技攻关项目，包括上海世博会园区、上海旗忠网球中心、上海源深体育馆、上海市磁浮交通示范运营线、江西奥林匹克体育场、湖南黄花机场新航站楼等。发表论文学术期刊150多篇，其中SCI收录16篇、EI收录70余篇。在国际权威学术期刊ASCE Structural Engineering、ASCE Bridge Engineering、ACI Structural Journal、ACI Materials Journal、Engineering Structures、Canadian Journal of Civil Engineering、PCI Journal、Composite Structures以及Constructional Steel Research上发表论文十多篇。撰写著作6部，其中独著“现代预应力结构设计”（中国建筑工业出版社，已8次印刷，共13200册），主编“先进纤维混凝土”（同济大学出版社）。获省部级一、二等奖4项。授权国家发明专利9项。作为主编或副主编，编制国家标准2部、上海市标准3部。

**叶列平** 清华大学教授、博士生导师。主要从事土木建筑结构工程领域的教学、科研和工程实践。在混凝土结构基本理论、钢骨混凝土结构、高强混凝土结构、预应力混凝土结构、高层建筑结构、结构抗震与减震理论、结构弹塑性动力分析、纤维增强复合材料、结构加固等领域进行了多项研究工作。负责承担国家自然科学基金、国家自然科学基金重点项目、国家自然科学基金国际合作项目、国家杰出青年海外青年学者合作基金（国内合作者）、国家高技术研究发展计划（863计划）各1项、国家十一五支撑计划项目子课题2项、中国工程院咨询项目1项、建设部国际合作项目1项和30余项其他科研项目。参加《混凝土结构设计规范》和《钢骨混凝土结构设计规程》等7本国家标准和中国工程建设标准化协会标准的编制。近年来在国际国内核心刊物和学术会议共发表论文350余篇，其中SCI收录16篇，EI收录73篇，独立编著和参加编著8本专著和教材，译著2本。获1995年北京市优秀青年教师、2000年宝钢教育优秀教师奖、2003年教育部自然科学一等奖（排名1），2007年中冶集团科学技术奖特等奖（排名3），2007获得北京市教学名师称号，2009年获得国家科技进步二等奖（排名6）等奖励，所主讲的《混凝土结构》获得国家级精品课程。现任国际土木工程FRP复合材料学会理事，2008年当选该学会资深会员；中国工程建设标准化协会混凝土结构标准技术委员会委员；中国土木工程学会混凝土与预应力混凝土分学会理事；中国建筑学会抗震防灾分会常务理事；中国振动工程学会结构控制专业委员会常务委员。

**赵君黎**

中交公路规划设计院桥梁技术中心副主任，教授级高级工程师。曾任标准规范研究室主任，现任桥梁技术中心副主任，主管标准规范研究工作。目前担任《公路工程结构可靠性设计统一标准》、《公路桥涵荷载标准》、《公路桥涵设计通用规范》、《公路钢结构桥涵设计规范》、《公路大跨径梁桥设计施工技术指南》、《公路组合结构设计施工技术细则》、《公路桥涵抗撞防撞设计指南》、《公路长大桥隧风险评估管理办法和指南》、《公路特殊桥梁设计文件编制办法》等数十个行业标准规范编制任务的执行、主管负责人。同时还担任工程结构可靠度委员会委员、茅以升科技教育基金会桥梁委员会副秘书长、环境保护部环评中心常聘专家等。参与编制了国家标准《工程结构可靠性设计统一标准》和多个公路建设项目的环境影响评价报告的技术评估工作。在公路行业标准规范方面，尤其是桥梁规范的宏观方向、桥梁规范及其风险等方面有较深的研究。

**郑文忠**

哈尔滨工业大学教授、博士生导师。国建教育部长江学者奖励计划特聘教授，“新世纪百千万人才工程”国家级人选，黑龙江省优秀中青年专家，教育部新世纪优秀人才，教育部宝钢教育基金会优秀教师，黑龙江省杰出青年科学基金获得者。他主持多项国家自然科学基金和国家科技支撑计划等重要课题。在预应力结构、结构抗火、结构改造与加固等方面取得创新成果。“5·12”四川汶川大地震后，第一时间赶赴灾区，牵头创造性完成国家国防科技局交付的技术任务。获中国高校科技进步一等奖1项，黑龙江省科技进步一等奖1项、二等奖3项，全国优秀建筑结构设计奖1项，黑龙江省重大科技效益奖1项。获得及申请国家发明专利技术10项。合作起草或审查《混凝土结构设计规范》、《混凝土结构工程施工规范》等国家及行业等标准10余本，SCI和EI收录论文100余篇。培养毕业博士24人，硕士58人。任《建筑结构学报》、《建筑结构》、《工程加固》、《混凝土与标准化》等刊物编委及多个学术团体的理事。